ELECTRONIC INSTRUMENT DESIGN

Electronic Instrument Design

Architecting for the Life Cycle

KIM R. FOWLER

New York Oxford
OXFORD UNIVERSITY PRESS
1996

Oxford University Press

Oxford New York
Athens Auckland Bangkok Bombay
Calcutta Cape Town Dar es Salaam
Delhi Florence Hong Kong Istanbul
Karachi Kuala Lumpur Madras Madrid
Melbourne Mexico City Nairobi Paris
Singapore Taipei Tokyo Toronto

and associated companies in
Berlin Ibadan

Published by Oxford University Press Inc.,
198 Madison Avenue, New York, New York 10016

Library of Congress Cataloging-in-Publication Data
Fowler, Kim R.
Electronic instrument design : architecting for the life cycle /
Kim R. Fowler.
p. cm.
Includes bibliographical references and index.
ISBN 0-19-508371-7 (cl)
1. Industrial electronics.
2. Electronic instruments—Design and construction.
I. Title.
TK7881.F68 1996 95-52711 621.3815í4—dc20

2 4 6 8 9 7 5 3 1

Printed in the United States of America
on acid-free paper

CONTENTS

PREFACE

AIMS

Education is a piecemeal process of collecting and synthesizing information. Often the application of newly acquired knowledge is left haphazardly to the student. This book attempts to fill the gap by providing a coherent and integrated picture of the design process. The material connects engineering principles to real applications from the overall perspective of a systems solution.

SCOPE

The book provides a framework for developing electronic instrumentation, from handheld devices to consoles of equipment. It offers practical design solutions and the interactions, trade-offs, and priorities encountered and then uses specific details or situations as examples. The methods may be applied to single prototypes as well as mass-produced devices. The applications are not technology dependent, so the principles presented will be useful for years and will not be outdated by the next generation of hardware or software.

The focus is on engineers who manage and participate in small projects in small or medium-sized companies, but many of the principles apply to larger projects as well.

GENERAL APPROACH

Developed from real-world experience, the material outlines some general methods of systems engineering and then covers individual issues. I present more general subjects first, followed by more specific topics. Examples are used frequently, but the material does not examine every possible electronic component and situation. My intention is to provide an introduction to specific topics; the references point to in-depth materials on each subject.

INTENDED AUDIENCE

This book will serve as a reference handbook for practicing engineers. It can also be used by junior and senior students in electrical or industrial engineering. The material assumes a diverse background of fundamental courses in electrical engineering: circuit theory with Laplace transforms, digital and analog electronic design, classical electromagnetics, and introductory computer science.

ACKNOWLEDGMENTS

Any book is the culmination of the efforts of many people. The Johns Hopkins University Applied Physics Laboratory supported this work with a Janney Fellowship and provided the composition and typesetting. I am grateful for the fine effort by the TIR publications group: Bill Pullin, Murrie Burgan, Patrice Zurvalec, and Nancy Conklin. The art department produced all of the figures; thanks to Gloria Crites, Libby Lee-Falk, and Ruth Novick. EEI copyedited the manuscript; thanks to Gayle Dahlman and Keith Ivey for the good job.

I am indebted to the following reviewers: Brian Alvarez, Murrie Burgan, Guy Clatterbaugh, Oonagh Fowler (my wife), Mars Gralia, Stan Kaveckis, Steve Lammlein, Dave Lohr, Mike Souders, Mark Tilden, and Tom Van Doren. Thanks to Oxford University Press for providing the comprehensive technical review.

I am particularly grateful for Jack Ganssle, who reviewed the entire manuscript and helped me from his wealth of experience, technical insight, and writing skills. And finally, thanks to Steve Zeise, my friend, who gave some honest and insightful criticism. You are a friend indeed.

If you have any comments or questions about this book, I welcome and appreciate your contacting me:

Ixthos
741-G Miller Drive
Leesburg, VA 22075
Tel.: (703) 779-7800
Fax: (703) 779-7805
e-mail: k.fowler@ixthos.com

Kim R. Fowler
June 1995

ABBREVIATIONS

AC	alternating current
ADC	analog-to-digital converter
AFD	arc fault detection
AGC	automatic gain control
AM	amplitude modulation
ANSI	American National Standards Institute
ASIC	application-specific integrated circuit
ATE	automatic test equipment
AWG	American Wire Gauge
BGA	ball grid array
BIT	built-in test
BTL	back-plane transceiver logic
CAD	computer-aided design
CAM	computer-aided manufacture
CASE	computer-aided software engineering
CCD	charge-coupled device
CFC	chlorofluorocarbon
CMOS	complementary metal-oxide semiconductor
CMRR	common-mode rejection ratio
CMS	Colorado Memory Systems
CPU	central processing unit
CSMA/CD	carrier-sense, multiple-access with collision detection
DAC	digital-to-analog converter
DC	direct current
DFE	design for environment
DIP	dual in-line package
DMA	direct memory access
DMM	digital multimeter
DNL	differential nonlinearity
DoD	Department of Defense
DUT	device under test
DVM	digital voltmeter
ECL	emitter-coupled logic

EISA	Extended Industry Standard Architecture
EMI	electromagnetic interference
EPLD	electrically programmable logic device
EPP	expanded polypropylene
ESD	electrostatic discharge
ESL	equivalent series inductance
ESR	equivalent series resistance
ESS	environmental stress screen
FCC	Federal Communications Commission
FDA	Food and Drug Administration
FET	field-effect transistor
FFT	fast Fourier transform
FPGA	field-programmable gate array
FM	frequency modulation
FMECA	failure mode, effect, and criticality analysis
FTA	fault-tree analysis
GOMS	goals, operators, methods, selection rules
GPS	global positioning system
HP	Hewlett-Packard
IC	integrated circuit
ICE	in-circuit emulator
IEEE	Institute of Electrical and Electronic Engineers
INL	integral nonlinearity
I/O	input/output
ISO	International Standards Organization
ISR	interrupt service routine
LAN	local area network
LCC	leadless carrier chip
LCD	liquid crystal display
LEAP	Light Exoatmospheric Projectile
LED	light-emitting diode
LISN	line impedance stabilization network
LRU	line-replaceable unit
LSB	least significant bit
LVDT	linear variable differential transformer
MCM	multichip module
MLV	memory-loader-verifier
MOSFET	metal-oxide semiconductor field-effect transistor
MOV	metal-oxide varistor
MSB	most significant bit
MTBF	mean time between failures
MTU	magnetic tape unit
mux	multiplexer

NAS	Native American Services
NC	normally closed
NO	normally open
NRE	nonrecurring engineering
NSS	Neurological Stimulation System
OOP	object-oriented programming
PAL	programmable array logic
PC	personal computer
PCB	printed circuit board
PERT	program evaluation and review technique
PGA	pin grid array
PGA	programmable gate array
PID	proportional-integral-differential
PLC	programmable logic controller
PLD	programmable logic device
PLCC	plastic leadless chip carrier
PP	prediction processor
QFP	quad flat package
R&D	research and development
RAM	random-access memory
RC	resistance-capacitance
REP	range extraction processor
RF	radiofrequency
RFI	radiofrequency interference
rms	root mean square
ROM	read-only memory
RTOS	real-time operating system
SCSI	small computer system interface
SEM	standard electronic module
S/H	sample-and-hold
SMT	surface-mount technology
SNR	signal-to-noise ratio
TCE	thermal coefficient of expansion
TCR	temperature coefficient of resistance
TDR	time-domain reflectometry
TTL	transistor-transistor logic
TTU	tape transport unit
UHF	ultrahigh frequency
UL	Underwriters Laboratories
UPS	uninterruptible power supply
VCR	videocassette recorder
VHF	very high frequency
VLSI	very large-scale integration
VME	VersaModule Eurocard

ELECTRONIC INSTRUMENT DESIGN

1

SYSTEMS ENGINEERING

Prevent stupidity, control ignorance.
—Eddie Fowler (my dad)

1.1 INTRODUCTION

Electronic instruments abound in our everyday life (Fig. 1.1). The sheer diversity is mind boggling, but the development of electronic products can be both systematic and controlled. This book aims to provide you with that systematic insight.

I introduce the concepts of systems engineering and design architecture for electronic instruments in the first two chapters and then delve into specific areas that affect product design in the remaining chapters. Each chapter begins with general principles and then focuses on specific issues. Examples and case studies illustrate specific points and principles. Throughout each chapter, I try to relate the specific subject to the overall development of an electronic product.

1.2 OVERVIEW OF SYSTEMS ENGINEERING

Every electronic instrument is a system of components and interactions. Each instrument is a part of a larger system. And each component is a system unto itself. Consequently, the development of electronic instruments should be a systematic effort called *systems engineering*. Systems engineering is an approach, an attitude, and a loose set of methods for solving complex problems.

Systems engineering requires a broad approach that applies general principles to promote understanding across disciplines and clarify interactions between system components. Moreover, systems engineering helps bound the unknowns within, between, and around components. (System components include hardware, software, environmental influences, and human operators.)

Systems engineering provides a framework to develop a product from concept, through design and test, to delivery and documentation. That path, from concept to documentation, delivers the best product at the lowest cost.

Fig. 1.1 Examples of electronic devices, appliances, and instruments that you may design and develop.

Through systems engineering you apply knowledge, reason, and creative judgment to arrive at a solution. That solution is your product. Once you have defined the interactions between system components and specified their uses and controls, you have defined your solution. In unpredictable circumstances where the interactions cannot be absolutely defined, you must bound and reduce the unknowns through experimentation. Though unpredictable circumstances are normal and lead to mistakes and failure, you can learn from them. *In other words, prevent stupidity through learning, and control ignorance through reason and creativity.*

The discipline of systems engineering strives for a global perspective of a problem, even though a single best solution generally does not exist. Systems engineering evaluates numerous parameters, balances priorities among requirements, synthesizes a solution, and measures the effectiveness of the solution. Fig. 1.2 illustrates aspects of systems engineering. Feedback loops within the block diagram demonstrate the iterative nature of systems engineering. Iteration refines the perspective of the problem, corrects mistakes, and reduces unknowns.

The techniques of systems engineering vary according to the problem. A general framework of methods suits most problems, but the framework is flexible and may be drastically altered to fit a particular problem. For example, if you develop

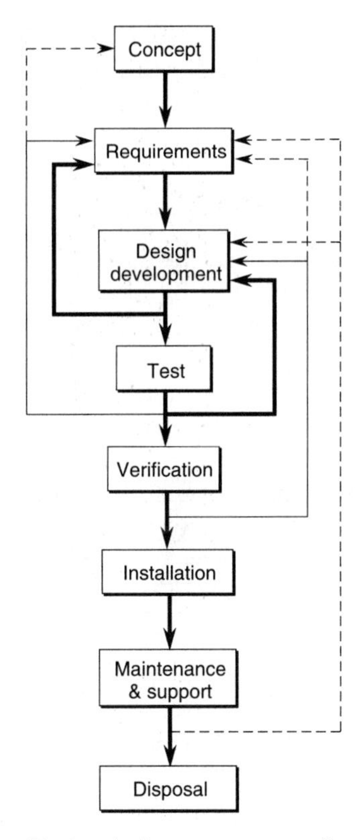

Fig. 1.2 Top view of a project. The heavier lines represent more frequent iteration; the dashed lines, less frequent feedback.

a medical instrument you will emphasize software quality assurance to reduce product liability, whereas if you develop an electric food mixer you will focus on manufacturability to reduce cost. Clearly, different problems require different methods, and you must tailor your approach accordingly.

Any approach must have a good foundation. The foundation in systems engineering is integrity, responsibility, and teamwork, which you must relentlessly pursue ethically and with excellence. Doing any less or expecting someone else to pick up the loose ends will haunt you and your organization for the life of the product.

Finally, systems engineering does not ensure an optimal solution; often an optimal solution is not possible. A fighter aircraft, for example, is never a perfect solution to a military requirement. Compromises and trade-offs among current technology, threat, mission, speed, maneuverability, range, weapons loads, and reliability prevent a solution that completely satisfies the need. But this is a characteristic of systems engineering—compromise and trade-off.

In general, systems engineering incrementally improves the design or implementation of an instrument. Paradigm shifts in product design are unpredictable and impossible to control. Witness the phenomenal development of the personal computer and computing power; who could have predicted the fortunes and failures within the computer industry for the last two decades? Yet in the course of innovation, systems engineering can provide some sanity in managing product development.

1.3 THE SYSTEMS PERSPECTIVE

Systems engineering maintains a broad, long-term view of the constraints and the product life cycle. In addition to operations, logistics, and measurable parameters, your perspective must encompass some intangible attributes such as the culture of your customer—whether corporate, social, economic, political, or ethnic—to fit the solution to the problem. Moreover, you must be committed to continual improvement to satisfy the long-term requirements within the life cycle of the systems solution.

Constraints peculiar to each situation shape the methods that you apply. Some basic constraints are found in most projects:

1. *Functionality*: Does the product fulfill the need?
2. *Cost*: Is the cost as low as possible?
3. *Safety:* Is the product safe enough?
4. *Reliability*: How long will it function?
5. *Maintainability*: How easy is it to fix?
6. *Utility*: How easy and obvious is it to use?
7. *Time*: How long will it take to develop and produce?

Constraints force you to measure both the progress of development and the conformance to requirements. Clearly, you must understand and establish the constraints in every project before you outline a plan of action.

Within such constraints, a product passes through a life cycle that includes development, evaluation, maintenance, and disposal. The life cycle describes the evolution of the system bounded by the constraints. Table 1.1 lists the main areas within the life cycle of a product.

Table 1.1 Definition and costs within the life cycle of a product

Area	Activities
Start-up	Company R&D, concept development
Acquisition	Research, design, construction
Software	Design, maintenance, upgrades, error recovery
Documentation	Technical data generation, distribution, storage, plans, test results
Production	Manufacture, distribution
Operations	Resources, power consumption, facilities
Training	Users, operators, maintenance
Maintenance and repair	Equipment, staff, service calls
Inventory	Facilities, stock
Legal	Litigation, staff, regulations
Disposal	Dismantling, inventory, regulations

1.4 DOCUMENTATION

Documentation pervades systems engineering and is central to good engineering. Understanding the customer and the requirements, measuring progress and test results, creating drawings by computer-aided design (CAD), producing instructions, and presenting the product in a user manual all require communication and records. Concise, clear, and complete documentation will provide a critical avenue of communication and record progress effectively.

Documentation serves several functions within engineering. It records progress during development. These records indicate how well you are satisfying the requirements. Documents help establish the legal liability you and your company have for a product. Finally, documents are an integral part of every product, from instruction pamphlet to repair manual. *A well-prepared manual will enhance the customer's perception that yours is a quality product—and perhaps provide the important differentiation that makes your company successful.*

Documentation, like all other aspects of systems engineering, should be tailored to the need. Brief, neat notes in an engineering notebook may be all that are necessary for a particular project. Always strive for simplicity and clarity in communication and writing. Include only the necessary; avoid the extraneous that just adds to the height of the document pile.

All documentation fits into an overall scheme called *configuration management* that ensures a match between the delivered product and its documentation and that captures a history of the development effort. Toward these ends, the plan for configuration management must define the product complexity, list resources, define the work structure, set the level of documentation, schedule the development process, and inventory the results.

1.5 CONCEPT DEVELOPMENT

Systems engineering begins with defining the problem. No matter how superb your ability, you can't solve a problem if you don't understand it. Therefore, definition is an absolutely necessary first step. You must establish the following:

1. Customer objectives
2. User needs
3. Mission or regions of operation
4. Constraints
5. Regulations and standards

Defining these elements will provide the *who, what, where,* and *when* of a problem. Corporate vision will explain *why* this problem is important and why it should be solved. The remaining techniques described in this book will help establish the *how*.

After definition, concept development enters an iterative cycle of analyses to refine candidate solutions. The analyses include functional analysis, modeling, and feasibility. Beginning with functional analysis, you determine what must be done by illuminating the operations flow. All systems have some sort of operations flow—for example, data flow, energy distribution and consumption, monetary transactions, or passenger and material transport. Once functional analysis defines the operations, you may discover that numerous operations or solutions exist for a particular problem. You then model each solution to characterize it and perform a feasibility study to reveal trade-offs. After you define the operations, model the solutions, and test their feasibility iteratively, the concepts you develop can be used in determining the requirements. Fig. 1.3 illustrates the elements and the iterative nature of concept development.

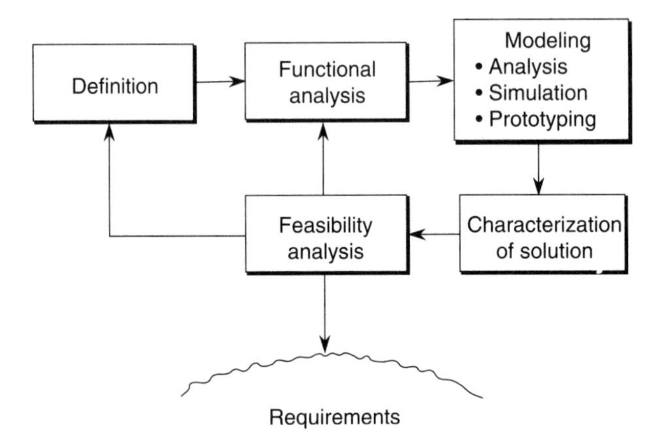

Fig. 1.3 Progression of elements within concept development.

Functional analysis breaks down operations into successively more detailed functional components of what must be done. Each level uses a diagram composed of functional blocks, each representing a specific function. The block diagrams aid the logical development of a system and ensure that necessary functions are not forgotten. Fig. 1.4 illustrates an example of part of a functional analysis for a hypothetical irrigation system. A systems designer or architect would expand higher-level concerns into progressively more detailed functions to illuminate the elements of a problem: the *who, what, where,* and *when.*

The iterative and fluid nature of concept development is reinforced in Fig. 1.4, in which the feasibility study may occur before the functional analysis. The important consideration is that all three analyses—functional, modeling, and feasibility—are sufficiently exercised to select a system solution that leads to the definitions of requirements. As the problem complexity grows, the analyses become significantly more necessary, detailed, and time consuming.

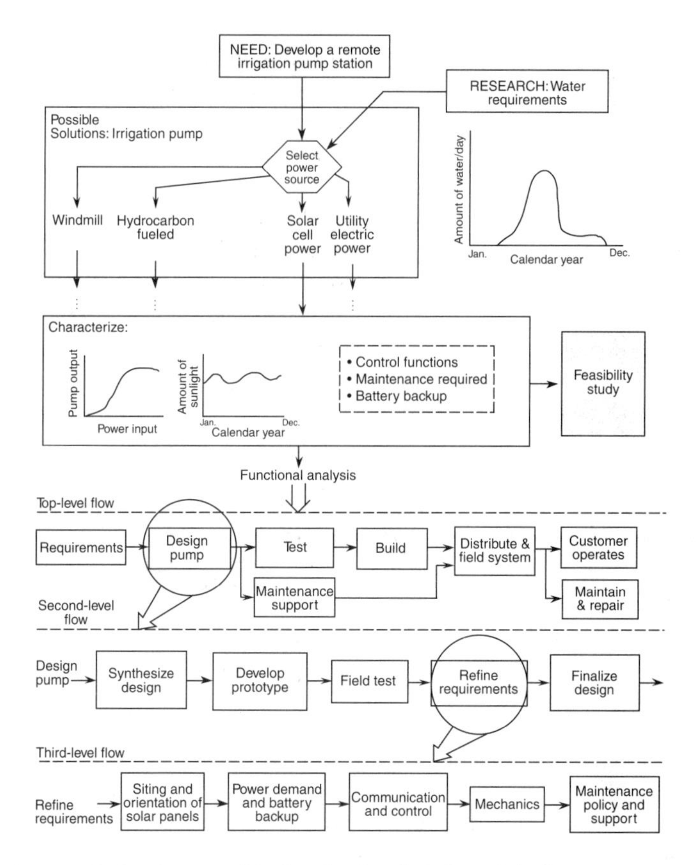

Fig. 1.4 An example development of the functional analysis for an irrigation system.

After the functional analysis, you model problems and solutions in three ways: mathematical analysis, simulation, and prototyping. A mathematical analysis is a paper study that strives for an analytical solution but generally can only bound part of the problem. Simulation provides some insight into the hypothetical interactions between components and people within the system. Component variations and "what if" situations may be considered with simulation. Prototyping produces a subscale vehicle or subset of the solution for in-depth study and refinement. Each technique is different, and each provides valuable insights for different aspects of a problem, depending on time, money, and potential payoff. Developing a satellite or space station, for example, will rely on a great deal of mathematical analysis and simulation, since prototyping a space vehicle would be prohibitively expensive. On the other hand, a prototype of an irrigation pump station (as in Fig. 1.4) may reveal operational concerns more clearly than mathematical analysis or simulation.

Modeling characterizes possible solutions so that you can study their feasibility and compare their outcomes. A feasibility study should uncover the limitations and shortcomings of each proposed solution. From the example in Fig. 1.4, you might compare the available amount and cost of power, for instance, to assess the feasibility of each approach to developing an irrigation pump station.

1.6 REQUIREMENTS

Requirements crystallize the function, operation, and performance of the system. The requirements result from the various efforts of concept development: analysis, simulation, and prototyping as well as functional and feasibility analyses. The requirements establish metrics for quantifying progress during development of the system solution. Requirements define the *what* of system function (but not the *how*).

However, requirements can change. *The mere existence of a developing system solution will spark change in the requirements by causing customers and potential users to envision possibilities not previously considered.* As the designer, you will have to strike a balance between useful options and those features outside the scope of the project. Software developers will recognize the problem of *creeping featurism*: How often have they heard "Just add a menu here" or "Let's add another button there—after all, it's only software; it can be changed"? You will have to find a balance between flexibility and firmness while navigating the fulfillment of requirements during development.

Creating the requirements will sometimes suggest a schedule and milestones. (Appendix A outlines those milestones.) You must prepare a development plan in concert with the requirements, and both plan and requirements should be mentioned in the configuration management plan. Include the following elements:

- Schedule and milestones
- Deliverables
- Design reviews
- Test reviews

Document each step and element as it is completed. This documentation will represent your efforts to fulfill the requirements and will give you confidence in preparing plans for future projects.

Requirements have attributes that vary from hard metrics with readily measurable properties to subjective qualities that are difficult to evaluate (Table 1.2).

Table 1.2 Outline of requirements—types and examples

General types of requirements	Specific examples of parameters
Performance	• Range • Speed • Throughput • Error rates • Size • Weight • Power consumption • Efficiency • Test levels - Electromagnetic interference - Vibration and shock - Thermal cycle
Reliability and maintainability	• Mean time between failures • Failure rate • Maintenance downtime
Human factors and user interface	• Response latency • Number of operations per sequence • Expertise required • Intuitive operation • Ease of use
Safety and failure modes	• Failure mode, effect, and criticality analysis • Fault-tree analysis • Hazard analysis
Operational regimes and environment	• Duty cycle • Location • Temperature extremes • Stress range
Logistics support	• Maintenance intervals • Personnel expertise • Maintenance-task analysis

1.7 DESIGN DEVELOPMENT

Design synthesizes the *how* of component interaction within the system (Blanchard, 1991). It molds the *what* from requirements into functional relationships to derive an integrated whole. Most of this book is devoted to design—optimizing and integrating components into a system.

All parts of systems engineering revolve around the synthesis of the solution. Concept development proposes various approaches to design. The requirements structure the framework on which the solution hangs. Integration and testing determine how well the solution fits the requirements. The logistics of the life cycle directly depend on the components defined by the design.

Various standards may apply to your design process. These standards serve as guidelines for developing the *how* of design. For military projects in the United States, Department of Defense (DoD) Standard 2167 is the complete road map for development and documentation. You may customize it to ensure careful and complete development in commercial environments. International Standards Organization (ISO) 9000 standards are more flexible procedures to certify quality in your products. The Institute of Electrical and Electronic Engineers (IEEE) and the American National Standards Institute (ANSI) both provide standards for individual components, devices, and systems.

Besides standards and procedures, close-knit teamwork is necessary to implement a successful design. The synthesis of functionality, testability, manufacturability, and marketability requires an interdisciplinary team throughout the product development. This is a parallel effort, not a sequence of events in which one group passes along the project, or "throws it over the wall," after they have added their input. Communication is central to design. A favorite buzzword for this process of interdisciplinary cooperation is *concurrent engineering*.

Within the necessary standards and teamwork approach, development may proceed along one of several lines or methods:

1. *Top-down*: The requirements completely drive the design.
2. *Bottom-up*: The solution is synthesized from current designs and available technology.
3. *Outside-in*: System interfaces drive the design.
4. *Inside-out*: The design is driven by developing technology.
5. *Hybrid*: A combination of approaches is used.

Chapter 2 will explore these methods, discussing some of their strengths and weaknesses.

The actual design process involves identification, trade-offs, and specification of different approaches to components and component interaction. For example, you might weigh temperature stability versus cost in selecting a resistor type in circuit design. The steps that define interactions occur in the following sequence:

1. Identify parameters (performance, cost, etc.).
2. Select the evaluation techniques necessary to analyze each choice.
3. Perform a sensitivity analysis for uncertain parameter values.
4. Identify the attendant risks.
5. Recommend the preferred approach.

Sensitivity analysis uses the extreme values for a parameter to calculate the range of possible outcomes. For example, manufacturing variation in a capacitor affects its final value in a filter. Choosing the maximum and minimum expected capacitance allows you to calculate the minimum and maximum cutoff frequency of the filter. Knowing the range of filter responses and how it factors into the system allows you to estimate the total system response.

Design reviews help close the feedback loop in design. These reviews can and should range from casual discussions and informal communication to carefully planned, formal arrangements that mark milestones in development. Finally, documentation provides an audit trail for all design development.

1.8 RAPID PROTOTYPING AND FIELD TESTING

Rapid prototyping and field testing provide two more ways to close the feedback loop in design. They can pull the design toward the intent of the requirements and fit the system to the customer's desires. Specifically, they *validate* a design.

Early in the project, rapid prototyping presents various functions to a user who then assesses each of the possible modes. This helps frame the range of each requirement. (Remember, the mere existence of a developing system may cause changes in the requirements.)

Rapid prototyping is most useful for addressing issues related to the human interface:

- Presentation
- Ease of use
- Intuitiveness
- Response latency

Developing menu screens for a process controller on an assembly line is an example where rapid prototyping is very effective. You can prepare a series of formats for the various screens without generating all the control functions. Selected users then evaluate the effectiveness of each mode. This interaction between designer and user not only improves the utility of the product, but also improves its acceptance because the users feel that they own a part of the solution.

Rapid prototyping is characterized by a short duration, or *latency* between mode comparisons. The prototype usually implements a subset of the system functions to give the user a feel for its operation.

In contrast, a field test implements a more finished product and exhibits longer durations between updates or alterations. Field testing concentrates on the longer-term effects and attempts to shake out the more obscure bugs, flaws, and inadequacies.

Flight testing a new aircraft is an example of field testing. The tests characterize the control system responses and allow the tuning of gains and parameters of the control laws to satisfy requirements. However, a prototype occasionally crashes in such a flight test when instability is found in a new region of the system response.

Setting up your prototype at various customer sites for initial evaluation is another form of field testing. It is called *beta testing*.

Finally, bench tests and breadboard evaluations of circuits and devices fall somewhere between rapid prototypes and field tests. Breadboards can help verify concepts and critical performance parameters, as well as allow experimentation with various approaches.

1.9 VALIDATION, VERIFICATION, AND INTEGRATION

Formal tests certify the fulfillment of the requirements and the system goals. Validation determines how well the requirements suit the intent of the system solution. Verification evaluates how well the system satisfies the requirements. Integration is the process of assembling the components and subsystems and performing the acceptance tests of validation and verification.

You can perform a number of tests to qualify your design during integration. The specific tests chosen depend on the purpose of your instrument and the criticality of its function. Validation is a qualitative description of the purpose of the instrument. Table 1.3 lists some tests that help validate a design.

Table 1.3 Tests that may be used during validation

Test	Development phase	Function
Hazard analysis	Early and mid-design	Find areas where deviations can generate hazards, and determine the consequences.
Challenge tests	Late design	Confirm the results of the hazard analysis.
Event-tree analysis	Design, integration, operation	Identify potential accident as a consequence of an initiating failure or error.
Fault-tree analysis	Design, integration, operation	Track backward from postulated accidents to possible initiating failures.
Failure recovery and error correction	Late design, integration	Confirm that recovery and correction occur predictably (similar to challenge test).
Human factors evaluation	Early and mid-design	Identify problems and risks of human interaction.
Failure mode, effects, and criticality analysis	Design, integration, operation	Identify all failure modes of each component and their potential effects on the system.

Verification quantifies system performance and compares it with the requirements. Verification occurs later in the development process and tends to be more extensive than validation. Table 1.4 lists the areas within verification.

Integration may proceed along one of several courses:

- Modular
- Spiral development
- Expanding envelope

The *modular* approach adds each subsystem sequentially, followed by a prescribed set of tests. It works well for integrating a very modular system such as a rack of

Table 1.4 Areas within verification

Area	Specifics
Environmental	Temperature range, shock, vibration, humidity, corrosion
Electromagnetic	Conducted and radiated interference and susceptibility
Reliability	Burn-in, life testing
Maintainability	Time and sequences, skill levels, diagnostics, test equipment, procedures
Technical data	Operating procedures, maintenance procedures, supporting data
Performance	Size, weight, power, capacity, throughput, speed, etc.
Software	Performance, failure modes, user acceptance

test equipment. *Spiral development* prepares and tests the major functions first, followed by lower-priority functions. Some software programs follow a spiral development model. Finally, the *expanding envelope* approach is used for systems that must have a critical mass or a majority of assembled components and subsystems before meaningful tests can be performed. This kind of integration gradually loosens constraints on testing parameters to confirm performance. Flight testing of aircraft, for example, employs tests that expand the performance envelope incrementally.

Testing actually reduces cost during development. The earlier a flaw is discovered, the cheaper it is to fix (Fig. 1.5).

You will have to plan, prepare, and complete the appropriate documentation to record the results of validation and verification. This documentation proves that you have qualified the system. I cannot stress this point enough; documentation is the heart and soul of good testing and integration.

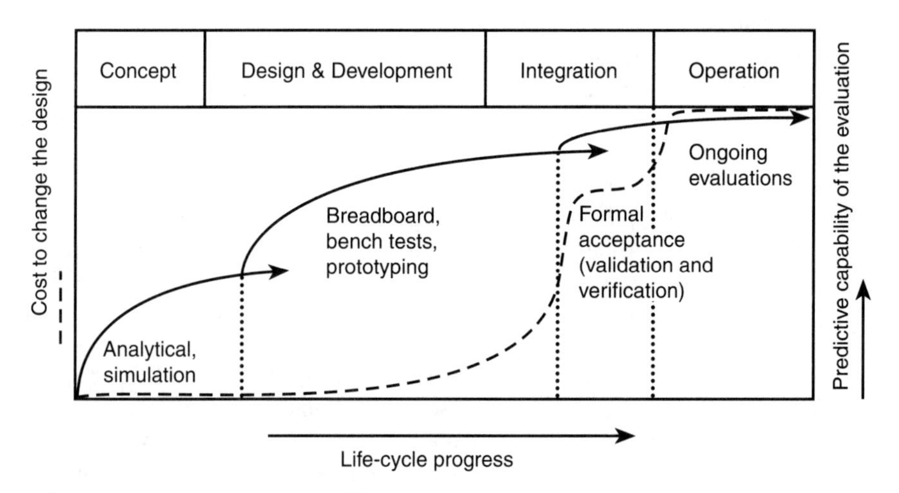

Fig. 1.5 The cost of design changes. The dotted vertical lines demarcate the beginning of a new series or type of test or a crossover point for a large cost increase.

1.10 MAINTENANCE AND LIFE-CYCLE COSTS

Many designers fall short in planning for the maintenance within the life-cycle costs of a system. Consequently, maintenance becomes a bag of trial-and-error tricks that can be time consuming, inefficient, and costly. Maintenance must be carefully considered to build success into a product.

Maintenance is either planned or unscheduled. Planned maintenance is the replenishing of resources within the system—for example, battery replacement or lubrication. Unscheduled maintenance, or repair, returns the system to operation after a failure. General maintenance and repair must have a philosophical basis to orient the work, a diagnostic procedure, and an administrative scheme.

The philosophical basis assigns a level of functionality and availability to the system during either a failure or a replenishment procedure and simultaneously requires a specific level of expertise from the repair personnel. The five levels of functionality, generally arranged in decreasing order of availability or in increasing order of expertise required for repair, are as follows:

1. *Fail operational*: Redundant architecture allows uninterrupted operation.
2. *Cold spare*: Switch in a redundant system to minimize interruption.
3. *Line-replaceable unit*: Remove the failed circuit board or module, and plug in a spare.
4. *Component replacement*: Diagnose and repair the specific component failure.
5. *Disposable*: Throw away the failed device and get a new one.

Once you establish the maintenance philosophy, you will need to consider the diagnostic procedure. Diagnostic support may be entirely included within the system, as a *built-in test* (BIT) or it may require specialized external equipment, such as an oscilloscope or digital multimeter, to localize the problem. Ways of diagnosing a problem run from an automatic self-test to stimulating special test connectors or pins that scan a circuit to probing components with a meter. The expected level of expertise of the maintenance personnel, the sophistication and cost of test equipment and circuitry, reliability of operation, and system availability are all factors in determining the appropriate diagnostic procedures.

Diagnostics are usually reserved for locating problems and failures during repair; administrative schemes, however, are a part of every maintenance, whether replenishment or repair. As you plan maintenance, you must consider the steps and components of each administrative action. Among the components are the expertise of the repair or maintenance personnel, availability of spare parts, and accessibility of the system. Obviously, a full-featured sewing machine requires a different set of repair skills, tools, and procedures than a handheld hair dryer. You will need to understand the unique applications of your system before prescribing the maintenance procedures.

Thorough documentation should describe and record all procedures for diagnosis, repair, and replenishment. Good maintenance records help solve those inevitable mysteries that arise whenever a system is fielded. The maintenance documentation includes the following manuals:

- Replenishment
- Diagnostics
- Repair
- Records

Depending on the device or system, some manuals may be combined; maintenance instructions and records for an automobile, for instance, are often combined with the owner's manual.

Finally, life-cycle costs figure into all considerations of operation and maintenance. These costs are difficult to measure or even estimate, but they directly affect your customer and consequently the future acceptance and success of your device or system. A reasonable and clearly defined set of life-cycle costs increases the customer's confidence in you and your product; it also provides a gauge for the effectiveness of future product developments. Life-cycle costs fall, rather arbitrarily, into the categories given in Table 1.5.

Table 1.5 Life-cycle costs with some examples

Category	**Example components**
System at delivery	• Amortized development costs and nonrecurring engineering • Material and construction labor
Operation	• Personnel • Power consumption
Consumable resources	• Fuel • Lubrication • Batteries • Auxiliary supplies (e.g., paper for printers)
Training	• Operators and users • Maintenance personnel
Maintenance and repair	• Labor • Spares • Inventory • Downtime cost to customer
Upgrades	• Hardware • Software • Capability and complexity

1.11 FAILURE, ITERATION, AND JUDGMENT

Petroski (1985, p. 62) says, "No one wants to learn by mistakes, but we cannot learn enough from successes to go beyond the state of the art." Learning by experimentation, iteration, and failure advances technology and engineering development. The goal of learning is to perfect your engineering judgment because "nearly all engineering failures result from faulty judgments rather than faulty calculations" (Ferguson, 1993, p. 20).

The collapse of the Tacoma Narrows Bridge (Fig. 1.6) on November 7, 1940, which spawned research into aerodynamic effects on structures, is a poignant example of failure precipitating an advance in the state of the art. Furthermore, the collapse is a case of ignorance in design rather than stupidity. Research revealed two identical failures more than half a century before: the Wheeling Bridge in West Virginia and the Brighton Chain Pier in England. However, the cause of these failures was not generally known—that is, the engineering profession was ignorant—before the disaster over the Tacoma Narrows of Puget Sound.

Ultimately, you want to engineer a successful product and reduce failure. Experimentation and design iteration can reduce the magnitude of failure due to erroneous assumptions—at the same time you learn and expand your experience. Thomas Edison was famous for his practical approach to problem solving; he

Fig. 1.6 The Tacoma Narrows Bridge collapse was a failure that advanced the engineering of bridge design (photograph courtesy of University of Washington).

experimented prodigiously, testing more than 1000 different filaments before demonstrating a successful incandescent lamp. Don't be afraid to experiment. Don't fear failure; learn from it.

1.12 SUMMARY

The systems engineering of real electronics is fundamentally different from the pure, analytical science portrayed in many university courses. First, systems engineering is a closed loop activity with much iteration through design and test. Every possible condition cannot reasonably be accounted for by the initial concept and requirements; only test and experimentation can begin to bound the operation of your product. Second, engineering is more art than science; judgment is a critical skill. Ferguson (1993, p. 18) writes, "A successful new design combines formal knowledge and experience and always more judgment than certainty." Enhance success by reducing failure, but failures will remain a part of our experience because our knowledge and judgment are limited.

1.13 RECOMMENDED READING

Beam, W. R. 1990. *Systems engineering: Architecture and design.* New York: McGraw-Hill.

Blanchard, B. S. 1991. *System engineering management.* New York: Wiley.
These two textbooks provide a thorough introduction to systems engineering and general treatment of systems beyond the scope of this text.

Kamm, L. J. 1991. *Real-world engineering: A guide to achieving career success.* New York: IEEE Press.
Delightful and insightful reading for the practicing engineer.

Middendorf, W. H. 1990. *Design of devices and systems.* 2nd ed. New York: Dekker.
Interesting reading on ethics, liability, and invention.

Petroski, H. 1985. *To engineer is human: The role of failure in successful design.* New York: St. Martin's.

Petroski, H. 1994. *Design paradigms: Case histories of error and judgment in engineering.* Cambridge: Cambridge University Press.
These two books are readable essays on failure and experimentation and very good. Most of the content is mechanical and architectural engineering, but it still provides a good perspective for electrical engineers.

1.14 REFERENCES

Blanchard, B. S. 1991. *System engineering management.* New York: Wiley.

Ferguson, E. S. 1993. How engineers lose touch. *Invention and Technology* 8(3):16–24.

Petroski, H. 1985. *To engineer is human: The role of failure in successful design.* New York: St. Martin's.

2

ARCHITECTING AND ENGINEERING JUDGMENT

Where there is no vision, the people perish.
—Proverbs 29:18

2.1 GOOD ENGINEERING

Good engineering is a mix of analytical, logical, experimental, and judgmental skills. Chapter 1 introduced the general principles and philosophy behind designing a good product. This chapter builds on those principles with heuristics and examples.

First, good engineering requires integrity and good communication. Integrity provides the foundation of character that motivates right actions and responses. Good communication must follow integrity to provide synergy—that is, effort exceeding the sum of its parts. Rational knowledge places a distant third by providing the tools of the trade. Soon the technical content of this text will be obsolete, but integrity and communication will always be necessary to implement the current technology.

Second, good engineering plans for the long term. Architecting will help you develop long-term vision, goals, and plans. Throughout the book I will interject concerns about testing, maintenance, liability, and ecological impact—all long-term issues—when examining specific examples of product development.

Third, good engineering employs technology to give customers understanding and control. Cause and effect must be clear; people won't use what they don't understand. (How many people do you know who fear to touch a desktop computer yet will drive a new car with complex embedded computers for engine control without a second thought?) Consequently, validation of purpose should continue throughout development of an electronic instrument.

2.2 QUESTIONS TO ASK

When you begin any new project, you must answer a number of questions to form an understanding of the problem. The purpose of this section is to get you thinking of the appropriate questions for your particular project; it is not meant to be an exhaustive list of questions that will define any problem. (Appendix B lists some questions that you can use in design reviews that will help define the problem.)

Determine the type of problem:

- Experimental—one of a kind, custom
- Specialty or limited-run production
- Large production run

What market will your product serve?

- Military
- Industrial
- Medical
- Commercial

(Sometimes a commercial product has to withstand worse abuse than a military product, but it generally doesn't have to deal with the environmental extremes.)

Who will use your product and what control interface does it need?

- Skills and education of users
- Culture

What is the function of your product? What is its capacity?

- Data throughput and communications
- Memory
- Energy consumption
- Size

How will your product be manufactured? Will it need maintenance or is it a throwaway? If it needs maintenance, what level of expertise is needed for repair? How much will it cost the customer?

To what standards will your product conform?

1. Market
   a. Military
   b. Industrial
   c. Medical
   d. Commercial
2. Operational and safety
3. Hardware
   a. Electromagnetic compatibility
   b. Reliability
4. Software
5. Human factors
6. Environmental and ecological

How complex is your product? How much time do you have to develop and deliver it? Where can you optimize? Often a single factor drives optimization, and it depends on the market (Kaplan, 1987):

- Consumer: cost
- Avionics: weight
- Medical: safety
- Military: reliability

How does it fit into the organizational culture of your company?

2.3 ARCHITECTING

The purpose of architecting is to conceive, build, test, certify, and operate a system that maintains its integrity and performance. It is a process that matches, balances, and compromises function and form; it helps make designs flexible enough to accommodate changes caused by time and circumstances (Rechtin, 1992).

To understand architecting, you need to define systems first. Eberhardt Rechtin (1992, p. 66) defines a system as "a collection of different things so related as to produce a result greater than what its parts, separately, could produce." A primary concern in architecting is to understand the interactions between the components and how they contribute to the overall behavior of the system. Rechtin goes on to say, "First, all systems have subsystems and all systems are parts of larger systems. . . .Thus, complex systems cannot be architected, built, or operated in isolation." You should always look for the "big picture," or global perspective, when developing a product; it is a system that fits within a larger system.

You can use architecting in one or a combination of four different approaches (Rechtin, 1992):

1. The *normative* approach includes standard practices and quantitative solutions, such as communications protocols and design handbooks.
2. The *rational* approach uses quantitative analysis and algorithms; an example is the scientific method of data collection, hypothesis, and testing. Optimization through analytical solutions is a goal in the rational approach. Most university courses teach this approach.
3. The *argumentative* approach uses broad participation and aims for consensus. It requires good team dynamics and group commitment to a common goal. Brainstorming sessions are a good example of the argumentative approach. Its weakness is design by committee.
4. The *heuristic* approach employs common sense and rules of thumb that are based on the experience and judgment of the architect. This approach is best for predicting and avoiding pitfalls and for recalling lessons learned. An architect can quickly discard unreasonable options through heuristic judgment to ensure the integrity of the product. Heuristics provide bounds on design solutions but not necessarily an optimal solution.

A product or system can be implemented by any of four design methods, or tools, which represent combinations of the four architecting approaches described above:

1. *Top-down* (constraint driven): normative and rational approaches
2. *Bottom-up* (built from current designs): argumentative approach
3. *Outside-in* (interface driven): rational, argumentative, and heuristic approaches (can be a superset of top-down design)
4. *Inside-out* (technology driven): heuristic approach (zero-mass design) (Wallich, 1987)

Zero-mass design starts by asking, "What does this product do?" It is a nihilistic approach to building a working product; "the technique is at least partly predicated on failures. 'If your first design works, . . . then it's overbuilt'" (Wallich, 1987, p. 38).

Regardless of the design method (that is, *how* you do it), you also have a process model, or schedule, that explains *when* you're going to do it. The process model quantifies the effort by laying out the steps you take to solve a problem. Software management tools are available that can help you prepare a realistic schedule.

Here are some of the important steps you must schedule. Beginning with available technology, use analysis to drive the requirements. When you have to break new ground, provide latitude for changes in the conceptual design. Test and characterize components before using them in your design. Be sure to validate the design specifications so that the product actually satisfies the customers' desires. Test and verify the performance of the design. Finally, manufacture, deliver, and maintain the device.

2.4 DESIGN CONCERNS AND HEURISTICS

During architecting of electronic instruments, you must understand certain issues to make effective trade-offs and produce good designs:

- Standards
- Defensive design
- Interface definition
- Concurrent engineering
- Heuristics

I finish this section with some cautions to temper the application of these principles.

2.4.1 *Trade-offs and Optimization*

Most products have a critical factor that limits usefulness; usually it is one of the following (Wallich, 1987):

- Cost
- Weight or size
- Performance
- Safety
- Power
- Reliability

You should first optimize the design for that critical factor. (The more sophisticated the system and its application, however, the less likely that it has a single critical factor.)

You must select the philosophy of design for your product; in particular, consider *modular* versus *custom* design. A modular architecture allows parallel effort in design, test, and fabrication and consequently can be quicker to build and verify. On the other hand, a custom design can optimize a single function (such as performance, weight, or size) for a particular application more easily than the modular approach can. Generally, custom design is better for single-purpose devices. Larger, more complex systems require modular architectures to control the escalating cost of testing, verification, manufacture, and maintenance.

While modularity concerns the relationships between components, *commonality* concerns the relationships between models of similar products. Commonality of design within a market can reduce redesign time, maintenance, inventory, and training. Example 2.4.1.1 gives an extreme but illustrative example of commonality in design.

Example 2.4.1.1 The Airbus A330 and A340 airliners share a common design for the airframe (wings, fuselage, landing gear, and avionics). The difference is that the A330 has two jet engines while the A340 has four. Commonality in design between the two models reduces inventory and allows for common assembly stations. It also improves the utilization of the aircrews, because the flight controls and characteristics are similar (Bak, 1993).

Other areas for consideration and optimization include the following (Evanczuk, 1989):

- Testability
- Manufacturability
- Maintainability
- Reliability

Sophisticated systems require greater capability for testing to measure performance and ensure function than simpler devices do. Similarly, more complex systems need greater access for maintenance. Furthermore, producing many units increases the importance of design for manufacture to reduce cost. Reliability can reach into many areas of product design, including thermal analysis; fault-tree analysis (FTA); and failure mode, effects, and criticality analysis (FMECA).

Reduce the parts count for higher reliability and lower cost. Fewer parts usually take less time to assemble, so labor costs are lower. Fewer parts also have fewer potential locations for failure. Of course, all this may be offset by complex components that are finicky and expensive if you substitute them for larger circuits that use simpler, cheaper, more reliable components. Use standard parts—such as microprocessors, buses, form factors, operating systems, and communications protocols—wherever possible, for higher reliability and lower cost (Wallich, 1987).

Table 2.1 illustrates some of the steps where trade-offs and optimization can occur in the development cycle.

2.4.2 *Standards*

Standards help ensure quality and interoperability with other equipment. Besides guidelines that are specific to your market (they will be mentioned throughout the book), you will probably encounter the ISO 9000 quality standards. These are a set of standards to certify your design and production process. ISO 9000 provides a road map for the series of standards 9001 through 9003. ISO 9001 is the broadest and covers design, manufacturing, installation, and servicing. ISO 9002 is a subset of 9001 and covers production and installation; it is meant for commodities where little design is done. ISO 9003 targets distributors, not engineering or manufacturing, and is a subset of 9002; it covers final inspection and test. ISO 9004 gives guidelines for managing quality.

2.4.3 *Defensive Design*

You are responsible for the effectiveness, quality, and safety of your product. Legal liability means that a manufacturer that negligently causes injury to a person should pay the victim's losses. A manufacturer is liable for the following (Bell, 1987):

- Misrepresenting the safety of the product
- Faulty manufacture
- Failing to warn of dangers
- Failing to provide instructions for safe use
- Failing to design an adequate safety margin for the intended use

A device can be ruled defective if it is poorly designed, poorly manufactured, or inadequately marked with warnings.

You and your company can reduce the risk of liability. First, provide appropriate and obvious warnings. Next, try to uncover and compensate for the unusual or abusive use of your product, such as the "woman who electrocuted herself by putting an electric heater in the bathtub with her to heat the water" (Bell, 1987,

Table 2.1 Trade-offs in the development cycle of an electronic product (adapted from Wallich, 1987, pp. 36–37)

Concept	Development	Product	Marketing and Maintenance
Conceptual design Nearly all costs stem from the initial design. Spend more time here and less in later changes. Software design has similar concerns. Balance modularity with customization and commonality with uniqueness. Identify the critical factor.	**Components** Consider testability, reliability, and automatic insertion among the trade-offs between simpler, less expensive parts and more complex and integrated parts.	**Cost** Cost depends on the application and involves a trade-off with performance and function.	**Sales and service** Availability and utility are better than capability. Meet the needs of the customer. Good documentation and adherence to industry standards are important considerations.
Specifications Specs constrain the design and set the performance and costs. Balance flexibility with defining a clear-cut course of action. An experimental system tends to need less rigid definition than a mass-produced or military system. Make sure specs can be verified in later testing.	**Assembly** Automated assembly is good for large production runs, but it needs to operate to capacity for lowest cost. Contract manual labor may be appropriate for small or custom lots. Careful layout of components can improve quality and lower cost during assembly and improve maintainability and testability.	**Performance and power** Generally, higher performance requires greater power consumption. Thermal dissipation goes up with power consumption. Internal communications within a chip are faster and less power hungry than external communications between chips (on a circuit board).	**Maintenance** Reliability, testability, and maintainability often go hand in hand. Maintainability depends on the application and involves a trade-off between production cost and effort to diagnose problems.
Analysis Thorough analysis and simulation saves time and money. This is the place to examine most trade-offs and optimizations.	**Testing** Testing during manufacture reduces rework, speeds diagnostics, and defines quality. It is necessary to verify performance against specifications. Balance the extent of testing with the consequences of failure and liability.	**Size and weight** Less is almost always better. Reducing components reduces size and weight and improves reliability. Size and weight are proportional to power consumption.	**Inventory** Consider the shipping environment: shock, vibration, and temperature. Size and shape will affect the ease of packaging.
Technology Understand the state of the art to identify potential challenges. Off-the-shelf components can reduce risks in design, testing, and reliability but may result in larger systems and higher manufacturing costs. Integrating many logic functions into a gate array or application-specific integrated circuit (ASIC) may significantly reduce size, power, and failures.	**Prototypes** Validate the need and purpose of the product. It must be quick and inexpensive to debug, change, and use. There is a trade-off between simulations and prototypes: Prototypes are good for more complex systems with many unknown interactions.	**Reliability** Reliability depends on the application and involves a trade-off with performance. Often circuit integration onto an ASIC improves reliability over circuits with discrete components. Simpler designs usually are better than more complex designs.	**Disposal** There is a trade-off between purchase cost and ecological impact for disposal of batteries, recycled enclosures, and manufacturing by-products.
Tools Good tools can speed development but are expensive and may require a significant learning curve. Balance current capability with needs for the project. Facilitate communication between team members.	**Software** There is a trade-off between extensive use of software with hardware. Software is flexible in developing complex algorithms, but hardware is faster in simpler operations. Use hardware to limit the effect of faults in software.	**Human interface** Utility depends on obviousness. Fewer, more understandable functions are better than many or complex operations.	

p. 39). Carefully document your design to establish a record of following standards with due care and foresight. Have a legal review of your product's design to ferret out possible risks. Finally, get liability insurance.

2.4.4 *Interface Definition*

The outside-in method first defines interfaces between components before specifying them; it is appropriate and effective for many product designs. It provides modular, well-defined interfaces so that various teams can design different subsystems concurrently (Reinertsen, 1992). Failure to define interfaces early forces excessive communication between teams and may cause later rework of the modules when the definition is finally set.

Example 2.4.4.1 A company that builds and sells products for local area networks (LANs) used the outside-in method for developing a bridge/router. Their effort was described as follows: "It's important to define the problem, partition the problem and set up the interfaces correctly before you do each of the modules. . . . [We] spent a lot of time up front doing that. Without that, we would never have been able [to] create a system this complex"(Child, 1993, p. 101).

Interfaces allow assignment and testing of individual functions as separate modules and reduce the complexity of system interactions. They have several configurations and purposes such as loose or tight coupling.

Loose coupling is an interface configuration in which modules are not highly dependent on each other. It provides several advantages over other configurations (Reinertsen, 1992):

1. It tolerates changes, thereby making the system more robust.
2. It reduces rework for redesign.
3. It increases reuse in other designs and projects.
4. It aids concurrent and independent development of multiple modules.
5. It adapts more easily to market variations.
6. It eases the test burden because modules can be more fully tested before integration, thereby reducing the uncertainty in system performance and speeding the time to market.

Loose coupling among interfaces provides more margin, but it can produce a system that is more complex, bigger, heavier, or more power hungry than absolutely necessary.

On the other hand, *tight coupling* makes modules highly dependent on each other. Its characteristics include the following (Reinertsen, 1992):

1. It reduces the initial cost of manufacturing.
2. It optimizes performance.
3. It makes redesign more difficult because redesign affects the entire system.
4. It requires more extensive testing of the system.

The primary purpose of interfaces is to hide complexity. As mentioned already, good interface design should reduce the burden of both testing and verifying system performance. Even more important, a good interface design should reduce complexity in capturing, organizing, and communicating information. Nowhere is this more true than in the user interface. In general, you should keep the user interface simple and intuitive (see Chapter 4). Bob Brunner, manager of industrial design for the Apple Newton, says, "Technology should hide complexity, not overwhelm you" (Gottschalk, 1994, p. 72). Often users won't use the full capability of an instrument if the configuration and interface are too complex. A designer of microprocessor emulators remarked, "If a feature isn't understood, it doesn't exist" (D'Alessandro, 1993, p. 91).

2.4.5 *Concurrent Engineering*

Concurrent engineering is a recent development that redefines good engineering practices. Concurrent engineering is the simultaneous development of various facets of a product that demands communication between different disciplines, such as design, engineering, manufacturing, and marketing. Its tenets are built on sensible notions of human conduct and collaboration to build better products. Carter and Baker (1992), in their book *Concurrent Engineering: The Product Development Environment for the 1990s,* proposed five components of concurrent engineering: technology, tools, tasks, talent, and time.

Concurrent engineering takes advantage of the latest *technology* and industry standards to produce quality products. Companies should create a long-range, global vision that integrates R&D with development and overcomes the "not-invented-here" syndrome.

Tools should be compatible, integrated, and automated. An integrated database and a network make parallel efforts possible. The downside is that engineers usually need retraining to use the tools.

Concurrent engineering defines and divides up *tasks* to concentrate power, improve communication, and manage complexity. Upper management needs to shepherd the company *talent* through vision, continuing education, preparation for change, commitment, and granting authority. The final element, *time,* should be measured for future improvement in the development cycle.

Product development in concurrent engineering follows these steps (Carter and Baker, 1992):

1. Define a common vocabulary.
2. Agree on a common purpose based on customer requirements and company vision.
3. Agree on priorities.
4. Couple responsibility with authority.
5. Agree on the scope of the development (see domains below).

Carter and Baker give the scope of development one of four labels: task, project, program, or enterprise. Software consultant Timothy Lister uses four slightly different labels for scope or design domains: ad hoc, intuitive, process, and system (Jensen, 1993).

The *ad hoc* design domain is similar to the *task* scope. It is a simple design completed by one engineer and is usually experimental. An example of an ad hoc development is a stepper motor control for laboratory equipment.

The *intuitive* design domain is similar to the *project* scope. It is a little more complex than an ad hoc design: a single, well-defined process that is completed by a small team. It may be an experimental or custom application. An example of an intuitive development is a telephone answering machine.

The *process* design domain equates to either a *project* or a *program* scope. It requires rigorous planning for success and uses a set of well-defined processes operating in parallel developed by multidisciplinary teams. It can be used to develop custom or commercial or military products. Monitoring and coordination occur through meetings and discussions. An example of a process development is a medical device.

The *system* design domain equates to either a *program* or an *enterprise* scope. It uses more processes and requires a master plan of development with multidisciplinary teams. It develops complex or mass-produced equipment. An example of a system development is a satellite with its attendant coordination between launch services, range control, ground support, and subcontractors.

Most projects in this book fall into the intuitive or process domain, but the principles also apply to the system domain. As products become more complex, they move into the process and system domains. You must determine the scope of your project, identify its domain, and then communicate the domain and consequences to management. Strangely enough, people usually assume that project development is simpler than it really is.

In addition to the design domain, you must identify the processes within the project and the responsibility for handling them, specify the communication links, and set up metrics for reporting progress.

2.4.6 *Heuristics*

Heuristics are empirical methods for making sound engineering judgments. Each heuristic is gained from experience by practitioners in the field of endeavor and is communicated as a rule of thumb. For example, you may have heard of the heuristic called Murphy's law: "If anything can go wrong, it will."

Example 2.4.6.1 A friend who designs mechano-optical instruments reports a heuristic his company uses for estimating the feasibility of development: "A device is producible if the density of components (weight/volume) is equal to or less than the density of water."

Much of this book is based on heuristics. Each chapter has a unique set of heuristics for a specific area (such as enclosures, power, cooling, or software) within the development cycle of an electronic instrument. This chapter focuses on planning and executing project development. One rule of thumb, for instance, is that up-front planning sets the cost of the final product. Therefore, spending more time up front results in better final products. Table 2.2 lists a number of heuristics for developing electronic instruments.

2.4.7 *Cautions*

As with everything, there is a need for balance among heuristics. Some may be contradictory, or at least limited. For instance, "The customer is always right" is not true if customers rip you off or don't pay for your services. "The customer is usually right" is more accurate but lacks the punch of the original heuristic and allows too much latitude in judgment. I will introduce a few cautions here, related to the heuristics given previously. Ultimately, you will have to resolve which heuristics are appropriate for your project and company.

Unfortunately, *concurrent engineering* has become a fashionable buzzword that can be misused and abused if a serious commitment to real change is lacking. In particular, corporate inertia and company culture resist changes in the status quo and can derail the process (Huthwaite, 1993). Otherwise, concurrent engineering can be used to place blame and point fingers later; *trust is necessary.* For concurrent engineering to succeed, team members must be allowed to meet and have sufficient time to devote to the project.

Another buzzword that is often coupled with concurrent engineering is *time-to-market.* While timely introduction of the finished product into the marketplace is often the most critical competitive factor, hasty decisions made to satisfy the time-to-market window can lead to mistakes: as Proverbs 19:2 says, "he who makes haste with his feet errs." Careful planning does not have to be slow—just careful.

Metrics are necessary to gauge progress and improve development. But no universal design tool or single metric can measure progress, just as a speedometer does not describe the entire state of an automobile. You need a suite of metrics tailored to your application to describe progress adequately.

Table 2.2 Heuristics for architecting, designing, and building electronic instruments (adapted from Rechtin, 1991; Strassberg, 1992a, 1992b)

Category	Heuristic
Planning	Spend more time up front in design to reduce costs and delays later. Define the need clearly. Limit changes after definition, but don't be inflexible. Murphy's law: If anything can go wrong, it will. Occam's razor: The simplest solution is usually the correct one. Less is more (smaller, lighter, less complex, less power consumed). Expect the unexpected. Predicting the future is impossible; ignoring it is irresponsible.
Quality, management, and morale	Get management commitment to approach. Include design engineers in application and field research. Build loyalty through individual responsibility, ownership of solution, and initiative. Involve all team members early in project. Don't forget peripheral organizations (purchasing, publications, and customer service) and customers during development. Use consistent project-management tools throughout project. Don't do what others do better. Build in quality; it can't be tested in. Strive for milestones in quality. Demand good documentation (complete, concise, and correct). Don't fear mistakes.
Fresh approaches	Encourage team members to take risks. Learn from failures. Don't immediately reject off-the-wall ideas. Don't surrender to opposition unless there are very strong reasons to do so. Give the team additional challenging and useful goals, but don't expect to meet all of them. *How* you make technological, social, and cultural change is more important than *what* you do.
Good communication	Provide for immediate feedback on new ideas and suggestions. Locate team members in one area; have them rub elbows. Provide ready access to e-mail and networks. Document design and progress; transfer this knowledge to later projects. Make sure that everyone who needs to know knows; otherwise someone will foul up.
Customer service	Make the product serviceable. Consider documentation as integral to your product (brochures, data sheets, price lists, installation and maintenance manuals, user manuals).
Validation	Build simulations and prototypes. A model is not reality.
Interfaces	Relationships between components and subsystems determine the value of the overall product. Partition components and subsystems for loose coupling if possible. • Minimize communication between subsystems. • Local activity is high speed; global activity is low speed.

Table 2.2 Heuristics for architecting, designing, and building electronic instruments (continued)

Interfaces (continued)	Match functional operation with physical structures. Hide complexity. • Make operation intuitive and obvious. • High internal complexity is okay, but external complexity should be low. Don't tack on "features" just because they are available.
Building, testing, diagnosing, and repairing	Pay attention to component placement; it affects all later operations. The failure rate of a component or system is proportional to its cost. Performance, cost, and schedule are not independent. Find and fix problems at the source. Cost of repair increases exponentially with development. "Quick-look" analyses are often wrong. Uncover the failure mechanism and eliminate all other explanations of failure before declaring a successful recovery. Knowing that a failure has occurred is more important than the actual failure. The probability of failure is proportional to the level of upper management in the vicinity. Fault avoidance is far better than fault tolerance. Multiple, simultaneous failures happen and often are linked.
Culture	No product is an optimal solution. ("You can't fool all the people all the time" or "You can't satisfy all the people, ever.") Don't assume that being good at one thing implies being good at another. Reduce the number of people and sequential operations required to accomplish a project. (The probability of implementation is an inverse function of the number of people in the chain and depends on their understanding and communication of the idea.)

Another caution: Too much emphasis on customer feedback will lead to creeping featurism (the continuous addition of new capabilities to a product) as well as prevent the introduction of new products. Creeping featurism drags out development and complicates testing; this is particularly easy to do with software. At some point you have to solidify the requirements and product definition to finish development. Customer feedback can also slow innovation because customers cannot predict features yet to be invented; only you can imagine them. An editorial by Michael Slater (1992) in *Microprocessor Reports* states, "Customers often don't know what is possible, and if vendors just build what customers ask for, progress will be slow."

2.5 TEAMWORK AND TRUST

Teamwork is essential to any successful product. Fashions and fads in engineering will come and go, but teamwork will remain. It is the synergism and trust between different people with various capabilities working toward a common goal. Teamwork includes the user, client, management, designers, manufacturing, procurement, and vendors. Table 2.3 illustrates the people and phases involved in development.

Table 2.3 People involved in the product life cycle (adapted from Kaplan, 1987)

	Product maturity →													
	Concept	Conceptual design review	Analysis	Specifications	Preliminary design review	Design	Prototypes	Critical design review	Validation/verification	Production	Sales	Training/customer help	Maintenance	Disposal
Management	•		•		•			•		•				
Design architect	•	•	•	•	•	•	•	•	•	•		•	•	
R&D	•	•	•	•	•	•	•	•						
Electrical engineering	•	•	•	•	•	•	•	•	•			•	•	
Software engineering	•	•	•	•	•	•	•	•	•			•	•	
Mechanical engineering	•	•	•	•	•	•	•	•	•			•	•	
Industrial engineering	•	•	•	•	•	•	•	•	•			•	•	
Manufacturing	•	•		•	•	•	•	•		•				
Purchasing	•	•		•	•	•		•		•				
Publications	•			•		•		•		•	•	•	•	
Marketing	•	•	•	•	•	•	•	•			•	•	•	
Service	•	•		•	•	•	•	•					•	•
Financial			•							•				
Legal				•	•			•						
User and customer	•						•				•	•	•	•

2.5.1 *Commitment*

Teamwork requires commitment from all members. Vision, responsibility and authority, and inclusion all help in gaining that commitment. Charles Rogers, a manager with IBM, says that a "focus on the project rather than on an individual's functions" results from a responsible, committed team (Cortes-Comerer, 1987, p. 43). His claim is reinforced by Wolfgang Feix of Siemens U.S.A., who says, "When it's all your [the group's] responsibility, . . . motivation reaches levels you never imagined possible" (p. 44). Furthermore, Dan Strassberg (1992b, p. 105), in an editor's analysis, writes, "Create an atmosphere in which manufacturing people are not afraid to express their ideas. Listen when they talk. Incorporate manufacturing suggestions into your designs." And learn from previous failures. Bart Huthwaite (1993, p. 72), director of the Institute for Competitive Design, writes, "Call on those who have suffered from previous design flaws to help you avoid them in the new design. Give them a *pro-active* role, not one that simply asks them to review your design."

2.5.2 *Management*

Working together as a team involves numerous concerns, such as an understanding of each other's problems, meetings where all participate, responsibility and authority to work together, and individual support for the decisions made by the group. Encouraging teamwork is a function of management, who must provide the vision, select the right people, and emphasize cooperation: "a 'contract' between top management and the project team that spells out the schedule, budget, resources, product performance, and time-to-market, plus the team authority to carry it out" (Gautschi and Goldense, 1993, p. 76). The contract should be written if it is to mean anything.

2.5.3 *The Customer*

The customer plays an integral part on this team. Hewlett-Packard believes that "exposure to customers, as well as to sales representatives who have to sell the product, broadens an engineer's outlook" (Cortes-Comerer, 1987, p. 46). Keithley Instruments "rotates design engineers into applications-engineering assignments lasting approximately six months. Answering dozens of phone calls each day about how to accomplish specific tasks using instruments you helped to design (and whose operation you probably thought was self-evident) changes your perspective" (Strassberg, 1992a, p. 65).

2.5.4 *Communication*

If people are the necessary components of teamwork, then communication is the glue that holds them together. Face-to-face discussions, informal walk-throughs, formal design reviews, networks and e-mail, on-line database and configuration management, and documentation all serve in good communication.

Documentation is integral to the architecture of the product and plays a vital role in communication. Documentation includes engineering notebooks, drawings, flowcharts, interactive monitoring and tracking, records of all decisions, and memos. One final heuristic: Define acronyms early and stick to them; doing so reduces confusion and demonstrates clarity of thought and action.

2.6 A COMMON PROBLEM: REAL-TIME CONTROL

While this chapter is a general introduction to the heuristics of development, many of you will encounter real-time control in one form or another. Consequently, I will examine some technological issues that often occur in embedded control.

Control has two basic configurations: open loop and closed loop. *Open-loop* control systems have no indication of whether the control action really occurred. The system is unaware of whether the output responded to the input or not. Examples of open-loop systems are light-emitting diodes (LEDs), displays, solenoids, and relays. *Closed-loop* control systems sense the actuated output and return a result (*feedback*) to the input. Typically, one or more output parameters—such as position, velocity, acceleration, angle, or rotation—are monitored for feedback.

Feedback sends a portion of the output signal to the input for comparison with the input command. Feedback closes the loop and provides the system with a sense of how well the output responded to the input signal. It makes output responses more predictable and less sensitive to variations in the system. Fig. 2.1 illustrates a general block diagram of a closed-loop control system.

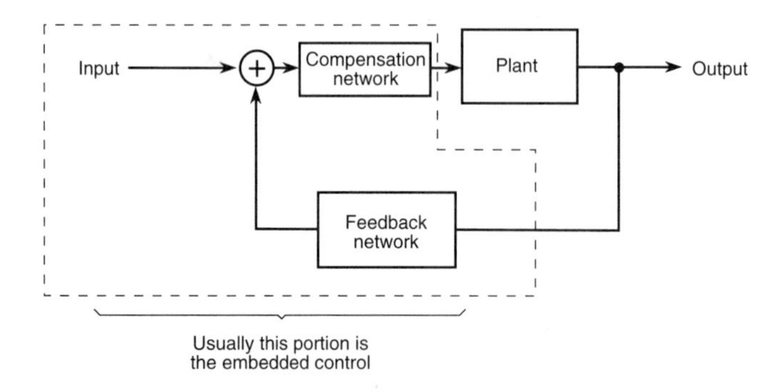

Fig. 2.1 General outline of a control system.

Example 2.6.1 The cruise control in an automobile uses feedback to maintain speed. Setting the desired speed provides the input command. The rotational speed of a wheel is the output. Feedback sends the rotational speed (a count of wheel revolutions per unit time) to the cruise-control computer, which converts it into vehicular speed and then compares it with the input setting. If the comparison shows the car's speed to be lower than the setting, the cruise control opens the throttle further to speed the car up. If the speed is higher, the cruise control closes the throttle proportionally to slow the car down. Consequently, feedback reduces the sensitivity of the car's speed to hills and valleys and maintains the desired speed.

The *plant* is the device or component to be controlled. In the example, the plant is the engine and drive train, but it could be a motor, a vehicle, an amplifier, or something else. The feedback converts the physical property of the output—such as shaft speed, vehicle position, or load current—to an electrical signal. The embedded control system compares the feedback with the desired command at the input and adjusts the plant accordingly.

Feedback provides two functions:

1. It reduces sensitivity to parameter variations in the plant.
2. It rejects disturbances injected into the plant.

Feedback has side effects such as oscillations and overshooting, as well.

The concerns about closed-loop control are response, accuracy, stability, and plant characteristics. For response, you are interested in how long the system takes to reach steady-state equilibrium. Once at steady-state equilibrium, the accuracy should be good, meaning the amount of error that remains is small. For stability, you usually want no wild oscillations and a rapid decay of any that occur.

The concerns about plant characteristics are energy storage, energy loss, and motion and signal delays. Energy storage can be inertia, springiness, inductance, or capacitance. Energy loss usually occurs through friction or electrical resistance. These factors all affect how the system responds to any signal. Ultimately, energy storage and signal delays shape the frequency response of the system.

2.6.1 *Potential Problems*

Aside from understanding the plant characteristics, you must understand how digital processing can affect system operation. Digital control can degrade, if not destroy, the operation of a system through low sampling rate, harmonic distortion, low resolution, or long processing time (Jones, 1993).

A low rate of sampling will fold high-frequency components of the signal into the low-frequency band, through a phenomenon called *aliasing*. The *Nyquist limit* sets the absolute minimum rate of sampling at twice the highest frequency. In practice, sampling at greater than 10 times the highest frequency in the system is far more reliable and reduces the possibility of aliasing. Furthermore, filtering the signal input to remove high-frequency components will relax constraints on the sampling rate. These types of low-pass filters are also called *antialias filters.*

Analog outputs from digital control systems must undergo a digital-to-analog conversion. The conversion process necessarily produces sharp transitions between discrete values that translate into high-frequency components called *harmonic distortion.* A low-pass filter on the output of the digital-to-analog conversion will reduce harmonic distortion. Even then, the digital system must have enough bits of resolution to provide an adequate signal-to-noise ratio (see Chapter 12 for more detail).

Signal delay due to processing time within the embedded controller can contribute to system instability. The longer the delay for processing, the more difficult the compensation. Furthermore, variable durations for processing make compensation nearly impossible.

2.6.2 *Four Types of Control*

Four classes of control that you may encounter are linear, nonlinear, fuzzy logic, and neural networks. (Obviously, many more exist, but these will give you some feeling for the variety available.) Each has advantages and particular areas of application.

Linear control is the most well developed and understood. It can be difficult to use in real-time control because it requires a linear system and an accurate model of the plant.

Nonlinear, or complex, systems often require unique designs that tolerate a wide range of conditions. Understanding the operation of the system and its environment is critical to designing a successful nonlinear control system. Often, designers force nonlinear systems to operate in an approximately linear region so that they can apply classical linear control. On the other hand, fuzzy logic and neural networks are robust and can sometimes simplify the design of real-time control for nonlinear applications.

Fuzzy logic is useful when exact mathematical descriptions do not exist; linguistic descriptions of the system suffice to design the control. Fuzzy logic is good for complex situations where an expert system provides the control; the system can give the reason for each decision it makes. Conversely, an expert system is only as good as the expert that programs it. The more numerous and complex the decisions, the more computational power it requires.

Neural networks are inherently parallel in their decision-making structure and give fast response. They are good for pattern matching. They are "trained" on subsets of real data rather than programmed, so they do not require expert programming for each application. But neural networks cannot necessarily provide the reason for each decision they make.

2.6.3 *PID Control*

Proportional-integral-derivative (PID) control is a common configuration used in linear real-time control. Applications include thermostats, servos, and cruise control for vehicles. PID control has a well-known theoretical background (an advantage), but it is limited to linear systems. Also, you must adjust the proportionality constants for each application; that, in turn, requires a very good understanding of the linear models for each system.

Fig. 2.2 illustrates the general form for PID control. The proportional term adjusts out the difference, or error, between the input and the output. The integral term reduces the steady-state error. The derivative term increases stability by adding damping.

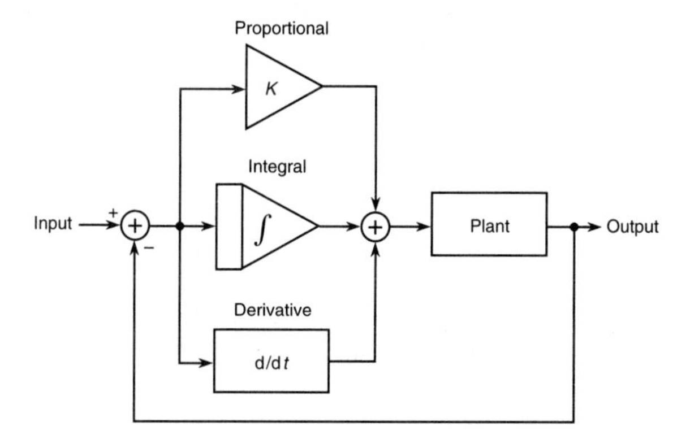

FIG. 2.2 General form of a PID control system.

The proportional term essentially provides an amplifier with adjustable gain to reduce the difference between the input and the output. It has no time dependence because it does not have energy storage or frequency dependence. A system with only proportional control can be unstable, however, and have offset errors in the steady state. Fig. 2.3 illustrates how a simple proportional control responds to a step change in the input.

The integral term integrates the difference between input and output to reduce steady-state error. Integral control improves the transient response, but it may lead to oscillatory behavior (instability). Fig. 2.4 illustrates the response to a step change in the input for a control system with proportional and integral terms.

The differential term anticipates the error and provides early correction, but it is effective only during input transitions. Derivative control increases stability by

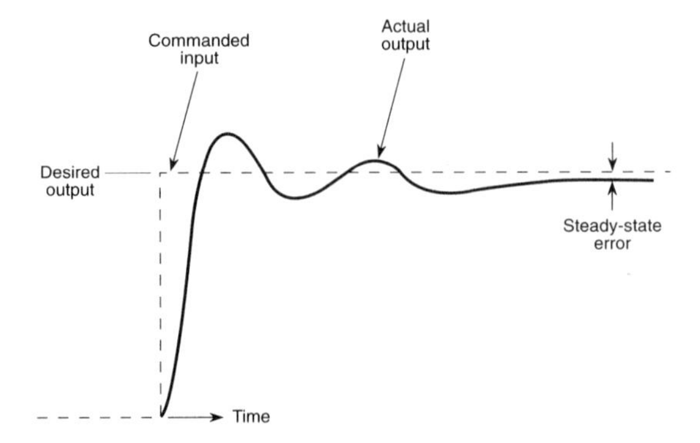

Fig. 2.3 Response to a step change in the input for a proportional-only controller.

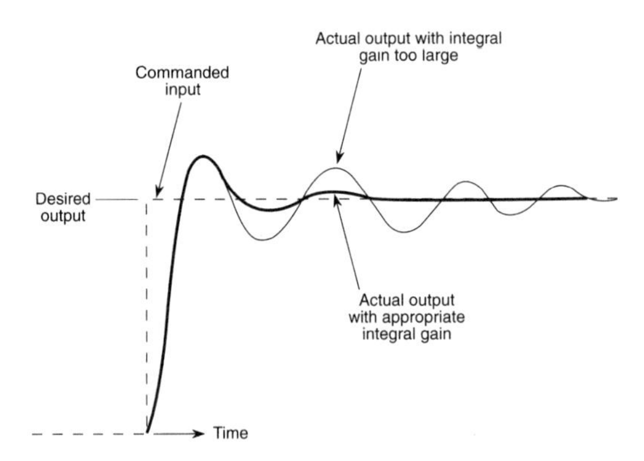

Fig. 2.4 Response to a step change for a proportional-integral controller.

adding damping that permits a large proportional gain. Too large a derivative term adds significant damping that slows the response. Derivative control also amplifies noise signals and is never used alone. Fig. 2.5 illustrates the response to a step change in the input for a PID control system with all three terms—proportional, integral, and derivative—operating.

Ogata (1970) provides good traditional approaches and mechanical models for PID control. Vanlandingham (1985) is outstanding for implementing PID control with digital systems.

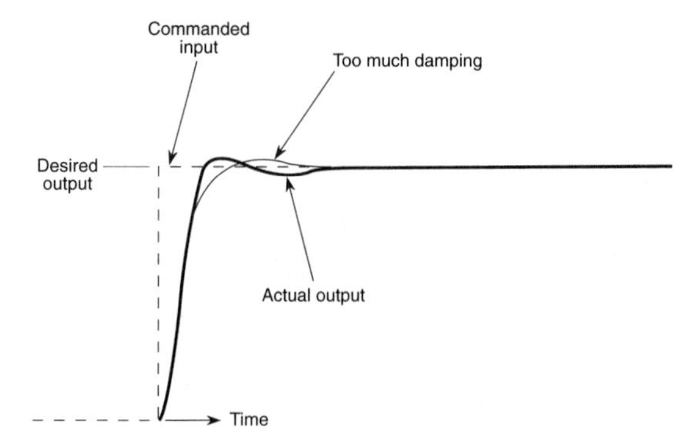

Fig. 2.5 Response to a step change for a tuned PID controller.

2.6.4 *Fuzzy Logic*

Fuzzy logic is rapidly finding a niche in real-time control, particularly for discrete, nonlinear, or complex situations. Applications include clothes washers, concrete kilns, train controls, automatic transmissions for automobiles, camcorder focus controls, and commercial dough mixers.

Fuzzy logic is a formulation of logic that deals with imprecise terms such as *very*, *most*, and *few* by rigorously describing the terms according to fuzzy set theory. Fuzzy logic works better than predicate logic and probability-based models when dealing with imprecise information because it (1) uses approximate reasoning to deal with the imprecise human concepts, and (2) provides a systematic framework for dealing with linguistic quantifiers (such as *many, infrequently, about*) to ease implementation of expert systems. There is nothing really "fuzzy" about fuzzy logic.

Fuzzy logic uses linguistic descriptions and *if-then* rules to provide the control action. It has a constrained operating space, or *universe of discourse*, that is robust and tolerant to failed sensors and inputs. The constraints on operation to a stereotyped format should simplify the design of real-time control for many applications and make software validation and verification more accessible in complex systems.

A system with fuzzy logic uses expert knowledge in an inference engine to generate control actions (Fig. 2.6). The expert knowledge base provides the specific knowledge for the application through membership functions and rules. The inference engine implements the knowledge and generates actual control actions. The inference engine acts like a compiler and assembles the necessary operations with three distinct components: fuzzification, rule evaluation, and defuzzification.

A distinguishing feature between fuzzy logic and probability is the membership function; it takes linguistic descriptions and assigns a "degree of member-

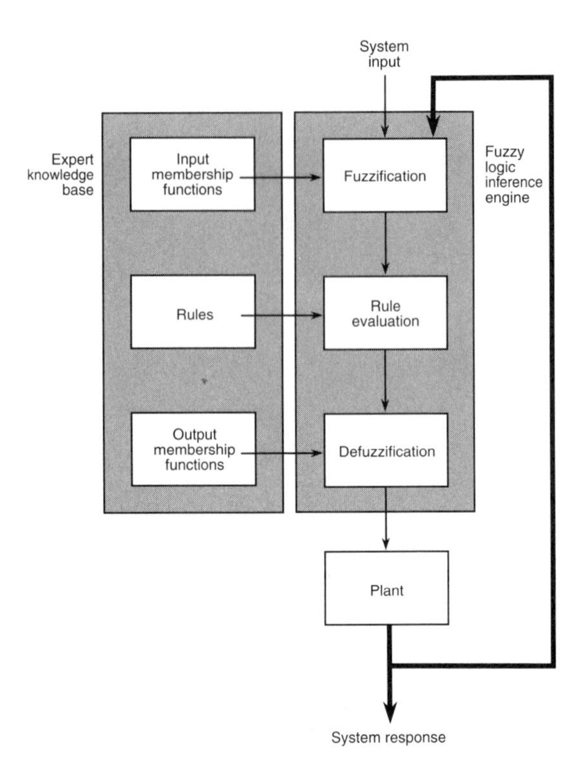

Fig. 2.6 A fuzzy logic system. Fuzzy logic uses an inference engine that draws on a knowledge base to fit each application. The fuzzification and defuzzification units convert between real signals and linguistic variables. The rule evaluation unit uses linguistic variables.

ship" to a variable that ranges between 0 and 1. This degree of membership describes how "truthful" or accurate the variable is. Describing room temperature as "hot" becomes more true (closer to 1) when the temperature approaches 25°C (77°F). Hotness decreases to false, or 0, as the temperature drops toward 15°C (59°F). Likewise, the temperature is "cold" (close to 1) as it approaches 5°C (41°F). Coldness decreases to false or 0, as the temperature rises toward 20°C (68°F). Fig. 2.7 graphs the membership functions for room temperature.

The x-axis is the *universe of discourse* and represents the possible values for one input to the system. The y-axis is the *grade of membership* and ranges from 0 (false) to 1 (true). Each overlapping outline is a membership function for the labeled *linguistic variable,* such as "hot" or "cold." The dashed lines indicate the degree of truth for two different linguistic variables: One is 0.7 true, while the other is 0.3 true.

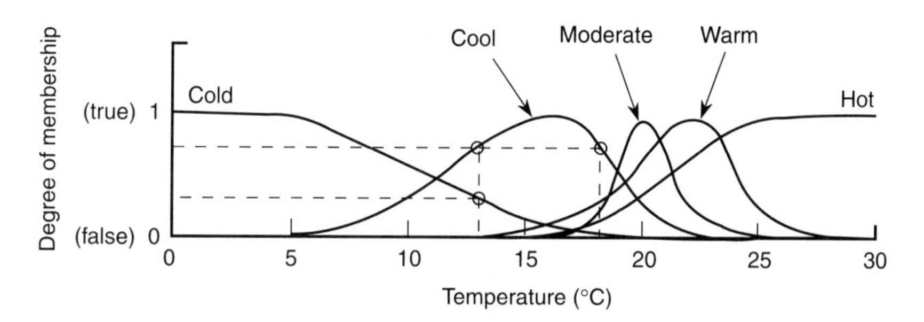

FIG. 2.7 An example of membership functions that describe temperature in a room. Each label (*cold, cool,* etc.) is a linguistic variable for one membership function. Each variable has varying degrees of membership. Cold is completely true, or 1, below 5°C and then tapers off as temperature rises; it is only 0.3 true at 13°C. Cool is true, or 1, at about 16°C but only 0.7 true at both 13°C and 18°C.

In contrast, probability describes the likelihood that a variable is either completely in a region (its membership is 1) or completely outside it (its membership is 0). Fuzzy logic allows a smooth gradient of membership over a range of values.

An inference engine uses these membership functions to map, or *fuzzify,* an input into the system. Then the inference engine evaluates the rules according to the knowledge base to infer outputs from the inputs. A thermostat based on fuzzy logic might evaluate some rules like this:

1. If temperature is hot, then command maximum cooling.
2. If temperature is warm, then adjust to medium cooling.
3. If temperature is moderate, then make no change.
4. If temperature is cool, then adjust to medium heating.
5. If temperature is cold, then command maximum heating.

For more complex rules that have overlapping outcomes, you must combine the outputs in one of several different methods to derive the final output. One inference method scales the output membership function so that it peaks at a value less than 1; another inference method clips the output membership function so that it has a "flat-top" value less than 1. Once the inference engine has prepared an output, it must convert the linguistic variable or fuzzy value to a single, crisp value. This step is called *defuzzification.* Fig. 2.8 illustrates an example of combination of rule outputs and defuzzification.

Fortunately, fuzzy logic is not complicated to use; software systems and compilers are available to make it easier. You only have to define the variables (in the membership functions), give the rules, and specify the inference method. The compiler guides you through each of these steps and keeps all the details.

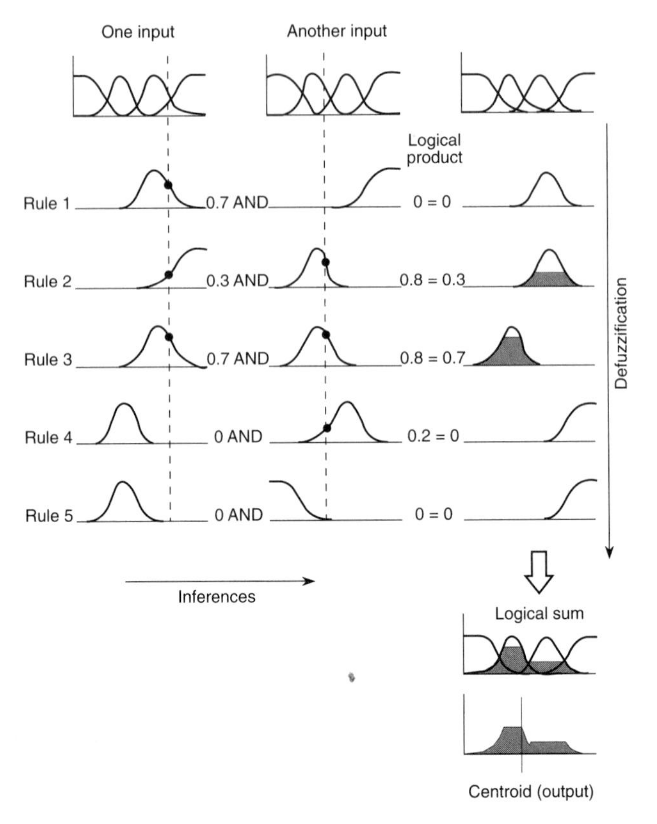

FIG. 2.8 A simple example of what might happen behind the linguistic variables. The inference method, logical product, takes the minimum value resulting from combination of the components within each rule, and sets a maximum output. All the outputs are combined, and the centroid provides the crisp output value.

2.6.5 *Neural Networks*

Neural networks use arrays of simple, nonlinear elements, called *neurons*, connected together. The connections between neurons may have varying effectiveness in coupling or *weights*. These weights change as you train the neural network. Signals from multiple neurons are scaled and summed by each neuron, and then a threshold function determines when the neuron will fire its own output signal. Fig. 2.9 gives a general outline of a neural network and a constituent neuron.

Once trained, neural networks have a high speed of response. The accuracy of response depends on how representative the training subsets are.

Neural networks are robust and withstand failure of individual elements well and failure of groups of elements fairly well. Their performance degrades gradually with failure.

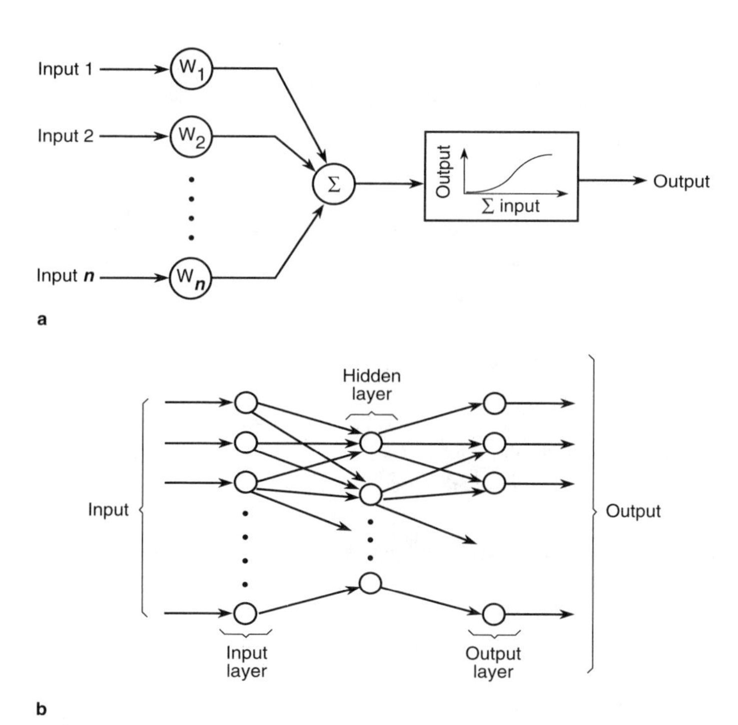

FIG. 2.9 General outline of neural networks. (a) A single neuron. (b) A neural network with three layers of neurons. Each circle represents a neuron.

2.7 CASE STUDIES

Any important point is best made with good examples. These case studies represent the use of some of the heuristics laid out in this chapter. You will find common problems and issues, but you will also find unique solutions in each of these studies.

2.7.1 *Video Gain Control for a Satellite Camera*

The Delta 183 project was an experiment for the Strategic Defense Initiative Organization in the U.S. government. The intent was to study the possibility of orbital interception of ballistic missiles. Specifically, the project developed a satellite in a very short time to observe missile plumes during boost.

I helped develop video cameras to view the visible and ultraviolet spectrum of space. Each camera had a video intensifier on its front end to amplify the brightness of the scene. A video intensifier is a bundle of millions of coated glass capillaries, each acting like a photomultiplier tube. When a photon strikes the inside surface of a capillary, it starts an avalanche of electrons that strike a phosphor screen that emits a dot of light into the camera video tube. Each capillary roughly equates to a picture element, or *pixel.*

The gain of each capillary is controlled by high voltage imposed between the front and back ends of the bundled capillary tubes. The intensified images have a wide dynamic range of brightness, up to 1 million; not only that, the intensity gain is exponentially related to voltage. The control, therefore, is discrete and nonlinear. Analysis showed that no analytical, closed-form solution existed.

I implemented the gain control with a microcontroller that generated discrete steps of gain. The circuit counted the number of pixels that exceeded a low-intensity (or dim) threshold and the number that exceeded a high-intensity (or bright) threshold and then decided whether to increase, decrease, or maintain the gain. If too many pixels exceeded the high-intensity threshold, the circuit decreased gain on the intensifier. If too few pixels exceeded the low-intensity threshold, the circuit increased gain. It had a fixed slew rate because it changed gain only by a fixed increment after each video frame. Fig. 2.10 shows the block diagram of the video gain circuit, and Fig. 2.11 gives the conceptual model of the operation of the gain circuit.

The critical factor for this circuit was stability in the intensity of the scene. (A previous design for an earlier project, called Delta 180, oscillated terribly. I was quite embarrassed to see the scene of an approaching rocket blooming and bursting—that is, oscillating from very dark to very bright—on national television.) Part of the problem is that a camera in orbit can view a wide variety of scenes:

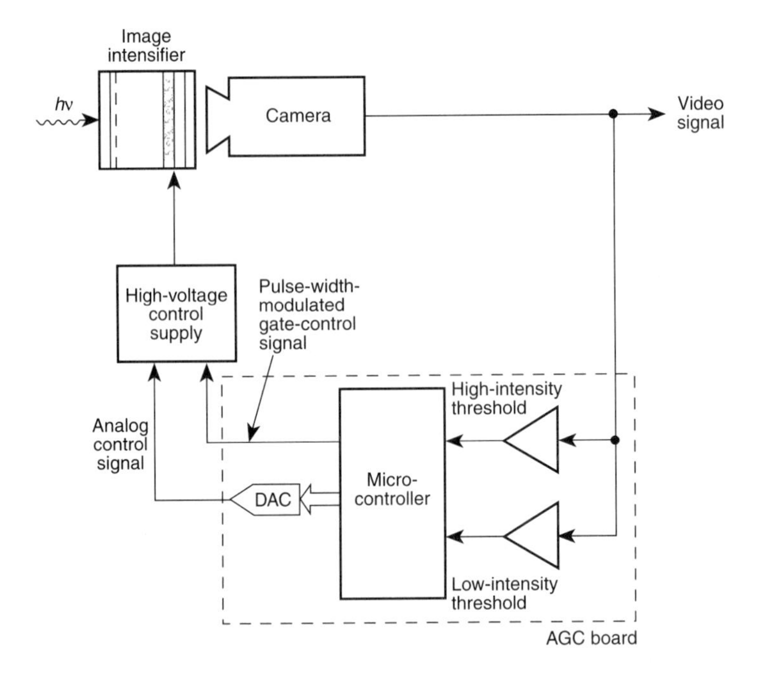

Fig. 2.10 Block diagram of the video gain control for an image intensifier. (AGC = automatic gain control; DAC = digital-to-analog converter.)

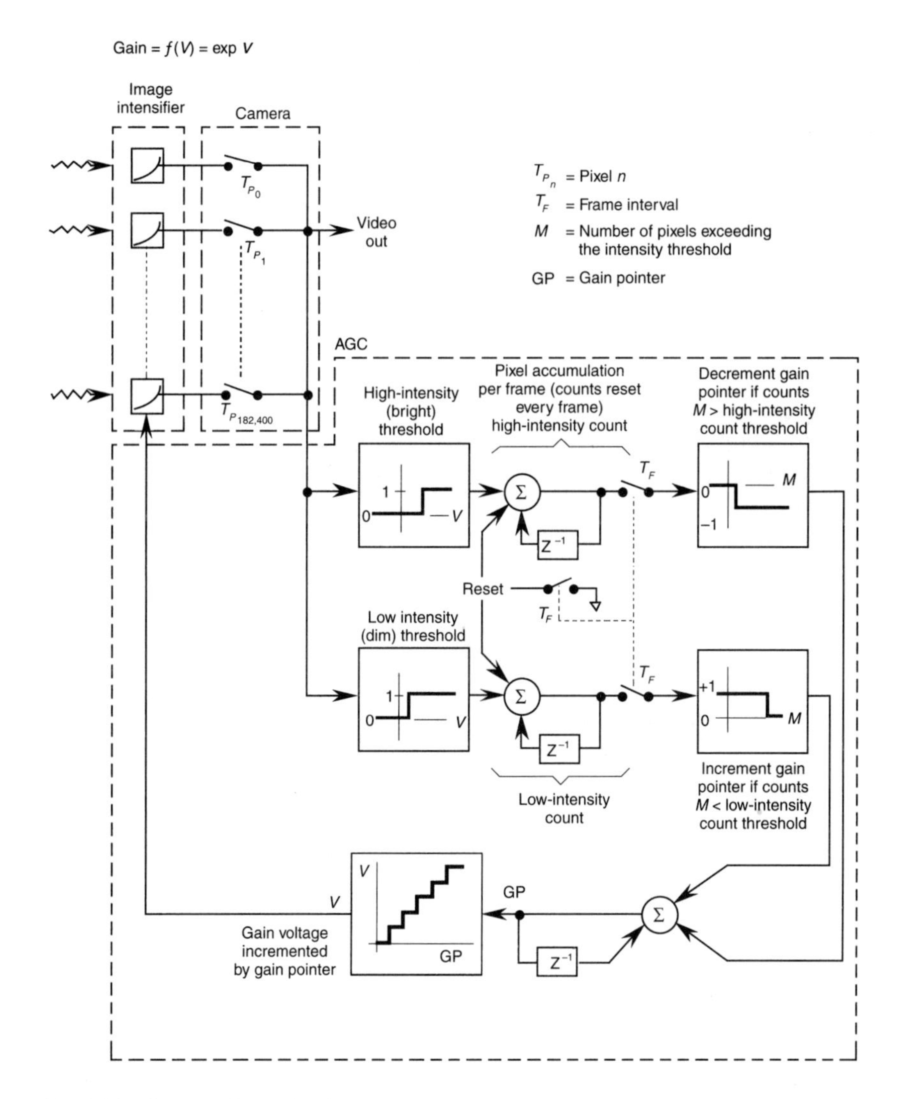

Fig. 2.11 Conceptual model of the video system.

blackest space, the earth's limb, the sun, and stars (bright point sources). Furthermore, individual pixels within a charge-coupled device (CCD) camera head will bloom (saturate on bright point sources) and exacerbate the stability problem.

A small team of engineers simulated and experimented with various sources to develop a robust algorithm. We avoided oscillation with a combination of values for gain step and threshold that allowed ample margin for variations in parameters.

This was a typical small project, a single well-defined circuit board within a much larger, complex system. It was an experimental project that had to be designed and built quickly—it used wire wrap on the circuit boards and non-space-qualified components throughout the satellite. The design domain would be closest to intuitive in scope because of the defined teamwork and the rigor of space-qualified design.

The system was nonlinear. A PID control system would have had trouble with the wide dynamic range of the camera and with remaining stable without blooming. By experimenting with different ranges of inputs, we tailored the nonlinear control system to remain stable for all inputs.

Scope: intuitive or project
Duration: 9 months
Number of people involved: 4 engineers, 2 technicians, 1 draftsman
Level of effort: full time
Lessons learned: The development of a nonlinear gain control for intensified cameras succeeded for two reasons:

1. Previous experience from a very similar project.
2. Careful simulation and experimentation at the beginning of the project provided a robust design.

This complex and nonlinear circuit could have used fuzzy logic for even faster development.

2.7.2 *Neurological Stimulation System (NSS)*

Electrical stimulation of the spinal cord can block or reduce chronic pain. Several companies serve a small but important market by manufacturing and marketing implantable devices for stimulating the spinal cord. A neurosurgeon threads electrodes into the patient's spinal column and then implants a receiver under the skin. After release from the hospital, the patient wears a transmitter on a belt that powers the receiver through radio frequency (RF) coupling and encodes it with the desired stimulation parameters. Fig. 2.12 illustrates the general configuration.

Adjusting the stimulator for each patient is extremely tedious because of the many possible combinations of parameter values. (Eight electrodes represent 6050 combinations of polarity; 16 electrodes have over 62 million possible combinations.) The NSS eases the adjustment burden by providing an interface between a computer and commercial stimulators. The software helps optimize treatment by selecting appropriate subsets of stimulation parameters from the wide variety available; the interface controls the stimulators to deliver the selected stimulation. The NSS has a graphics pad and high-resolution screen that allows patients to describe and tailor their own treatment. Fig. 2.13 illustrates the block diagram of the NSS; Fig. 2.14 shows one configuration of the equipment.

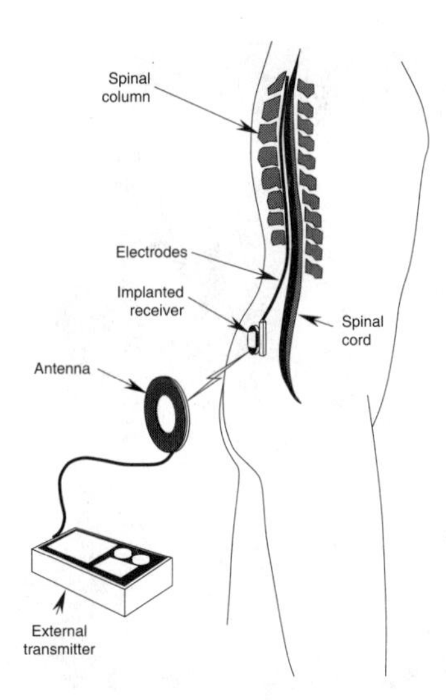

Fig. 2.12 Equipment for electrical stimulation of the spinal cord to relieve chronic pain in the lower back and legs.

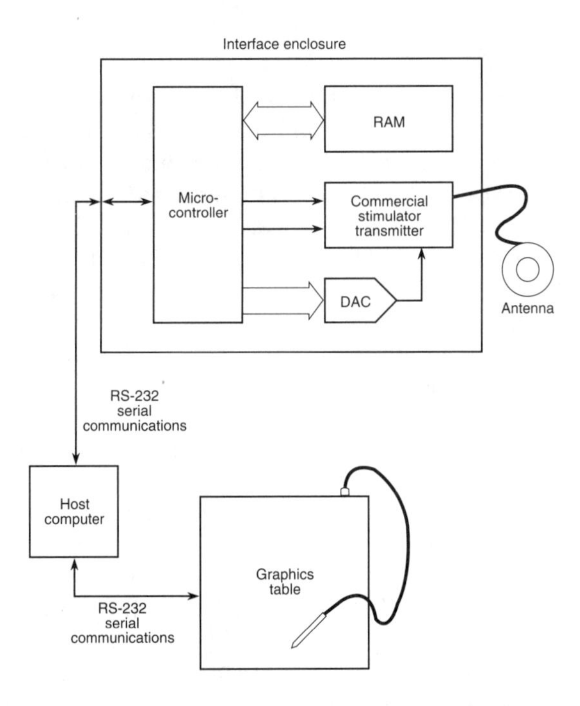

Fig. 2.13 Block diagram of the NSS. The microcontroller runs the mundane tasks of sequencing and timing the stimulation and frees the host computer to provide the patient interface, data collection, and data analysis.

Fig. 2.14 One configuration of the NSS equipment.

At the beginning of the project, we wished to avoid the extensive approval cycle by the U.S. Food and Drug Administration (FDA). I designed the hardware interface to drive the commercial transmitters with little modification to their hardware. That way we operated under the FDA approval for each device, since we were not exceeding their envelopes of operation.

The design emphasized utility and an intuitive human interface. Teamwork was important in developing the NSS; I worked directly with physicians, patients, and commercial manufacturers to design a useful instrument. We spent a great deal of time preparing the software and the human interface so that it was easy to use even for the most computer illiterate. We chose QuickBasic Professional for ease of software development. We used pull-down menus and screen buttons to make the interface more graphical and intuitive. Chapter 4 goes into more detail on the human interface for the NSS.

Extensive clinical studies have proved the quality and objectivity of the data derived from the NSS. Now, a physician or an assistant can sit for a few minutes with a patient to provide instruction on the basic use of the graphics tablet. (The patient never touches the computer keyboard, answering questions only on the screen by either drawing outlines or touching YES/NO squares on the tablet.)

In review, the NSS has had an involved and challenging development history but has proven to be a useful instrument. Time spent by medical personnel to adjust stimulators has dropped from hours to minutes, the data are more reliable, and the patients are intimately involved in their own treatment.

Scope: intuitive
Duration: second-generation device, 3 years
Number of people involved: 1 engineer, 1 technician, 2 software developers
Level of effort: one-quarter time
Lessons learned: This project had several common faults and some important lessons:

1. We underestimated the time and effort needed to program the software; the problem lay in lack of definition and planning for configuration in the beginning of the project.
2. Complex features and selections of parameters proved to be the biggest source of delay in finishing the software.
3. The human interface and its hardware packaging proved to be important in the success of the instrument (see the first case study in Chapter 4).

2.7.3 *EF-111A Magnetic Tape Unit*

The EF-111A Raven is a U.S. Air Force jet that jams radar. The number of possible emitters and locations worldwide is very large. Consequently, changing the parameters for jamming is a time-consuming manual task for the electronic warfare officer, who must type the changes at a small calculator-styled keyboard in the cockpit, often en route to a mission.

The Johns Hopkins University Applied Physics Laboratory developed a database of radars and jamming parameters for the EF-111A. Moreover, we developed software and a tape unit that retrieved the appropriate parameters from the database and loaded the data onto a cassette, or tape transport unit (TTU), designed for military avionics. The software uses heuristics and logical algorithms to assemble the jamming countermeasures. The magnetic tape unit (MTU) provides the interface between the computer with the database and the avionics TTU; it downloads the countermeasures data to the cassette tape. The ground crew then loads the data in the TTU into an aircraft through a memory-loader-verifier (MLV). Fig. 2.15 gives an overview of retrieving data from the computer database and loading jamming parameters into an EF-111A.

The MTU was a custom project that had a small production quantity (eight were built) and adhered to a number of military specifications and procedures. I designed a microcontroller-based interface (similar to that for the NSS in the previous case study), as shown in Fig. 2.16. Fig. 2.17 shows photographs of the MTU hardware.

We developed the software for the MTU by using the EF-111A simulator at Warner-Robbins Air Force Base to verify its operation. The simulator is the closest thing to the actual aircraft and an excellent test bed. (This is an example of prototyping.)

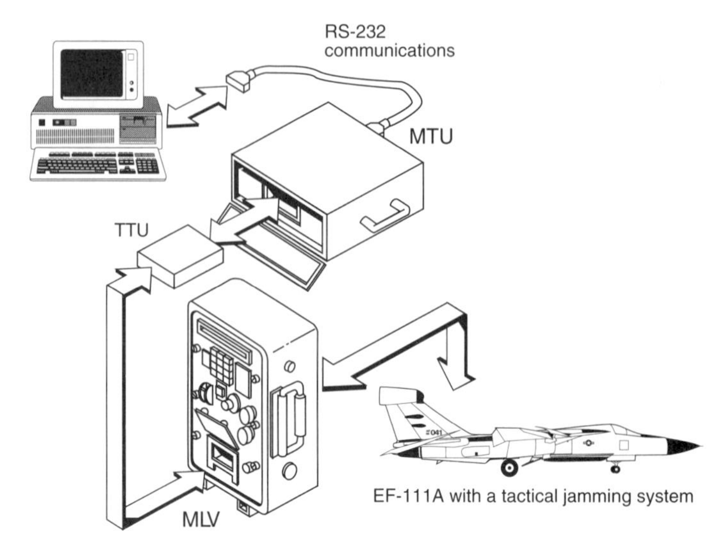

Fig. 2.15 Summary of the operations that transfer data from a computer to an EF-111A.

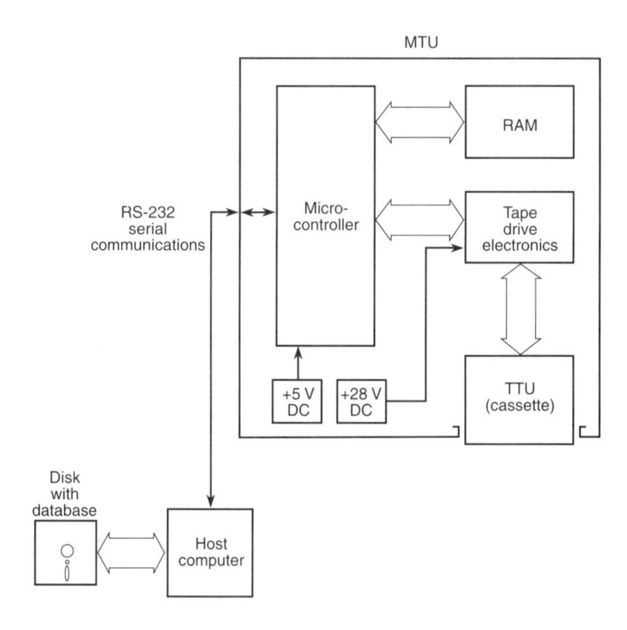

Fig. 2.16 Block diagram of the MTU for loading jamming parameters onto a tape cassette.

The military standards required protection against electromagnetic interference (EMI) and radiofrequency interference (RFI), so we contracted for testing the MTU according to MIL-STD-461C. The sealed front panel and EMI/RFI gaskets

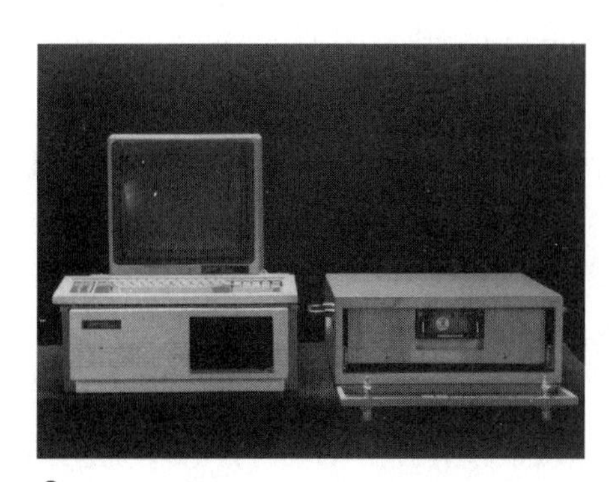

a

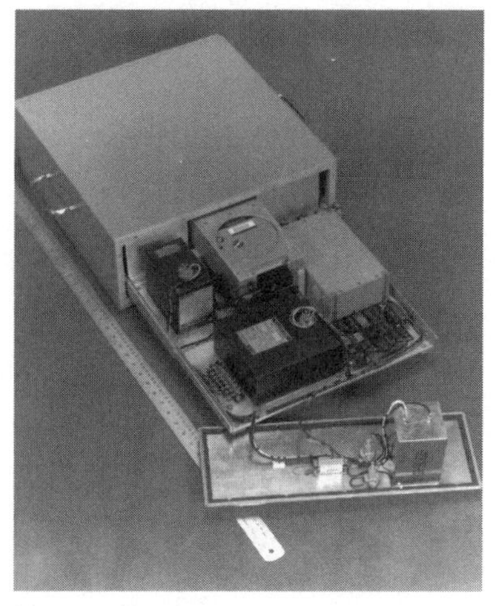

b

Fig. 2.17 Computer system with database of jamming parameters for generating mission-specific settings. (a) Host computer with MTU. The MTU has a custom enclosure that is welded and sealed with gaskets to prevent electromagnetic interference. (b) Electronic components in the MTU slide out for maintenance or repair. Some components and modules were purchased and some, like the circuit board, required custom design. These are some of the choices that you face in developing instrumentation.

around the connectors on the back panel provided an electromagnetically tight box; consequently, the MTU passed the tests with flying colors.

Finally, the project demanded extensive documentation: description of the hardware, description of the software, repair manual, and user manual.

Scope: process or program
Duration: second-generation device, 1 1/2 years
Number of people involved: 1 engineer, 1 software developer, 1 technician, 1 draftsman
Level of effort: half time
Lessons learned: The MTU had a good technical design, but the project suffered from three major failings:

1. Logistic support was lacking. The air force did not follow up with training for either the electronic warfare officers or the ground crews to load tapes.
2. Tape loading proved clumsy and time consuming for the ground crew. The MLV, which loaded the data into the EF-111A, had cables and electrical connectors with pins that bent easily.
3. I did not spend enough time with the electronic warfare officers to develop the MTU design; they needed to "own" the solution to ensure its success. Chapter 4 will cover this in more detail.

2.7.4 *Hyperbaric Chamber Control*

Deep-sea divers who go to great depths must breathe special mixtures of gases such as nitrogen, oxygen, helium, and hydrogen. Part of their training includes sessions in hyperbaric chambers, where they experience high barometric pressures and various gas mixtures. Controls for hyperbaric chambers vary the pressure of the gases and must have safety interlocks for sustaining life and avoiding faults or mistakes. All controls are manual.

The U.S. Navy asked for a proposal to automate the controls to a hyperbaric chamber. They wanted to simulate deep diving conditions for navy divers in a large pressure chamber. This would have been a custom experimental project in the process design domain with only one unit built. Concerns included life support mechanisms and human interface for the controls. Fig. 2.18 outlines the system to be controlled.

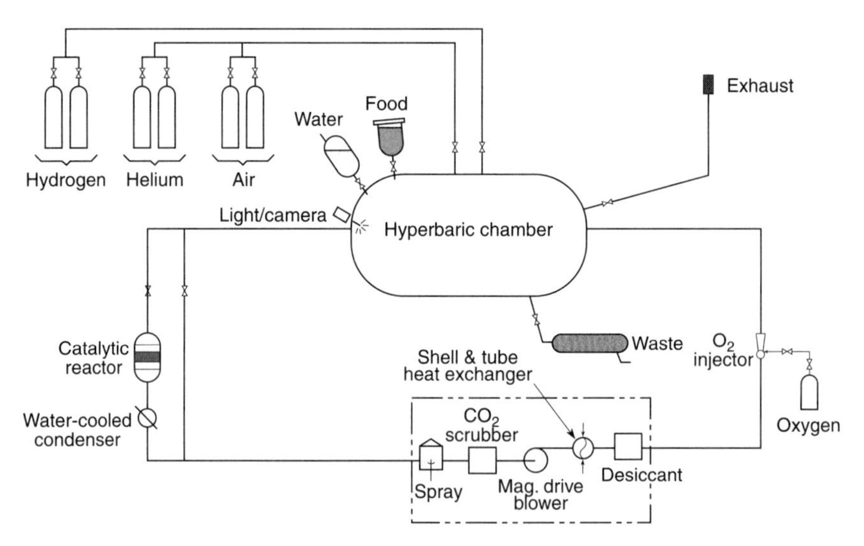

FIG. 2.18 Schematic outline of the gas, water, food, and waste lines controlled in a hyperbaric chamber.

The proposed solution used dual redundant programmable logic controllers (PLCs) for reliability and "hot swap" of circuit boards for maintenance. (A *hot swap* is the replacement of circuit boards while power is supplied to the system.) We would also have to write software for the display screens to control the human interface. Finally, we incorporated a design for mechanical overrides for failure recovery and fault tolerance.

Besides the technical design, we also estimated the cost and time to develop the control system. I used a spreadsheet to calculate the effort, including meetings, design time, reviews, software programming, and integration. I used project-scheduling software to organize a PERT chart with specific items of development and their milestones. The results of these exercises gave a realistic estimate of the size, effort, and expense for the project that surprised the sponsors; ultimately, they decided not to go ahead with the project to automate the controls.

Scope: would have been either intuitive or process
Duration: 1 week
Number of people involved: 1 engineer
Level of effort: full time
Lesson learned: Careful thought and preparation up front revealed the extent of the project and prevented a disaster in cost and schedule overruns for the sponsor.

2.7.5 *Arc Fault Detection*

Navy vessels occasionally suffer arcing faults in the switch gear that distributes electrical power. Dust, dirt, or contamination forms a conducting path that initiates an arc between relay contacts or bus bars. Arcing faults can quickly burn up switch gear and disable a vessel. Circuit breakers do not trip and extinguish the faults, because arcs are "high-impedance" short circuits with currents below the trip point of a breaker.

The Johns Hopkins University Applied Physics Laboratory developed a system that detects arcs by sensing simultaneously both the bright flash and the pressure of the shock wave from an arc. Then it signals and trips the appropriate circuit breaker to extinguish the arc. The system has four types of circuit boards: cabinet, zone, relay, and BIT. The functions of the circuit boards followed the functional division of the power distribution system within a vessel. A cabinet board filters signals from sensors with timing criteria to prevent false positive alarms and avoid false negatives (missing an arcing fault). A zone board combines outputs from multiple cabinet boards (up to eight) and signals the relay board. The relay board actually trips the circuit breaker to open it. The BIT board exercises the other three boards to ensure their proper function. Fig. 2.19 shows the block diagram of the arc fault detection (AFD) system. Fig. 2.20 is a photograph of the AFD equipment.

The AFD system was a military project for installation in submarines to prevent damage from arcing faults. The design and development was to full military specification, construction, and documentation. The system underwent several qualification tests, including shock, vibration, and EMI/RFI. Finally, we turned over the production of the final units to a commercial firm.

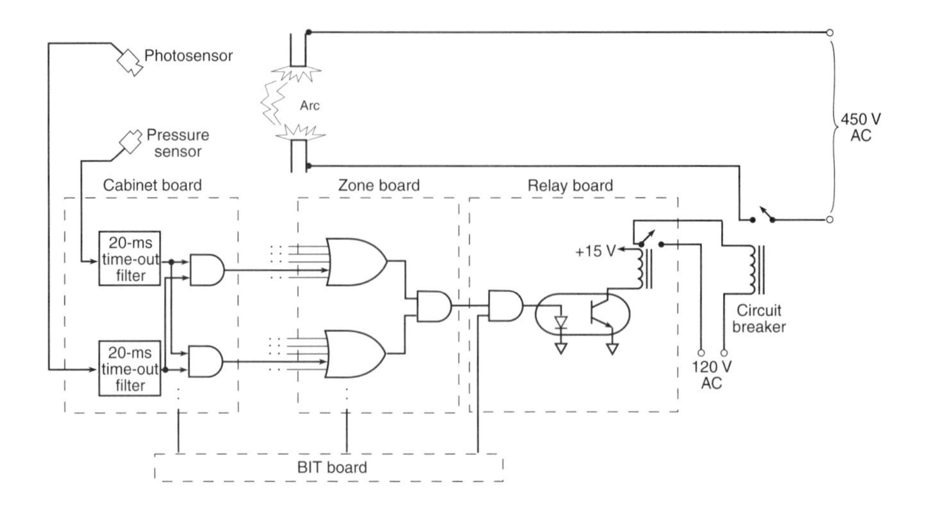

Fig. 2.19 A block diagram of the AFD system.

Our involvement began with research into the detection of arcs using magnetism, light, sound, and pressure. The navy then asked for a prototype of a system to detect arcs. After initial design and development by other engineers, I was asked to revamp the electronics and finish the design. Several of us tried but failed to persuade the project management to replace the design in Figs. 2.19 and 2.20 with a triply redundant system that would use multiple copies of a single type of circuit board. This kind of design would have reduced size, power, and maintenance (integrated BIT for immediate diagnoses) and provided for hot swap.

Scope: system or program
Duration: 6 years (after initial period of 4 years for research)
Number of people involved: 2 project managers, 4 engineers, 3 technicians, 3 draftsmen, shop personnel to fabricate enclosure
Level of effort: full time
Lessons learned: I think that I could have persuaded management to use a better design if I had done several things differently:

1. Prepared more effective arguments, with studies of trade-offs between the current design and the triply redundant design.
2. Not allowed the pressures of time to development to seem more important than a superior design (not always true but it was in this case).
3. Increased credibility by experience (a kind of "chicken-and-egg" problem since I was only a young junior designer).

Fig. 2.20 One configuration of the AFD equipment. This is an example of a completely custom design. The enclosure, circuit boards, back plane, and sensors (not shown) were all designed and fabricated for this specific application; only the power supply was purchased.

2.7.6 *Light Exoatmospheric Projectile (LEAP)*

The LEAP project was a large experiment to demonstrate the feasibility of intercepting a tactical ballistic missile. The target was a modified second stage from a Minuteman intercontinental ballistic missile. It emulated a tactical ballistic missile like a Scud (a Soviet derivative of the old German V-2). The interceptor was a U.S. Navy standard missile modified with a third-stage booster and a kinetic kill vehicle. Fig. 2.21 outlines the experiment's configuration to launch and intercept a tactical ballistic missile.

Analysis showed that the navy ship radars had insufficient resolution, so high-resolution radars at Wallops Island, Virginia, had to track the target. The analysis studied the cumulative errors from the radars, telemetry, and dynamics of both the interceptor missile and the kinetic kill vehicle to determine the resolution required

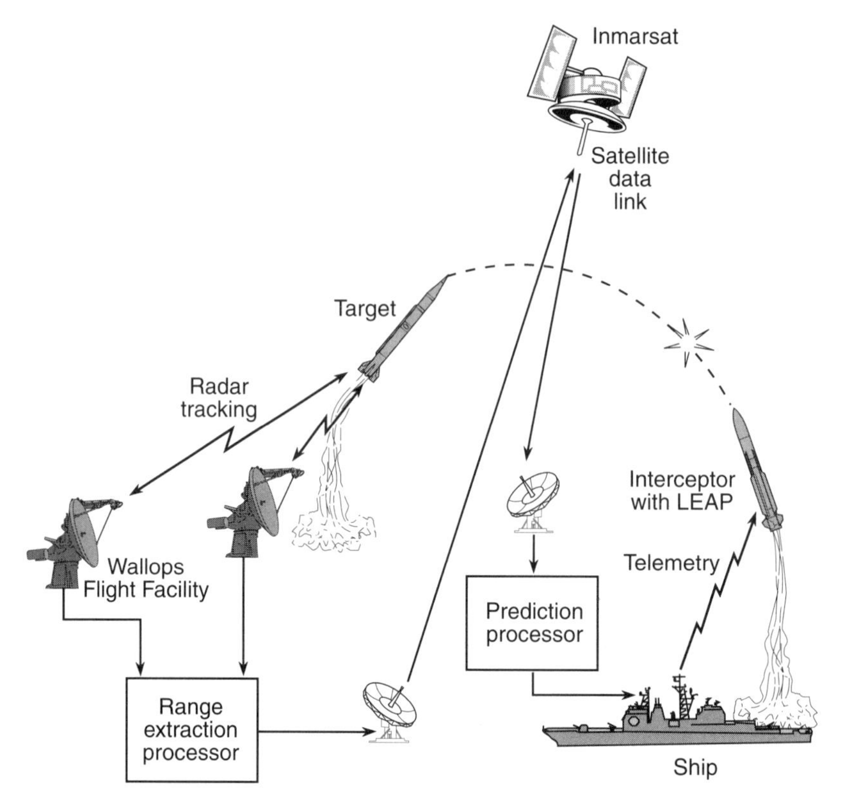

FIG. 2.21 Outline of the LEAP experiment, in which a modified U.S. Navy standard missile intercepts a ballistic target.

and margins deemed necessary. Further concerns included redundancy in systems components for fault tolerance, software management, data analysis, and documentation. The LEAP project had high visibility in both the public and political arenas.

I was involved in the system design of the processors and communications link between Wallops Island and the ship. The processor at Wallops Island, called the range extraction processor (REP), collected data from the radars, filtered the target track with a six-state Kalman filter, and transmitted the track to the ship. The processor on the ship, called the prediction processor (PP), received the data, predicted the intercept time and point, and indicated when to launch the interceptor missile. Fig. 2.22 outlines the REP and PP. Fig. 2.23 has photographs of each processor.

We designed the processors around a VMEbus chassis and programmed the software in Ada, a high-level language developed for military projects. The devel-

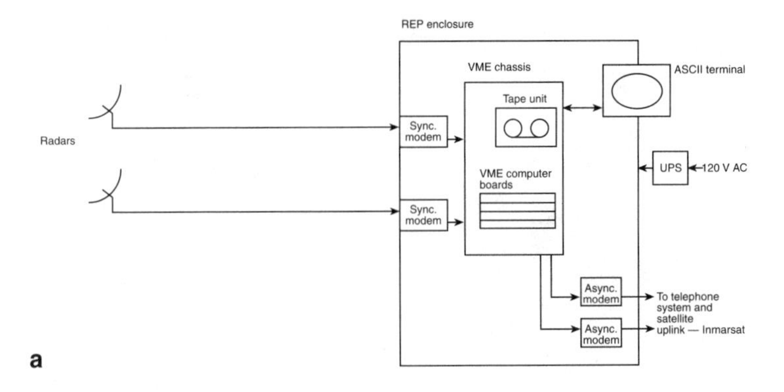

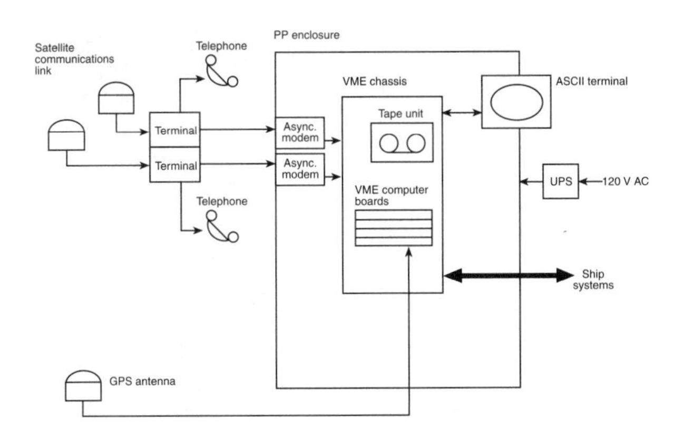

Fig. 2.22 Block diagrams of processors in the LEAP experiment. (a) REP at Wallops Island, Virginia. (b) PP onboard the ship that launches the LEAP interceptor. (GPS = global positioning system; UPS = uninterruptible power supply.)

opment environment included workstations on an Ethernet network; it was large and expensive and required expertise in both Ada and Unix environments. A different architecture could have avoided the expense of a large, expert programming team. We could have based the processors on a ruggedized version of a powerful desktop computer and programmed in another language environment, instead.

Furthermore, procuring components caused unavoidable delays. Several manufacturers of necessary components were weeks to months late making delivery. We could have ordered components earlier to allow more margin in the calendar for delivery only if the design had been finalized earlier. Again, a chicken-and-egg problem: How could we have known, except in hindsight, that we needed more time for delivery? The only answer is to use this experience in the future.

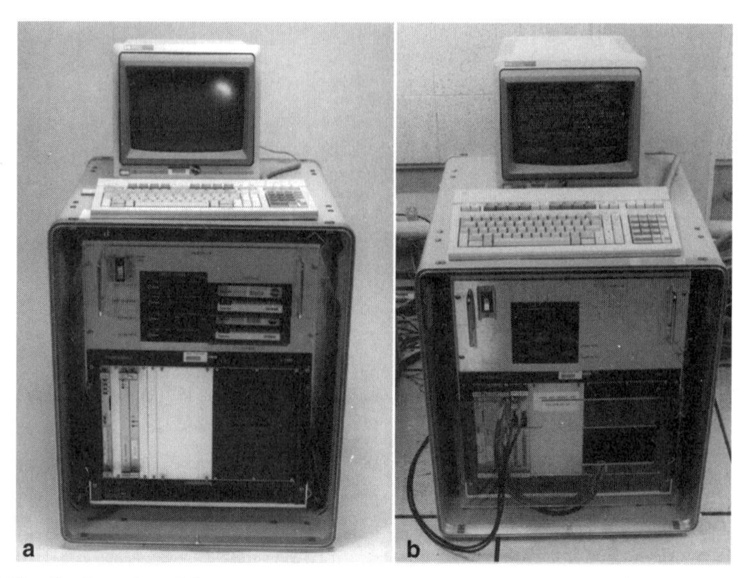

Fig. 2.23 Configuration of the two processors that track the target and calculate the interception. All the electronic components were purchased, integrated together, and then controlled by software written for the specific application. Only the sheet-metal tray and some cables had to be fabricated. (a) The REP receives radar data through synchronous modems and transmits the filtered results through asynchronous modems to the satellite communications link. (b) The PP receives the satellite communications through asynchronous modems, calculates the predicted intercept, and notifies the ship systems for launch.

Scope: system or enterprise
Duration: 2 years
Number of people involved: (The entire project involved more than 40 companies and organizations.) Immediately involved in the REP and PP were 2 project managers, 2 engineers, 7 software developers, 1 analyst, and 2 technicians.
Level of effort: full time
Lessons learned: While the LEAP project was technically challenging, teamwork proved to be the biggest problem. Some interpersonal relationships, in particular, could have been better if, early on, I had done the following:

1. Defined responsibility rather than assuming others' functions.
2. Had reviews for the conceptual design of each processor rather than wait for a later review and introduce delays from the requested changes.
3. Clearly communicated the experimental nature of the project and hence the expectation for changes in scope and requirements throughout the project.

2.7.7 *Helicopter Remote Control (Wesolowski, 1992)*

A remote control system built by Native American Services (NAS) of Huntsville, Alabama, enables a ground-based pilot to fly a helicopter. Remote control provides a more realistic target for testing weapons by allowing rotary-wing maneuvers such as terrain following, hovering, and pop-up.

The remote control was an experimental program with several basic requirements:

1. Maintain the stability of the helicopter by preventing maneuvers that cause a crash.
2. Provide controls as easy as those of a video game, not requiring a flight-qualified helicopter pilot.
3. Transmit sight and sound from the cockpit of the helicopter to the ground-based operator.
4. Provide instrumentation with continuous feedback of altitude, airspeed, attitude, position, direction, rate of climb or descent, fuel level, manifold pressure, and oil pressure.

The emphasis of this case study is on quick development. A VMEbus chassis houses the remote control system. It is a modular design that integrates off-the-shelf components and permits rapid development (see Fig. 2.24). NAS managed the software and limited its capability (features) to expedite development into something workable that produced rudimentary maneuvers. NAS also spent considerable effort developing the operator interface, understanding helicopter controls, and testing.

FIG. 2.24 Remote control system for piloting a helicopter (photograph courtesy of Native American Services).

The operator wears a specially equipped helmet with video eyepieces and speakers (Fig. 2.25). Head movement by the ground-based operator causes the dual video cameras in the cockpit to slew accordingly. The system must limit the lag between the head movement and servo response of the cameras; to do so, it generates 60 pan-and-tilt commands per second. One camera sends images to one video eyepiece and presents a narrow-field view, with high resolution, to the dominant eye so that the operator can distinguish distant objects. The other camera presents a wide-field view to the other eye for good peripheral vision.

NAS had to learn the helicopter controls to incorporate them into the remote control system. The controls were the collective for vertical maneuvers, the cyclic for pitch and roll, the throttle, and the tail rotor for counterforce to the torque of the main rotor. NAS engineers also had to learn and incorporate into the remote control system the helicopter flight characteristics, such as ground effect and the fact that the helicopter begins to act like a fixed-wing aircraft for speeds exceeding 55 to 75 km/h (30 to 40 knots). After understanding the characteristics and controls of

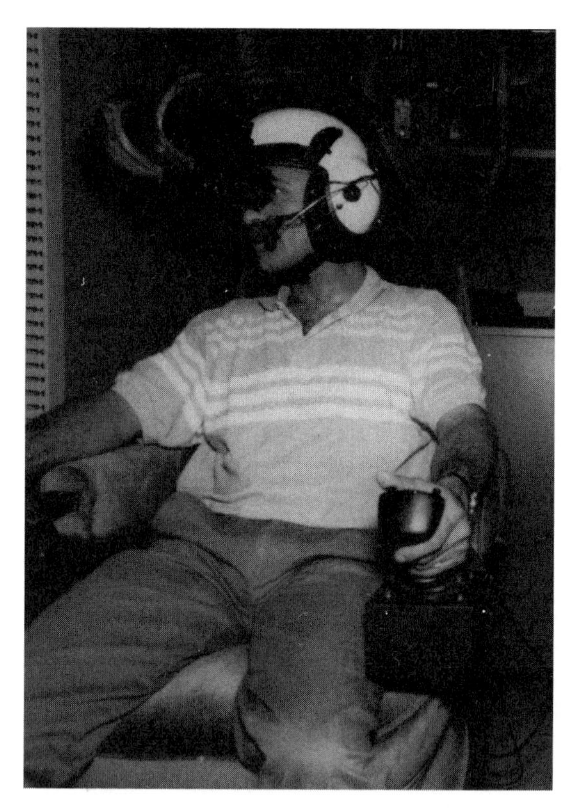

Fig. 2.25 Operator station for helicopter remote control (photograph courtesy of Native American Services).

a helicopter, they determined that the flight control task must update the joystick positions at 20 commands per second for stability and that the system should download status information every half second. They also designed for three levels of override: (1) the test pilot physically wrests control, (2) the pilot presses a button to release the actuator clutches, and (3) the pilot pushes the kill button for the autopilot power.

One important aspect of this case study is that NAS engineers prepared and spent a good deal of time in testing the control system throughout development. The application code was modular; they partitioned it into independent tasks for ease of testing (an example of loose coupling). They developed detailed simulations to prepare the control laws for the system. A real pilot performed maneuvers that were recorded on a digital tape; they compared the real maneuvers with simulated responses and adjusted the control laws accordingly.

To test the remote control system, they bolted the helicopter skids to a wheeled platform and had the remote control system move it about a runway apron. These trials with ground movement tested the RF data link for both reliability and orientation and verified the translation of the operator's joystick movements into cyclic control. For the first test flight, a human pilot initiated flight and then allowed the remote control to take over; after it performed some prescribed maneuvers, the pilot landed the helicopter.

The testing revealed some interesting concerns. First, the tape drive that collected data proved unreliable during the sharp changes in ambient temperature as the helicopter ascended. (You will find that temperature transients can cause all sorts of problems for a wide variety of electronic products.) Second, the GPS antenna location proved very sensitive to multipath reflections from the rotor blades; this became obvious in the prototype; no simulation would have predicted it.

Scope: process or program
Lessons learned: The helicopter remote control provided good examples of concurrent effort and loose coupling in the interfaces. Testing figured into the development early in the design and proved necessary to the final success.

2.7.8 *Pegasus Avionics Suite (Steffy, 1990)*

This case study focuses on the avionics in a commercial rocket booster built by Orbital Sciences Corporation. The avionics form a complete subsystem within the larger project. Many of you will work on projects partitioned similarly—hence the usefulness of this example. It resides somewhere between the process and system design domains.

The Pegasus rocket is a commercial rocket booster that lofts a satellite of about 300 kg (660 lb) into a 370-km (230-mi) polar orbit. A B-52 bomber or large commercial jet carries the Pegasus under its wing and launches it at 12,000 m

(40,000 ft) altitude, as shown in Fig. 2.26. This arrangement significantly reduces the size of the rocket booster.

The main selection and design criteria for the rocket attempted to optimize development:

1. Use good aerospace practices, not military specifications.
2. Maintain a low recurring cost to allow small-payload users to afford launches.
3. Lower nonrecurring cost to promote reasonable commercial development.
4. Shorten the schedule. Launch within 14 months after order.
5. Simplify testing—for example, by not using many BITs.
6. Maintain safety by using standard practices for ordnance control, airborne operations, and range safety.
7. Design for the environment, to withstand shock, vibration, and temperature changes with rocket flight and to reduce weight.

Fig. 2.26 Pegasus, a commercial booster for lofting small satellites, being prepared for launch from under the wing of a B-52 bomber (photograph courtesy of the National Aeronautics and Space Administration).

This case study is a good example of heuristic design and development by a small team. The basic design uses a star network from a VMEbus flight computer to control actuators, fire thrusters, and gather telemetry with remote units based on microcontrollers. A VMEbus flight computer resides inside a sealed enclosure with extensive EMI filtering. Communications between modules are serial and asynchronous, and microcontrollers receive and validate commands from the flight computer and then actuate the appropriate control. For faster development, the designers depended on experience rather than detailed trade-offs. Two of the heuristics were "RS-422 offers good noise immunity" and "Serial interfaces will be easy to test" (Steffy, 1990, p. 330).

The design team tested components and modules with common personal computers to reduce development delays. The architecture allowed early and partial testing of the distributed processors. The display and ground support equipment used personal computers to maintain commonality and availability.

The skunk works organization of a small team of multipurpose engineers responded well to rapidly shifting changes in the design. (This approach requires commitment from engineers, management, and customer to work well.)

The design team used a commercial operating system and wrote most of the software in the C programming language. Personal experience, however, led to some Forth and assembly language in the embedded microcontrollers. The team had no full-time software developers; rather, each engineer was involved in several aspects of the project and maintained a system perspective.

Testing, as already mentioned, was important throughout the development of Pegasus. The team followed a philosophy of testing that permitted simple debugging, required a minimum of ground support equipment, and allowed simple field operations. Consequently, team members were able to exercise the preliminary hardware and software early and often. Testing had several phases:

1. Lab bench tests, which limited the troubleshooting gear to common personal computers, power supplies, oscilloscopes, digital voltmeter (DVM), and cables.
2. Simulation.
3. Mobile test bed, in which "hangar" testing wrung out problems in the avionics, procedures, and wiring.
4. Static firing of the motor to allow integration of the telemetry unit.
5. Captive flight to check out the navigation unit, exercise fin actuators, test power transfer, and validate the countdown procedures.

Scope: system or program
Duration: 3 years
Number of people involved: 3 engineers
Lessons learned: The Pegasus avionics suite is a good example of a small team developing a robust design through heuristics, loose coupling, and commonality.

2.7.9 *Hospital Monitoring System (Reiche, 1991; Westerteicher, 1991; Westerteicher and Heim, 1991)*

This next case study is a commercial medical instrument by Hewlett-Packard (HP) whose design architecture is a good example of utility, modularity, flexibility, and expandability. It is a patient monitor for medical observation and diagnosis. Typically, it monitors parameters like electrocardiogram, blood pressure, pulse oximeter (SaO_2), and respiratory gases. Fig. 2.27 is a photograph of the patient monitor.

Modularity and expandability allows HP to configure each monitor to the customer's application and to install future upgrades that measure new parameters from the patient. The hardware and software building blocks form functional modules that tailor the system to the application and the patient. In a good example of loose coupling, HP clearly defined interfaces early, thereby making each module (both software and hardware) easier to design and integrate. Expandability is important because hospitals invest in new equipment only every 7 to 15 years, so they want something that is easy to upgrade to new technologies (Westerteicher, 1991).

The patient monitor has three main subsystems: a parameter module rack, a computer module, and a display. Up to eight parameter modules can fit into one

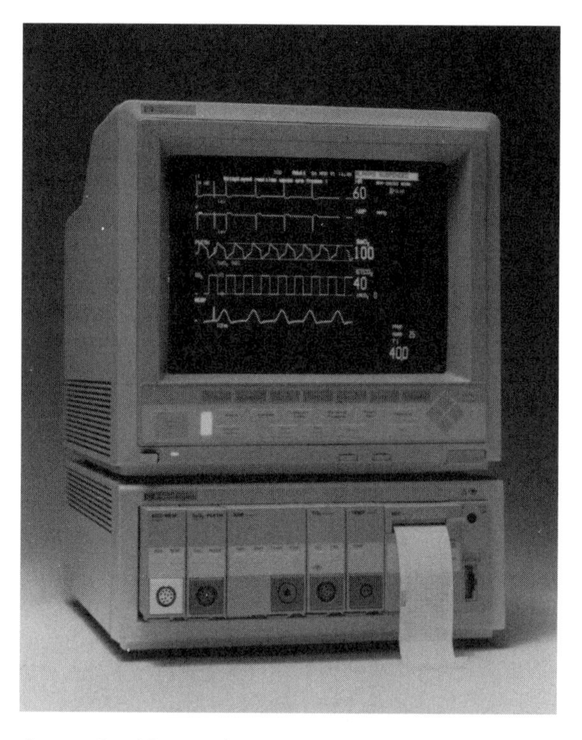

Fig. 2.27 HP patient monitor (photograph courtesy of Hewlett-Packard).

rack, and each computer module can support up to four parameter racks. This configuration allows placing the parameter modules near patients, thus reducing the length of cable and the tendency to have cable "spaghetti" draped over patients (Westerteicher, 1991; Westerteicher and Heim, 1991).

HP designed the interface for ease of use by trying to make the display and its operation intuitive for doctors and nurses. Medical personnel can operate the instrument through either a keyboard or the touch panel on the display.

The computer module has cards of a single size that plug into a back plane. It uses a message-passing bus that allows both arbitrary locations of the cards on the back plane and arbitrary functions for the cards. Each card in the computer module has a chip—an application-specific integrated circuit (ASIC)—to control communications with the message-passing bus; it decouples the function of the card from the bus. Inserting the appropriate cards into the module customizes the unit to the application.

The software has a decentralized architecture built on decoupled, standardized, and formalized interfaces. It too has modularity: Each application-specific module is self-contained and exchanges information with other cards only via the message-passing bus. It uses a concept called the *virtual processor* that makes the current configuration invisible to all applications. Each application does not "know which modules are assigned to which [virtual] processor, how many processors are available, which interface cards are present, or where messages come from [or] where they go. Therefore, software developers must not make any assumptions as to where their applications will be executed. The only way modules are allowed to communicate with each other is by exchanging messages" (Reiche, 1991, p. 15). Furthermore, the modules link together only at boot to allow independent configuration and ease in maintenance.

> *Scope:* system or enterprise
> *Lessons learned:* The hospital monitoring system is an excellent example of loose coupling at its best. The advantages of modularity include the following:
> 1. Adding functionality by inserting new hardware modules and updating the software (if necessary).
> 2. Technological updating by replacing modules of hardware or software.
> 3. Simplified manufacture (each component is assembled and tested separately; a complete system is not needed for thorough testing).

2.7.10 *Manhattan Traffic Control System (Heurikon Corp., 1992)*

JHK & Associates of Norcross, Georgia, recently redesigned the traffic control system for the city of New York. Their work replaced an obsolete computer system with an expandable and maintainable system that had the following goals:

- Control up to 12,000 intersections.
- Tie in Manhattan to improve traffic flow and decrease pollutants.
- Reduce long-term maintenance costs.

This project is a good example of the system design domain, where you would be a member of a team that develops one aspect of the overall design.

JHK chose VMEbus for the architecture of the traffic control system because of the wide range of circuit cards available. In general, proprietary systems are not as flexible. They strove for a survivability of 20 years, with a potential for 50% expansion in the number of intersections that needed control. Their development included testing to verify performance and function at each step along the way.

JHK designed the hardware for fault tolerance, maintenance, and expansion. Each equipment rack has an alarm panel for immediate diagnosis of faults. A system watchdog timer reboots the system if it ceases proper operation. The power distribution included line filters and power factor correction at each rack to reduce interference and susceptibility on the power lines. Each rack had redundant power supplies, and each supply had a 150% capacity to supply an entire rack and allow for more circuit cards. Maintenance personnel may insert circuit cards while power is on the rack; the design for live insertion specified that the ground pins are the longest, power pins next in length, and signal pins the shortest. Every harness, lug, wire, and connector was labeled for ease of maintenance. Each cable was labeled every 30 cm (12 in) with its function and the connections on each end. JHK routed the cables through cable guides and avoided sharp edges on the racks to prevent abrasion and cuts in the wires. The design also incorporated many different configurations for connectors to reduce accidental and wrong connections.

Scope: system or program
Duration: 3 years
Number of people involved: 31
Lessons learned: The traffic control system is a good example of robust design for fault tolerance, maintainability, and expandability:

1. Fault tolerance is important for continuous operation that does not disrupt traffic flow.
2. Good practice, such as labeling the cables, eases maintenance, reducing downtime and possible foul-ups.
3. Live insertion, or hot swap, allows expansion within the system that extends its useful life and decreases the life-cycle cost to the city of New York.

2.7.11 *Crown Lift Truck (Maloney, 1992)*

The design of a lift truck may appear to be highly mechanical and outside the scope of this book, but many principles used in designing and building the Crown lift truck apply to electronic products that you may design. It is a good example of the system design domain.

Crown Equipment Corporation of New Bremen, Ohio, builds equipment for materials handling. During the mid-1980s the corporation decided to get into the highly competitive market for four-wheel counterbalanced lift trucks with a new product. Its entry, the Crown FC, is shown in Fig. 2.28.

One of the most important aspects of the successful design was that Crown engineers began by seeking customer feedback and studying lift truck operations. They studied trucks of competitors. They met with purchasing agents, warehouse managers, lift truck operators, and service mechanics to discuss features. And they observed lift trucks in operation around warehouses and manufacturing plants, even videotaping them to study the habits of operators. David Smith, an industrial designer who has worked with Crown for years, says, "It's a big mistake not to let engineers and industrial designers out into the field. Crown has gotten its design teams heavily involved in this type market research from the company's earliest days" (Maloney, 1992, p. 47).

FIG. 2.28 Crown FC series lift truck (photograph courtesy of Crown Equipment Corporation).

The design concerns for the new lift truck included ergonomics, wear and reliability, maintenance, and utility. Crown designed the lift truck with several ergonomic features:

- A wide, low step and a tilt steering wheel for ease of entry and exit.
- A direction-control lever that responds to a finger flick so that the driver's hands can stay on the steering wheel.
- An adjustable, contoured seat with a padded place to rest an arm when backing up the truck.

The goal for reliability was that parts with wear points must stand 15 years of abuse.

Crown designed the control panel in the cab for ergonomics and maintenance. The panel gives status on the operation of the lift truck such as motor temperature, brush wear, and battery charge. Furthermore, it can display in any of six languages for the international market. For maintenance, the panel has BIT and stores a history of faults.

The electric motor propulsion can run from batteries with 36, 48, or 80 V (for the European market). The battery compartment has rollers that allow a quick exchange of batteries for the next shift in a 24-hour-per-day work schedule. Maintenance personnel can remove the rear panel for easy access to the power electronics.

Crown used concurrent engineering to develop the lift truck. The team included electrical, mechanical, and hydraulic engineers; CAD specialists; industrial designers; and manufacturing personnel. This approach worked well; for instance, "design engineers originally wanted to make the metal part from one stamping. . . . This would require . . . an expensive new stamping machine. Instead, manufacturing suggested welding and grinding techniques that, along with improved styling, would achieve the desired fit and finish with unique pressbrake techniques [on equipment that they already owned]" (Maloney, 1992, p. 50).

Extensive use of models, mock-ups, and prototypes improved the design. Models of interior compartments, for example, helped the team make sure that the components could be easily serviced. "The company produces operational 'hard' prototypes, which are actually tested in the field before any volume production is started. Even these prototypes will be changed—based on feedback from customers and equipment operators" (Maloney, 1992, p. 52).

One way Crown ensures design excellence and manufacturing quality is through vertical integration. Crown manufactures most of the necessary components for its products—such as hydraulic cylinders, gearbox and gears, sheet metal for bodies, and lift assemblies—and Crown will soon build its own DC motors. (Other people argue that the alternative is to include vendors and suppliers in design instead of using vertical integration; the decision depends on company culture, capability, and preference.)

Scope: system or enterprise
Duration: 5 years
Lessons learned: Crown attributes the success of its lift truck to four factors:

1. Field research on customer and operator needs by the design team.
2. Concurrent engineering to speed development.
3. Modular design for quality and ease of change.
4. Industrial design for ergonomics and style.

2.8 SUMMARY

This chapter focused on some of the heuristic, or "squishy," aspects of good engineering—those elements that are hard to define but essential to success. The application of these rules varies from company to company, individual to individual, and project to project, but several principles emerge:

1. Commit to a common vision.
2. Commit to working together.
3. Communicate.
4. Plan ahead.

The chapter covered architecting, some of the common concerns and heuristics, and teamwork. It then introduced several examples of design: real-time control and case studies of some successful and not-so-successful products. The remainder of the book will build on these principles and examine specific areas in detail.

2.9 RECOMMENDED READING

Carter, D. E., and Baker, B. S. 1992. *CE, concurrent engineering: The product development environment for the 1990s.* Reading, Mass.: Addison-Wesley.
A readable introduction to concurrent engineering.

Rechtin, E. 1991. *Systems architecting: Creating and building complex systems.* New York: Prentice-Hall.
An excellent book for learning about the heuristics in design and architecting a design. Though it aims at large-scale projects, the principles apply to the development of any electronic products.

Skytte, K. 1994. Engineering a small system. *IEEE Spectrum*, March, pp. 63–65.
An excellent introductory article for developing electronic products.

2.10 REFERENCES

Bak, D. J. 1993. One size fits all at Airbus. *Design News,* March 22, pp. 62–68.
Bell, T. E. 1987. Defensive designs minimize hazards. *IEEE Spectrum,* May, pp. 39–40.

Carter, D. E., and B. S. Baker. 1992. *CE, concurrent engineering: The product development environment for the 1990s.* Reading, Mass.: Addison-Wesley.

Child, J. 1993. Bridge/router design takes best of builds and buys. *Computer Design,* January, p. 101.

Cortes-Comerer, N. 1987. Motto for specialists: Give some, get some. *IEEE Spectrum,* May, pp. 41–46.

D'Alessandro, J. 1993. The value of usability. *Hewlett-Packard Journal,* April, p. 91.

Evanczuk, S. 1989. Back-end engineering moves to the front. *High Performance Systems,* November, pp. 18–25.

Gautschi, T. F. and B. L. Goldense. 1993. Eliminating the "old boy" network. *Design News,* July 5, pp. 74–80.

Gottschalk, M. A. 1994. Newton: Apple's eye on the future. *Design News,* January 3, p. 72.

Heurikon Corp. 1992. Unique 600 MIPS real-time system manages New York City traffic: 200 Heurikon VME computers controls [*sic*] 12,000 intersections. Brochure.

Huthwaite, B. 1993. Do's and don'ts of concurrent engineering. *Design News,* July 5, pp. 71–74.

Jensen, R. 1993. Managing embedded designs. *Embedded Systems Programming,* January, pp. 30–40.

Jones, D. 1993. Avoiding control system pitfalls. *Embedded Systems Programming,* November, pp. 26–56.

Kaplan, G. 1987. On good design. *IEEE Spectrum,* May, pp. 27–28.

Maloney, L. D. 1992. Crown puts design on a pedestal. *Design News,* July 20, pp. 46–52.

Ogata, K. 1970. *Modern control engineering.* Englewood Cliffs, N.J.: Prentice-Hall.

Rechtin, E. 1991. *Systems architecting: Creating and building complex systems.* New York: Prentice-Hall.

Rechtin, E. 1992. The art of systems architecting. *IEEE Spectrum,* October, pp. 66–69.

Reiche, M. 1991. Component Monitoring System software architecture. *Hewlett-Packard Journal,* October, pp. 13–18.

Reinertsen, D. G. 1992. Use product architecture to slash design time. *Electronic Design,* December 3, pp. 59–62.

Slater, M. 1992. Limits to integration. *Microprocessor Report,* June 17, p. 3.

Steffy, D. A. 1990. The Pegasus avionics system design. *Proceedings: IEEE/AIAA/NASA 9th Digital Avionics Systems Conference, October 15–18, 1990, Virginia Beach, Virginia,* pp. 326–332. New York: Institute of Electrical and Electronics Engineers.

Strassberg, D. 1992a. Design it right: Part 1. *EDN,* October 1, pp. 60–70.

Strassberg, D. 1992b. Design it right: Part 2. *EDN,* October 15, pp. 97–105.

Vanlandingham, H. F. 1985. *Introduction to digital control systems.* New York: Macmillan.

Wallich, P. 1987. How and when to make tradeoffs. *IEEE Spectrum,* May, pp. 33–39.

Wesolowski, C. 1992. How to use a VMEbus system to fly a helicopter. *VMEbus Systems.* 9(5):29–42.

Westerteicher, C. 1991. Introduction to the HP Component Monitoring System. *Hewlett-Packard Journal,* October, pp. 6–10.

Westerteicher, C., and W. E. Heim. 1991. Component Monitoring System hardware architecture. *Hewlett-Packard Journal,* October, pp. 10–13.

3

DOCUMENTATION

Of making many books there is no end...
—Ecclesiastes 12:12

3.1 DON'T SKIP THIS CHAPTER!

I am astounded at the general lack of good documentation in products. Many overlook or downplay its necessity, but documentation is communication, and communication is vital to good engineering (Fig. 3.1). This chapter should raise your awareness of the importance of documentation and give you helpful hints, tips, and techniques to create and maintain good documentation.

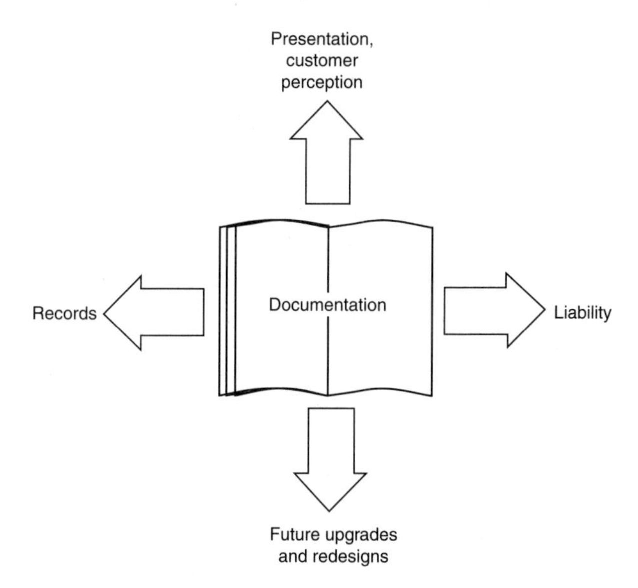

Fig. 3.1 Documentation pervades many aspects of a system.

Documentation is integral to any product. It provides the foundation for understanding the product and promotes its capability and utility. Documentation is necessary for operation, maintenance, repair, upgrades, redesigns, fixes, and recalls because it explains why the product is useful, how it may be controlled, and how it works. Documentation also reduces the burden of legal liability by warning of hazardous operations, modes of failure, and limits of operation. Finally, good documentation creates a favorable perception of your product. Equally, poor documentation slights your product by creating a perception of lesser quality.

3.2 TYPES OF DOCUMENTATION

Documents have many different forms; each fulfills a specific role within the system. Engineering notebooks, memos, drawings, diagrams, photographs, software listings, and technical presentations all record the progress and development of a product. Manuals explain the operation of a product, while proposals and advertising brochures introduce the product to customers and users. Table 3.1 outlines the types of documents that are part of any product.

Table 3.1 Types of documents and their use and format

Category	Specific Type	Use and Format
Introduction	Owner's manual Videotape instruction	Explains the product to the user with clear, simple language. Should be professionally prepared.
Instruction	Operations manual	Tailor to users—are they owners or expert operations personnel? Should be easy to use (e.g., spiral binding that lies flat). Format usually is company specific.
Records	Hardware description Memos Drawings, photographs Software listings Electronic files	Company-defined format. Engineering notebooks must be written in ink and signed and dated for legal authenticity.
Promotion	Technical presentation Advertising brochures Videotape presentation	Company-defined formats. Advertising needs professional preparation.
	Proposals	Granting agency defines rules for the format.

3.3 RECORDS, ACCOUNTABILITY, AND LIABILITY

Most of your effort in documentation will be to record the development, account for progress, and establish the liability of your product.

3.3.1 *Records*

Documentation speeds modifications, upgrades, fixes, and recalls for your product when they occur—and they will occur—by helping you to remember the how and why of the original development. Moreover, these records provide a basis to help customers and users when they encounter unforeseen problems. In both situations, good documentation helps you to avoid reinventing the wheel while you solve new problems.

3.3.2 *Accountability*

Beyond maintaining a project memory, documentation also records the progress toward satisfying requirements and provides an audit trail of the development. To measure progress, your configuration management plan must lay out a road map for completing documents that help you ensure the direction and performance of the product. Moreover, documents help establish your rights to inventions that originate in development.

3.3.3 *Liability*

Finally, good documentation goes a long way toward establishing and limiting product liability. You can reduce the possibility of lawsuits by labeling warnings clearly, presenting concise instructions, and listing necessary details about the use of your product. A thoughtfully considered plan for documentation also supports rigorous testing, validation, and verification to give you more information about your operation, reduce the chances of failure, and decrease your liability.

3.4 AUDIENCE

Once you understand and accept the importance of good documentation, you must determine who will read it and how they will use it. Because your documentation will educate customers and help users to overcome problems, you must understand who the users are, what jobs they perform, and why they will read your manuals. They may be the owners of the product, operators, service personnel, or dealers and distributors.

Know the users. Will they be the customers who actually buy your product? Will they be part of the untrained general public or sophisticated and trained personnel? Are they dealers and distributors who rely on the documents and manuals to sell and service your products? To fulfill their needs, you must understand their jobs and the information they use (Horton, 1991).

1. How well do the users understand your product? Are they learning to operate a new piece of equipment and need step-by-step instructions? Or are they familiar with similar devices and need only cursory guidance?
2. What decisions do they make? Are they in control of every aspect of a task and therefore need detailed descriptions? It is pointless to explain the groundwork for making decisions when the user doesn't make them.
3. Are they motivated? Do they view their work as exciting or boring? Do they want the details or just enough to accomplish the immediate task?

Users range from the naive to the sophisticated—and few will read the owner's manual until they have trouble. Often they are under pressure to get work done and will read only enough of the manual to get by the initial problem. No one reads manuals for literary fulfillment; people want to get the information and be done with it. Usually people read manuals to find answers to problems, fixes for failures, and references to explanations. Occasionally, dedicated users read the documentation to educate themselves about the technology or your product.

For international markets you must consider the differences in culture (Horton, 1991; Schoff and Robinson, 1991; Woelfle, 1992). Figures, shapes, numbers, and colors can have different connotations in different countries, and you should be sensitive to them.

You will need to account for these situations in preparing your documentation by analyzing the users through one or more of the following: questionnaires, informal interviews, beta-site observations, and videotape task analysis. (You will do similar analyses for developing the human interface of your product, as described in Chapter 4; in fact, documentation is part of the human interface.) Tailor your documents by matching these elements to the users' needs (Schoff and Robinson, 1991):

- Level of reading comprehension (simple, short, and concise is always best)
- Layout of text and graphics
- Intuitive format or instructional flow
- Warnings
- Updates
- Appropriate detail

Fig. 3.2 is a checklist of questions that will help you understand the users and match these elements to their needs (Schoff and Robinson, 1991).

User capabilities
- ❑ How familiar are users with your product?
- ❑ Do they use similar products?
- ❑ Are they literate?
- ❑ Are they trained personnel?
- ❑ Do they use and understand technical language?
- ❑ Do they understand charts and diagrams?

Information needed
- ❑ Basic introductory material for operation (owner's manual)?
- ❑ Educational material for new products?
- ❑ Routine maintenance?
- ❑ Troubleshooting and repair procedures?
- ❑ Explanation of the technology?
- ❑ Are the users in control of the task?
- ❑ Do they need descriptive detail?
- ❑ Are they performing upgrades or modifications?

Cultural concerns
- ❑ Do the users have a work hierarchy that prevents certain actions?
- ❑ Do they have any familiarity with technology?
- ❑ What figures, shapes, or colors have favorable connotations?
- ❑ What figures, shapes, or colors are objectionable?
- ❑ Do the users have a preferred writing style?

How the manual will be used
- ❑ Will it be used to set up and operate your product?
- ❑ Will it be used only for breakdowns or problems?
- ❑ How often?
- ❑ Do users need to read the entire document?
- ❑ Or are they likely to read a section here or there?
- ❑ Will the manual lie flat when opened to any page, so that both hands can be used for repairs?
- ❑ Does it need to resist water, grease, mud, or food?
- ❑ Will the lighting be good?

FIG. 3.2 A checklist of questions for defining characteristics of the users and how the documentation will be used.

3.5 PREPARATION, PRESENTATION, AND PRESERVATION

Now that you better understand the importance of documentation and the potential audience, how do you organize the various documents, prepare and present the information, and maintain the necessary records? This section lists documents and organizational techniques that will help you generate appropriate and complete records throughout the development cycle of a product.

Basic documents such as the engineering notebook, drawings and schematics, software source listings, and memos form the foundation of any record-keeping effort. Various manuals (such as operations, maintenance, and repair), presentations, proposals, and advertising brochures are a more visible portion of the documentation that exposes the product to customers and users. Finally, archival systems such as filing cabinets, electronic records and diskettes, and microfiche preserve your development work for future access.

3.5.1 *Engineering Notebook*

The engineering notebook is the foundation of any engineering task. It contains the reasons for your designs and the results of your experiments and tests; they are the rational groundwork for developing a successful product. You will return often to the notebook to remind yourself of the reasons for design decisions when you make future modifications to the product and prepare manuals for users.

The most important aspect of an engineering notebook is that you write one—moment by moment, day by day. Basically, it's a journal, and it should detail the *what*, *when*, *where*, *who*, *why*, and *how* of your work. That means you should record the circumstances and equipment used in each experiment and the results *exactly* as they occurred. You should write in ink for permanence and date and sign each page. Have someone cosign pages to witness important events. Fig. 3.3 shows an example page from an engineering notebook.

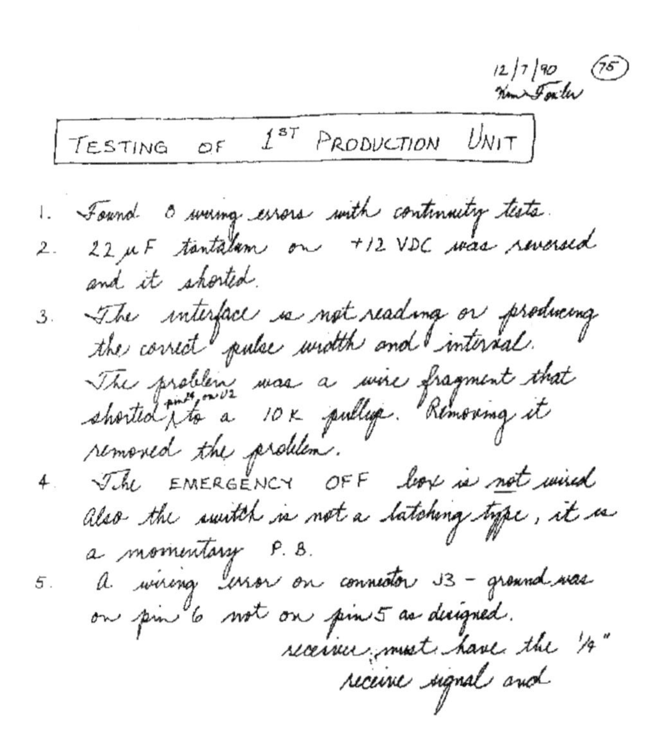

12/7/90 (75)
Kim Fowler

TESTING OF 1ST PRODUCTION UNIT

1. Found 0 wiring errors with continuity tests.
2. 22 μF tantalum on +12 VDC was reversed and it shorted.
3. The interface is not reading or producing the correct pulse width and interval. The problem was a wire fragment that shorted (pin 14, on U2) to a 10 K pullup. Removing it removed the problem.
4. The EMERGENCY OFF box is not wired. Also the switch is not a latching type, it is a momentary P. B.
5. A wiring error on connector J3 - ground was on pin 6 not on pin 5 as designed.

receiver must have the 1/4" receive signal and

Fig. 3.3 An example of a page from an engineering notebook. It's not elegant, but it's useful and very necessary.

3.5.2 *Drawings and Schematics*

Along with the engineering notebook, drawings and schematics are important records in project development. They should be continually maintained much as a notebook is. Drawings describe the *what*, *when*, *who*, and *how*. They should have title boxes that list the project title, page title, date, drawing and revision number (and the reason for the change), and the responsible engineer (Fig. 3.4).

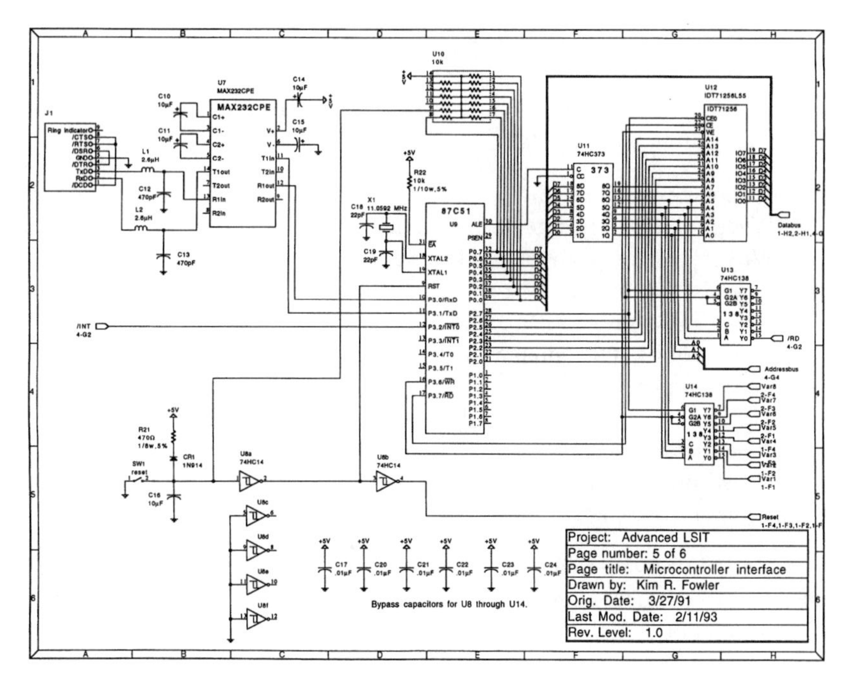

FIG. 3.4 An example of a schematic that might be used in documentation.

3.5.3 *Software Source Listings*

Software development is key to the evolution of many products; consequently, I have devoted Chapter 11 to software. Maintaining records of its development is as critical as the engineering notebook is for laying a rational foundation of development. Chapter 11 covers reasonable formats and methods for recording and preserving the development of software.

Although not actual code, programmable array logic (PAL) files should be maintained as necessary documentation. They should be clearly and consistently laid out, as should software listings. All software and PAL listings should have consistent sets of comments that indicate function and operational flow.

3.5.4 *Memos*

Memos serve as a formal record of product development and provide a chronological audit trail. Like the engineering notebook, memos should contain the *what*, *when*, *where*, *who*, *why*, and *how* of the development. Company policy will set your memo format.

3.5.5 *Manuals*

Manuals are an integral part of any product because they specify *what*, *why*, and *how* to customers and users. All manuals share some basic characteristics. Good manuals are simple, clear, and easy to use; these characteristics are achieved through attention to the following points:

- Basic content
  - What: Prominently note the model number and version.
  - Why: Clearly explain the need or theory.
  - How: Detail the operation.
- Clear organization
- Concise writing style (simple and to the point)
- Modular format
- Clearly illustrated figures and tables
- Detailed schematics
- Table of contents, index, and cross-references
- Appropriate binding

An introductory section titled "Getting Started" often facilitates a user's understanding of the operation. Many times it is the only section that users read. Prepare it carefully.

3.5.6 *Brochures*

Brochures and advertising literature introduce your product to customers. The format is specific to the market, company, and product. You should get help from a professional design consultant to prepare this kind of literature because every line, jot, and tiddle has significance to marketers and customers. These documents form the first impression in customers' minds and lay the foundation for their perception of your product.

3.5.7 *Presentations*

Presentations are frequent forms of communication and, like memos, cover all the particulars (*what*, *when*, *where*, *who*, *why*, and *how*) of a project. Simplicity should rule the organization of a talk. Follow the outline in Fig. 3.5.

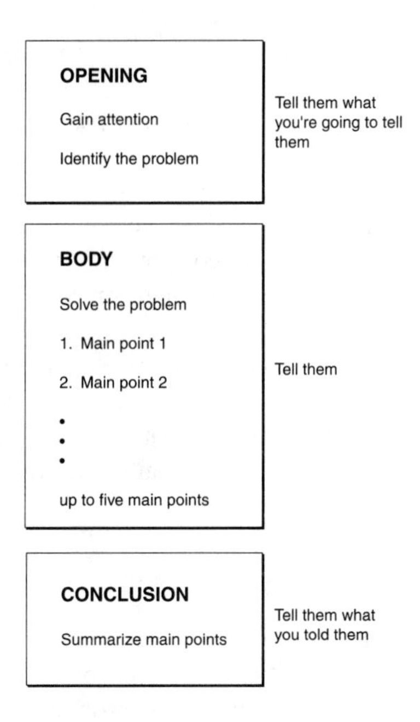

Fig. 3.5 Format for giving a talk or technical presentation.

3.5.8 *Proposals*

Proposals are carefully prepared pieces of logic and persuasion to get funding for a project. One of the most important aspects of writing a proposal is to follow the rules of the funding agency. They will usually specify the elements and format of the proposal, including the following (Meador, 1991):

- Cover letter
- Abstract
- Introduction
- Statement of the problem
- Objectives of the project
- Description of the project
- Schedule
- Budget
- Personnel and organization
- Future use or application
- Appendixes
- References

3.5.9 *Electronic Networks and Files*

Communications, computers, and networks have pushed engineering and business toward the "paperless" office. Electronic files now constitute a significant amount, if not the majority, of correspondence, business records, software programs, drawings, and illustrations. Although they provide quick and easy access, these files also require precautions to avoid corrupting or losing both records and data through power failure, misplacement, or illegal entry.

Backing up your files on diskettes, magnetic tapes, or optical disks and storing copies in separate places is your first and best defense. Using an uninterruptible power supply (UPS) for your computers further protects your data from power failure. Locking up removable disks with your security-sensitive records and using antiviral programs to check your files can reduce losses from theft and malicious destruction.

3.6 METHODS

Several universal principles stand out for preparing good documentation. First, strive for consistency and simplicity in the format of the manuals, the layout of the figures, and your style of writing; these will promote the fastest understanding among readers. Consistency and simplicity reduce surprise and confusion.

Second, use all available resources in planning and executing a document. These include photographs, figures and drawings, charts, tables, videos, and references. (Note that most of these resources are graphics, so the next section is devoted to visual techniques.)

Third, keep records of both the time per page for writing and the time per figure for preparing visuals to build up statistics for planning future documents. You may be surprised to learn that you complete only one or two pages or figures per day after a document has been through the full cycle of composition, rough draft, editing, layout, proofing, and production. Start early!

Beyond these general methods, three particular areas of documentation and communication warrant further consideration: technical presentations, proposals, and foreign translations.

3.6.1 *Technical Presentations*

Technical presentations are generally brief and persuasive talks, lasting 15 to 40 minutes, that use graphics to illustrate and reinforce points. They should be marked by verbal and visual simplicity, appropriate timing of graphical aids, and clarity in thought and speech. Technical talks usually require extensive planning, preparation, and rehearsal. Preparing a storyboard of the figures and talking through the presentation flow will aid your preparation and ensure consistency and clear communication. Never read to your audience. Tell them a story. Be prepared so that you can speak from a short list of notes or the slides.

Fig. 3.6 outlines the steps to giving a technical presentation. Section 3.7 details techniques for preparation of the visuals.

Purpose:	What do you want to accomplish?
Audience:	Who are they, and what do they want?
	General thoughts on the audience:
	• They don't know as much as you do about the topic.
	• They usually have a particular interest, but don't assume you know it. Ask or find out what it is.
	• They want you to succeed because they want to take away something from your talk that they can use.
Key points:	What do you want the audience to remember?
Outline:	Prepare opening, body, and conclusion.
Visual aids:	Keep them legible and simple.
Storyboard:	Organize sequence of points and flow of talk.
Rehearse:	Polish your talk, get critique from colleagues, and gain confidence.

Fig. 3.6 Outline of method to develop good talks.

3.6.2 *Proposals*

Proposals succeed through starting early with long-term planning, working with the reviewers, knowing your strengths and weaknesses, and understanding your competitors. According to proposal consultant R. N. Close (1989), there are seven common failures in proposals:

1. Late start
2. Not following the rules
3. Proposing to the wrong customer
4. Not knowing your company's weaknesses
5. Poor proposal management
6. Surprising the customer with innovation or change
7. Inept negotiating tactics

Using a storyboard to lay out the proposal will give a consistent format to your preparation. Debriefing with the sponsors after the award or rejection of your proposal will help you to prepare more effective proposals in the future.

3.6.3 *Foreign Translations*

Preparing manuals for the international market has many pitfalls. Most important, spend the money to get a *good* translation, because the document is important to your product!

Example 3.6.3.1 I once built a wooden ship model, from a foreign manufacturer. The instructions were poorly translated, for example: "Smooth the surface. . . taking care in particular of the rounding off of the strakes 20 in correspondence of the upper bending of the stern frames; the same strips, beginning from the middle of the hull, must be adjusted so as to form a continue surface with the ramaining planking." The misspellings and incorrect words are quoted directly, and I am still not sure I know what it means.

Aside from accurate and correct grammar, attention to cultural differences is important. Remember that some shapes, colors, and gestures have significant or offensive connotations. Fig. 3.7 is a checklist of some techniques to develop an effective manual for international markets (Fitch, 1990; Horton, 1991; Schoff and Robinson, 1991; Woelfle, 1992).

- ❑ Show rather than tell—use graphics.
- ❑ Standardize vocabulary—limit its scope.
- ❑ Avoid abbreviations, jargon, slang.
- ❑ Avoid attempts at humor, especially puns (they don't translate).
- ❑ Use simple and direct grammatical constructions.
- ❑ Use relatively formal writing style.
- ❑ Avoid using people in photos or drawings—if required, reflect the culture.
- ❑ Define color and use it only in a technical context.
- ❑ Use modular format.
- ❑ Keep side-by-side translations on same page, and use identical page numbers and figures.
- ❑ Leave space for translation (e.g., English may expand 20% in French, or 50% in Russian).
- ❑ Have a local distributer proofread.

FIG. 3.7 A checklist for developing effective manuals for the international market.

3.7 VISUAL TECHNIQUES

Graphics should make a statement or reinforce a point; their purpose is to clarify the message of the document. Good visuals are supremely important. Don't most people leaf through a book or manual and look at the pictures before they read it? It's easier to show people how to do something than to tell them. But visuals are useful only if they are well executed. This section provides some tips and techniques of good graphic design.

First you need to know what types of visuals are available. They have several forms; each has its purpose and associated technique:

1. Drawings are instructional and emphasize specific points.
2. Photographs are informational and give perspective.
3. Charts give a qualitative feel and allow a grasp of the big picture.
4. Tables give a quantitative measure and fill in the details.

Technical talks and presentations contain variations and combinations of these four basic visuals. Viewgraphs are combinations of visuals and text to reinforce points in a talk. Video and animated cartoons can relate time, motion, and sequences. (Remember to get written permission if you use someone else's photographs, diagrams, or text.)

Multimedia presentations can provide a combination of slides and animation and follow the same rules for good format. But they can become too busy and fast moving or can be amateurish if not prepared and handled carefully. Consider getting professional help to develop multimedia presentations.

3.7.1 *Line Drawings*

Line drawings are prevalent in manuals. By omitting extraneous details or showing hidden components, they allow the reader to focus on the subject (see Fig. 3.8). Line drawings reproduce easily and require less memory to store in a computer than photographs.

3.7.2 *Photographs*

Photographs are easily understood because the realistic setting puts the subject in context. Photographs tend to be more difficult and costly to produce than drawings, and they photocopy poorly. Also, details in photographs can clutter or even obscure the main point of interest. Fig. 3.9 is an example of how a photograph can give immediate perspective and context to a piece of equipment that otherwise would need many words and drawings to describe.

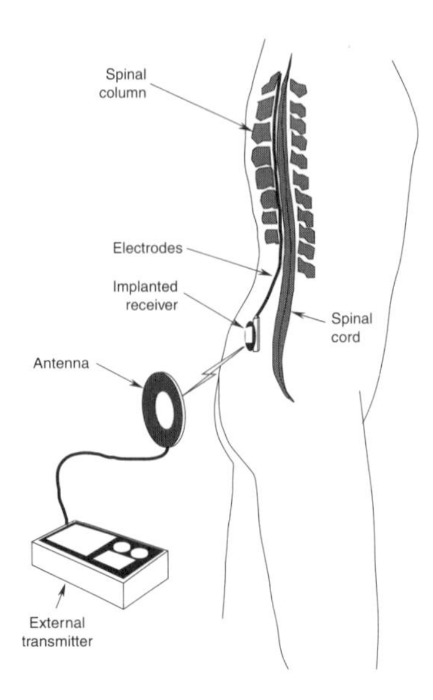

Fig. 3.8 An example of a line drawing that shows hidden components and deletes extraneous detail.

Fig. 3.9 A photograph that gives immediate context to the relationship among components.

3.7.3 *Charts*

Charts can be either qualitative or directive. (*Qualitative* techniques provide a context for comparison or subjective feel.) Bar charts, pie charts, and line graphs are qualitative and invite comparison of elements. Flowcharts are directive and define the sequence of operations and actions. Both forms are abstracted from the actual system and emphasize specific points, and both can become useless and cluttered if too many details are included. Figs. 3.10 and 3.11 illustrate the utility of charts if the detail is not overdone.

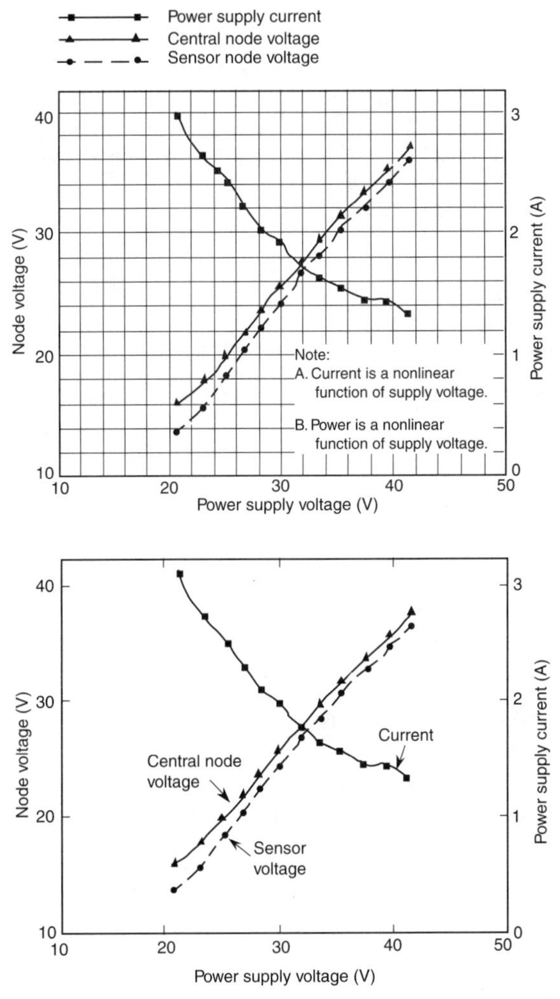

FIG. 3.10 Removing extraneous detail from a graph. The upper graph is too cluttered. The lower graph is simpler and a better visual.

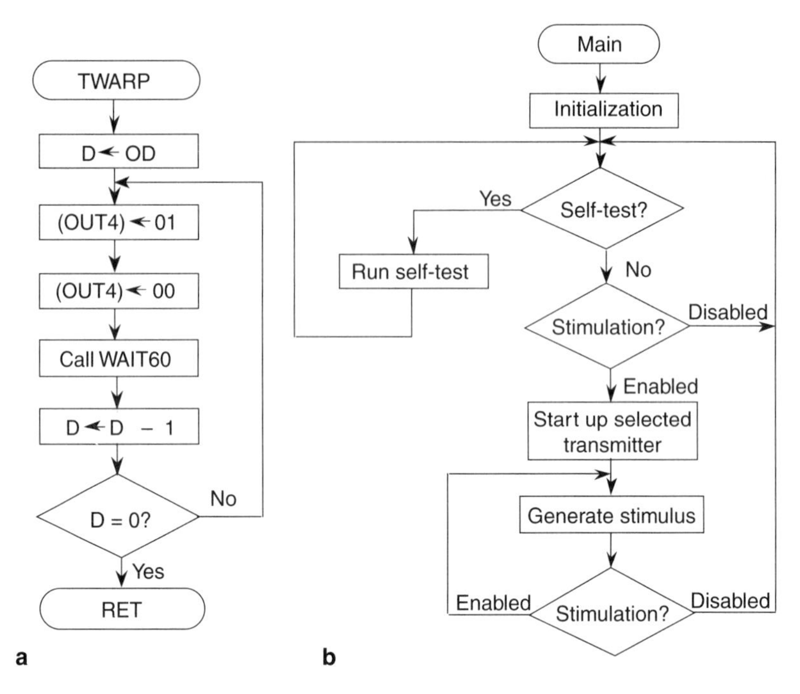

Fig. 3.11 Removing extraneous detail from a flowchart. Flowcharts should not detail each line of code in a program as the left-hand chart (a) does. (Exception is when flowchart gives a sequence of instructions to an operator.) (b) The right-hand flowchart is more abstract and easier to understand.

3.7.4 *Tables*

Tables, on the other hand, are quantitative. (*Quantitative* techniques are objective and provide precise information.) Tables arrange comparative information in a compact form. They give exact detail, usually numbers, and are more quantitative than charts. There are several rules for tables (Schoff and Robinson, 1991):

1. Arrange the data so that they can be compared vertically.
2. Align numbers on the decimal point.
3. Put the units of measure in the headings.
4. Use lines judiciously to divide rows and columns—they can clutter and confuse otherwise.

Figs. 3.12, 3.13, and 3.14 give examples for arranging tables.

Performance comparisons between systems

Model	Processor speed (MIPS)	Data channel (Mbps)	Bus width (bits)	Memory size (kbytes) RAM	EEPROM	PROM	Software	Support equipment
V	3	1.28	8/16/32	512	352	64	C, C++	Desktop computer
W	1	2.5	16	704	704	32	Ada	Workstation
X	0.3/1.5	?	16	64	0	32	Proprietary	Workstation
Y	8	3.0	32	1000	0	128	C, C++	Workstation
Z	2.5	1.7	16	512	0	128	C, C++	Desktop computer

Performance comparisons between systems

	V	W	X	Y	Z
Processor speed (MIPS)	3	1	0.3/1.5	8	2.5
Data channel (Mbps)	1.28	2.5	?	3.0	1.7
Bus width (bits)	8/16/32	16	16	32	16
RAM (kbytes)	512	704	64	1000	512
EEPROM (kbytes)	352	704	0	0	0
PROM (kbytes)	64	32	32	128	128
Software	C, C++	Ada	Proprietary	C, C++	C, C++
Support equipment	Desktop computer	Workstation	Workstation	Workstation	Desktop computer

Fig. 3.12 Two versions of the same table, showing that data comparison is easier by columns (upper table) than by rows (as in the lower table).

System comparison

Model	Cost (million $)	Weight (lb)	Size (in^3)	Power (W)
V	?	26	938	20
W	1.4	> 4	246	< 5
X	1.25	22	695	8
Y	2.0	12	477	15
Z	0.78	8	200	7

System comparison

Model	Cost (million $)	Weight (lb)	Size (in^3)	Power (W)
V	?	26	938	20
W	1.4	> 4	246	< 5
X	1.25	22	695	8
Y	2.0	12	477	15
Z	0.78	8	200	7

Fig. 3.13 Two versions of the same table, showing that white space (left table) provides good separation between columns and rows, whereas lines (right table) create clutter.

Radiation hardness

Model	Total dose (krad)	SEU (errors/bit-day)	Latch-up
V	10	> 0.23	None found in tests
W	> 15	CPU: 5.3×10^{-5}/device-day Memory: 2×10^{-7}	Immune
X	300	$< 10^{-10}$	Immune
Y	100	CPU: 4×10^{-5}/device-day	None found in tests

Radiation hardness

Model	Total dose (krad)	SEU (errors/bit-day)	Latch-up
V	10	> 0.23	None found in tests
W	> 15	CPU: 5.3×10^{-5}/device-day Memory: 2×10^{-7}	Immune
X	300	$< 10^{-10}$	Immune
Y	100	CPU: 4×10^{-5}/device-day	None found in tests

Fig. 3.14 Adding rules. Lines can be useful when the spacing between columns is irregular. Tables like the upper one can be confusing without lines like those in the lower one.

3.7.5 *Viewgraphs*

Viewgraphs combine text and visuals to reinforce the message of a technical talk. Joseph P. Cillo (1989, p. 33) writes, "WYSIWYTH. What you see is what you take home." Simplicity rules in viewgraphs because the viewer's short-term memory saturates at somewhere between five and seven facts. Here are some general rules for laying out effective viewgraphs:

1. Use correct grammar and spelling.
2. Put the most important information at the top of the visual.
3. Use a simple, sans serif typeface for legibility.
4. Lay out for legibility from the back row of the audience.
5. Don't crowd—no more than 10 or 20 words per visual:
   a. Use phrases.
   b. Space appropriately.
   c. Leave generous margins.

6. Generally avoid the following:
    a. Long bulleted lists
    b. All capital letters
    c. Mixing type families and sizes
    d. Fancy typefaces
    e. Italics (use only to *emphasize* a word or phrase)

Fig. 3.15 shows examples of bad and good viewgraphs.

Cellular Logic Operator

Operation: Each CLO ASIC operates on a universe of 510 x 510 cells plus a one-cell wide border. This implementation does not permit wrap-around from line to line but does permit a fixed, arbitrary border. The input neighborhood for CLO processing is the nine site Moore neighborhood (3 x 3) and the output neighborhood is a single cell, the center one.

ASIC Design: Each CLO has the following circuits: a 1027-bit shift register, a 32 x 16-bit RAM, two 18-bit counters, and various control and interface circuitry. It is laid-out for a 6.4 x 4.8-mm die and uses a 2-μm CMOS process.

ASIC Performance: The CLO ASIC will operate at a 30-MHz clock rate. This corresponds to a throughput of 33 ns per pixel or 8.7 ms per image transformation. Pipelining of CLOs requires only 514 clock cycles per additional CLO. Estimated power consumption is 400 mW.

Cellular Logic Operator

- **Operation**
    - 510 x 510-cell universe
    - 3 x 3 neighborhood
- **Design**
    - 1027-bit shift register
    - 2-μm CMOS
- **Performance**
    - 8.7 ms/image transform
    - 400 mW

Fig. 3.15 Bad and good viewgraphs. The upper viewgraph is too detailed; it will lose the audience. The lower viewgraph is much better for conveying the message.

Simplicity is a key for making good viewgraphs. Think of visuals as billboards; you can make your point only through a short and clear message. Don't use a schematic when a block diagram will do (see Fig. 3.16). For complex concepts, use three or four visuals, each building on the previous one to deliver the message.

Color is being used more and more often in viewgraphs, but you must handle it with great care. Color's primary function is to enhance viewers' understanding by focusing attention, linking and categorizing elements, emphasizing, and highlighting points. Strive for consistency by standardizing color between groups for identification. Use paler tints for backgrounds in viewgraphs or slides to keep from

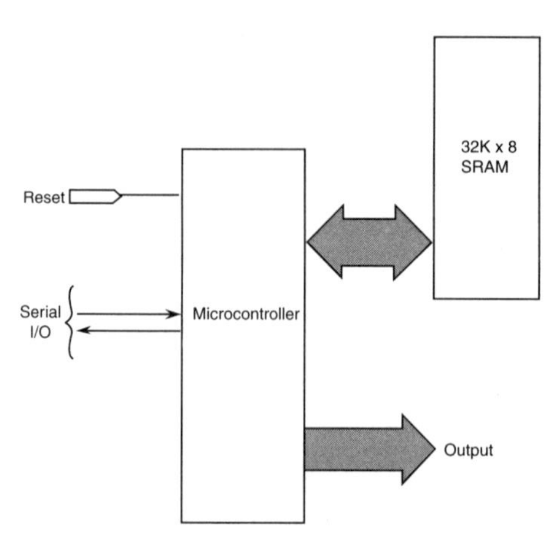

Fig. 3.16 Simplifying diagrams. Don't use more detail than necessary to communicate your point. This block diagram adequately describes the schematic in Fig. 3.4 for many purposes.

obscuring the main point of attention. If you employ graded shades in the background, it is safest to shade from dark at the top to light at the bottom. Above all, be sparing and careful with colors—particularly since software packages for generating them are readily available and easy to use. Fancy software does not make you a competent graphic artist.

All visuals should have a purpose. You need to define that purpose and examine the visual to see whether it fits the purpose. The checklist in Fig. 3.17 will help you evaluate visuals (Horton, 1991). The recommended reading at the end of this chapter provides excellent, in-depth instruction for preparing good visuals.

- ❑ Does it have an immediately obvious point?
- ❑ Does the main point stand out?
- ❑ Are the details necessary?
- ❑ Does every line, symbol, and word contribute to the main point?
- ❑ Can it be reproduced easily?
- ❑ Will it be legible under most expected lighting?
- ❑ Is it consistent with the text and the document format?
- ❑ Is it an appropriate graphic, or would another visual be better?
- ❑ Is it useful?
- ❑ Is the caption meaningful? Does it relate the graphic to the spoken or written word?

Fig. 3.17 A checklist for evaluating a visual.

3.8 LAYOUT

Preparing the format of a document requires the same kind of consideration as preparing the visuals described in the last section. Just as an artist carefully composes a picture from many components, so you must balance all the elements of a document when laying out each picture, page, and section. Clarity and consistency are watchwords in preparing documents; they unify the elements to produce a coherent message.

Three common complaints about documentation all concern lack of clarity (Gudknecht, 1982):

1. The language is vague and misleading.
2. The manual's organization is confusing.
3. The data are sketchy.

Clarity and consistency begin with the organization of the document. It should have a sensible arrangement of subjects that a reader can follow easily. Most documents use these elements:

- Table of contents
- Introduction
- General description
- General operation
- Limits of operation
- Maintenance
- Troubleshooting and diagnostics
- Appendixes
- Index

Notice that the flow is from the general to the specific.

3.8.1 *Organization*

3.8.1.1 Headings

Within each chapter are sections, subsections, and paragraphs to delve into deeper detail. Headings and subheadings reinforce the organization of the document. Chapter titles may have larger, boldface type and all uppercase letters, while subheadings within the chapter have smaller type and a mix of uppercase and lowercase letters. (Standards for some sorts of publications, including military documents, may require numbering of every chapter, section, subsection, and paragraph. The numbers make cross-referencing easier, but they can be confusing and make for slow reading. Consider carefully whether they are necessary in your documentation.)

3.8.1.2 Composition

Headings are one portion of a modular format that suits clear and consistent style. Careful use of white space and margins in a simple, uncluttered format, moreover, will best serve the reader. Finally, make sure that the chronology, or sequence of events, is correct. Don't give a warning, for instance, about a critical step in the instructions after the fact (see Fig. 3.18).

Once you have chosen the headings, you can begin to organize the text and visuals. This is a difficult task, and only experience and many good examples can help you. The alternation between text and visuals should have a smooth flow and rhythm for the reader. Section 3.7 detailed the preparation of specific types of visuals. Now you have to decide which ones fit each section of your document.

1. Clean the circuit board with detergent and deionized water.
2. Air dry for 1 hour.
3. Brush conformal coating onto the circuit board.
4. Allow the coating to cure on the circuit for 24 hours.

Note: Coating must not be brushed repeatedly after application.

NOTE: The conformal coating will form voids and not adhere to the circuit board if it is brushed for more than 5 minutes.

1. Clean the circuit board with detergent and deionized water.
2. Air dry for 1 hour.
3. Brush conformal coating onto the circuit board.
4. Allow the coating to cure on the circuit for 24 hours.

FIG. 3.18 An example of the importance of the chronology of instructions. The warning in the left-hand box is too late for some readers.

3.8.1.3 Drawings versus Photographs

Drawings are instructional and emphasize specific points by showing detail selectively. Photographs are informational and can show lots of detail. They can give accurate context and perspective to a subject. Fig. 3.19 shows how a drawing can eliminate the background clutter and focus attention. In contrast, Fig. 3.9 shows how a photograph reveals contextual detail better than a drawing might.

3.8.1.4 Charts versus Tables

Charts and graphs give a qualitative comparison of data and allow the reader to grasp the big picture. But they generally can handle only one type of data and are

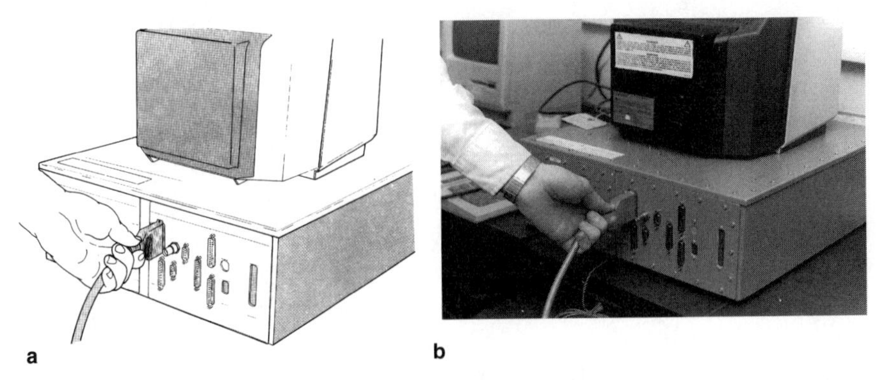

Fig. 3.19 Using a line drawing (a) to eliminate distracting detail and background from a photograph (b).

good for comparison on small amounts of data. Tables give a more quantitative comparison of data and fill in the details. In addition, tables can compare a variety of data types and handle large amounts of data simultaneously (see Figs. 3.12, 3.13, and 3.14). Conversely, Fig. 3.20 shows how a chart can be more immediately descriptive than a table.

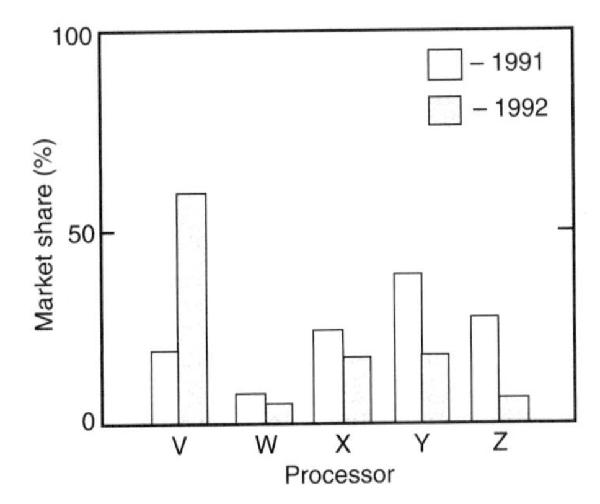

Processor	Market share (%)	
	1991	1992
V	17	62
W	5	3
X	20	14
Y	35	15
Z	23	6

Fig. 3.20 A comparison shown as a bar chart and as a table. The bar chart makes qualitative comparisons immediately more obvious than the table.

3.8.2 *Implementation*

Moving from the organization to the implementation of a document, you must establish guidelines for the text and visuals. The typography, color, and even the binding contribute to the message of your document.

3.8.2.1 Typography

The style and size of font affect your readers in subtle ways. Simpler styles and a mix of uppercase and lowercase letters make the words easier to read. Boldface fonts and all uppercase letters are tiring and more difficult to read. Fig. 3.21 illustrates some rules for good typography.

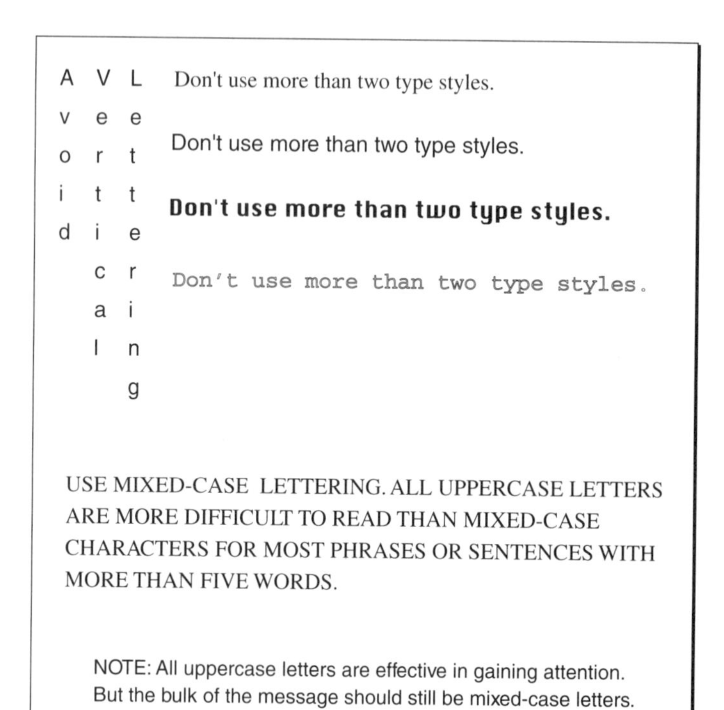

FIG. 3.21 Some rules for typography and layout.

3.8.2.2 Significant Figures

Be careful with the number of significant figures you assign to a value. Calculators and spreadsheets have increased the apparent number of significant figures in many calculations. Why use 0.1764705 when the circumstances support only three significant figures, or 0.176? This is a matter of integrity, as well as accuracy.

3.8.2.3 Color

Assign color deliberately. Decide what the reader should understand, and apply color for the following purposes:

- Focusing attention
- Leading the eye
- Explaining relationships
- Coding or linking elements
- Categorizing
- Emphasizing
- Highlighting

Don't use color simply for decoration. It should have utility and meaning. Standardize color to identify groups, or use it to signal changes in the direction of a presentation.

Beware of certain effects of color. Complementary colors, such as orange and blue, red and green, yellow and purple, can cause an unpleasant vibrating sensation. Bright colors tire the viewer's eye quickly, so use pale tints for large areas and backgrounds.

3.8.2.4 Binding

Even the choice of binding can affect the user's impression of the document and ultimately of your product. For an instructional manual, a three-ring or spiral-bound notebook that will lie flat when it is open is easier to use. For an archival document that contains records, a flat configuration will ease storage, but a three-ring notebook allows the quick insertion of updates and may be preferable for that reason alone. Finally, electronic files and archives may provide the easiest and most compact form of documentation for your particular work.

3.9 WRITING WELL

Writing well is a challenge for every author. This section summarizes some elements of good writing, but a style guide such as Strunk and White's *The Elements of Style* (1979) can prove indispensable.

To write well, be brief and simple. The active voice is shorter and more interesting than the passive voice. Use of the first and second person will add vigor to your writing. For instance, "I did . . ." is better than ". . . was completed." Furthermore, you should omit some common expressions or replace them with a simple word or phrase (Table 3.2).

Engineers tend to have peculiar problems with their writing that hinder comprehension. Jargon, acronyms, long strings of nouns, dropping the article "the," and subject-verb agreement plague many of us. Here are some examples to clarify or avoid:

Acronyms
GPS
CPU
I/O
GUI

Long strings of nouns
voltage control oscillator signal waveform
logic analyzer timing diagram
computer monitor screen readout
human factors interface engineering data

Table 3.2 Phrases to avoid and acceptable substitutes

Avoid	Use instead
along these lines	(omit)
at the present time	now
by means of	with
despite the fact that	although
due to the fact that	because
during the course of	during
during the time period of	during
for the purpose of	for
in addition	also, too
in order to	to
in the event that	if
in the vicinity of	near
it is felt that	(omit)
it should be noted that	(omit), note that
-ize	(reword)
in view of the fact that	because, since
is comprised of	comprises
on the part of	by
personally	(omit)
the capability of the product is	the product can
utilize	use
-wise	(reword)

3.10 SUMMARY

You may think that I have a hidden delight in writing documentation. Well, yes and no. Yes, because it's an integral part of any product. No, because the job of preparing documentation never ends. However, you can produce good, even excellent, documentation if you believe that it is essential to your product and plan for it accordingly.

Even though documentation can never be perfected, it can match the quality of your product. *Good documentation can't help a poor product, but poor and inadequate documentation can destroy a good product.* Remember these principles:

- Plan each document.
- Do it early.
- Do it often.
- Strive for simplicity and clarity.

3.11 RECOMMENDED READING

Horton, W. K. 1991. *Illustrating computer documentation: The art of presenting information graphically on paper and online.* New York: Wiley.
A well-written text for preparing good visuals and graphics.

Schoff, G. H., and P. A. Robinson. 1991. *Writing and designing manuals: Operator manuals, service manuals, manuals for international markets.* 2nd ed. Chelsea, Mich.: Lewis.
This short and simple text is very well written and easy to read; it should be on every engineer's shelf.

Strunk, W., Jr., and E. B. White. 1979. *The elements of style.* 3rd ed. New York: Macmillan.
The standard for good style.

These two IEEE Press books will serve you in writing and delivering effective presentations:

Beer, D. F. 1992. *Writing and speaking in the technology professions: A practical guide.* New York: IEEE Press.

Woelfle, R. M., ed. 1992. *A new guide for better technical presentations: Applying proven techniques with modern tools.* New York: IEEE Press.

3.12 REFERENCES

Cillo, J. P. 1989. Some fundamentals for presenters. *Macintosh Business Review*, January, p. 33.

Close, R. N. 1989. Selected Vugraphs from the proposal win strategy seminar/workshop. Rockville, Md., R. N. Close Associates.

Fitch, D. 1990. Producing foreign language presentations. *Audio Visual Communications,* August, pp. 17–20.

Gudknecht, A. R. 1982. Documentation usefulness rises with a little fine tuning. *EDN*, February 3, pp. 161–162.

Horton, W. 1991. *Illustrating computer documentation: The art of presenting information graphically on paper and online*. New York: Wiley.

Meador, R. 1991. *Guidelines for preparing proposals*. 2nd ed. Chelsea, Mich.: Lewis.

Schoff, G. H., and P. A. Robinson. 1991. *Writing and designing manuals: Operator manuals, service manuals, manuals for international markets*. 2nd ed. Chelsea, Mich.: Lewis.

Strunk, W., Jr., and E. B. White. 1979. *The elements of style*, 3rd ed. New York: Macmillan.

Woelfle, R. M. 1992. Considerations for international presentations. In *A new guide for better technical presentations: Applying proven techniques with modern tools*, edited by R. M. Woelfle. New York: IEEE Press.

4

THE HUMAN INTERFACE

Perfect simplicity is unconsciously audacious.
—George Meredith

4.1 MAN-MACHINE DIALOGUE AND INDUSTRIAL DESIGN

The interaction between a user and an electronic instrument is complex. Look, feel, operational flow, perception, and even documentation compose a dialogue between user and machine. It's more than display layouts. It's more than style and packaging. It's the entire instrument communicated to the user, from the front panel to the last nuance of program flow. That dialogue is the human interface of the product and defines the set of interactions, encompasses the user's view and understanding of your instrument, and affects the entire design of a product.

The human interface should influence everything that you do during development of a product. It is communication, and it establishes the message and the dialogue. The discipline of industrial design can help you define the message, establish the man-machine dialogue, design the human interface, and ultimately design the product. Industrial design is a multidisciplinary process that uses many concepts to refine the design of the human interface. It involves the following (Couch, 1993):

1. The design team: engineers, software developers, industrial and graphic designers, model makers, and writers
2. Stakeholders: end users, influencers, customers, and clients
3. Repeated discovery, development, and delivery
4. Early model concepts

In general, don't just design features; design a product with its whole purpose and function in mind. Don't fall in love with your ideas; users can have very different ideas about what is important. Consequently, you must integrate users into the design process to understand their needs, wants, and desires.

User acceptance and satisfaction help measure the success of an instrument. According to Marcus and Van Dam (1991, p. 52), "designers have learned that a good product must be easy to produce, market, learn, use, and maintain." Furthermore, subjective issues such as "look and feel" and attractive packaging are equally important to users and your customers.

This chapter introduces guidelines and principles that drive the design of human interfaces for devices, appliances, and instruments. It starts with the needs and desires of the user, goes on to define the elements of design in a human interface, and ends with some specific concerns. *You should carefully define the interface before code is written, circuits are designed, or packages are built.*

4.2 USER-CENTERED DESIGN

Understanding users is essential to successful design. For instance, McDonald and Schvaneveldt (1988) cited a study that showed pilots could access the information from a cockpit display more efficiently when the menu structure was derived from judgments of the pilots rather than developed by a design team.

Collaboration with the user during design will improve the chance of success in the final product. Including users in the process helps eliminate false assumptions and estimations. Consulting with users during design also encourages them to "own" the design solution and enhance their acceptance of the instrument.

Example 4.2.1 Don't take a solution or some interesting technology and go looking for a problem. A former officer in the U.S. Coast Guard explained that when programmable calculators became widely available a number of people developed programs for calculating values in celestial navigation. But crews did not use the programs and calculators because the manual preparation of the charts was the only departure from the monotony of the cruise and they looked forward to the calculations each day.

4.2.1 *Users, Influencers, and Customers*

First and foremost, establish who will use, influence the use of, and buy your product. You must know the difference between the customer (buyer) and the actual user. A *user* is the predominant operator of your instrument. Generally users are the most familiar with its function, though they are not always knowledgeable about its operation. The *customer* is the purchaser of the device. An *influencer* exercises considerable control over the purchase and use of a product and may have considerable knowledge about its operation.

Successful companies meet customer expectations. Site visits, resident knowledge, user profiles, focus groups, and concept descriptions will help you

Example 4.2.1.1 Consider a portable medical device for ambulatory patients. The patient is the user. The family may frequently purchase refills for the device and therefore acts as a customer and possibly an influencer. Certainly the doctor who prescribes the device is an influencer with much control over the use of the product. Nurses and medical assistants who train the patient can also exert considerable influence over the use of the product.

understand the population of users, customers, and influencers and their capabilities and expectations. The following three sections describe processes, analyses, and tests that define the users, customers, and influencers and refine the design concept of a product with their interests in mind.

4.2.2 *Prescription and Process*

You need to know the purpose for your device before knowing how to implement it. The desires, capabilities, and expectations of users, customers, and influencers will define the purpose for your device. Don Norman (1986) gives the following prescription for design:

- "*Do user-centered design: Start with the needs of the user.* From the user's point of view, the interface *is* the system. . . . User-centered design emphasizes that the purpose of the system is to serve the user, not to use a specific technology, not to be an elegant piece of programming" (p. 61).
- "*Take interface design seriously as an independent and important problem.* It takes at least three kinds of special knowledge to design an interface: first, knowledge of design, of programming, and of technology; second, knowledge of people, of the principles of mental computation, of communication, and of interaction; and third, expert knowledge of the task that is to be accomplished. . . . A truly user-centered design . . . will have to be done in collaboration with people trained in all these areas" (p. 60).

The prescription states *what* should be done; the following process tells *how* it is done. Applying this process will identify the major elements of the user interface (Card et al., 1983):

1. Analyze user needs.
2. Specify the following:
   a. Performance requirements
   b. Tasks
   c. Methods to do the tasks

3. Design alternative methods for a task to be clear to the user and easy to apply.
4. Design for error recovery.
5. Analyze the sensitivity of predicted performance and compare with assumptions. Use prototype and field testing.

Documented standards, such as DoD Standard 2167 for the U.S. military, help catch some obvious problems, but they can provide only a rough outline for the human interface. A successful design is not the domain of a smart designer or even a team of designers; rather, it is the synthesis of insights, opinions, and preferences of a group of interested parties, including the client, the users, and the design team.

4.2.3 *Analysis*

Refining the human interface requires analyzing the task, understanding the users, and testing prototypes. Analysis may take the following forms:

- Task analysis
- Questionnaires
- Informal and formal interviews
- Focus groups
- Alpha test sites
- Beta test sites

Begin development of the human interface by performing a task analysis. Get to know potential users through questionnaires and interviews. Early on you can use informal sessions to define the interface. Later in development you will want to formalize the interviews to compile useful statistics.

Focus groups are a good source of primary research to set direction for product development. A focus group has a moderator leading a group of potential users in a free-form discussion of the product's utility. An impartial moderator will have wide latitude, steering away from the *how*, and focusing on the *what*.

Alpha test sites allow close interaction between designer and potential users to refine designs. Beta test sites require potential users to evaluate the function of devices in nearly finished form, with little or no interaction with the designer.

Example 4.2.3.1 A company collaborated with an industrial design firm to develop lunch boxes for school children aged 6 to 12. It extensively studied children, how they used lunch kits, and concepts for lunch boxes. The analyses took several routes; they asked children to do the following:

- Sketch lunch boxes.
- Color in line drawings of lunch box concepts.

- Place color chips on lunch box concepts.
- Indicate what characters they wanted on the side of the lunch box.

The company also built models of the concepts and asked parents and children to evaluate the concepts. (Note: The users were children, the customers were their parents, and the influencers were peers and parents.)

The researchers found that children wanted different compartments for thermos, sandwich, and fruit. Interestingly, they also found that the children had ecological concerns and wanted environmentally safe materials. The new lunch kits sold very well (Couch, 1993). (While the lunch box was not an electronic product, the principle of investigating user desires is still applicable.)

Obviously, involving users throughout development is integral to these analyses, but you should follow several maxims when dealing with users. First, strive to understand users; your knowledge will always be incomplete. Second, don't necessarily try for consensus; accept and work with differences. Third, you will have to balance the wants of the user against your risks to develop a minimum product with maximum benefit.

4.2.4 *Rapid Prototyping and Field Testing*

Rapid prototyping and field testing are two methods that implement the interaction and collaboration between designers and users. Rapid prototyping allows the interactive development of rudimentary functions by the designer and user in an operational environment. A graphical interface may be simulated with screens that represent, but do not implement certain functions, thereby speeding development and isolating problems early. Field testing involves more mature products, takes longer, and allows users to examine functional products.

Example 4.2.4.1 I drove a car that had a control knob for the dashboard lights mounted on the upper edge of the dash near the rim of the steering wheel. When I made left turns, my knuckles would occasionally rake across the knob and shut off the dash lights even though my hands were gripping the steering wheel properly. Thorough prototype and field testing would have revealed that the knob location was poor and needed to be moved.

Rapid prototyping and field testing are absolutely necessary to the successful development of an instrument. Rapid prototyping and field testing can speed the development process by revealing problems and illustrating applications. David

Moriconi, president of IDE Inc. in Scotts Valley, California, stated, "I'd have to say that the prototyping probably has more impact on a successful program than any other phase" (Fraser, 1990, p. 196). He went on to say, *"The actual users, in some cases, gave us just the opposite feedback from what we expected"* (p. 198).

A report by the Defense Science Board Task Force on Military Software (1987) emphasizes both rapid prototyping and field testing for developing software: "Users cannot, with any amount of effort and wisdom, accurately describe the operational requirements for a substantial software system without testing by real operators in an operational environment, and iteration on the specification. . . . [I]t is simplest, safest, and even fastest to develop a complex software system by building a minimal version, putting it into actual use, and then adding function, enhancing speed, reducing size, etc., according to the priorities that emerge from actual use."

You use rapid prototyping during informal sessions with individual users and focus groups and at alpha test sites. Field testing applies to beta test sites. You can collect data through interviews and questionnaires with the users and through measurements and observations of the prototype.

Prototype testing requires flexible, repetitious design cycles to fit users' perceptions and habits with the goals of generating more consistent interface models and useful, enjoyable tools. Achieving these goals helps users own the design solution and improve their final acceptance of the instrument.

4.3 FIVE ELEMENTS OF SUCCESSFUL DESIGN

Five interdependent elements contribute to good design: cognition, ergonomics, utility, image, and ownership. These five elements define the human interface once you have established the purpose for your product and understand the user population. They ensure a useful and functional dialogue.

A user's view of an interface (and the product) differs from yours. Both user and designer form conceptual models of how the instrument works. Unfortunately, those conceptual models often differ vastly. Your goal as a designer is to create one conceptual model of the product, or at least to accommodate both models.

Fig. 4.1 illustrates the basis from which users and designers view a product through the five elements of design. The conceptual model of a product is built on these five elements.

If you don't make provision for all five categories—called *product interface levels* by Robert Case (1993)—your product is deficient. *Cognition* involves the mental tasks and computations involved in operating a device and relates to the intuition or expectations of the user. *Ergonomics* deals with the traditional concerns of human factors. *Utility* measures ease of use and is a differentiating factor for a product in a competitive market. *Image* represents more than the styling of a product; it includes the user's perceptions of the instrument and its operation. *Ownership* is the level of commitment that a user exercises to use your product.

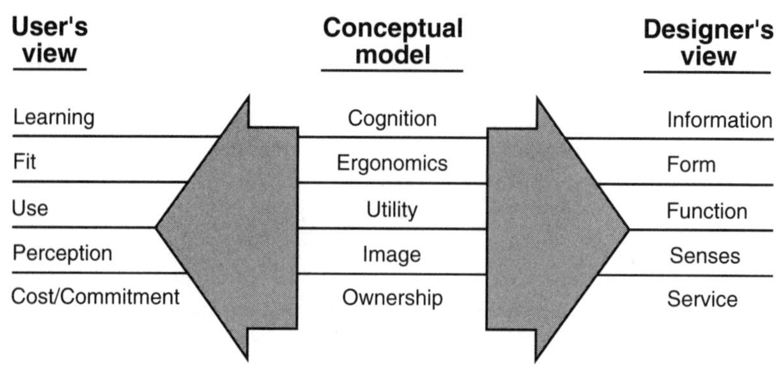

Fig. 4.1 Five elements of a human interface as viewed by the user and the designer.

These five elements intertwine to provide a basis for a conceptual model of a product. The cognitive complexity for performing a particular task ties directly in to the utility of a device. Image can often enhance ownership, while ergonomics may place buttons where a user might expect them, thus simplifying a task.

Users want to understand a product and have control over events. If your instrument is difficult to use, people won't use it in spite of its capability. Have you tried to use a remote control for some consumer electronics or program a VCR? Most people never learn to use the full range of functions because of the complexity of use. Perhaps you, like me, have a VCR that continuously blinks 12:00 because you have not figured out how to set the clock. Fig. 4.2 shows a remote control for a stereo system; it has 32 identical-looking buttons. Who is ever going to remember all of them?

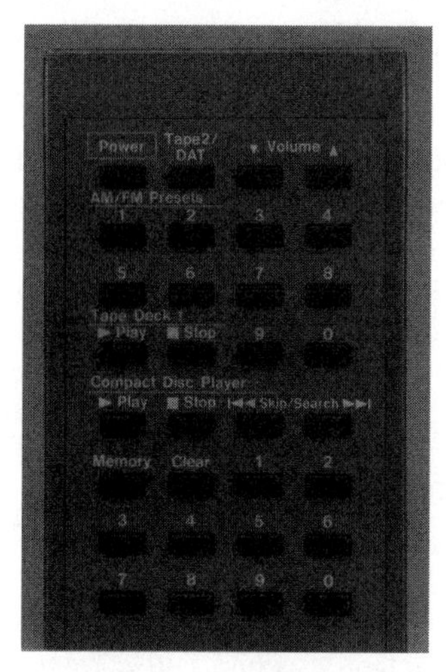

Fig. 4.2 Panel from a remote control for a stereo system. The buttons are small, identical in shape, and regular in layout. Consequently, operation is confusing, and mistakes are frequent.

4.4 COGNITION

The cognitive aspects of instrument design have many ingredients, including learning, memory, organization, and consistency. Most of these ingredients come to the fore in the graphical user interface that provides the dialogue between the user and the instrument. Marcus and Van Dam (1991, p. 49) write, "The purpose of the user interface is to facilitate user-computer communication by enveloping hardware and software, particularly the semantics of applications, in a dialogue. This dialogue hides the structure of input/output devices. . . . The user interface remains embodied in these essential aspects: one or more underlying metaphors or ideas, the conceptual organization of data and functions, navigation techniques, appearance characteristics, and interaction sequences."

4.4.1 *Learning*

Nearly every device requires learning by users before they can operate it. Some devices are more obvious or intuitive and therefore require less time to learn. Novices usually need more time to learn than experienced people that have used similar tools. (Sometimes, however, habits established with similar devices can impede learning of a new instrument that is somewhat different.) For complex systems, learning can be a significant investment in the ownership of the product, so follow these rules (Fischer and Lemke, 1988):

1. Provide incremental instruction for inexperienced users so that they can accomplish useful results even when they know only a small part of the system.
2. Make experts more efficient by allowing them to build on their knowledge and use short cuts around simple, time-consuming operations.
3. Give users a relevant context so that they can choose the next steps. This is simply good organization.

Use examples in instruction. Complete examples allow extrapolation to other actions through relevant context (as long as the format and operation are consistent). Beyond examples, you can enhance the cognitive aspects of communication with metaphors, exaggeration, repetition, and associated symbols that serve in an intuitive dialogue between the user and the instrument (Marcus and Van Dam, 1991). Icons, windows, menus, buttons, and knobs may use these techniques of communication within a graphical user interface. Unfortunately, if they are poorly done, they can obscure the obvious.

4.4.2 *Memory*

Memory recall and cognitive processes play an important role in ease of use. Simplifying the amount, content, structure, and learning of knowledge will improve the ease of use (Kieras and Polson, 1985).

The user's short-term memory connects one event to the next during interface dialogues. It has a capacity of about five items. Don't require the user to remember sequences of operations, items, or locations that burden short-term memory.

Memory recall has two psychological components—recall-directed searches and recognition-based scanning—that must be considered to simplify retrieval of information. Recall-directed searches use remembrances to retrieve the desired information—for example, combinations of keystrokes to initiate operation. Recognition-based scanning looks for familiar landmarks to navigate a decision tree and retrieve information—for example, finding appropriate icons in completing an operation. An expert user will use recall-directed searches more often; a novice will use recognition-based scanning (Lansdale, 1990).

Finally, response latency of the interface affects memory recall. Longer latencies will significantly impede understanding of the instrument, increase the cognitive complexity, and reduce its ease of use (Norman, 1986).

4.4.3 *Organization*

Organization puts information where users expect to see it. For instance, categorical menus are better than alphabetical or random organizations (McDonald and Schvaneveldt, 1988). Organization should be simple and concise, with few components. Too many colors, windows, overlays, and character sizes will distract the user and hide the relevant information (Van Der Veer et al., 1990). Ultimately, the user should have the final say in the organization of the interface. According to McDonald and Schvaneveldt (1988, p. 318), "menus organized according to users' empirically derived cognitive structures are superior to . . . alternatives (e.g., alphabetical, random, and subjective organizations)."

4.4.4 *Consistency*

Consistency is the hallmark of intuitive operation. Consistency in form, color, and operation promotes understanding, training, and acceptance by the user. Each user forms a mental model of the instrument. A good mental model is necessary for the user to employ the device effectively (Norman, 1986). Consequently, consistency "helps a user to stick to a simple mental model," while inconsistency "prevents the user from constructing a comprehensive mental model," according to Van Der Veer et al. (1990, p. 148). Consistency within and between applications of the user interface significantly reduces training time (Polson, 1988). Increasing complexity within instruments and systems requires greater learning time and fault tolerance in the design; consistency plays an important role. Ziegler et al. (1990, p. 27) write, "The more the functionality of a system is broadened, the more the ease of learning is determined by the degree of consistency of use between the different functional domains."

4.4.5 *User Capabilities*

The abilities of the user population must be considered. Users vary in experience, frequency of use, and whether they are "imagers" or "non-imagers." Since novices don't know the terminology or "syntax of commands, they should be allowed to choose from (semantically well-structured) menus or to work in a form-filling mode. After some hours of experience, they will need the possibility to turn to command mode. If novices are not sure about semantics, well-designed icons may be helpful" (Van Der Veer et al., 1990, p. 147). Occasional users "need the possibility to ask for information about the present mode of interaction, about previous situations and actions, and about possibilities to proceed" (p. 147). Finally, you must incorporate the cultural, professional, and personal preferences of the user. All these concerns result in trade-offs between the elements of design; intuitive operation does not imply minimum operations, for example, and consistent metaphors may require repetitious rather than minimal operations.

4.5 ERGONOMICS

Studies in human factors have traditionally concentrated on the physical layout of interfaces. A classic example is the arrangement of stove top controls: How easily do they map to the heating elements? Human factors (ergonomics) focus on accessibility, arrangement, and fit to make a product usable.

Ergonomic data provide a rough baseline or starting point for interface design. Even so, Fraser (1990, p. 198) writes, "Despite the aid of ergonomic data, designers can still be surprised when they send prototypes out into the field for testing." Hence the continued need for testing with users.

4.5.1 *Anthropometric Data*

Statistics on size, weight, and reach of general populations exist in numerous sources (Matisoff, 1990; Woodson et al., 1992). These statistics constrain the physical dimensions of a workplace. Fig. 4.3 illustrates some of the human measurements used for anthropometric data.

The capability of the user population varies with age, sex, size, mobility, dexterity, coordination, reaction time, training and experience, and cultural background. Don't just apply these statistics to an "average" user and assume that your work is done. Each task needs refinement to suit the particular group of users.

4.5.2 *Workplaces*

Your instrument may be used in a wide variety of situations. Equipment arrangement, posture, and seating make up an inexhaustible array of workplaces. Fig. 4.4 just begins to sample some of these configurations.

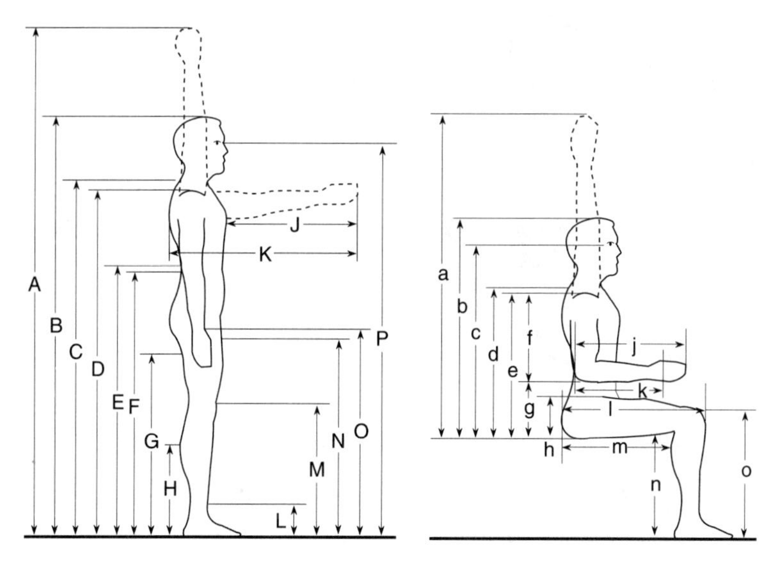

Fig. 4.3 Some measurements found in anthropometric data (modified from Woodson et al., 1992).

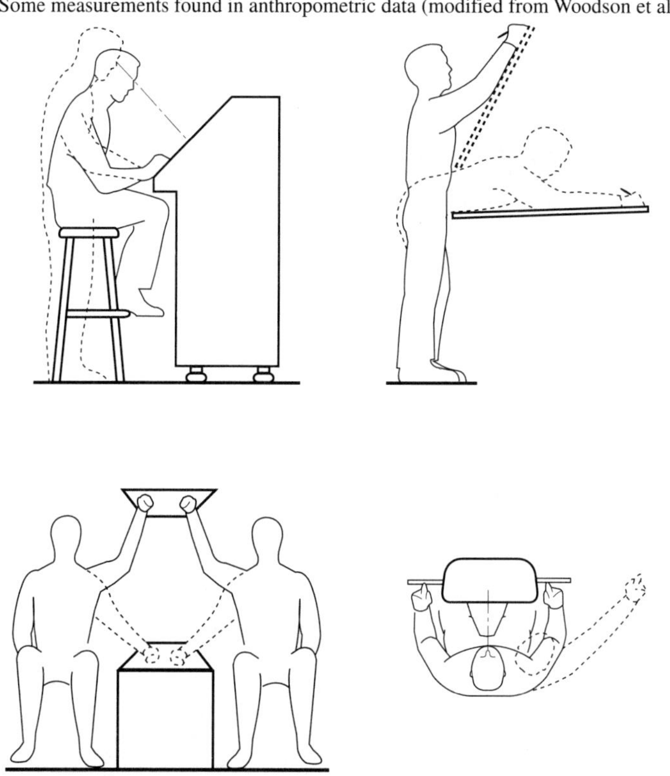

Fig. 4.4 Equipment arrangements, positions, and postures found in some environments. Each has limitations on reach, angle, and comfort (modified from Woodson et al., 1992).

Access and viewing add to the complexity and considerations of ergonomic design. Fig. 4.5 illustrates several accessibility concerns.

In general, you should design workplaces with these rules in mind (Woodson et al., 1992):

- Avoid awkward positions.
- Use normal limb movement and reach limitations.
- Minimize fatigue.
- Reduce hazards.
- Reduce distracting noise.
- Control temperature and humidity.

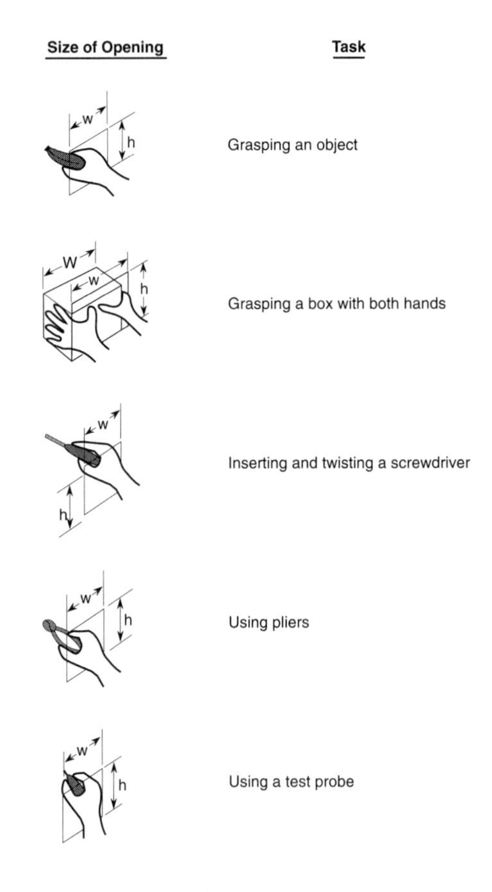

Fig. 4.5 Some examples of accessibility in maintaining equipment.

Example 4.5.2.1 A company proposed to the U.S. Air Force a redesign of a large loader for cargo aircraft. They employed an industrial design firm to help redesign the cab of the loader. The original cab used gauges, switches, and knobs laid out according to military specifications, but it was still difficult to use. They built cardboard and plywood models of the cab and attached the controls with Velcro to allow quick rearrangement. Air force personnel evaluated the different layouts of the cab. The designers found that the most important issue was a good seat, because operators spend entire work shifts in it. The seat became the focus of their design (Couch, 1993).

My colleague Steven Lammlein (letter, June 2, 1994) predicts that legal issues will increasingly affect interface design:

> Interface design can in many cases have very important consequences, with accompanying liability implications. Indeed, one frequently reads of people suing manufacturers and designers for allegedly defective product design. These suits sometimes yield lucrative settlements or decisions even where the products seem to have been used in a clearly uninformed or irresponsible manner. The potential liability consequences of interface design, and the fact that an engineer may be called upon to defend his/her interface design in court before a hostile audience, should be sobering thoughts to those who may be tempted to take interface design and error checking lightly. Perhaps interface designers may wish to remember that the words *user* and *suer* consist of mere transposes of the same letters.

Another concern is government regulations, such as the Americans With Disabilities Act, which makes it very difficult for an employer to discharge or refuse to hire a person with disabilities. Consequently, businesses must accommodate disabled persons with alternative ways of doing a job. The implication for interface design is that interfaces may have to address more sensory modalities and require novel approaches for persons with disabilities.

4.5.3 *Lighting*

Illuminating the instrument and workplace requires attention too. Lighting intensity, angle, and uniformity of coverage are important concerns. Glare can be distracting or even hazardous if it obscures important information on computer screens and panel displays. You need to know the environment where your product will be used to select appropriate lighting.

4.5.4 *Cues*

Cues are the indicators of function. Cues may be visual, auditory, or tactile (even olfactory if the equipment malfunctions and begins to smoke). Each type of cue has unique advantages in the operation of devices.

Visual indications provide rich cues that have spatial, context-sensitive, and quickly recognized attributes. Lamps give active visual cues. They can indicate the following:

- Whether a function is active
- Mode of operation
- Warnings

Many products use a small LED to signal when power is on. This cue is important when someone is trying to diagnose a malfunction. (Did you plug it in?) Using too many lamps can hide a particular event, because human senses accommodate to multiple conditions that are similar.

Once again, there are no absolutes in designing interfaces. Generally, you should not use very many lamps for indications, but occasionally many points of light can give a unique view into the operation of an instrument. Some people could read the patterns of flashing lamps on old minicomputers and actually ascertain the state of the computer.

Pictorial symbols describe a situation quickly and efficiently. Symbols also make instruments easier to use for people who speak other languages. A stop sign, for instance, is a familiar, everyday pictorial symbol. Such symbols suit certain displays, and some are illustrated in Fig. 4.6.

Beyond lamps and pictures, even shapes give visual cues to operation. The arrangement of door handles, for instance, can and should, but in reality does not always, cue the proper operation. A knob indicates a twist, but do you pull or push the door? Other cues are then necessary, such as how the door fits the lip of the jamb. A horizontal bar indicates pushing the door, while a vertical handle indicates pulling. Be careful to investigate the visual cues of your design; make sure that they simplify task flow and do not give contradictory or inconsistent signals.

Example 4.5.4.1 The door handles at the post office in my hometown are backward (Fig. 4.7). Horizontal bars are mounted on the exterior of the doors, and vertical handles are inside. This arrangement cues people to push the doors when they enter and pull them when they exit—but the doors swing outward, not inward. Fire codes in the United States require all exterior doors to open outward, but the visual cues from the bars and handles override this fact and people don't remember it when they use those doors.

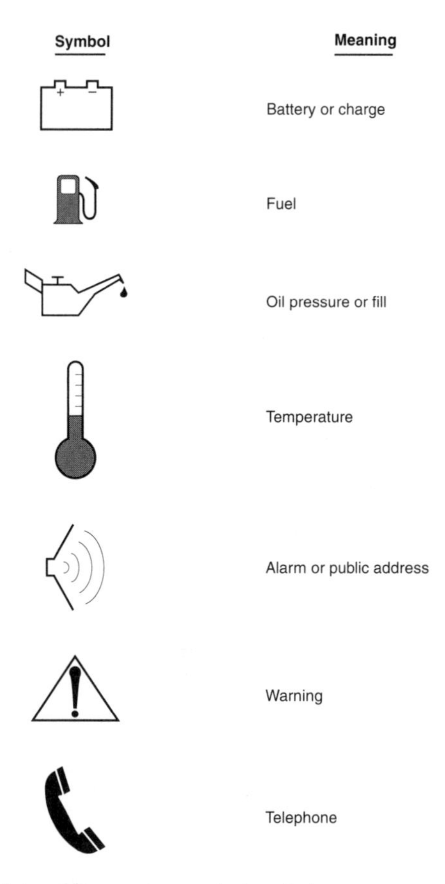

Fig. 4.6 Examples of pictorial cues using standard symbols.

Auditory cues are excellent for signaling unique and infrequent conditions. They have low information density and are good for warnings, but too many different sounds are confusing and eventually ignored. How many times have you been confused (or annoyed) by an automobile with multiple alarms? Is that buzzing for keys in the ignition, or the headlights left on, or what?

Be careful to set the volume appropriately: too loud and you will startle and upset users, too quiet and they will not notice. "Appropriate" means fitting the level of concern; some warnings must be insistent for hazardous operations.

Tactile cues can use size, shape, force, and texture to differentiate functions. They are good for controls that an operator manipulates without looking away from another task.

FIG. 4.7 Doors mounted backward so that the handles give the wrong visual cues. The exterior horizontal bars beg to be pushed, while the interior vertical handles indicate a pull—both backward from the way the doors actually swing.

Example 4.5.4.2 I have a video camcorder that has a fade option. The button for fade is larger than other buttons, but, more important, it has small bumps on its surface that easily distinguish it from the other buttons. I can find the button with the touch of a finger and control fade without having to look for it. This feature allows me to keep my eye on the viewing screen during filming.

Tactile cues have size and shape limitations; they can't be very small or subtly shaped. Cues must usually be large enough to grasp with several fingers and have significantly different shapes to allow differentiation of functions. Different shapes can provide an important safety factor by differentiating certain controls such as those found in an aircraft or a power plant.

4.6 UTILITY

Utility measures how well the expectations of users are met. Does the instrument do what the users want? How completely does the device accomplish a task? Utility affects primarily the feel of the user interface and includes simplicity, minimal operations, and easily remembered memory processes.

Utility relates closely to cognition and ergonomics. Simplicity and minimization often correlate with intuitiveness, but some simple, minimal operations may also be completely nonintuitive. For example, a software command may be initiated by either a keystroke sequence or a selection from a pull-down menu. The keystroke sequence will probably be faster and take fewer steps than the menu selection, which may be more intuitive and instructional. An expert user may prefer a keystroke sequence for speedy execution, whereas a novice may prefer the menu selection for understanding the process.

4.6.1 *Simplicity*

Simplicity involves graphical expression, user understanding, and instrument operation. Simplicity in expression reduces the task of locating information. Simplicity in understanding speeds both learning and acceptance of the instrument by the user. Simplicity in operation reduces mechanical movements, the number of buttons and knobs, and the psychological calculations. A push button, for example, is better than a three-way knob on a lamp if all you want is on/off operation.

4.6.2 *Minimal Operations*

Minimal operations will speed the execution of a task. Card et al. (1983) have developed a model of cognitive structure called GOMS (for "goals, operators, methods, selection rules") to analyze tasks. The GOMS analysis allows comparisons between methods by quantifying performances in practice, memory, keystrokes, elapsed time, reaction time, matching, perception, and motor skills. Minimal operations may be counterintuitive, however, and trade-offs between these elements must be discussed with the users.

4.6.3 *Usability Testing*

Usability testing helps measure the utility of your product. It employs prototypes and potential users in a variety of situations to evaluate the designs. Study each prototype through task analysis, interviews with users, and focus groups. In developing tests for usability, follow these guidelines (Trower, 1993):

1. *Don't* use members of the development team as subjects.
2. *Don't* draw conclusions from the results of one subject. Use an appropriate sample of users, including those from different cultures and countries if you are designing for the international market.
3. *Do* test for whether the outcomes are appropriate—that is, whether testing accomplishes what it is supposed to.
4. *Do* test early and often.

No hard and fast rule exists for the number of times you'll have to repeat tests.

4.7 PRINCIPLES FOR APPROPRIATE OPERATION

The first three elements of design—cognition, ergonomics, and utility—are rather artificially divided; in fact, the issues are interdependent. The goal of these three elements is to help you build good conceptual models of interfaces. Two different views may help you see how these elements may intermingle and contribute to good design.

4.7.1 *Visibility, Mapping, and Feedback*

One view, expressed by Don Norman (1988), is that an interface provides three attributes for building a good model:

1. *Visibility* shows the user the available options and the state of the system. It relates to the elements of cognition, utility, and ergonomics. It helps the user understand the system.
2. *Mapping* gives the user the relationship between controls and their results. It encompasses the elements of cognition and ergonomics and enhances user control and understanding by meeting expectations. A natural mapping is intuitive; it fits the user's culture and expectations. Mapping occurs on various psychological levels: between intention and possible action, between actions and effects, and between the state of the system and what is perceived. Mapping usually requires visibility.
3. *Feedback* informs the user about the action performed and the result accomplished. It is used in cognition to clarify operation and reduce mental tasking. Feedback aids visibility, for example, by providing tactile feedback in push buttons, audio cues, and visual cues.

4.7.2 *Control, Predictability, and Economy*

The other view, from Marcus and Van Dam (1991), sums up some guidelines for intuitive operation:

1. *User control:* Users should feel that they have direct control and can do useful things. (Beware. Automation can take away control if you are not careful.)
2. *Predictability:* Strive for consistency, or the "rule of least astonishment."
3. *Economy of expression:* Terse and appropriate expression of users' wants, needs, and actions makes users feel they are working with an intelligent partner.

4.8 IMAGE

Image is very important to the success of a product. Marcus and Van Dam (1991, p. 50) write, "An attractive look and feel in a product's user interface can be as important to its success as its functionality." Unfortunately, image is also the most fickle element of design.

4.8.1 *Dimensions*

Image has several dimensions that determine its impact on the user: visual composition, physical size, and operation. Visual composition, while subjective and difficult to define, is the most important component of image. The visual component comprises color, materials, textures, and proportions, along with style and fashion. Entire colleges are devoted to training professionals in commercial design; a technical designer should draw on the expertise of professionals in graphic and commercial product design to develop an attractive package for an instrument. Even highly advanced customers and users may alter their judgments of an instrument because of an attractive package. (The first case study at the end of this chapter is an example of subjective judgments from doctors and engineers based on image alone.)

In general, a designer should strive for compactness; big is seldom better. Finally, the operation, or feel, of the instrument should be smooth, giving a clear indication of progress with little delay in response.

4.8.2 *Perception*

The user's first impressions, expertise, and culture all factor into developing an attractive look and feel for the instrument. A user's first impression of an instrument is a powerful determinant of acceptance. An engineer and industrial designer said, "Most of the consumers out there are not engineers. They really don't know whether a product is well-engineered or not. All they can see is the outside, the housing. If the housing communicates to them the idea that a product is high quality, then they'll have confidence in buying it. For general consumers, the appearance is the only means they have of determining whether they want to purchase a product or not" (Fraser, 1990, p. 195).

Documentation, particularly user manuals, forms a portion of the product's image. Effective manuals are clear and concise, with drawings, indexes, and cross-references.

Example 4.8.2.1 Impressions affect professionals as well. A colleague related a story about how appearance prevented the acceptance of a computer system. He had designed a minicomputer with a graphics terminal in an instrument rack. Since the sides of the rack were exposed, he covered them with beaverboard (Masonite paneling). The computer system worked fine, but the only comment he ever heard about it was a disparaging reference to "that thing with beaverboard sides."

4.8.3 *Operation*

The user should perceive the operation of your product as logical, appropriate, and consistent. One event should naturally lead to the next in a smooth flow. Operation should also be pleasing to the eye, ear, and touch.

4.8.4 *Impact*

Novice users and experts are equally affected by the image and package. New products in an international market require particularly careful attention. Even an inappropriate name can prevent successful market penetration. A good image enhances a product, *but a poor image can utterly destroy a product, regardless of its utility.*

4.9 OWNERSHIP

Ownership is the commitment to a product by users, customers, and designers. Users invest time, effort, and psychic energy to learn and use the device. Customers spend money and time to investigate and purchase the device. Designers invest their energy, time, and emotions to develop the device. All these represent a considerable commitment to using your product. Have you made it worth their while?

> **Example 4.9.1** A developer of software wrote about an experience testing a program: "The software asked the user whether it could modify a user's file. Surprisingly, no user hesitated or declined the change, even though in the real situation it could have significant impact. When debriefed afterward, several users said they had responded that way because it wasn't really their machine" (Trower, 1993, p. 16). Talk about a classical problem of ownership!

Image and ownership are closely linked by psychic, emotional, and perceptual aspects of a design. Like image, ownership is difficult to measure. The elements of cognition, ergonomics, and utility all contribute to a user's commitment to a product. Other users can be powerful influences toward ownership. If they really like the product, they can sway other people to use and accept it. Owner referral and repeat sales are the most cost-effective methods of marketing a product.

4.9.1 *Cost*

Obviously, monetary pricing is important in ownership. The cost must be commensurate with the capability of the device and the service that it renders.

Customers are happy to pay the appropriate cost if the device meets or exceeds their expectations.

Furthermore, the cost over the life cycle of the device affects ownership. Designers often overlook its value in customer and user commitment. The costs of power consumption, maintenance, and repair all factor into life-cycle cost.

4.9.2 *Good Design*

Careful and thorough application of the other elements of design (cognition, ergonomics, utility, and image) will go a long way toward influencing users to commit to your product. An intuitive, easy-to-use device that provides significant control and feedback in accomplishing a task begs for ownership by users.

Designers also need a measure of ownership to perform at their best. Service planned and provided by the designer or design team indicates their measure of ownership.

4.9.3 *Instruction*

Appropriate training and instruction will improve user commitment. Users need to understand the full range of power and capability of your device to truly own it. You can use a number of techniques and methods to instruct ownership:

- Intuitive operation
- Metaphors
- Instruction by example
- Training manuals
- Training classes

Advertisements and sales brochures influence customers and can instruct users.

Ease of training is an important issue with employers because many people do not read instructions well—or at all. Consequently, some businesses will emphasize ease of training when choosing equipment.

4.9.4 *Specialty Equipment*

The ownership problem is easier to tackle with specialty equipment than with mass-produced devices. With specialty equipment, a small, fairly homogeneous group of users usually operates it and has a good working knowledge of its function. Often tasks are complex and operators are sophisticated in using the equipment. You can accelerate commitment by involving users early so that they have a significant hand in the design process and own a portion of the design of the product.

Example 4.9.4.1 I designed a tape unit (see the case study in Chapter 2 on the MTU) for the U.S. Air Force to speed data transfer between desktop computers and aircraft avionics. Unfortunately, the units were not used by the aircrews for two reasons. First, I did not spend nearly enough time with the aircrews in rapid prototyping, particularly developing software for the computer screens. Second, the aircrews did not get consistent follow-up training in using the computer or the tape unit. The lack of involvement and training nearly eliminated the feeling of ownership by the aircrews. Consequently, the computers and tape units ended up collecting dust instead of providing useful service. (This experience also became an example of a failure that increased my expertise and taught me what not to do. The moral: Learn from your failures.)

4.9.5 *Influencing and Marketing*

Finally, sales and service influence ownership. Conscientious and truthful marketing will build a faithful following of customers.

4.10 PRACTICAL APPLICATIONS AND SYSTEM IMPLICATIONS

This section gives some practical tips for designing screen displays on computer-based instruments. It covers screen layout, input devices, how to prepare user interviews and questionnaires, and the implications for hardware and software. The applications contain some of the principles and elements of design presented in the preceding sections.

4.10.1 *Some Display Applications*

Understandability is a customer expectation and often a driver for use. Therefore, promote understanding and control. Specifically, the design of computer displays should fit the operator's expectations, experiences, and culture. In general, you can enhance understandability if you follow these rules (Dinan, 1992):

- Input data to the left or above.
- Output data to the right or below.
- Justify columns.
- Indicate only exceptions instead of all test data.
- Identify page numbers.
- Use consistent formats.

Fig. 4.8 illustrates some good and bad examples of these guidelines.

The last point, consistent format, is an issue of visibility and deserves some elaboration. Operator *perseveration* leads to errors when formats are inconsistent; that is, users automatically continue doing what they are accustomed to do, even

Bad Form

	Test Type							
Unit	A	B	C	D	E	F	G	H
1	P	F	P	P	F	F	P	P
2	F	P	P	P	P	P	P	P
3	P	P	F	P	P	P	P	P
4	P	P	P	P	P	P	P	P
5	P	P	P	P	F	P	P	F
6	F	P	P	P	P	P	P	P
7	P	P	P	P	P	P	P	P

Better

Unit	Test Failures
1	B, E, F
2	A
3	C
4	
5	E, H
6	A
7	

Bad form:

870.3
2.5
1500
3550.2
45.76
27.3
108.9

Better:

870.3
2.5
1500.0
3550.2
45.76
27.3
108.9

Bad form:

frequency: 870 MHz

pulse width: 2.5 μs PRF: 1500 Hz

radiated power: 3550 W

angle of coverage: 45 direction: NE

Better:

frequency:	870	MHz
pulse width:	2.5	μs
PRF:	1500	Hz
radiated power:	3550	W
angle of coverage:	45	
direction:	NE	

FIG. 4.8 Examples of good and bad organization of information.

when situations change. You can accommodate this human tendency by making displays and inputs consistent within each screen and between screens. Consistent formats will help highlight operational changes.

Appropriate design in the expression of the interface dialogue will simplify the task of locating information. Table 4.1 gives some rules for simplifying expression (Plauger, 1993; Reid, 1984). Good design will make the information more understandable. Table 4.2 gives some rules for improving displays (Reid, 1984).

Users need feedback to understand device operation. Table 4.3 shows how you can give operators suitable guidance.

Apply appropriate levels of feedback. A caution should alert users to reduced performance but allow them to continue operations: a low battery indication, for instance. A warning may preempt operation momentarily to force a response; for example, saving a new computer file with the same name as an existing file could reasonably evoke a warning to help the user avoid erasing the existing file. An

Table 4.1 Rules for simplifying expression

Rule	Comment
Don't overfill the screen.	If it is more than 25% full, the user's ability to locate and recognize information begins to diminish.
Show only changing data.	Leave out static background.
Limit complexity.	Hide the nonessential details and controls.
Put exceptions in upper right quadrant.	Users are least sensitive to changes in the lower right quadrant of the screen.
Use mixed-case lettering.	ALL UPPERCASE (OR CAPITAL OR BLOCK) LETTERS ARE MORE DIFFICULT TO READ THAN MIXED-CASE CHARACTERS FOR MOST PHRASES OR SENTENCES WITH MORE THAN FIVE WORDS. NOTE: All uppercase letters are effective in gaining attention. But the bulk of the message should still be mixed-case letters.
Reduce head, eye, and hand movements.	Design the layout of the screen and instrument so that the user's eyes fall naturally on the next item.
Blinking attracts attention.	Use sparingly because blinking becomes very distracting.

Table 4.2 Rules for improving displays by simplifying understanding

Rule	Comment
Color coding	Good for labeling; better than size, angle, or shape for identification; easier to locate than alphanumerics. • Arbitrary assignments of color can be counterproductive. • Avoid unpleasant visual effects, such as afterimages (intense colored patterns) and vibration (apparent movement due to complementary colors). • 8% of males and 0.4% of females are color blind. • Use a maximum of five distinct colors. • Brightness and texture coding are poor substitutes for color.
Numeric data in tabular format	Justify column on the decimal points.
Alphanumeric information in linear and horizontal format	Vertical lettering is difficult to read.
Position coding	Effective only for low-information displays.
Length coding	Good for ratio comparisons, poor for identification. Use only four to six levels.
Angle coding	Use up to 24 levels; can be confusing.
Pictures—better than words	Quick comprehension.

Table 4.3 Components of design that guide operators and reduce errors

Principle	Specifics
Visibility	Make the program operation visible. • Use spinning beach ball cursor. • Indicate (with a symbol) when operator input is locked out. • Show acceptance of data and resulting action. • Highlight selection.
Feedback	Give feedback on operations and results. • Use active voice in descriptions. • Give list of options. • Allow audio alarm to be turned off, but leave visual reminder.
Reduce errors	• Easy-to-learn dialogues lead to fewer errors. • Use "smart" shortcuts. They could be a source of errors, but consider how they may speed execution for an advanced user. • Test execution on naive users with a single-thread problem to find sources of errors.
Specify error recovery	• Design for error. Prevent slips; otherwise, detect errors and correct. • Make actions reversible; make irreversible actions difficult.

alarm may suspend all activity and force the user to select a necessary operation for life-threatening situations—a vital signs monitor in a hospital, for instance.

Response latency is important. When a user commands a function, the machine response should be appropriate to the task. If the machine delays more than a fraction of a second, it should indicate that something is happening; feedback such as a clock face with moving hands in place of the cursor fills this role. Even cursor flashing is an important consideration; too high a frequency or too bold an indicator will distract and irritate users.

4.10.2 *Input Devices*

While the display consumes much effort in developing the human interface, the input needs your attention also. Input devices can be as simple as a push button or as complex as a full keyboard on a touch-panel display. Input devices also include mice, joysticks, and graphics tablets. Sherr (1988) gives an in-depth treatment of input devices.

Keyboards allow the rapid entry of large amounts of alphanumeric data. The keys need enough size and spacing to permit efficient use and should give tactile feedback. Table 4.4 gives some ranges of parameters for keyboard design (Wiklund, 1992b).

The keys should provide tactile feedback. Generally, they should not wobble or have a "squishy" feel. If the keyboard is used in an industrial environment, it should be rugged enough to withstand a poke from a screwdriver. You should also test its suitability for users with various skills and in various postures, such as sitting or standing.

Table 4.4 Recommended parameters for keys on a keyboard

Criteria for keys	Recommendation
Minimum height	0.25 cm (0.1 in)
Minimum width	1.2 cm (0.47 in)
Spacing	
Vertical	1.8–2.1 cm (0.71–0.82 in)
Horizontal	1.8–1.9 cm (0.71–0.75 in)
Travel	0.20–0.40 cm (0.08–0.16 in)
Force	0.5–0.6 N (0.11–0.13 lbf)

Pointing devices (joysticks, graphic tablets, light pens) usually direct a cursor on a visual display and tend to be used for special purposes. Frequency and duration of use in each application, together with training effects, help determine the appropriate input device. Table 4.5 lists some of the capabilities of these pointing devices (Wiklund, 1992a).

Touch screens are intuitive and good for tasks where users give only partial attention or are under stress. They structure the user interaction and are good for first-time or infrequent users. Because they do not provide tactile feedback, however, they are not suitable for extended or frequent use.

Table 4.5 Usefulness of some pointing devices

Operation	Mouse	Trackball	Joystick	Graphic tablet	Touch screen
High resolution	+	+	+	+	–
Minimum training	0	0	0	0	+
Drawing	0	–	–	+	–
Tracking moving objects	+	+	+	–	–
Eye-hand coordination	0	0	0	0	+
Small space requirements	0	+	+	–	–
Unobstructed display	+	+	+	+	–
No parallax	+	+	+	+	–

Note: + = a positive capability; – = a drawback; 0 = neutral or in-between.

4.10.3 *Questionnaires, Interviews, and Observations*

Studies support the initial decisions for the direction of the product. Tests occur later in the process and help evaluate the design of prototypes. Designers can use questionnaires, interviews, and observations to collect data from users on product function and utility. Table 4.6 outlines some points to use in designing questionnaires, interviews, and observations (Bailey, 1982).

Don't just copy someone else's formats or methods when you design a questionnaire. Define your purpose, decide what information you need to address that

Table 4.6 Concerns in the design of questionnaires, interviews, and observations

Methods	Specifics
Questionnaires	Provide large cross section of data economically. • Use words that are simple and familiar. • Make instructions clear and complete. • Make questions understandable. • Make categories clear and distinct. • Set instructions apart from questions with a good layout. • Provide adequate space to write answers. • Provide adequate time for completion. • Avoid suggestions for response. • Don't embarrass respondents with inappropriate wording. • Use words with precise meaning. • Sequence elements in a logical, easy manner. • Design for an appropriate length and completion time.
Interviews	Allow greater interaction between user and designer by providing immediate feedback and correction of misunderstandings. • Face-to-face interviews give the best response rates with greatest information, but they are the most expensive and time consuming. • Telephone interviews are shorter.
Observations	Provide unobtrusive insight into how users operate devices. Avoid interruption because presence of observer affects user actions.

purpose, and design a questionnaire to collect it in a reliable and valid fashion. Beware that many people throw questionnaires away and some deliberately fill in the wrong information. Getting professional help in preparing questionnaires is a good idea.

You will have to sample users for inclusion in studies of your product. You can use random, systematic, or subjective sampling for testing. Random sampling helps remove bias from the testing but requires more resources and effort. Systematic sampling can reduce the effort by selecting classes of users to cover most situations. Subjective sampling expedites testing but runs the very real risk of wrong or skewed outcomes.

Tests help ensure success in the development of your product, but they are a means to an end. Don't collect data for its own sake. Always test for the specific purpose of improving your product.

4.10.4 *Implications for Hardware*

The principles of design have particular application to hardware. Intuitive or natural mappings, visibility, visual cues, utility, and control all relate to the arrangements of switches, knobs, gauges, panels, and screens. Ergonomics, tactile cues, work flow, and image affect the size, shape, and layout of switches, knobs, and enclosure. You need to refine the arrangement and operation carefully before committing to a production model; rework is expensive.

Example 4.10.4.1 In the United States most power switches should be oriented vertically, with ON being up and OFF being down, not horizontally. (Room light switches in the United Kingdom operate just the opposite: ON is down, and OFF is up. This is an example of cultural bias affecting natural mappings.) Even better, backlight the switch to indicate that power is on, thus providing visibility and feedback.

4.10.5 *Implications for Software*

The software for a human interface is often more difficult to develop than the core calculating program. Keep the program modular and separate the interface from the core, so that changing one does not affect the other.

4.11 SOME SOURCES OF ERRORS

Errors arise from many sources. The designer may have overlooked some modes of operation. The user may forget functions or sequences. A janitor may splash cleaning fluid onto the keyboard or through a cooling vent onto sensitive circuitry. All of these problems could induce errors in operation.

Errors fall into several broad categories:

1. Cognitive: designer or user
2. Physical abuse: mechanical, chemical, or electrical
3. Need for refill or recharge

Table 4.7 lists some problems and why they occur.

Table 4.7 Examples of mistakes, errors, and problems when people use devices, appliances, and instruments

Problem	Reason or comment
Operator-induced problems	
Pushing the wrong button	• Accidental Perseveration, remembering another task or event Misunderstood, counterintuitive operation • Setting manual or book on top of equipment or switches • Mischievous or random key presses • Incorrect design or instructions
Force fit or breaking	• Wrong or absent visual cues • Counterintuitive operation • Wide manufacturing tolerance makes some components too big
Drops	Smaller and transportable devices are more likely to be dropped
Spills	• Coffee on computer keyboard • Cleaning fluids (soap and water) seep into device
Incorrect battery changes	• Put in backward, sockets not polarized or marked • Batteries don't last long enough for application • No gauge to indicate hours left in battery
Keys in pocket with 9-V batteries	Key short-circuits across terminals and kills battery
Design oversights and hurdles	
Inappropriate operations	• Counterintuitive, not predictable, poor mapping • Lack of visibility and feedback • Incorrect instructions • No differentiation (color-coded steps may help)
Inappropriate alarms	• Indistinguishable or unintelligible • Incorrect • Inconvenient
Refill or charge size too small	Causes inconvenience to users
Manufacturing inconsistency	Parts that don't fit well or operate smoothly
Lack of coordination	• Components don't work together • Refills from other manufacturers aren't compatible

4.12 INTERFACE DESIGN SPECIFICATIONS

You should prepare interface design specifications early in product development. They will reduce costly modifications by identifying goals and user needs and will balance the expectations of users with your capability to build the product. Rapid prototyping will help refine the interface specifications by revealing user desires and actions. After you have designed and tested the initial prototype, the interface design specifications will help identify the design goals for validation.

The specifications should include user performance requirements and general and specific interface specifications. The contents should follow this general outline (Eisenberg, 1993):

1. Introduction
   a. Scope
   b. Overview
   c. List of definitions
   d. References
   e. Conventions
   f. Role of prototype
2. General description
   a. Information about system
   b. Types of users
   c. User performance
3. General user interface specification
   a. Strategy
   b. Guidelines
   c. Definition of the major functional components
   d. Viewing distance
   e. Fonts, color, visual and auditory feedback
4. Specific user interface specification
   a. Interpretation of information
   b. Basic interface logic and layout
   c. Common interface elements
   d. Specific functions, tables

If you work for a small company, you may be the author of the specifications. If so, try to include relevant input from several sources: users, design consultants, engineers, and marketing. In a large company a human factors engineer will write the specifications with help from users, design consultants, software and hardware engineers, quality assurance teams, writers of manuals, and marketing.

Sometimes the specifications may be included in the software requirements. Then the document may have several authors and be more difficult to use. Separate documents for interface design specifications and software requirements will have redundant information and need careful configuration control to track changes in each and assess the impact on the other.

You should write the interface design specifications after the system requirements are under way; after some prototyping, concept testing, and usability testing; and before the software requirements are written. The detail of the specifications should be high level and practical and should promote understanding among the members of the design team. Use text, pictures, and prototypes to describe the specifications, and leave little to chance. Evaluate the specifications with both document review meetings, informal and formal, and reviews by potential users of the document.

4.13 CASE STUDIES

4.13.1 *Neurological Stimulation System*

I designed an instrument in collaboration with Richard B. North at The Johns Hopkins Hospital (see the case study in Chapter 2 on the NSS) that provided neurological stimulation. The computer-controlled system optimizes stimulation for treating chronic pain. It has two modes: clinical and research. In clinical practice, the system provides for patient interaction while reducing the time required from health care professionals to make adjustments to stimulators. Applied to research, the system can deliver arbitrary and unique patterns of stimulation far beyond the current capabilities of commercial neurostimulators (Fowler and North, 1991a, 1991b).

North used a prototype instrument for four years before we redesigned the system. The prototype and redesigned systems provided valuable insight into developing a successful instrument. First, the package's image significantly influenced acceptance of the instrument. Second, the population of users were physicians, physicians' assistants, and patients and included a wide range of understanding and expertise. Third, field testing and rapid prototyping aided the development of the software.

The differences in packaging can be seen in the photographs in Fig. 4.9. The differences in image (outlined in Table 4.8) had a dramatic impact on user acceptance. A physician observed the prototype and called it a toy. Eighteen months later the same physician attended a slide presentation that covered the redesigned system and expressed great enthusiasm for using it. The interesting point is that the prototype and redesigned instruments operated identically in the clinical mode,

a

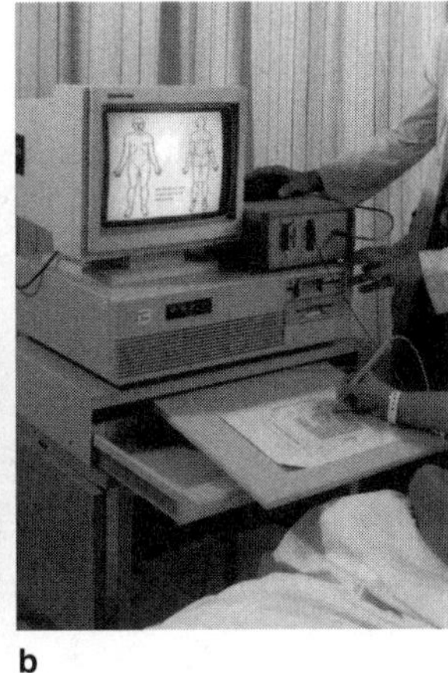

b

FIG. 4.9 First and second generations of the NSS. (a) First generation. (b) Second generation. Note the wide differences in packaging (or image) and graphical resolution.

Table 4.8 Differences in presentation between the prototype and redesigned systems

Presentation	Prototype	Redesign
Package	• Contact paper over sheet-metal box • Button placement was done at convenience of machinist	• Attractive commercial case • No knobs or buttons
Output: monitor screens	• Monochrome; black background • Question-and-answer format	• Color display; blue and white backgrounds • Windows, menus, and mouse buttons
Input: mechanical interface	• 3 push buttons • 2 rotary dials • Paper overlay on a small sketchpad; scratchy feel and heavy pressure	• Graphics tablet; smooth operation on a plastic overlay • Mouse
Size	Large	Compact
Operation	Custom hardware prevented migration of interface to other computers; upgrades and modifications difficult	Standard communications allow migration path to pen-based computers; facilitates future upgrades; can be used with a laptop computer

the mode that he would primarily use. Even more interesting was the response of some engineers who received a breadboard of the instrument; they expressed disappointment over not getting the commercial enclosure, even though they did not need it to do their development work with the device.

We used frequent communication, observation, and field testing with physicians and medical assistants to develop the graphical user interface. Primarily, we developed a smooth flow of operation through the windows and menus that was intuitive to medical personnel. For example, we used a cursor-drawn slash through a bar to represent a subjective rating of pain by the patient rather than a cursor-dragged color bar. A slash is the traditional method of rating according to the Palmer Pain Questionnaire, even though for most applications the colored bar seems more intuitive.

Scope: intuitive
Duration: first-generation device, 5 years; second-generation device, 3 years
Lessons learned:

1. Packaging image is very important for user acceptance.
2. The user interface is peculiar to the application and requires careful consideration.

4.13.2 *DesignJet Plotter User Interface Design: Learning the Hard Way about Human Interaction*[*]

How many times have you picked up a product and found it easy to use? If a product is easy to use, it probably was no small task to make it that way. With some products, even simple ones, I can be frustrated because I can't make it do what I expect it to do. Either I don't know how or it doesn't have the capability. The manual may be within reach but I have no interest in consulting it. With any product, users hope to combine their own intuition with external clues to determine the machine's capabilities and correct operation. What a confidence builder it is when a person can walk up to a new machine and operate it correctly!

We engineers, with our generally analytical minds, think that the simple solution is to publish a manual with step-by-step instructions. Although users expect such a manual, at least 50% will attempt operation without even opening it. Apparently users do not find it pleasant to build their mental model of a machine's operation by using only diagrams and text from written instructions.

Although the cause is noble, achieving intuitive operation is not so easy. A designer, familiar with each intricacy of the mechanism, will express an opinion and come up with the initial user interface design. Is this going to work for all users? I think not, but it is a place to start.

Our brain is parallel processing input from five senses and filtering it through past experiences. No amount of analytical thinking in a serial fashion can possibly predict the human response. The designer's experience with the development precludes any useful help in the user interface area. Perhaps, we might think, an experienced person could help us determine how people will react. Mistake! No one person can determine the best user interface that will appeal to the most people. This is a very difficult concept. When your manager tries out your design and has a certain difficulty, that can seem like the highest-priority problem, but it may just be a corner case. User testing is the key.

The DesignJet plotter started life as a gleam in the eyes of an architectural team. Prototypes were built, and along with proving the functionality came early user testing of some of the concepts the designers were concerned about. For example, to load roll media, the user needed to preselect roll format on the front panel, place the roll correctly on the spindle (it can go four ways), insert the media into the load slot, and lift the pinch roller release lever to align. Was the roll-to-spindle orientation intuitive? Would people lift the lever when instructed by the front panel?

The designer's opinion was that we needed to redesign for automatic lever lift; we should not be asking people to raise the lever. Mistake! No one person, especially the designer, can determine user reaction. A week's worth of investigation resulted in an unacceptable impact to cost and schedule and so the idea was reluctantly shelved. Later user testing showed that people had no problem or frustration when the display read "Lift lever." They simply lifted the only lever on the machine.

---

[*] From Wield (1992). Copyright 1992 Hewlett-Packard Company. Reproduced with permission.

The same initial user testing showed that all users intuitively oriented the roll on the spindle backwards. This sent us into another redesign, this time for two months, resulting in a prototype that allowed the roll to be installed this way. Mistake! Users look for clues and parallel process all information. This new prototype displayed a new set of problems of both function and ease of use. Going back to the original design, a more subtle change was made. A graphic label installed under the roll cover was the clue that users were looking for. It was only in the absence of any other information that they showed the opposite preference for roll orientation.

With the initial concerns addressed by user testing, we thought we could continue with hard tooling. Mistake! Not all user interface problem areas can be predicted. The entire system must be tried by typical users not familiar with the product. Testing after hard tooling revealed that people will not select sheet or roll mode before trying to load the media. Initially a button on the front panel toggled two light-emitting diodes that displayed the type of media the DesignJet plotter expected. From the previous problem, we had already learned that the DesignJet did not need to be redesigned; we simply had to provide more clues. Now, after the load is initiated, the plotter pauses and asks the user, by means of the front-panel display, to select which format of media has been loaded. This tested well and has the added benefit of providing the user with time to stop and decide (with hands off) if the plotter has a good, even grip on the media before continuing. Unfortunately in this case, the hard tooling needed to be modified as an unexpected expense.

At the same time, it was found that first-time users almost always created a paper jam while trying to load. This was, of course, unacceptable. Observation of user tests showed the various techniques people were using to load. The DesignJet plotter grabs the media after a set amount of time when the media passes a sensor in the insertion slot. People did not know that they needed to wait for this time and then let go of the media so that it could move into the mechanism. We needed to provide some clues. An audible click, which was part of the original concept, was not enough. Almost by accident, it was discovered that if we moved the media quickly at the start, users would instinctively let go.

Another problem was that some people were using a line on the side of the machine to adjust the media squareness. People complained that there needed to be a guide on the side to align the media. It wasn't until we thought about the comment that we realized this was how they expected to keep the media square. They were frustrated because the DesignJet would reject that load as misaligned. It was not possible to use the marks on the side to adjust the squareness. Our other plotters use a side surface to reference the media and these people were accustomed to that type of system. The correct DesignJet method is to push the media against the pinch rollers which are, unfortunately, out of sight. The lines on the side were only an approximate left/right location reference. Within 1/4 inch of the side line would have been fine. Our solution was to reduce the length of the line on the right so it didn't look like something to align to and to add a label in the area that explains in six different languages to "Push media against rear stops." A better solution might have been to make the pinch rollers visible to the user.

The position selected for the label is not very eye-catching but has the benefit of not cluttering the appearance of the machine. While looking at this solution, the general comment was, "No one will ever see it." Mistake! Don't ask people to predict how others will perceive. User testing of first-time users showed that they were unsure of how to load and were actively searching for clues. Almost all noticed and comprehended the label. A bail lift timing change bought us some extra design margin which, combined with the improved user interface, gave us a vastly improved and acceptable design.

The design might be improved further now that we have gained a clearer understanding of DesignJet user perceptions. But to give an accurate model of user perception, user testing requires a complete product, and a complete product is not receptive to many changes. This creates a minor dilemma neatly expressed as "In every product's development, there comes a time when you must shoot the engineer and go into production." In my case the wound was not fatal and I'm recovering nicely.

Scope: system, program, or enterprise
Lessons learned:
1. User testing is necessary. A designer or single experienced user cannot completely specify the interface of a mass-produced instrument. The entire system must be exercised by a number of typical users.
2. Users look for multiple cues and process the information in parallel.
3. A combination of cues can sometimes help users operate instinctively.

4.14 SUMMARY

The primary ingredient for success is an interactive team effort between designers and users in prototyping and field testing. During development, instrumentation should undergo continuous redesign to implement changes suggested during testing and evaluation. The instrument must have an adaptable architecture that allows for experimentation and change.

Often users make mistakes not because they are stupid but because the product and its human interface are deficient. The aim of this chapter has been to help you design better products by reducing errors in the interfaces. If you satisfy the elements of cognition, ergonomics, utility, image, and ownership in the human interface and give the user visibility, feedback, predictability, and control, your product has a much better chance of success.

The end result is summed up by Norman (1986, p. 49): "I want a system that is enjoyable to use. This is an important, dominating design philosophy, easier to say than to do. It implies developing systems that provide a strong sense of understanding and control. This means tools that reveal their underlying conceptual model and allow for interaction, tools that emphasize comfort, ease, and pleasure of use."

4.15 RECOMMENDED READING

Norman, D. A. 1988. *The design of everyday things*. New York: Doubleday.
A delightful and enjoyable book that exposes the follies and occasional good design of many common appliances and devices that we use.

4.16 REFERENCES

Bailey, R. W. 1982. *Human performance engineering: A guide for system designers.* Englewood Cliffs, N.J.: Prentice-Hall.

Card, S. K., T. P. Moran, and A. Newel. 1983. *The psychology of human-computer interaction.* Hillsdale, N.J.: Erlbaum.

Case, R. 1993. Streamlining medical product industrial design. Tutorial at the Sixth Annual IEEE Symposium on Computer-Based Medical Systems, June 13, Ann Arbor, Mich.

Couch, J. 1993. Panel discussion at the Sixth Annual IEEE Symposium on Computer-Based Medical Systems, June 15, Ann Arbor, Mich.

Defense Science Board Task Force on Military Software. 1987. *Report of the Defense Science Board Task Force on Military Software.* NTIS Accession Number AD-A188 561/5/XAB. Washington, D.C.: Office of the Under Secretary of Defense for Acquisition.

Dinan, J. A. 1992. Human-interface rules reduce test-program operator errors. *EDN*, February 3, pp. 95–98.

Eisenberg, P. 1993. Computer/user interface design specification for medical devices. *Proceedings of the Sixth Annual IEEE Symposium on Computer-Based Medical Systems,* June 13–16, Ann Arbor, Michigan, pp. 177–182. Los Alamitos, Calif.: IEEE Computer Society Press.

Fischer, G., and A. C. Lemke. 1988. Constrained design processes: Steps towards convivial computing. In *Cognitive science and its applications for human-computer interaction,* edited by R. Guindon. Hillsdale, N.J.: Erlbaum.

Fowler, K. R., and R. B. North. 1991a. Computer-optimized neurological stimulation. *Annual International Conference of the IEEE Engineering in Medicine and Biology Society* 13(4):1692–1693.

Fowler, K. R., and R. B. North. 1991b. Computer-optimized neurostimulation. *Johns Hopkins APL Technical Digest* 12(2):192–197.

Fraser, J. 1990. An engineer's guide to industrial design. *EDN*, August 20, pp. 195–200.

Kieras, D., and P. G. Polson. 1985. An approach to the formal analysis of user complexity. *International Journal of Man-Machine Studies* 22:365–394.

Lansdale, M. 1990. The role of memory in personal information management. In *Cognitive ergonomics: Understanding, learning and designing human-computer interaction,* edited by P. Falzon. London: Academic Press.

Marcus, A., and A. Van Dam. 1991. User-interface developments for the nineties. *Computer* 24(9):49–57.

Matisoff, B. S. 1990. *Handbook of electronics packaging design and engineering.* 2nd ed. New York: Van Nostrand Reinhold.

McDonald, J. E., and R. W. Schvaneveldt. 1988. The application of user knowledge to interface design. In *Cognitive science and its applications for human-computer interaction,* edited by R. Guindon. Hillsdale, N.J.: Erlbaum.

Norman, D. A. 1986. Cognitive engineering. Chap. 3 in *User centered system design: New perspectives on human-computer interaction*, edited by D. A. Norman, and S. W. Draper. Hillsdale, N.J.: Erlbaum.

Norman, D. A. 1988. *The design of everyday things*. New York: Doubleday.

Plauger, P. J. 1993. Enhancing displays. *Embedded Systems Programming,* May, pp. 89–92.

Polson, P. G. 1988. The consequences of consistent and inconsistent user interfaces. In *Cognitive science and its applications for human-computer interaction,* edited by R. Guindon. Hillsdale, N.J.: Erlbaum.

Reid, P. 1984. Work station design, activities and display techniques. Chap. 8 in *Fundamentals of human-computer interaction,* edited by A. Monk. London: Academic Press.

Sherr, S., ed. 1988. *Input devices*. Boston: Academic Press.

Trower, T. 1993. The human factor: A column on interface design, *Microsoft Developer Network News,* May, p. 16.

Van Der Veer, G. C., R. Wijk, and M. A. M. Felt. 1990. Metaphors and metacommunication in the development of mental models. In *Cognitive ergonomics: Understanding, learning and designing human-computer interaction,* edited by P. Falzon. London: Academic Press.

Wield, P. J. 1992. DesignJet plotter user interface design: Learning the hard way about human interaction. *Hewlett Packard Journal,* December, p. 12.

Wiklund, M. E. 1992a. Choosing an effective pointing device. In *Designer's handbook: Medical electronics*. Santa Monica, Calif.: Canon Communications.

Wiklund, M. E. 1992b. Keyboards: User satisfaction and performance capabilities. In *Designer's handbook: Medical electronics*. Santa Monica, Calif.: Canon Communications.

Woodson, W. E., B. Tillman, and P. Tillman. 1992. *Human factors design handbook: Information and guidelines for the design of systems, facilities, equipment, and products for human use.* 2nd ed. New York: McGraw-Hill.

Ziegler, J. E., P. H. Vossen, and H. U. Hoppe. 1990. Cognitive complexity of human-computer interfaces: An application and evaluation of cognitive complexity theory for research on direct manipulation-style interaction. In *Cognitive ergonomics: Understanding, learning and designing human-computer interaction,* edited by P. Falzon. London: Academic Press.

5

PACKAGING AND ENCLOSURES

Always design a thing by considering it in its next larger context.
—Eliel Saarinen

5.1 PACKAGING'S INFLUENCE AND ITS FACTORS

Packaging and enclosure drive the design of most electronic instruments. The physical constraints of size and weight, for instance, are dictated by customer needs, system requirements, cooling design, and the operational environment. Additional physical and environmental factors will restrict the shape of the enclosure and the materials used in it, thus constraining the design even further. Consequently, the packaging and enclosure will define the extent of functions like processing capacity, power density, cooling, and, ultimately, the development of the product.

Packaging is the mechanical structure, support, and orientation of components within an electronic product. Packaging your electronics and selecting an enclosure involves many considerations:

- Cost
- Size
- Shape
- Weight
- Mechanisms
- Materials
- Finishes
- Appearance
- Ergonomics
- Serviceability
- Reliability
- Regulations and standards

Different markets for electronic products call for different packaging and enclosures. Cost tends to be the prime driver for consumer products, while industrial instrumentation focuses on service support and reliability. The medical market demands rugged and safe devices (but image is still important; see Chapter 4). Military equipment requires ruggedness, reliability, and good performance in its packaging; cost is usually a lower priority.

Physical environment is the biggest factor in selecting packaging. Here are some electronic devices you might find and places where they are used:

- Home and kitchen appliances
- Farm machinery
- Automobiles
- Aircraft
- Medical instrumentation
- Spacecraft
- Robots and factory automation
- Military weapons
- Process control
- Petroleum exploration

Each of these applications has a physical environment that includes temperature, humidity, vibration, shock, and corrosion. Consumer appliances, for instance, must survive drops from a tabletop to a concrete floor, while medical instrumentation is regularly doused with liquid disinfectant.

Within these environments are specific factors that affect electronic instruments:

1. Temperature and cooling
2. Vibration and shock
3. Humidity and condensation
4. EMI
5. Salt spray and corrosion
6. Pressure and altitude
7. Vacuum outgassing
8. Radiation upsets and hardness

Industrial, medical, and military products have regulations and standards that specify ranges for these factors. Military electronics often have to operate at ambient temperatures between –55°C and +125°C (–67°F and +257°F), while industrial equipment has to operate between –40°C and +85°C (–40°F and +185°F). Usually the regulations specify two different ranges: operating range and non-operating (storage) range. You must know the applicable regulations and standards for your particular market to design an acceptable product. For example, Table 5.1 lists some of the regulations required by the U.S. military.

Table 5.1 Some standards that affect packaging and enclosures for the U.S. military

Standard	Description
MIL-STD-461D	Enclosure EMI/RFI—emissions, susceptibility, radiated, conducted
MIL-STD-810D	Pressure
MIL-STD-810E	Altitude, temperature, humidity, vibration
MIL-STD-883B/C	Integrated circuit screening and specifications for temperature and humidity
MIL-STD-901C	Shock testing

5.2 DESIGN FOR MANUFACTURE, ASSEMBLY, AND DISASSEMBLY

Enclosures and packaging account for most of the manufacturing in electronic products. Cost in manufacturing is based on materials, configuration, complexity, molding, machining, casting, assembly, and testing. Here are some ways to reduce the cost of both assembly and production (Bank, 1991; Daumüller and Flachsländer, 1991):

- Simple or automated assembly
- Reduction of adjustments and calibrations (to facilitate testing)
- Interchangeable parts
- Minimum inventory and parts handling
- Adaptation of existing manufacturing capabilities
- Minimum vendor count

Furthermore, modular construction may minimize manufacturing cost and facilitate disassembly for maintenance.

5.2.1 *Manufacturing*

The first question to answer is whether to buy or build. Many enclosures are available off the shelf. They will save you design and tooling costs, can be delivered quickly, and are cheaper for small or medium quantities. Otherwise, a custom enclosure can fit an application better and is cheaper in large quantities.

The materials used affect the price and performance of the enclosures. Many different plastics are available for small enclosures. Selecting one requires understanding the trade-offs between cost, dimensional stability, temperature tolerance, impact resistance, abrasion, and chemical resistance. Larger cabinets require steel or aluminum in sheets and frames for strength and rigidity. Generally, a molded plastic will be cheaper for small instruments manufactured in quantity, but only metal will be feasible for large racks of equipment.

5.2.2 *Assembly*

Parts count, more than anything, drives production costs for most products. *Reducing the number of parts will reduce the cost of assembly and ultimately the cost of the final product.* According to Robert Williams of Hewlett-Packard, "The key deliverable of any DFMA [design for manufacture and assembly] effort is a significantly reduced part(s) count" (Colucci, 1994, p. 21).

Attachment is an important concern in producing enclosures. Do you mold, weld, rivet, or screw parts together? Plastic molding can be the quickest and cheapest way to produce an enclosure in large quantities. (The initial cost for preparing the mold can be high.) Metallic enclosures may be riveted or welded. Riveting can be faster and cheaper than welding, but shielding for EMI, for instance, may dictate a continuous welded seam in a metal enclosure. Certain fasteners allow disassembly and may be snap-fit, screws, or nuts and bolts.

Snap-fit packaging uses molded plastic components that snap together. It reduces the pick-and-place problems in automated assembly with screws. But it has larger tolerances and sizes that take more space than screws. Also, plastic creeps and can lose its grip.

Screws provide tighter tolerances and smaller sizes than snap-fit molded pieces. The twisting action on screws, however, can gouge out tiny chips of plastic that could bind internal mechanisms. In addition, automated assembly requires that screws be designed to fit during feeding as well as in their final use (Teague, 1993).

Assembly issues go beyond the ease of manufacture and cost of raw stock. The weight, strength, and electrical and thermal properties of each enclosure are important, as is compliance with relevant standards and regulations in the marketplace. Your best bet is to work closely with an experienced packaging or mechanical engineer to design and fabricate an enclosure for your product.

5.2.3 *Maintenance and Disassembly*

Determine the need and priority for maintenance and disassembly before you design an instrument enclosure. A cheap appliance, for example, probably will not be repaired if it breaks, so design for disassembly is a waste of effort. Conversely, avionics require a highly modular and easily replaceable form factor to reduce turnaround time for maintenance.

Once you have established the need, priority, maintainability, and serviceability, you will have several other considerations in the design for disassembly. Consider these factors in the design of an enclosure:

1. Human factors: What is important for ergonomics, cognition, and image?
2. Reliability: How often will the enclosure be opened for repair?
3. Capabilities and qualifications of service personnel: Are special skills and knowledge needed?
4. Tools: Do regulations require specific operations for safety?
5. Accessibility and maintenance frequency: Can the fuse be easily and safely replaced?

Example 5.2.3.1 Clements (1993) writes that an industrial controller in a large, expensive process "demands the highest level of serviceability. Tradeoffs for cost that compromise serviceability cannot be made" (p. 63). "Once an industrial workstation has been designed and installed into a factory process it is rarely replaced or upgraded for reasons other than loss of support" (p. 64). These concerns for serviceability and support make design for disassembly a paramount issue for industrial equipment.

The cooling fans, for example, should have mounting attachments that allow replacement, since fan bearings rarely last the life of the equipment (10 years or more). Mounting options are numerous, including rack mount, mast mount, equipment stacking, and front or back reversibility. Regulatory compliance may require a tool to remove components and modules.

Example 5.2.3.2 For smaller instruments, packaging that disassembles easily with a single tool may fit your application. Fig. 5.1 illustrates such an enclosure. You simply push a screwdriver into the screw head in the lid, press down, and turn a quarter turn to the left to release each corner. Long wedges secure the walls of the box. Prying out the wedges with a screwdriver from each corner releases the wall for removal.

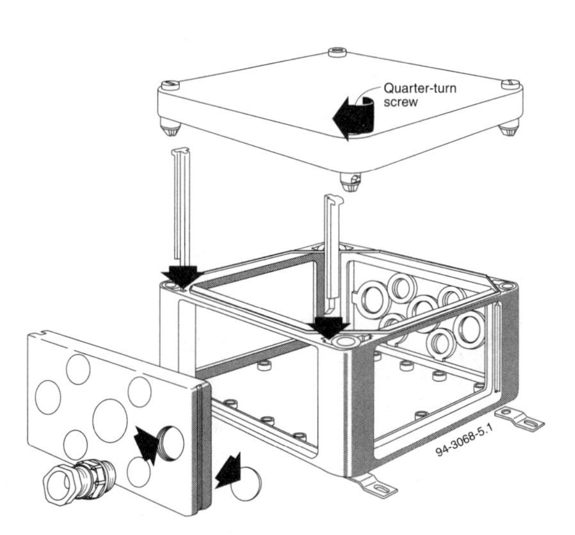

Fig. 5.1 Example of an enclosure that is easy to disassemble (fiberglass modular flange enclosure system by Rose Enclosures).

5.3 WIRING

Wiring and cabling play the vital role of interconnecting signals and power in electronics. *Mechanical failure of cabling, connectors, and solder joints is the biggest problem in most electronic systems.* Consequently, you should spend time designing your product's wiring.

Wiring has many applications and complex variations based on current-carrying capacity, mechanical strength, insulation properties, and shielding. Design concerns include the following:

- Vibration, abrasion, and shock
- Types of connectors
- EMI emission
- Wire routing
- Servicing
- Temperature, humidity, and fungus

5.3.1 *Cabling*

Most wire is soft, annealed copper because it is highly conductive, resistant to corrosion and fatigue, and easy to solder. Stranded conductor is preferable to solid wire because it is more flexible and less likely to break. Table 5.2 lists common conductor sizes.

Bundling wires together creates a cable. Careful routing of cables and tying them down reduces problems from vibration, shock, and mechanical flexure and

Table 5.2 Some common sizes of annealed copper wire, American Wire Gauge (AWG)

AWG	Diameter		Cross-sectional area		Mass or weight		DC resistance at 20°C (68°F)	
	in.	mm	in.^{2}	mm^2	lb/1000 ft	kg/km	Ω/1000 ft	Ω/km
10	0.1019	2.588	0.00816	5.261	31.43	46.77	0.9989	3.277
12	0.0808	2.053	0.00513	3.309	19.77	29.42	1.588	5.211
14	0.0641	1.628	0.00323	2.081	12.43	18.50	2.525	8.285
16	0.0508	1.291	0.00203	1.309	7.818	11.63	4.016	13.17
18	0.0403	1.024	0.00128	0.8231	4.917	7.317	6.385	20.95
20	0.0320	0.8118	0.000802	0.5176	3.092	4.602	10.15	33.31
22	0.0253	0.6438	0.000505	0.3255	1.945	2.894	16.14	52.96
24	0.0201	0.5106	0.000317	0.2047	1.223	1.820	25.67	84.21
26	0.0159	0.4049	0.000200	0.1288	0.7692	1.145	40.81	133.9
28	0.0126	0.3211	0.000126	0.0810	0.4837	0.7199	64.90	212.9
30	0.0100	0.2546	0.000079	0.0510	0.3042	0.4527	103.2	338.6
32	0.0080	0.2019	0.000050	0.0320	0.1913	0.2847	164.1	538.3
34	0.0063	0.1601	0.000031	0.0201	0.1203	0.1791	260.9	856.0
36	0.0050	0.1270	0.000020	0.0127	0.07568	0.1126	414.8	1361

prevents obstruction of mechanisms or service views. Routing and tie-down should optimize layout of the wiring for the shortest path, servicing, and EMI shielding. Color coding or wire identification sleeves aid the servicing of large cables that have many conductors. General guidelines for cabling are as follows (Matisoff, 1990):

1. Make it neat, sturdy, and short.
2. Permit easy inspection and testing.
3. Prevent damage to assembled parts and give clearance for moving parts.
4. Prevent strain on conductors, connectors, and terminals.
5. Space tie-downs 8 to 10 cm (3 to 4 in.) apart.
6. Use grommets (or rounded edges) on holes and penetrations to prevent abrasion.
7. Don't route over sharp edges, screws, or terminals.
8. Allow a reasonable bend radius (10 times outer diameter of cable if possible).

Fig. 5.2 shows examples of neat routing and tie-down that reduce problems with flexure. (Fig. 8.19 is another example.) Fig. 5.3 shows two different attachments of power cords. A long strain relief reduces flexure of the cord and subsequent fatigue and breakage of the conductors.

The weight and complexity of cabling can be important to an instrument. Spacecraft modules and aircraft avionics have strict weight limits, and cabling can be a sizable portion of the weight budget. Early in conceptual design consider using serial data transmission, multiplexed buses, or fiber optics to reduce the number of conductors and weight.

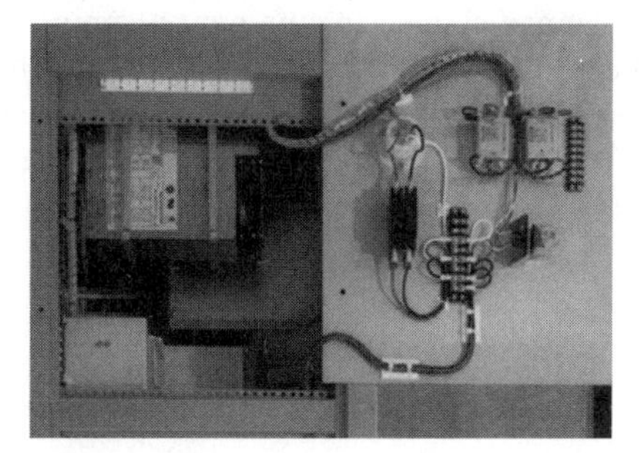

Fig. 5.2 Careful and neat routing of cables. Such routing reduces flexure fatigue, facilitates troubleshooting, and improves access for maintenance.

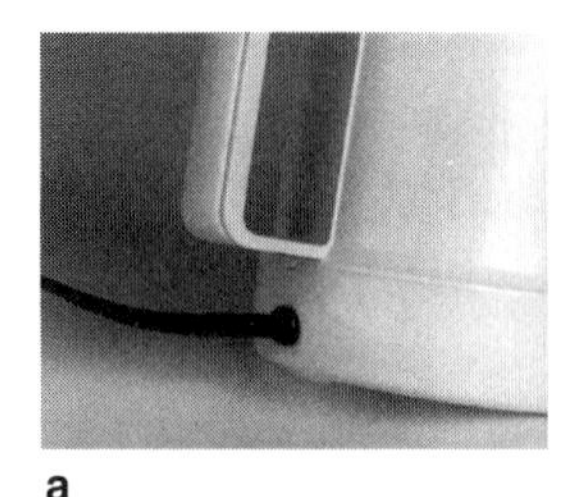

a

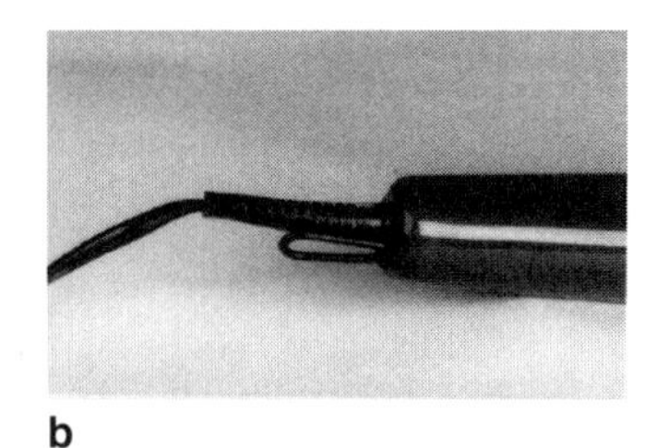

b

Fig. 5.3 Two examples of strain relief on a cable. (a) A grommet prevents abrasion but does little to reduce flexure of the power cord. (b) The long strain relief helps reduce flexure and the resulting mechanical fatigue of the wires.

5.3.2 *Connectors*

Connectors provide mechanical support for electrical connections. The metal-to-metal contact within connectors raises several concerns:

- Contact force and resistance
- Gastight fit
- Corrosion
- Life cycle, wear, and fatigue
- Keying (see Fig. 8.18)

Corrosion occurs in two regimes: atmospheric and galvanic. High moisture and temperature accelerate corrosion from atmospheric sources. Dissimilar metals cause galvanic, or electrochemical, corrosion. Therefore, carefully select the metals and alloys used in conductors and wires.

Plating contacts improves their resistance to wear, corrosion, and oxidation. Gold, nickel, and tin are used widely for plating contacts. Gold is an excellent conductor, chemically very stable, and best for corrosion resistance. You can use hard gold plating for contacts that experience many insertion/extraction cycles. Gold-over-nickel plating is less expensive and gives surface characteristics of gold while the nickel underplating prevents migration of the base metal. Tin is a good conductor and has excellent solderability. Unfortunately, tin wears quickly and corrodes easily and therefore should be used only for contacts that have few insertion/extraction cycles.

Example 5.3.2.1 Corrosion in connectors can cause baffling problems. A consulting engineer worked on a large display board in a professional sports stadium. Each pixel of the display was a small picture tube 25 mm (1 in.) in diameter. Occasionally a picture tube would not light. Pulling it out and testing it showed no failure. Eventually he discovered that the temperature and humidity around the exposed display board accelerated corrosion in the connector contacts to the picture tubes. He worked with the connector vendor to change the plating on the contacts to stop the corrosion.

Connectors and wire terminations have a huge variety of configurations. Fig. 5.4 shows a small sample of connectors.

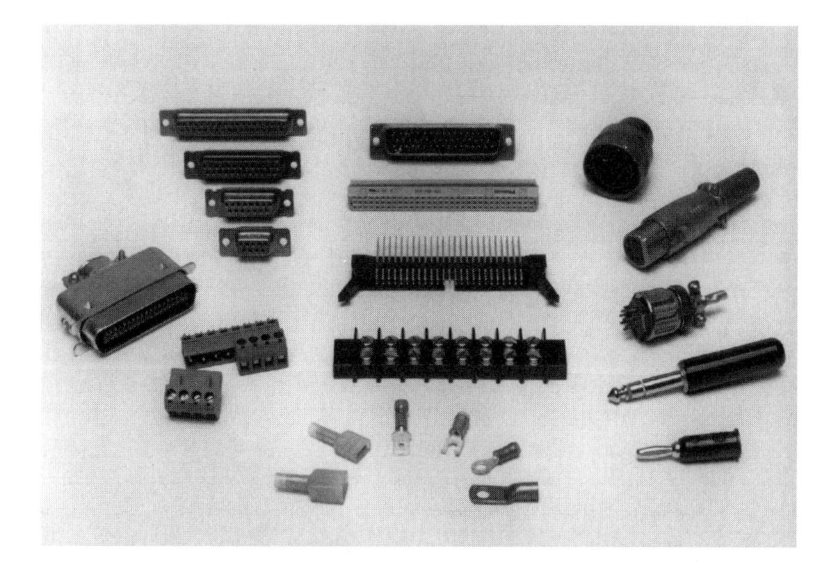

FIG. 5.4 Examples of connectors, plugs, and terminal blocks.

5.4 TEMPERATURE

Extremes in temperature stress electronics and cause failure. Mismatched coefficients of thermal expansion during temperature changes put stress on mechanical joints and accelerate failure (Benning, 1991). In fact, a large rate of change in temperature is often a bigger problem for reliability than extremes in temperature held in steady state. Unfortunately, most handbooks on reliability base their calculations on steady-state temperature. In some cases, you may be able to establish a baseline for reliability of the components and your design by testing them with many temperature cycles.

You can avoid thermally induced stress by controlling temperature with cooling (or heating for very cold environments). The goal is to minimize thermal transients and gradients in circuits.

5.5 VIBRATION AND SHOCK

Mechanical interruption causes many failures in electronic systems. Common problems include broken connections and components falling out of sockets. Flexure and fatigue breaks wires, solder joints, and component leads; cables are most susceptible. Problem areas include the following (Matisoff, 1990):

- Cables and connectors
- Flexure stress at junction of lead and component body
- Components shaking loose
- Transformers, batteries, large capacitors
- Mechanical relays chattering above 10*g* (acceleration) and 500 Hz

5.5.1 *Vibration*

Vibration kills electronic systems. Matisoff (1990, p. 163) claims that "vibration failures are four times as frequent as shock failures. Many components and systems can take up to 75*g* of shock yet cannot perform under as little as 2*g* of sustained vibration."

Circuit boards tend to vibrate in specific modes. Fig. 5.5 illustrates some of the modes. Most designers try to eliminate low-frequency, large-amplitude vibration by reducing the component count, reducing the mass at the center of the board, pinning the edges of the circuit board, and using stiffeners and a thicker dielectric in the printed circuit board (PCB). But you should take care that the stiffeners don't block airflow and bind card guides (Zubatkin, 1993). Also, thicker and stiffer component leads are less compliant, so solder fatigues sooner; dual in-line packages (DIPs) with longer leads are less likely to break their solder joints than are leadless packages. Fig. 5.6 demonstrates some good and bad practices for mounting components to survive vibration.

Here are some general guidelines for reducing problems with vibrations:

1. Avoid cantilevers and sliding joints.
2. Use stiff brackets (box, Z-shaped, and channel).
3. Mechanically damp heavy or large structures.
4. Firmly anchor large masses (e.g., transformers, batteries).
5. Add supports to long, flexible circuit boards.
6. Clamp large components.
7. Use short component leads.

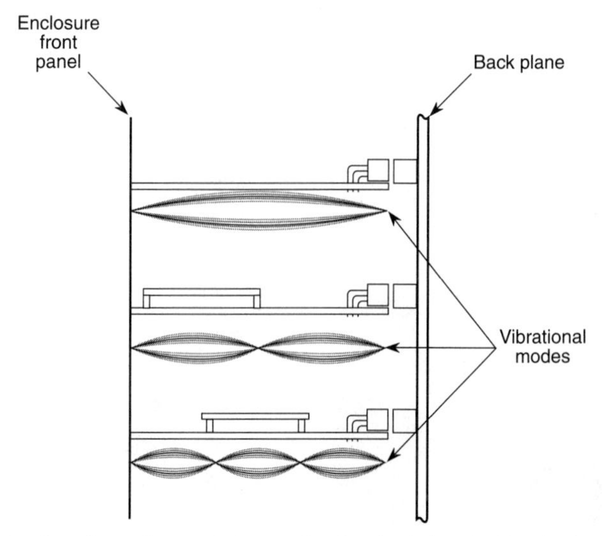

Fig. 5.5 Vibrational modes, or harmonic frequencies, for circuit boards. Piggyback cards alter the modes by adding mass and stiffness.

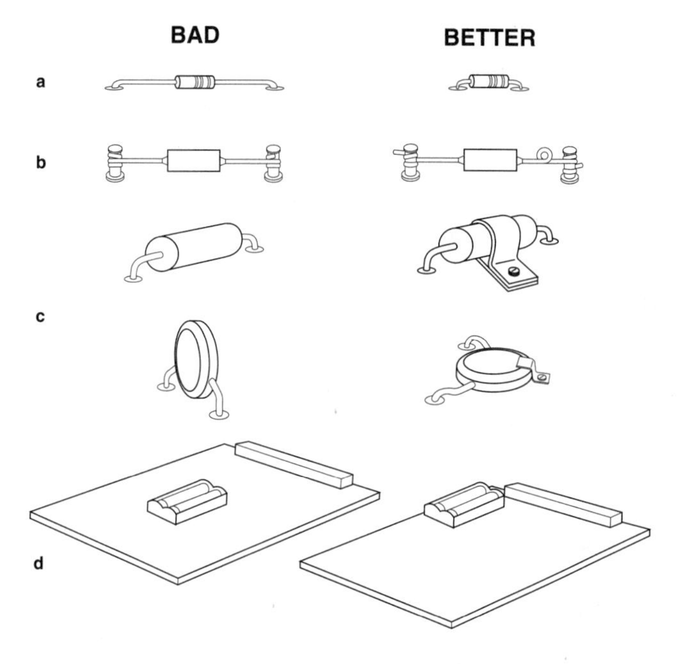

Fig. 5.6 Examples of how to mount components to withstand vibration. (a) Generally, leads should remain short to add stiffness. (b) Components mounted on posts should have a strain-relief loop. (c) Large components should be strapped or clamped down. (d) Heavy components should be mounted near mechanical supports.

5.5.2 *Shock*

Shock, which tends to be less destructive than vibration, occurs often enough in many applications. (How many times have you dropped an appliance?) The guidelines for reducing damage from vibration also apply to shock. Furthermore, special shock mounts can absorb or damp mechanical jolts as shown in Fig. 5.7.

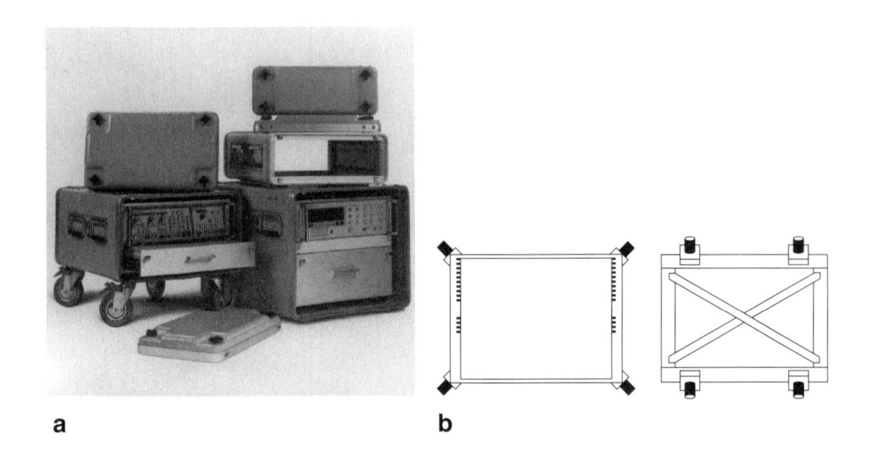

FIG. 5.7 Shock mounting within an equipment transit case. Shock mounts can damp jolts that otherwise might loosen or damage components. (a) Examples of transit cases with damping mounts (photograph courtesy of ECS Composites). (b) Equipment rack inside transit case with damping mounts near each corner.

5.6 RUGGED SYSTEMS

If you design equipment for the military or industrial markets, you will have to ruggedize your products. Even specialized products like medical instrumentation require attention to their particular forms of harsh environments. For rugged applications, you define the extremes of operation and then fit your design to the specified environment.

5.6.1 *Define the Environment*

First, determine the actual environment in which your product will perform: temperature, vibration, shock, humidity, and other characteristics. Don't blindly follow the standard regulations; they sometimes are insufficient. Investigate exactly how and where your equipment will be used.

5.6.2 *Temperature and Cooling*

Maass (1993, p. 5) writes, "It has been estimated that 65 percent of all military aircraft downtime is related to avionics problems and that about 50 to 60 percent of these problems are attributed to temperature or thermal problems." You need to determine the exact temperature profile for your application.

Example 5.6.2.1 Ground temperature in the desert can be 45°C (81°F) higher than the air immediately above it; when air temperature is 45°C (113°F), the ground can be around 90°C (194°F). At these temperatures, even thermoplastics in equipment transit cases begin yielding and failing.

Cooling can control temperature gradients and reduce thermally induced stresses. Cooling becomes imperative as circuits become denser and dissipate more heat. "The SEM-E module specification written in 1989 recommends a maximum power level of 35 W on a module. . . . It is not unusual to see 45 to 60 W conduction-cooled SEM-E [standard electronic module, E format] designs in present programs" (Maass, 1993, p. 5).

As always, you have trade-offs: Active cooling (fans, fluid pumps, and radiators) adds complexity that increases cost and decreases reliability. Fan bearings typically do not last as long as electronics and may need replacement. Liquid cooling can clog, and leaks can damage electronics—if they don't melt first. Thicker module and circuit board frames and quick fluid disconnects reduce circuit density and add significant cost to the system. On the other hand, you can integrate the thicker frames and the thermal conduction plates with stiffening to reduce vibration.

5.6.3 *Vibration and Shock*

Rugged systems must always survive some vibration and shock. As previously discussed, you can add stiffener bars or supports to circuit boards. Connectors must be both secured to prevent shaking off and strain-relieved to prevent conductor fatigue. Appropriately placed shock absorbers further isolate circuit boards from impact accelerations (jolts).

5.6.4 *Component Screening*

For rugged environments, integrated circuits (ICs) are tested more extensively to ensure that they will hold up. Generally, they are not built to be more rugged. (An exception is radiation hardness; Section 5.7.5 mentions the fabrication of radiation-hard ICs.) They also have ceramic packages, which seal out chemical assault better than plastic encapsulated components. You can contract for component testing to meet standards, such as thermal cycling for screening, if need be. Companies specializing in MIL-STD-883C (USA) can test components for you. Or you can buy components that are fully qualified to MIL-STD-38510 (USA), but some people question whether these components are any better; they argue that testing may actually reduce the life of the components (Zubatkin, 1993). Regardless, you will have to screen the components to at least establish a paper trail, if not remove early failures.

5.6.5 *Conformal Coating*

Dust, humidity, and condensation on circuit boards hasten their demise. Dust acts as both a thermal insulator that raises component temperatures and a gum to block mechanisms. Humidity and condensation short signals and accelerate corrosion that eventually opens circuits. A *conformal coat* is a polymer film that protects circuit boards and components by resisting corrosive invasion. Conformal coats can extend the functional life of the electronics at the expense of additional fabrication and material cost. Conformal coats do not protect connectors; you will have to take other measures to seal them.

Example 5.6.5.1 A manufacturer of home appliances had a problem with a model of steam iron. A small circuit board in the handle of the iron was sometimes shorting out after use because steam was condensing and saturating it. It was found that a thin parylene coating effectively sealed the printed circuit board and its components from moisture and prevented the short circuits.

5.6.6 *Future Directions*

Rugged systems tend to be less sensitive to cost than consumer products and therefore can take advantage of advanced technology such as materials, optical and electronic interconnections, component packaging, and high-capacity cooling. Some of the new materials include carbon-metal composites, graphite epoxy, aluminum alloys, and plastic foams. Optical interconnections can decrease weight and increase the bandwidth of signal transmission. Dense packaging includes both multichip modules (MCMs) and new forms of interconnections (see Section 5.7). New back planes have low dielectric constants, low capacitance, and matched impedance. Future systems will dissipate between 100 and 400 W per module; such high dissipation requires liquid flow-through cooling of the circuits (Benning, 1991). Fig. 5.8 illustrates a circuit card module with liquid flow-through cooling.

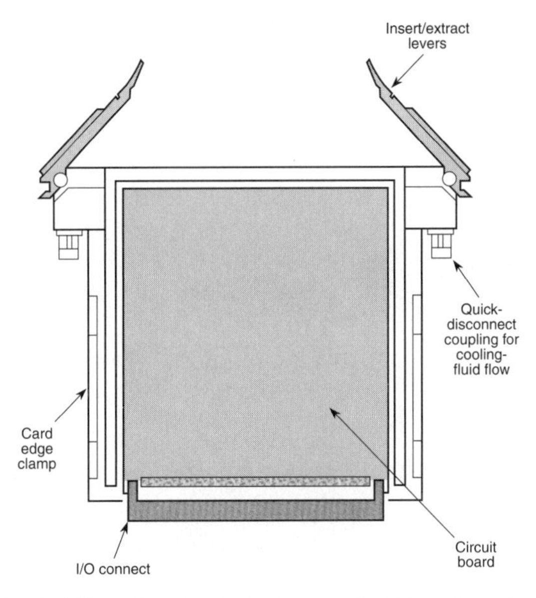

Fig. 5.8 Outline of an SEM-E module for avionics that uses liquid flow-through cooling.

5.7 COMPONENT PACKAGING

Faster clocks and data buses are fueling the drive for better packages. Currently, chip-to-chip interconnections negate the performance advances being made by ICs above 50 MHz. Component lead and wire bond inductances limit performance and create speed differentials between printed circuit boards and integrated circuit dies. Further concerns include thermal management, weight, reliability, cost, and standards (such as military specifications).

5.7.1 *Types*

The types of packages for integrated circuits include DIP, surface mount such as quad flat package (QFP) and plastic leadless chip carrier (PLCC), pin grid array (PGA), hybrid, and MCM (see Fig. 5.9) (Charles, 1993). DIPs have leads with 2.5-mm (0.10-in.) centers and are generally limited to 64 pins. Surface-mount components (QFPs and PLCCs) have leads with 0.85-mm (0.03-in.) or even 0.5-mm (0.02-in.) centers. QFPs and PLCCs allow square packages with leads on all four sides. Some QFPs can have as many as 304 pins. PGAs have pin spacings of 2.5 mm (0.10 in.) in multiple rows across the bottom of the package. PGAs can have more than 400 pins. Hybrids and MCMs have multiple bare chips die-bonded to a substrate; their form factors vary considerably.

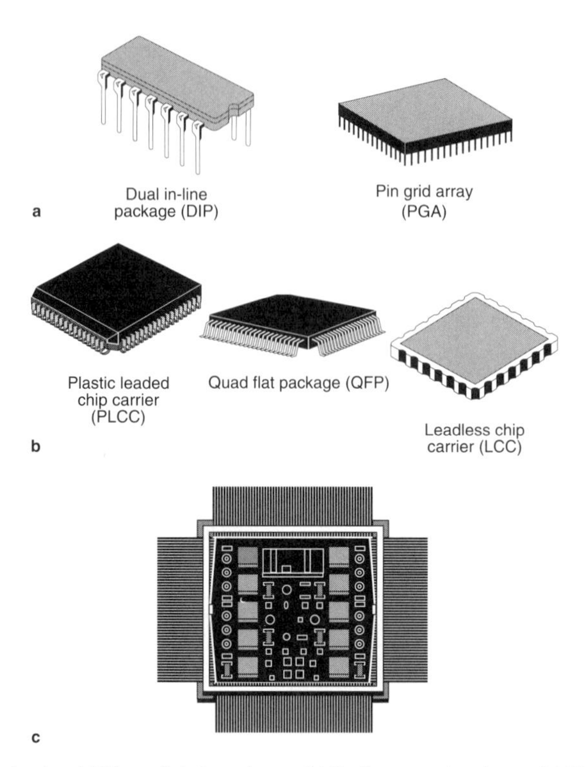

FIG. 5.9 IC packaging. (a) Through-hole packages. (b) Surface-mount packages. (c) Multichip module.

Working from the circuit die out, wire bonds connect the integrated circuit to the metal frame that is patterned to match the IC's pads to the PCB. The metal frame provides leads, using either through-hole or surface-mount technology (SMT), that act as a "space transformer" between the IC terminal pads and the traces on the PCB. Fig. 5.10 illustrates several types of IC packages. Any technology that shortens the wire lengths allows higher circuit density and clock rates.

Drilling holes very close together is very difficult and expensive. Hole size has limited the spacing of leads on through-hole components, such as DIPs and PGAs. Surface-mount technology (QFP and PLCC) bonds leads to pads on the circuit boards; leads can be much closer together since no holes are needed. Therefore, surface-mount technology has shrunk the size of components and circuit boards.

A recent development in packaging called ball grid array (BGA) is blending surface-mount technology with PGA packaging. A BGA uses solder balls with 1.5-mm (0.06-in.) spacing in place of pins on the bottom of the package. The spacing of the solder balls and the self-alignment of the solder bonds from surface tension makes the BGA package suitable for manufacturing (Gwennap, 1993).

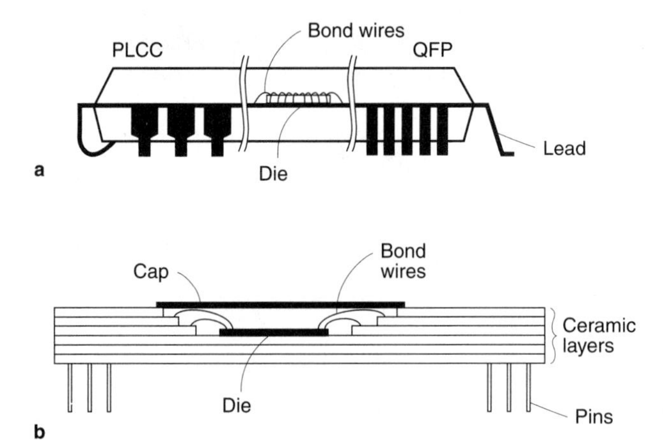

Fig. 5.10 Side views of some component packages. (a) PLCC and QFP. (b) PGA. (Modified from Gwennap, 1993.)

By combining multiple circuit dies onto a single substrate, MCMs and hybrids reduce the interconnection length between dies and increase the circuit density over standard IC packages on a circuit board. (An MCM has finer signal traces and greater functional density—that is, more logic gates—than a hybrid.) There are four types of MCMs, based on the substrate used (Klein, 1993; Maxfield, 1993):

1. MCM-L: Basically, MCM-L is a scaled-down PCB. The substrate is a laminate with copper traces and vias. It is the cheapest approach.
2. MCM-C: The substrate is either low-temperature cofired or thick-film ceramic. It may support up to 100 layers, but it has a relatively high dielectric constant and longer interconnection lengths. It is better for high-density rather than high-speed designs.
3. MCM-D: The substrate may be ceramic, glass, or metal and is formed by a thin-film IC process. It is good for high-speed designs but is expensive.
4. MCM-S: The substrate is a semiconductor, typically silicon. It uses optolithographic IC processes to form very fine signal traces. Transistors can be fabricated directly on the substrate. The coefficient of thermal expansion of a silicon substrate exactly matches that of the attached silicon chips. This is the most dense approach but the most expensive as well.

5.7.2 *Electrical Characteristics*

The package determines the electrical characteristics of the component and sets the performance limits. High-performance packages such as PGAs and MCMs have lower power and ground inductances and shorter interconnections than DIPs and consequently faster switching speed. These packages can also integrate capacitors between power and ground to lower the power supply impedance.

For the best performance, strive for the following:

- Impedance matching with the PCB
- Short signal traces
- Minimum inductances
- Low dielectric constants
- Low conductor resistance
- Minimal crosstalk (inductive or capacitive coupling)

5.7.3 *Mechanical and Thermal Concerns*

An IC package provides both mechanical and environmental protection for the chip during handling, assembly, and service. The physical packaging material may be molded plastic encapsulant, ceramic, or hermetically sealed metal. It conducts heat, resists moisture, and supports the electrical connections.

Plastic packages are much cheaper than ceramic, but ceramic packages can handle more signals, more heat, and larger chip die. Plastic packages usually dissipate less than 1 W of heat because they have a relatively large thermal resistance. They can be significantly lighter than ceramic packages. On the other hand, ceramic packages have a lower thermal resistance, and heat sinks can easily bond to them. They also seal out humidity better than plastic packages.

Hermetic seals and encapsulation provide moisture protection and chemical resistance for the IC die. Unfortunately, hermetic seals have higher thermal resistance and more interconnection bonds, providing another level of possible failures.

Packaging should provide for unimpeded heat transfer to reduce failures from thermal stress. The method and orientation of die attachment and the package (plastic or ceramic) largely determines the thermal resistance of the component package. Mounting the die upside down on the component lid shortens the thermal path to the heat sink and the outside world and improves heat transfer but is more expensive to fabricate. Research into substrate materials, such as polycrystalline diamond, that have both high thermal conductivity and high electrical resistivity will eventually improve packaging and circuit performance (McDonald, 1993).

Be aware that soldered components should have some allowance for expansion and contraction. Flexing in the leads relieves some thermal stresses on the component package. The long leads of a DIP flex to help relieve stress, whereas PLCCs do not have leads and are more brittle.

Finally, you may have to screen components with wire bond pull, die shear, and substrate attachment tensile tests. You will encounter these tests when using hybrids or MCMs in your designs.

5.7.4 *Trade-Offs*

DIPs of 64 pins are more than 75 mm (3 in.) long. Lengths of the circuit paths within the DIP can vary by a factor of 7, making control of lead parasitics, inductance, and capacitance difficult. A ceramic PGA dissipates heat better than a DIP and reduces the effect of lead parasitics; therefore, PGAs outperform DIPs but are much more costly.

The square packages of QFPs and PLCCs allow less variation in lengths of the circuit paths within the package, so control of lead parasitics, inductance, and capacitance is better than for DIPs. They have greater circuit density and better electrical performance than DIPs and some PGAs. QFPs and PLCCs usually are cheaper than PGAs, but matching the coefficient of thermal expansion between the plastic package and the circuit board is more difficult with them than with ceramic PGAs.

5.7.5 *Radiation Hardness*

If your application is space or military electronics, you will probably have to specify some form of radiation hardness in the selected ICs. Satellites are continually bombarded with cosmic rays that generate charges in the silicon substrate, while military components may experience fluxes of neutrons and ionizing radiation. The end results are either latch-up or bit flip errors.

Only specific processing techniques for the IC wafer fabrication can ensure radiation hardness. Since I am giving only the briefest of introductions, see Ma and Dressendorfer (1989) if you have to use radiation-hard components. Fig. 5.11 outlines the radiation environment for various systems.

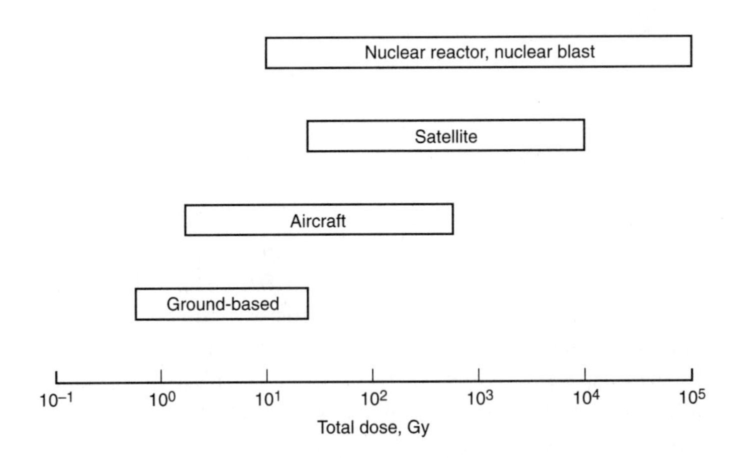

FIG. 5.11 Total dose exposure to radiation by application. Total dose refers to the cumulative exposure of components to ionizing radiation. (Adapted from Coe, 1990.)

5.7.6 *Trends*

Matta (1992, p. 68) writes, "Basic packages are simple and inexpensive. They meet most of the manufacturability and serviceability requirements, but electrically and thermally they are suitable only for nondemanding applications. Conventional high-performance packages, such as PGAs, provide significantly better electrical and thermal environments. However, they are fairly expensive and fail to meet most manufacturability and serviceability requirements." MCMs provide another advancement in performance over PGAs, but they are even more expensive and difficult to manufacture and service. You must understand the function, operation, performance, and cost of each project to choose the right component packaging.

Undoubtedly, innovations in packaging and manufacturing will extend the performance and lower the cost of DIPs, QFPs, PLCCs, and PGAs. Furthermore, we will be seeing more MCMs in all markets—specialized and military products first, but eventually in consumer appliances. According to McDonald (1993, p. 24), MCMs "offer the potential of 10 to 100 times improvement in density, up to two to three times reduction in power, a tenfold improvement in reliability, and lower cost."

5.8 OTHER MECHANICAL ISSUES

Several issues besides those already discussed need your consideration. Every project has some requirement or limit on size, weight, volume, and cooling. Materials, both the type and amount of raw stock, directly affect manufacturing costs. And safety can never be understated.

5.8.1 *Size, Weight, and Volume*

Every product has a requirement to fill a specified amount of space. Some equipment is more sensitive to size and weight requirements than others. Shaving a pound off the weight is much more meaningful for avionics, for instance, than for factory consoles. Many times it can't be too small or light. Who wouldn't want a computer that fits on a postage stamp and does everything that we want (unless, of course, it is easily lost)?

Some standards, particularly military ones, can add significant size and weight to an enclosure. The rigid supports and mounts that resist vibration and shock increase the size of the instrument chassis. A metallic shield for EMI will add weight to the enclosure. If your equipment needs operational flexibility, it may require expansion connectors and space for additional circuit boards. This potential for growth in capability will dictate a larger enclosure to accommodate additional circuit boards in the future. Computers with plug-in peripheral cards are an example.

Beyond simple size concerns, you must avoid obstructions between internal components and provide the appropriate tolerances in sizing your design. Even after careful preparation you should check the fit of the components within the enclosure. Until recently, mock-ups of components were constructed to check tolerances and possible obstructions. In some cases that may still be necessary, but software is available that generates three-dimensional models and checks the fit for you. Furthermore, the advent of virtual reality should ease the "walk-around" inspection of models and improve the thoroughness and reliability of checking the fit.

Vibration and shock affect the size and weight of instrument enclosures, too. As already mentioned, you may have to incorporate shock absorbers or special layouts of components to resist vibration and shock. Sophisticated software is available that performs the complex calculations of finite element analysis and helps you design vibration- and shock-resistant enclosures.

5.8.2 *Fans and Filters*

Most products must integrate the cooling design into the packaging and enclosure. They may have heat sinks, fans, ports, and plenums that you must situate carefully to allow heat transfer. Don't design a heat sink or a fan exhaust that may be blocked through casual use. Chapter 10 describes cooling of electronic equipment in detail.

Two more concerns are filters and negative pressure versus positive pressure. If the environment is dusty, you may need to use positive pressure and a filter to keep dust out of the electronics. However, if you can maintain a smooth airflow through the electronics, you may be able to use negative pressure and eliminate the filter. Clements (1993, p. 65) says, "Experience has shown that . . . filters do not get cleaned as often as required and lead to system reliability problems. Rather than filtering dust, the negative pressure design passes most dust through the product. The dust that does collect over time inside the product is far less detrimental than a clogged filter."

5.8.3 *Fire and Flood Survival*

Some simple arrangements of components can significantly reduce the damage from fire and water. *The primary concern for salvaging equipment is to prevent or reduce corrosion caused by the combination of sooty film, contaminants, and water.* Particularly if your equipment is expensive and difficult to replace, these tips can improve the prospect for reclamation after a fire. "On average, reclaiming equipment costs only 15 percent as much as replacing it, and usually gets it back in service sooner" (Kurland, 1992, p. 44).

First, avoid the possibility of soot, dust, and water accumulation. Mount circuit boards vertically to shed water quickly, and seal components with a conformal coat to prevent contaminant accumulation. Furthermore, you can angle vents in equipment enclosures downward and seal panels with rubber gaskets. Use screen or put holes in horizontal plates to drain any water. A simple precaution is to turn off unused equipment and cover it with a lightweight plastic dust cover (Kurland, 1992).

Second, avoid the use of corrosive materials such as polyvinyl chloride insulation and batteries. Polyvinyl chloride insulation releases highly corrosive chlorides when it burns. Situate batteries at the lowest point to keep acid from dripping on sensitive components. Fabricate the enclosure from corrosion-resistant materials such as anodized aluminum or galvanized steel.

Third, provide an emergency master switch to cut power and stop damaging electrolysis.

Finally, use modular and clearly marked components to aid disassembly and reassembly for cleaning. Fig. 5.12 summarizes some of these tips for fire and flood survival.

5.8.4 *Materials*

Get a good mechanical or packaging engineer for this one. Choosing appropriate materials for packaging and enclosures involves cost, weight, durability, corrosion resistance, wear, and life cycle. The materials also directly affect heat transfer and EMI shielding.

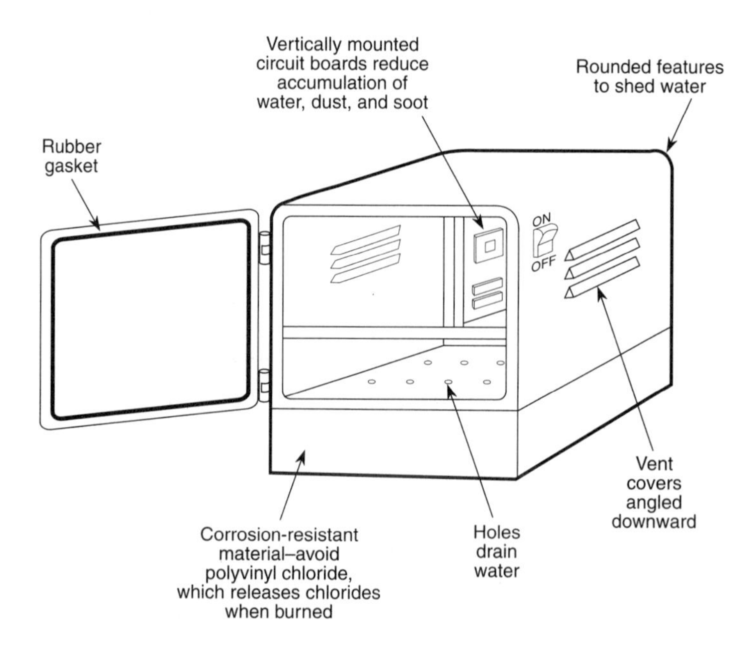

Fig. 5.12 Some features that help equipment survive fire and flood and make reclamation more likely.

5.8.5 *Safety*

Good packaging and enclosure design can improve the overall safety of an instrument. First, place warning labels in obvious places with clearly worded cautions. Second, use mechanical interlocks to protect against electrical shock to humans. Third, use shields and guards both to protect operators and to segregate manual functions.

Chapter 6 introduces grounding and fault interruption; beyond this, good mechanical design can protect operators and service personnel from shock hazards. Situating a fuse holder such that one lead of a fuse cannot be touched is one example. Another example is using interlocking switches on access panels covering high-voltage components.

Mechanical shields and guards can cover delicate mechanisms, high voltage and power, or sensitive circuits. Avoid sharp edges, on both the shields and the enclosure, that cause lacerations. Moreover, you should design the instrument enclosure and its human interface for proper operation that avoids or prevents mistakes (see Chapter 4). Recess switches and knobs to prevent accidental activation, for example (and allow clearance for fingers). Conversely, you can raise buttons with barriers in between. Critical operations should use two steps, such as pull-before-flip or push-before-turn; make sure to place instructions on a panel near the knob.

5.9 CASE STUDIES

5.9.1 *HP-PAC: A New Chassis and Housing Concept for Electronic Equipment*[*]

Business competition between PC and workstation manufacturers has resulted in shortened life cycles for computer products, faster development and production times, and steadily decreasing market prices. The Hewlett-Packard Böblingen Manufacturing Operation and its Mechanical Technology Center are faced with this trend, along with others, such as tightened environmental protection guidelines and take-back regulations.

These trends call for new concepts—environmentally friendly materials and matching manufacturing methods. Assembly and disassembly times for computer products have to be as short as possible. Assembly analysis of some HP products clearly showed the necessity to reduce parts as well as to improve manufacturing and joining techniques.

At the Mechanical Technology Center, these observations provided the motivation to look for a new packaging and assembly concept for computer products, one that would leverage existing techniques and incorporate new technological ideas.

[*] From Mahn et al. (1994). © Copyright 1994 Hewlett-Packard Company. Reproduced with permission.

Our objectives were to reduce the number of components and the number of different part numbers, to achieve considerable savings in the areas of logistics and administration, to save time in building a chassis, to automate the mounting of parts on the chassis, and to reduce overall chassis costs.

5.9.1.1 Genesis of an Idea

After we had critically weighed all of the technologies known to us—namely, producing enclosures and chassis of sheet metal or plastics—the only reduction potential seemed to lie in reducing the number of parts and using snap fits to save on fasteners and assembly times. However, in contrast to our expectations, we could not do this to the extent we had in mind.

Using snap fits is a disadvantage, since disassembly is time-consuming and can lead to destruction of the components or the enclosure. In the future, enclosures not only need to be assembled quickly but also need to be disassembled within the same amount of time to make recycling easier and cheaper.

We could not get out of our minds the idea of fixing parts in such a way that they are enclosed and held by their own geometrical forms. The idea is similar to children's toys that require them to put blocks, sticks, cards, or pebbles into matching hollows and at the same time keep track of positions and maintain a certain order at any time during the game. We applied this idea to our problem and thought about how our game collection would have to look in terms of composition and performance for us to be able to package and insert components for a workstation. It seemed most feasible to apply this idea at the assembly level, that is, to use the new method to fix conventional assemblies such as the disk, speaker, power supply, CPU board, and fan.

The only problem was what kind of material could we use to realize this idea. How could we achieve a form fit and not compromise on tolerances, feasibility, and price? Not to condemn the idea almost seemed impossible. It became obvious that we could no longer use conventional methods and standards to find the ideal material. We were forced to deal with a completely different field. The solution seemed to be to jettison everything we had learned before and direct our orientation towards something totally new.

The material we were looking for had to be pliable and bouncing—like foam, for example. Could foam be used for a form fit?

5.9.1.2 Raw Material Selection

After the idea had been born to use foam, we started our search for a suitable material. The goal was to embed all components necessary for an electronic device in one chassis made of foam synthetic material. We had plenty of material to choose from, including polyurethane, polystyrene, and polypropylene. The material had to be:

- Nonconductive to hold and protect electronic components
- Able to hold tolerances in accordance with HP standards
- Able to fix components without fasteners

With the help of our internal packaging engineers and an external supplier we soon found a suitable material: expanded polypropylene (EPP) with a density of 60 g/l.

In comparison to other foam synthetic material, EPP has the following advantages:

- Excellent mechanical long-term behavior
- Moisture resistance
- Resistance to chemicals
- Heat resistance
- 100% recyclability. Granules produced from recycled EPP can be used for manufacturing other parts such as packaging material and shock absorbers.

EPP foam parts can be produced in densities of 20 to 100 g/l. Lower density provides excellent shock absorption, while higher density offers tighter manufacturing tolerances. Thus design trade-offs are possible.

5.9.1.3 From the Idea to a Workstation

The next step was to apply this new concept to an already existing workstation. One workstation seemed suitable for the conversion. The existing concept (see Fig. 5.13), consisting of sheet-metal chassis (top and bottom), electrical components, sheet-metal enclosure, EMI liner, and plastic parts, was transformed into a foam chassis, electrical components, sheet-metal sleeves, integrated EMI liner, and

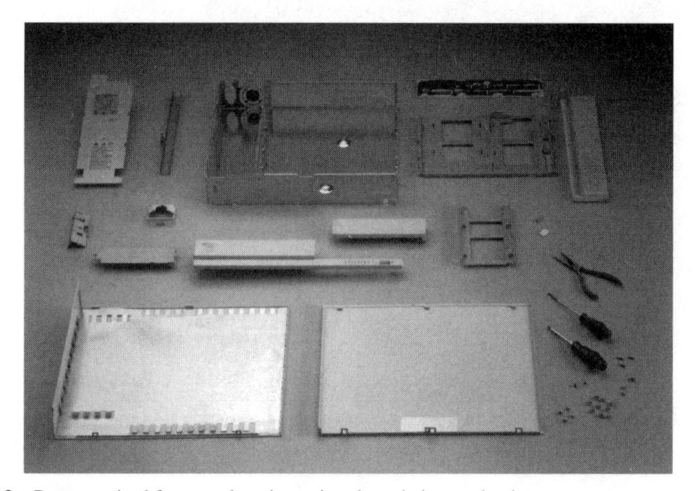

Fig. 5.13 Parts required for a workstation using the existing packaging concept.

modified plastic parts (see Fig. 5.14). In the new technology, all of the components are held by their own geometry in form-fitting spaces in the foam chassis. The connections between them are achieved through cabling held in foam channels (Fig. 5.15).

Time was saved by processing the foam chassis, the sheet metal, and the plastic parts in parallel. To obtain the foam parts quickly, we rejected the ordinary way of creating drawings with the help of a CAD system and instead created a 2D cardboard layout showing the placement of the components. A packaging company placed their sample tooling shop at our disposal for a few days. The first prototype was built step by step. We milled, cut, and glued, applying a lot of imagination.

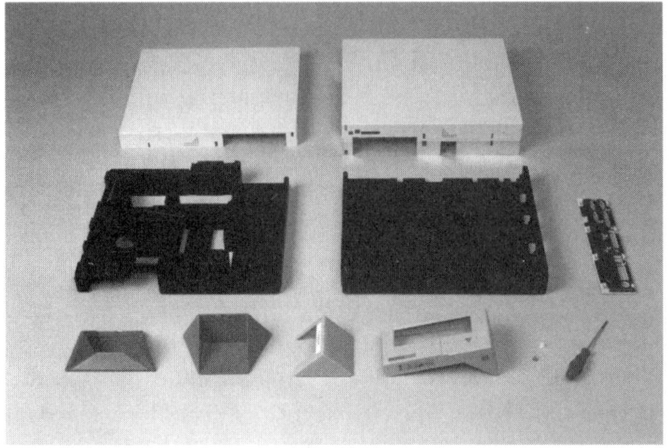

Fig. 5.14 Parts required for the workstation of Fig. 5.13 using the HP-PAC concept.

Fig. 5.15 Channels in the foam carry cooling air (shown) and cabling.

After two days the first prototype was nearly finished. Components were fixed in the necessary form fit and we were all aware that we had taken a step in the right direction. Back at HP we made a few minor changes to the EPP chassis and the remaining enclosure with the help of a knife and finished the prototype.

The major question now was, "Will it run?" We ran software on the workstation and started testing, surrounded by our production staff. The programs worked!

Next, temperature, humidity, and environmental tests were performed. Temperature problems were corrected by altering the air channels through cutting and gluing. HP class B2 environmental tests were passed (see Table 5.3 and Fig. 5.16).

Table 5.3 Thermal test results, CPU temperatures (°C)

Test point	HP-PAC	Original
UB7	29.1	32.1
UD15	33.0	37.5
UB20	36.6	44.8
UH25	35.7	46.7
TO-220	49.3	59.8
UM10	39.9	44.5
UM25	58.7	67.5
UR30	56.7	73.7
MUSTANG	70.0	82.9
CPU average	45.44	54.39

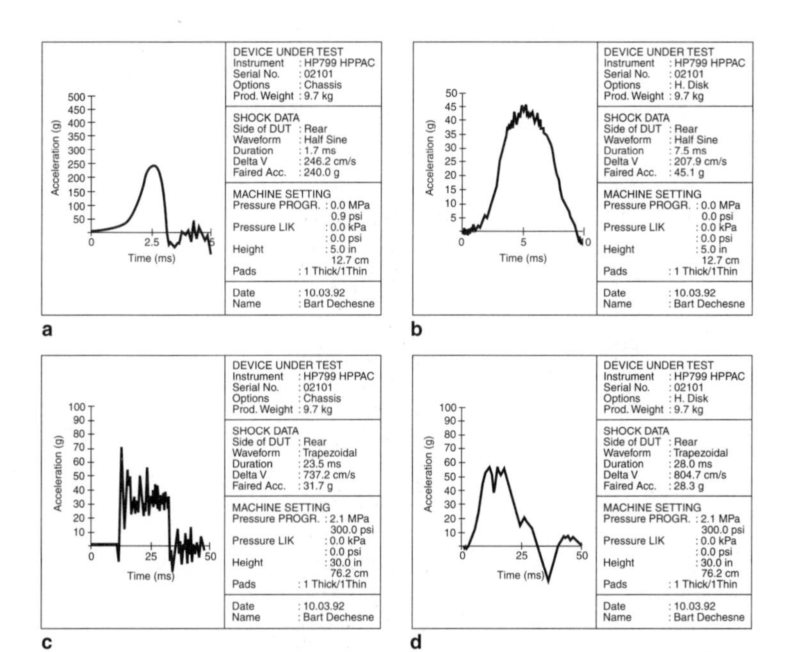

FIG. 5.16 Impacts transmitted to a hard disk drive by HP-PAC foam. In (a) and (c) the sensor is on the workstation. In (b) and (d) the sensor is on the hard disk drive held by HP-PAC.

Fig. 5.16 shows the impacts transmitted by HP-PAC to a hard disk for two different types of shocks (half sine, trapezoid). These impacts could be minimized by optimizing the design of the supports and form fits for the devices.

5.9.1.4 Savings and Advantages

Comparisons were drawn between a traditional HP workstation and an HP workstation in which the system components such as the CPU board, disk drive, and flexible disk were mechanically integrated using the HP-PAC concept. The HP-PAC workstation showed:

- A 70% reduction in housing mechanical parts
- A 95% reduction in screw joints
- A 50% reduction in assembly time
- A 90% reduction in disassembly time
- A 30% reduction in transport packaging
- A 50% reduction in time and expenditure for the mechanical development of the housing.

Compared to conventional chassis concepts, HP-PAC's advantages include:

- A reduction in the number of chassis parts.
- Separation between functionality and industrial design. The external enclosure is designed after definition of the mechanical interfaces between the enclosure and the chassis and between the enclosure and the components.
- One production step to produce molded parts.
- Simple, fast, and cost-effective assembly of the components (see Fig. 5.17). The assembly process is almost self-explanatory as a result of the indentations in the molded parts, and no additional joining elements and assembly tools are necessary. Assembly at the dealer's site is feasible.
- Reduced product mass because of the lighter chassis.
- Good protection against mechanical shock and vibration.
- On-the-spot cooling of components as a result of air channels in the foam.
- Cost savings during almost all working processes.
- Reduced transport packaging as a result of the good absorption of the chassis material. Also, less transport volume.

5.9.1.5 Impact on the Development Process

With HP-PAC, a 100% recyclable and environmentally friendly material is used for the construction of the chassis. Development of a chassis only means spatially arranging components within a molded part and adhering to certain construction guidelines (function- and production-specific).

FIG. 5.17 HP-PAC workstation assembly sequence. (a) Foam bottom chassis. (b) Loaded bottom chassis. (c) Partially loaded foam top chassis. (d) Open loaded chassis. (e) Assembled loaded chassis. (f) Lower enclosure added. (g) Upper enclosure added. (h) Enclosure completed.

The external enclosure is developed separately after definition of the interfaces. There are hardly any tolerance problems as a result of material flexibility. Changes are simple to perform by cutting, gluing, and additional grinding so an optimal solution can be reached quickly.

It is possible to perform all relevant environmental tests on the first prototype. Prototypes can also be used for the first functional tests. It is relatively simple to make changes during the tests, since industrial design and functionality are clearly separated and changes on the molded part can be performed in the lab. Once the design is complete, manufacturing of the molding tool is fast and cost-effective, and tool changes are not required.

The result is a short, cost-effective chassis development phase.

5.9.1.6 Material Development

HP-PAC places high demands, some of which are new, on the material used. Expanded polypropylene meets these demands in almost all ways.

So far EPP has been mainly used in reusable packaging and to an increasing extent in the automobile industry, where bumper inlays and side impact cushions for car doors are typical applications. Traditionally the automobile business has placed high demands on the quality of components. In terms of precision and long-term behavior these demands are identical to those of HP-PAC.

However, the situation is somewhat different for two new HP-PAC-specific material requirements: ESD (electrostatic discharge) suitability and flame retardant properties. We are working with raw material manufacturers in the U.S.A., Japan, and Germany to develop optimum raw material for HP-PAC in the medium and long term. For the short term, procedures had to be found and checked to meet these demands. In terms of ESD suitability this meant spraying or dipping parts in an antistatic solution. A suitable flame retardant was developed and patented together with a company specializing in flame retardants. Like plastic molding, treatment with flame retardant places an additional burden on the environment and impairs recyclability. However, a good product design renders flame retardants unnecessary. Three prototype HP-PAC products without flame retardant already have UL/CSA and TÜV approval.

We expect that the incentive for suppliers to develop suitable raw materials will steadily increase as the number of products using HP-PAC grows. Thus, in the future we hope to have custom-made materials available that will allow further improvements in quality at reduced cost.

5.9.1.7 EPP and Its Properties

EPP raw material is available on a worldwide basis. Some EPP manufacturers and their trade names for EPP are JSP ARPRO, BASF NEOPOLEN P, and Kaneka EPERAN PP. EPP comes in the form of foam polypropylene beads. Its chemical classification is an organic polymer, one of the class of ethylene polypropylene copolymers.

EPP contains no softeners and is free of CFCs. The product does not emit any pollution. Compressed air, steam, and water are used during the molding process. According to one EPP manufacturer, no chemical reactions take place during this process.

The specifications quoted here are for the JSP raw material we used. EPP from other manufacturers should vary only a little or not at all from these specifications.

Mechanical Properties. The following list shows the relevant mechanical properties of parts made of expanded polypropylene foam with a density of 60 g/l.

Density: 60 g/l
Tensile strength: 785 kPa
Compressive strength at 25% deformation: 350 kPa
Residual deformation after 24 hours at 25% deformation: 9%
Deformation under static pressure load (20 kPa): 1.2%
After 2 days: 1.3%
After 14 days: 1.4%

Thermal Properties. After the molding process, the parts are tempered so that the dimensions become consistent. Any further temperature influences will not result in significant contraction, expansion, or changes of mechanical properties between –40°C and 110°C. The coefficient of thermal expansion is 4.2×10^{-5}/°C from –40°C to 20°C and 7.5×10^{-5}/°C from 20°C to 80°C. Thus, a 100-mm length of foam at 20°C will be 100.375 mm long at 70°C.

The material changes state above 140°C. Thermal dissolution occurs at 200°C and the ignition point is 315°C.

There was no permanent deformation in a temperature loop test consisting of:

4 hours at 90°C
0.5 hours at 23°C
1.5 hours at –40°C
0.5 hours at 23°C
3 hours at 70°C and 95% humidity
0.5 hours at 23°C
1.5 hours at –40°C
0.5 hours at 23°C

Electrical Properties. EPP has good electrical insulation properties. This means that the foam parts can easily acquire an electrical charge. Consequently, methods are being developed to produce antistatic EPP. There is no noticeable interference between EPP material and high-frequency circuits with square-wave signals up to 100 MHz. Tests at very high frequencies (> 100 MHz) have not yet been conducted.

We can infer from solid polypropylene some of the electrical properties of expanded polypropylene:

- Dissipation factor of injection-molded EPP at 1 MHz: tan < 5×10^{-4}
- Breakdown voltage of injection-molded EPP: 500 kV/cm
- Surface resistance at 23°C and 49% relative humidity, untreated: 10^{11} to 10^{12} ohms.

Chemical Resistance. EPP has good chemical resistance because of its non-polar qualities. It is resistant to diluted salt, acid, and alkaline solutions. EPP can resist lye solutions, solvents at concentrations up to 60%, and alcohol. Aromatic and halogenated hydrocarbons found at high temperatures in grease, oil, and wax cause it to swell. When EPP is mixed with other substances, dangerous chemical reactions do not take place.

Reaction to Light. In general, EPP is sufficiently resistant to radiation at the wavelengths of visible light.

Reaction to Humidity and Water. Humidity has little or no effect on the mechanical properties of EPP. Water absorption is 0.1% to 0.3% by volume after one day and 0.6% after seven days. No changes are visible after 24 hours in water at 40°C.

5.9.1.8 Manufacturing Process

The raw material beads are injected in a precombustion chamber at a pressure of approximately 5 bar which reduces the pellet volume. The beads are then injected into the mold at a pressure of approximately 4 bar until a particular filling ratio is reached. The pressure is then reduced to normal so that the beads can reexpand and fill the mold. Once the mold is filled, steam at 180°C is injected into the mold through nozzles, warming the surface of the beads and fusing them together. This defines the foam part, which is left to cool down and then removed from the mold. Subsequently, by means of specified temperature cycles, controlled maturing and dimensional changes are induced in the part, resulting in its final form. This form remains constant within a specified temperature range.

5.9.1.9 Recycling of EPP

Polypropylene foam material can be recycled and used for manufacturing of other products. Manufacturers of polypropylene take back EPP waste free of charge.

EPP can be melted and fed back into source material polypropylene in thermoplastic form. Compression, melting, and granulation take place in gas extruders. The extruded recycled material can be used for polypropylene injection molded or extruded products. Recycling trials with a bumper system made of short glass fiber (approximately 20% of weight), EP rubber (approximately 20% of weight), and polypropylene produced a granule that can be used for complex injection molding.

5.9.1.10 Conclusions

To protect the HP-PAC technology in an appropriate manner we have filed for a patent under European patent application number 0546211.

It goes without saying that we will continue to develop the technology further. Efforts in which we are currently engaging are material development, prototype manufacturing, quality assurance, and marketing of HP-PAC.

We have not yet set any specific limits on user distribution. Possible areas for user application range from the electronics and electromechanical industries to home electronic equipment and transportation. At the Mechanical Technology Center, we offer various services ranging from consulting to complete solutions, not only for HP-PAC but also for sheet metal and plastic parts. We have experience in the computer, analytical, and instrument businesses and are in contact with others.

5.9.1.11 Acknowledgments

We would like to thank the Novaplast Company and the following HP entities: Böblingen Computer Manufacturing Operation, Exeter Computer Manufacturing Operation, Böblingen Manufacturing Operation Mechanical Technology Center, Böblingen Manufacturing Operation environmental test laboratory, Workstation Systems Group quality department.

Scope: system or project
Lessons learned: The objective for this packaging is to reduce the cost of manufacturing and use recyclable materials. Expanded polypropylene foam encases and holds electronic components; the design has these advantages:

1. It eases assembly, disassembly, and cost of manufacture:
   a. Fewer components and fasteners are needed.
   b. Assembly is nearly self-explanatory because of molded indentations.
   c. Mounting of parts in the chassis is automated.
   d. Molding is a one-step operation.
2. The foam is recyclable.
3. The foam has mechanical properties that support good packaging:
   a. Able to fix parts without fasteners
   b. Able to hold mechanical tolerances
   c. Nonconductive
   d. Shock absorbing
   e. Resistant to moisture, heat, and chemicals
4. Air channels cut in the foam tailor the airflow for cooling components.

5.9.2 *Rugged Designs for Tough Environments*[*]

Shock, vibration, radiation, dust, extreme heat and cold. Parts that survive for decades on the desktop might fail in hours when drafted for service in one of these hazardous environments. And, while electronic devices have become increasingly rugged, customers now insist that they must become even more so.

---

[*] From Gottschalk (1992). © Copyright 1992 Cahners Publishing Company. Reproduced with permission from *Design News*.

Users no longer baby portable computers. Instead, they toss them into suitcases and car trunks with abandon. Nowadays, military workstations must operate through near-miss explosions, sweltering desert heat, and rough vibrating terrain—but cost far less.

As they address these challenges, engineers have to take a more creative approach. Exterior and interior packaging, materials selection, and even additional protective sensors and electronics contribute to a good solution. The following examples illustrate some of these diverse design approaches.

5.9.2.1 Radiation Robots

Nuclear power plants present a unique hazardous environment. Concerns over occupational radiation exposure from the nuclear industry and government facilities have driven engineers to find alternative methods of working in certain areas. Humans shouldn't see more than 5 rads/year to remain healthy—500 rads in a short time could kill.

While the uncommon spill presents problems beyond the norm, even regular maintenance jobs—changing the filter cartridges for the reactor cooling water, for example—can expose workers to excessive radiation risks. Engineers at Odetics, Anaheim, CA, developed two robots to penetrate this environment (Fig. 5.18). ODEX III, a 900-lb, six-legged walker, can maintain remote areas originally designed for access on foot. And IRMA, the Integrated Radiation Mapper Assistant, relieves powerplant workers of the task of taking radiation surveys.

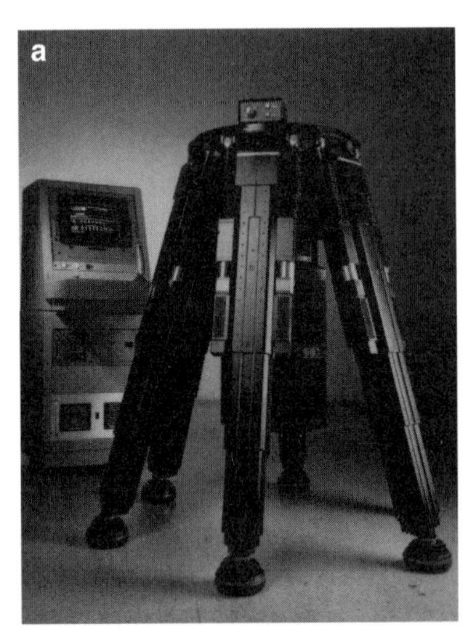

Fig. 5.18 Radiation rugged robots, ODEX III (a) and IRMA (b), can survive in environmentally unfriendly areas of a nuclear powerplant. Machined tungsten snout on IRMA shields the radiation sensor from all directions but the line of sight.

But even robots aren't immune to damage from extreme gamma-ray exposure. Radiation accumulation beyond 10^3 rads begins to break down microprocessors, causing bad signals, bad readings, lost data bits, and spurious signals. Approach 10^6 rads and electric motors lose power, plastics break down, rubbers degrade.

The classic solution might involve adding an abundance of lead shielding. However, these mobile maneuverers couldn't afford the extra weight. "You have to power this thing," notes project manager Steve Guzowski, referring to ODEX III. "Lead just isn't viable."

So engineers turned to more sophisticated radiation-resistant materials. They even reconsidered the need for absolute radiation immunity. "Do you design the robot to survive indefinitely?" questions Guzowski. "Or do you make it withstand a limited environment and replace degraded components after a period of time?" After comparing the costs and benefits of each approach, they went with a limited-life, interchangeable component design.

Though microprocessors exist that will withstand thousands of times the radiation of standard units, they cost much more to produce. Primarily low-volume designs developed for the military, their performance lags behind that of their non-hardened counterparts. So for ODEX III, engineers chose a hybrid design that combines standard Motorola 68020 processors with a single, radiation-hardened Intel 8086 chip. The 8086 unit also acts as a "watchdog," monitoring the degradation of the other critical system components, while the 68020s control the robot.

Teflon®, typically a top choice for wiring insulation, breaks down when irradiated and emits dangerous fluorides in the process. So engineers selected Tefzel®, another DuPont fluoropolymer that is in itself an irradiated product. To bond the primarily aluminum structure, the engineers applied a 3M aerospace adhesive. Limited radiation exposure assists in curing the material, initially increasing—not reducing—strength.

5.9.2.2 Dust-Free Design

"Contamination from radioactive dust is possibly a more severe environmental condition than radiation," claims Stephen Bartholet, senior staff engineer. "Like anything placed in a hazardous, radioactive environment, the robot must be free of contamination before being removed."

Typical cleaning processes involve wiping or even hosing off the object. But this cannot be done indiscriminately. Used water and cleaning solutions must ultimately be disposed of in expensive hazardous-waste containers.

Radioactive dust clings with a static electric charge that can be tenacious. To limit decontamination costs, engineers designed the robots not to get dirty in the first place. Tests inside a nuclear plant with hundreds of material samples showed that a special epoxy coating originally developed for low friction also attracted little dust. Engineers covered the entire exterior of ODEX III with this black, paint-like material.

As a bonus, the material resisted the sodium hydroxide containment spray used in nuclear reactors during spills. The corrosive solution would have eaten the bare aluminum structure and emitted explosive hydrogen gas.

On the smaller and less weight-sensitive IRMA, engineers chose a special electropolished stainless steel that also had excellent dust-resistive properties.

Even the payout system for IRMA's control cables limits contamination. It lays and picks up the cable instead of dragging it. This diminishes the amount of dust kicked up as the robot wheels by. "We designed to reduce the environment instead of trying to resist it," explains Bob Carlton, engineer on the IRMA project.

But sealing the robots to combat contamination creates another harsh environment—high temperature. Unable to actively cool the drive motors with outside, dust-laden air, engineers designed the primary structure of ODEX III as a heat sink. During normal operations, the robot might see temperatures of 300F. Luckily, the structural adhesive chosen for its radiation resistance also took the heat. Within a separate, closed compartment, fans circulate air in a closed loop across the circuit boards.

5.9.2.3 A PC That Takes Abuse

Portable computers pose special problems to designers. Ordinary use ranges from jostles and bumps inside a briefcase to an unscheduled fall to the floor. Users poke and scratch at the fragile display screen; sometimes they accidentally sit on it. To be acceptable, none of these ungracious acts should be fatal.

Engineers at Grid Systems, Fremont, CA, faced all of these ruggedizing issues when designing their new PalmPad, a 2-pound, 14-ounce pen-based PC (Fig. 5.19). Targeted for data collection markets like loading dock management, police accident reporting, and inventory control, "it was designed to survive a 3-foot drop onto a surface such as concrete," notes Steve Friend, mechanical CAD designer at Grid.

As with the Odetics' robots, designers decided prevention was the best cure. So the PalmPad incorporates a strap on the case that the user places a wrist through during use. "For both convenience and ruggedness, we made it wearable," explains Grid's Kate Purmal.

The display screen forms a weak link in many portable computers. To overcome this, Grid engineers designed a sandwich construction consisting of a bezel, digitizer overlay, LCD display, and a backlight diffuser with air gaps and neoprene gasketing to absorb shock (Fig. 5.20). Chemically strengthened, soda-lime glass, not plastic, protects the fragile LCD. "We tried acrylic overlays, but nothing would withstand what this glass can," says Friend.

As evidence, the face of the PalmPad brushes off repeated impacts from a 0.2-inch-diameter steel ball dropped from 6 feet. As a bonus, the glass is scratch-resistant—important in a pen-based computer.

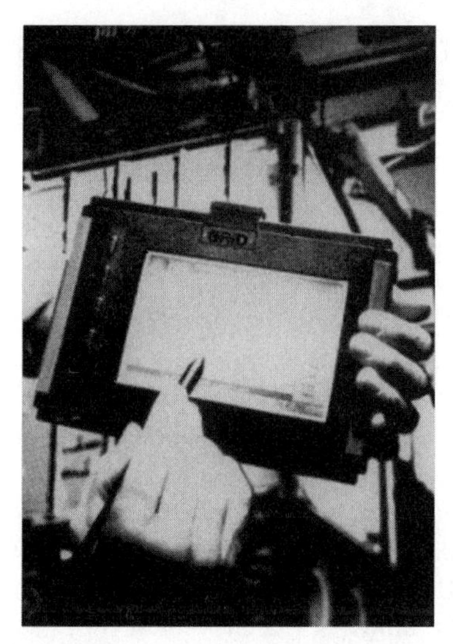

Fig. 5.19 Grid Systems' PalmPad emphasizes prevention as a means to ruggedization. The sturdy unit will withstand a 3-foot fall to concrete, but the wrist strap on the back should greatly reduce such risks.

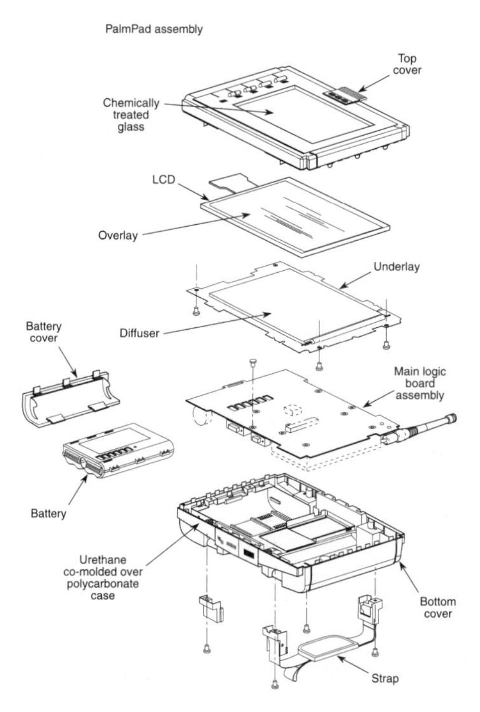

Fig. 5.20 Tight space constraints forced the PalmPad designers to use all available interior space. Chemically treated glass bezel and co-molded case contribute greatly to the unit's ruggedness.

For extreme shock resistance, the exterior consists of a co-molded composite of urethane over polycarbonate—a first for a PC of any type says Grid. The engineers worked with Trend Plastics, San Jose, CA, to develop the two-step process. First, the polycarbonate sections are formed and removed from the tool. Then the parts are placed into a second tool and a 0.040-inch-thick layer of urethane is molded over the top. Heat, not adhesives, fuses the two materials into a monolithic part.

In such a compact unit, dissipating heat became an issue. Initially, the cold-cathode fluorescent backlight caused local darkening of the LCD. Engineers fixed this by sinking the heat to a grounding shield behind the diffuser. Grid cites an operating temperature of 0 to 50C as evidence of the unit's environmental toughness.

For power drain and ruggedness, engineers chose to forego a hard disk for permanent storage. Instead, they selected the SunDisk from the Santa Clara, CA-based company of the same name. The solid-state memory storage card comes in capacities of 5 to 20 MB. Measuring the length, width, and three times the thickness of a credit card, the SunDisk matches the rest of the PalmPad's ruggedness, while remaining within the unit's tight volume constraints.

5.9.2.4 Hard Drive Senses Shocks

Seagate Technology hopes to get portable computer manufacturers like Grid to reconsider the hard disks as a rugged storage device. Probably the most complex mechanical mechanism in a computer, hard disks prove more sensitive to shock and vibration than the circuit boards with which they share space.

On a typical disk, tiny, regularly-placed magnetic guide marks, called servos, provide positional information for the recording head. Like lines on the highway, the servos help position the head over each of the thousands of tracks on the disk.

Problems arise when the computer is jarred while performing a read or write operation. Should the blow exceed the operating shock limit of the drive—typically around 2Gs—the positioning mechanism fails to maintain proper tracking and the head veers across adjacent tracks. During a read operation, this results in an irritating delay. During a write operation, it spells catastrophe. Both the written and the over-written data on the affected adjacent tracks will be lost. Should any servo mark be damaged, the drive could fail.

During tests to solve the shock problem, Seagate engineers discovered that the most harmless of impacts to a laptop computer can create severe G forces at the disk drive. For example: Dropping the entire computer four inches onto a rigid surface creates a peak force of 60 to 80Gs. Lifting and dropping just the front edge of the computer one inch generates 70Gs.

Unfortunately, laptops impose severe space constraints on designers, prohibiting shock-isolating mounts. To circumvent this problem, Seagate engineers incorporated a three-axis accelerometer and some processing circuits into three new drives—the ST9235AG, ST9144AG, and ST9080AG. This "shock sensor," a sensitive switch that detects shocks above a safe operating level, shuts off a disk write operation if necessary (Fig. 5.21). A buffer saves the data so that it can be written after the event has passed.

Generally, manufacturers rate the shock tolerance of their drives with three specifications: nonoperational shock rating or static damage boundary (100 to 150Gs), operating shock limit with recoverable errors (typically 10Gs), and operating shock limit without errors (typically 2Gs or less). "Our sensor extends the operating limit of the drive from the operating shock limit without errors up to the dynamic damage boundary," claims Seagate's Dave Tang.

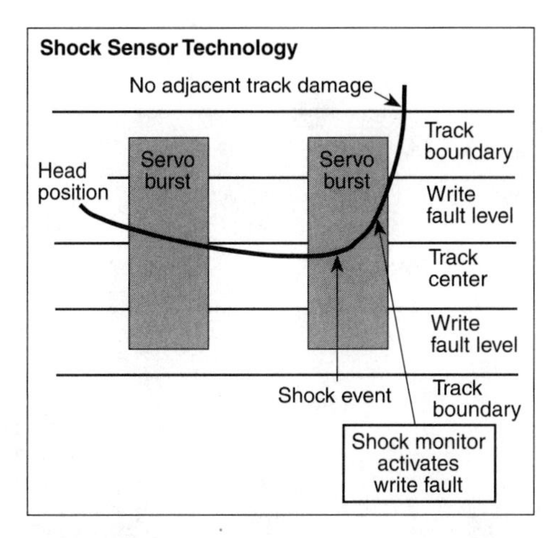

Fig. 5.21 Seagate shock sensor, shown in this exaggerated illustration, monitors the disk head, writing data along the black line near track center. Following a severe shock event that knocks the head off line, the shock monitor stops the drive from writing data. Later, when the head relocates itself, writing can continue.

5.9.2.5 Ruggedness on a Budget

With the military, rugged design is the norm, not the exception. So when Hughes Data Systems, Anaheim Hills, CA, set out to acquire the Navy's lucrative $172.3 million Tactical Advanced workstation Contract-3 (TAC-3), engineers felt that a different approach to ruggedness could win the contract. In a move mirrored with other government products, they included as many off-the-shelf components as possible, starting with Hewlett-Packard 9000-700 workstations. The benefit? A drastic reduction in the 200 to 300% increase over the price the military normally pays.

"The government no longer needs to control the design at the board level at an enormous cost disadvantage," notes Bruce Bromage, manager of rugged data systems at Hughes. Rapid advancements in commercial equipment, CPUs, and graphics allowed it to happen.

Specifying stock components, however, was the easy part. Hughes designers had to perform the deceptively difficult job of integrating the components into a single cabinet, then tuning the assembly to pass rigorous military shock and vibration specifications.

What the engineers ended up with was a design that combined a very stiff outer housing with the vibration isolation of internal components like disk drives. A coil-rope suspension supports the entire cabinet, isolating it from the 200G shocks specified in MIL-S-901C. The combination worked so well that they tested the unit to destruction for their own records (Fig. 5.22).

Fig. 5.22 TAC-3 workstation stands ready to receive a blow to its base of up to 300Gs. The test hammer strikes a blow upward from under the floor grating.

"The TAC-3 took 300G shocks—the maximum on the test machine—with only a broken on/off switch," says Hughes engineer Richard Chavez. It also passed MIL-S-167-1 vibration tests, during which the unit underwent two hours of 10G inputs at resonant frequency. All this, Chavez notes, occurred with the workstation operating and performing periodic I/O.

Fans pull 400 cfm of filtered cooling air through the cabinet's bottom. This, and a well-sealed design, allowed the TAC-3 to pass MIL-S-810E. It later exceeded the spec, surviving from 0 to 100C at 90% condensing humidity—while operating. "It was practically raining in the cabinet," jokes Chavez.

The design story is in the new approach to ruggedization. "The quality of commercial components is so good, all we really do is shock-isolate them," says Bromage. "We achieve ruggedization through the structure, not repackaging."

Lessons learned: These four applications of packaging and enclosures demonstrate the diversity of environments and particular solutions:

1. Radioactive environments: Design electronics to avoid the environment, not resist it.
   a. Use interchangeable components with a limited life instead of expensive radiation-hard components.
   b. Use a radiation-hard watchdog to monitor the CPU.
   c. Avoid Teflon insulation. It emits fluorides when irradiated.
   d. Use low-friction coatings to shed radioactive dust.
   e. Lay and pick up tethered cable instead of dragging it in the dust.
   f. Use the structure as a heat sink.
2. Rugged applications: Prevent problems rather than react to them.
   a. Use wrist strap on handheld computer to prevent dropping.
   b. Use shock-resistant materials: soda-lime glass, urethane over polycarbonate.
   c. Integrate heat sink with grounding shield.
   d. Use solid-state memory rather than a hard disk.
3. Survive shock: Sense and stop operation during impact.
   a. Dropping a computer several inches can generate 60*g* to 80*g* of force.
   b. Shock sensor shuts off disk write operation.
4. Rugged design does not necessarily have to meet military specifications. A good shock-isolated enclosure can protect commercial components.

5.10 SUMMARY

Designs begin and end with the package. It is the integrated function of all the individual components and represents the final product.

The enclosure often forms a major part of the human interface. It provides the mechanical, electrical, and functional inputs and outputs to the electronic circuits and protects them from severe or hostile environments. It also participates in the heat transfer and cooling of the electronics. The styling appeal of an enclosure may also be a major part of selling the product. Consequently, you must define the environment and consider packaging from the conceptual stages of any project. (Are you beginning to get the idea that all this stuff is interrelated?)

5.11 RECOMMENDED READING

These books are good references for all manner of electronic product packaging. Both are thorough, readable, well illustrated, and practical.

Ginsberg, G. L. 1992. *Electronic equipment packaging technology*. New York: Van Nostrand Reinhold.

Matisoff, B. S. 1990. *Handbook of electronics packaging design and engineering*. 2nd ed. New York: Van Nostrand Reinhold.

5.12 REFERENCES

Bank, L. 1991. A patient monitor two-channel stripchart recorder. *Hewlett-Packard Journal,* October, p. 27.

Benning, S. L. 1991. Progressive avionics packaging technologies. In *Proceedings: IEEE/AIAA 10th Digital Avionics Systems Conference,* October 14–17, pp. 158–163. New York: Institute of Electrical and Electronic Engineers.

Charles, H. K. 1993. Materials in electronic packaging at APL. *Johns Hopkins APL Technical Digest* 14(1):61–64.

Clements, B. 1993. Mechanical considerations for an industrial workstation. *Hewlett-Packard Journal,* August, pp. 62–67.

Coe, P. D. 1990. Rad-hard ICs. *Electronic Products,* October, pp. 39–42.

Colucci, D. 1994. DFMA helps companies keep competitive. *Design News,* November 7, p. 21.

Daumüller, K., and E. Flachsländer. 1991. Mechanical implementation of the HP Component Monitoring System. *Hewlett-Packard Journal,* October, pp. 44–48.

Gottschalk, M. A. 1992. Rugged designs for tough environments. *Design News,* November 9, pp. 78–86.

Gwennap, L. 1993. Packaging influences microprocessor cost. *Microprocessor Report,* September 13, pp. 12–16.

Klein, M. 1993. MCM design: A review. *Printed Circuit Design,* June, pp. 10–14.

Kurland, M. 1992. Surviving hell and high water. *IEEE Spectrum,* May, pp. 44–47.

Ma, T. P., and P. V. Dressendorfer, eds. 1989. *Ionizing radiation effects in MOS devices and circuits.* New York: Wiley.

Maass, D. 1993. Electronics packaging in avionics deserves more respect. *Military & Aerospace Electronics,* February 15, p. 5.

Mahn, J., J. Häberle, S. Kopp, and T. Schwegler. 1994. HP-PAC: A new chassis and housing concept for electronic equipment. *Hewlett-Packard Journal,* August, pp. 23–28.

Matisoff, B. S. 1990. *Handbook of electronics packaging design and engineering.* 2nd ed. New York: Van Nostrand Reinhold.

Matta, F. 1992. Advances in integrated circuit packaging: Demountable TAB. *Hewlett-Packard Journal,* December, pp. 62–77.

Maxfield, C. 1993. Multichip modules: Part 1. *Printed Circuit Design,* August, pp. 22–25.
McDonald, J. A. 1993. The brave new world of IC packaging. *Military & Aerospace Electronics,* February 15, pp. 21–24.
Teague, P. E. 1993. Polaroid breaks out of its box. *Design News,* January 4, pp. 52–60.
Zubatkin, A. 1993. Some techniques for ruggedizing commercial systems. *VMEbus Systems* 10(4):13–18.

6

GROUNDING AND SHIELDING

Men do not make laws. They do but discover them.
—Calvin Coolidge

6.1 FOUNDATIONS OF CIRCUIT OPERATION

Surely all of us have heard of a case of electrocution when someone drops a hair dryer into the water while bathing or seen static on the television when a kitchen appliance turns on. Consider how lightning can cause surges in power lines and damage a stereo system or induce a high-voltage transient in the sensor loops and damage a burglar alarm system (Doubek, 1991). A prominent consultant advertises seminars on electromagnetic interference control with a brochure that shows a car crashing through a highway barrier while an antenna radiates energy nearby. All these scenarios involve grounding and shielding (or the lack of them).

Good design of electronic circuits begins with grounding and shielding. Grounds and shields improve safety and reduce interference from noise. Properly connected grounds reduce dangerous voltage differentials between instruments. Shields minimize interference from noise by reducing noise emissions and noise susceptibility.

Simple models and schematics will help you understand the problems of safety and noise in each application. A cursory drawing of the paths of current flow from sources to receivers illuminates the problems and often suggests the solutions.

This chapter outlines the general principles of electromagnetic compatibility. It lays the groundwork for other chapters, in which specific examples of grounding and shielding are applied.

6.2 OUTLINE FOR GROUNDING AND SHIELDING DESIGN

The first step is to understand the safety and noise issues for your product. Next you need to know the possible mechanisms for energy coupling. Only then can you begin to define the necessary grounds and shields. This chapter follows the same general outline by first presenting the basics of safety and noise and then giving further details on grounding and shielding. Section 6.10 codifies these issues with some basic rules of design.

6.3 SAFETY

For safety's sake, reduce the voltage differential between external conducting surfaces. Usually the danger is conducted energy, low frequency (less than 1 MHz), and associated with power lines. (Microwave energy is not a shock hazard, but it does pose danger and demands special attention to shielding.)

Buildings and commercial equipment generally have been wired with a safety ground to provide a path for dangerous leakage currents and short circuits. A properly connected safety ground will reduce the voltage differential between external surfaces. Fig. 6.1 is a schematic of power and ground.

A safety ground must be a permanent, continuous, low-impedance conductor with adequate capacity that runs from the power source to the load. The *National Electrical Code* in the United States specifies the connection for a safety ground (National Fire Protection Association, 1993). Fig. 6.2 shows how outlets in a wall should be wired with a safety ground.

Impedance in the ground structure can force dangerous levels of current to flow through a bystander who is in physical contact with the instrument. Therefore, don't rely on a metallic conduit to form the conductive path for the safety ground; corrosion and breaks can open the circuit. Don't rely on building steel, either, because circulating currents can generate large and noisy ground potentials. A separate, dedicated conductor will avoid these problems. Fig. 6.3 illustrates one example of how important a correctly installed ground can be.

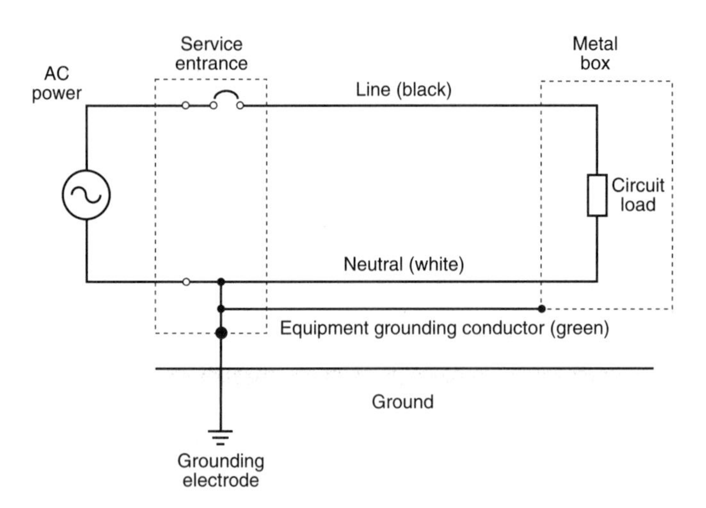

Fig. 6.1 General scheme for safety ground in buildings.

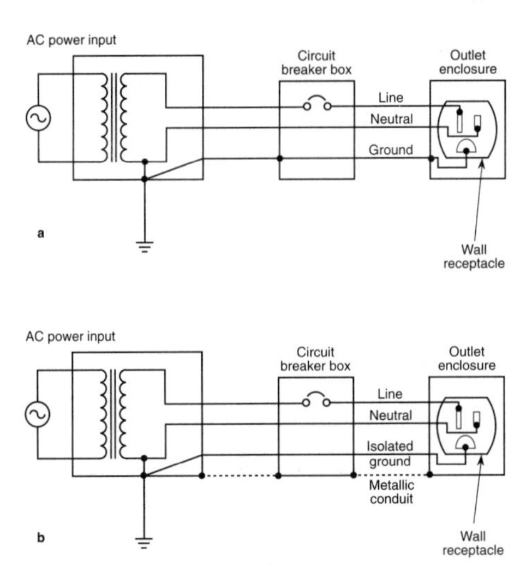

Fig. 6.2 General scheme for wiring wall receptacle in U.S. buildings. (a) If a metallic conduit is not installed, the ground must connect to both the circuit breaker box and the outlet enclosure. (b) The isolated ground must have low impedance (particularly inductance) and connect to the ground source, not to the metallic conduit.

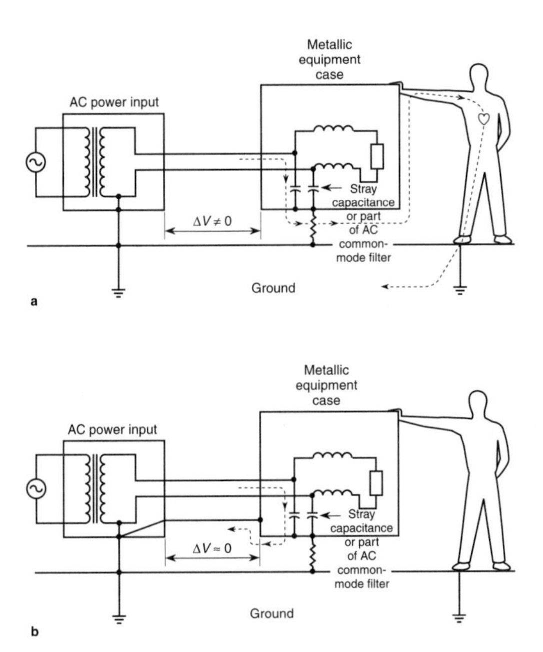

Fig. 6.3 How stray capacitance can route leakage currents to the equipment case. (a) An incorrectly grounded instrument or a ground connection with some impedance can allow dangerous current to flow through a human body. (b) An isolated ground with low impedance lowers the potential difference between the source and the instrument and reduces the threat of electrocution. (Adapted from St. John, 1992.)

Leakage currents, such as those in Fig. 6.3, will not trip protective circuit breakers. Circuit breakers open for large currents from short circuits between line and neutral or between line and ground, but leakage currents are orders of magnitude smaller than the trip currents. Consequently, ground-fault interrupters are needed for situations that are prone to dangerous leakage, such as the hair dryer in the bathtub. A ground-fault interrupter measures the differential in magnitude between the currents in the line and neutral conductors and opens the lines if an imbalance exists. Fig. 6.4 illustrates how a ground-fault interrupter can detect leakage current and protect against it.

Ground-fault interrupters are not a panacea, however. They neither control chassis voltages nor protect against transient ground voltages from lightning or power faults. They can be sensitive to nuisance trips (Van Doren, 1991).

Remember three things when you develop wiring for powering your instrument:

1. Consider the instrument and power mains as an integrated system (St. John, 1992).
2. Always draw your ground scheme to understand the possible circuit paths.
3. Don't blindly rely on building steel for a ground conductor.

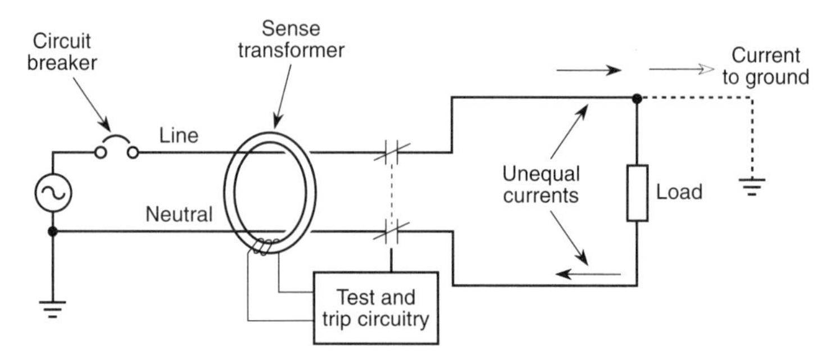

Fig. 6.4 A ground-fault interrupter. When it senses an imbalance between the line and neutral, it opens both lines. Unequal currents represent possible dangerous leakage to ground that may not be large enough to trip a circuit breaker.

6.4 NOISE

Noise is undesired electrical activity coupled from one circuit into another. Noise always includes three distinct components: a source, a coupling mechanism, and a receiver (Fig. 6.5).

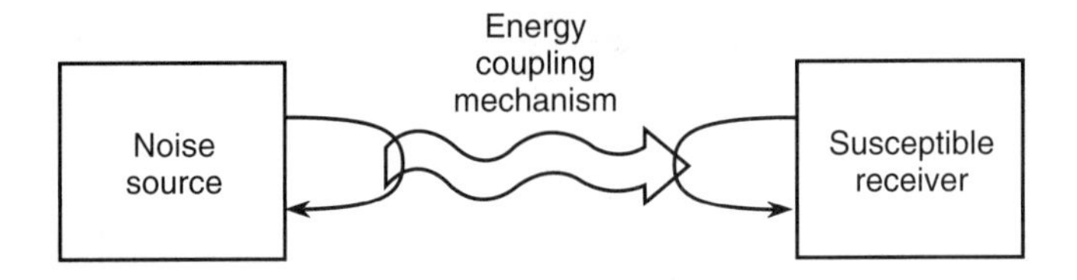

Fig. 6.5 Block diagram of noise disrupting a circuit.

6.4.1 *Sources*

Noise sources generate either a periodic signal or a transient pulse that disrupts other circuits. There are many types of sources:

- Power lines
- Motors
- High-voltage equipment (e.g., spark plugs, igniters)
- Discharges and sparks (e.g., lightning, static electricity)
- High-current equipment (e.g., arc welders)

6.4.2 *Energy Coupling Mechanisms*

Noise can couple by four mechanisms (Table 6.1): conductive, inductive, capacitive, and electromagnetic.

Table 6.1 Mechanisms that couple noise between sources and susceptible (receiver) circuits

Coupling mechanism	Frequency range	Comment
Conductive	DC to 10 MHz	Requires a complete circuit loop (really no upper limit to frequency)
Inductive	Usually > 3 kHz	Larger loop areas in circuits mean greater self-inductance and mutual inductance; associated with heavy currents. (Can get significant coupling from 50- or 60-Hz power.)
Capacitive	Usually > 1 kHz	Greater spacing between conductors reduces coupling; associated with high voltage. (Can get significant coupling from 50- or 60-Hz power.)
Electromagnetic	> 15 MHz	Needs antennas greater than 1/20 of wavelength in both the source and the susceptible circuits.

Current flow in conductive paths is probably the most common. Conductive coupling usually occurs at lower frequencies and is often caused by incorrect grounding. At higher frequencies circuit paths can have significant capacitive and inductive attributes that dominate coupling.

Changing magnetic flux can couple circuits, as shown in Fig. 6.6. The loop area of the circuit is the primary factor that determines the inductance and coupling.

Changing electric potentials can drive charge through stray capacitances as shown in Fig. 6.7. Appropriate grounding, shielding, and signal separation control the amount of capacitive coupling.

Electromagnetic coupling is a high-frequency phenomenon. It requires a transmitting antenna in the source and a receiving antenna in the susceptible circuit (Fig. 6.8). These antennas must be an appreciable fraction of the signal wavelength to couple effectively.

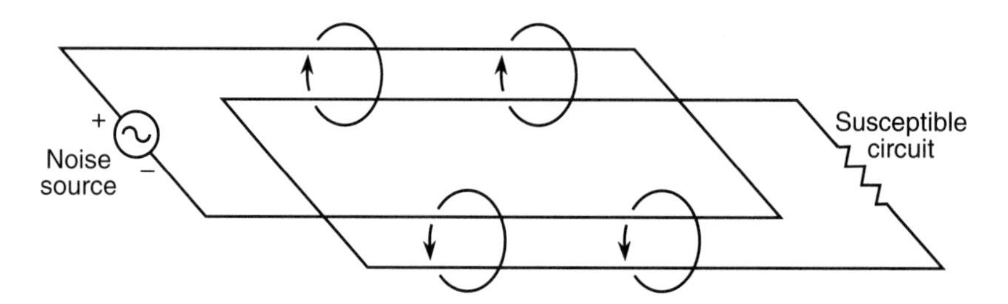

Fig. 6.6 Inductive coupling. Magnetic flux couples energy from the circuit of the noise source to the susceptible circuit.

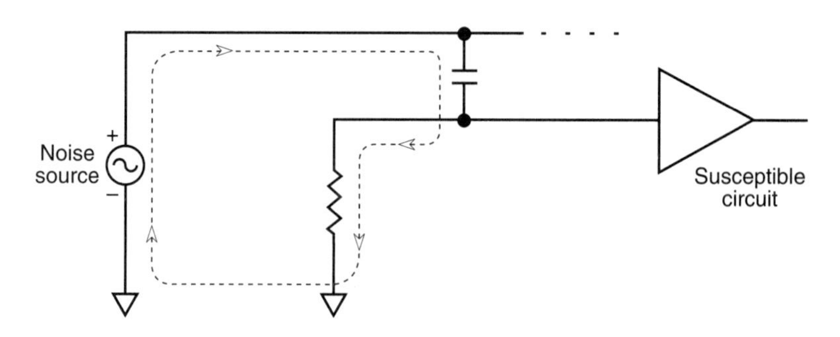

Fig. 6.7 Capacitive coupling. Coupling from stray capacitance completes the circuit of a noise source into a susceptible circuit.

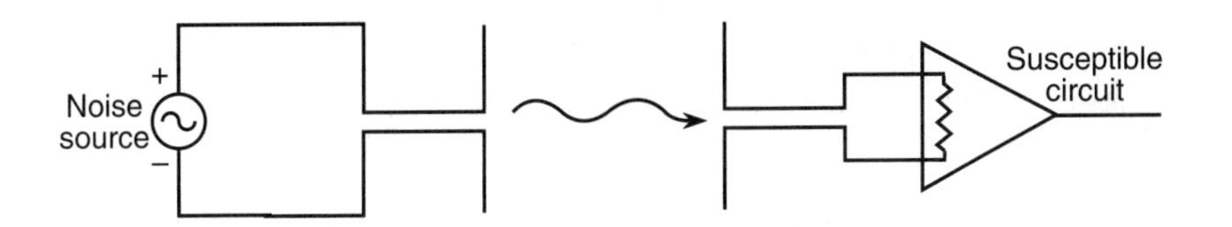

Fig. 6.8 Electromagnetic coupling. Radiated electromagnetic energy requires an antenna in both the noise and susceptible circuits. The antenna must be an appreciable portion of a wavelength, so such coupling is usually at high frequencies (> 15 MHz).

6.4.3 *Susceptibility*

The third component of noise is a susceptible receiver. Examples of susceptibility include crosstalk on inputs that leads to bit flips in digital logic, radio interference and crackle, and static discharge that destroys components. Susceptibility usually can be traced to incorrect grounding (or return paths) or long signal lines that are not properly shielded.

6.5 PRINCIPLES OF ENERGY COUPLING

To reduce noice, you must understand how electrical energy couples between circuits. Identifying the mechanisms of coupling will help you make wise decisions during the design process.

Current follows the path of lowest impedance, not necessarily lowest resistance. Consequently, charge follows the path of minimum inductive and maximum capacitive reactance for the lowest impedance. Current will flow in the path of minimum impedance, particularly for frequencies at or above the audio range (> 3 kHz). Remember that impedance is a function of resistance, inductive reactance, and capacitive reactance:

$$Z = \sqrt{R^2 + [\omega L - (1/\omega C)]^2}\ ,$$

where Z = impedance
R = resistance
ωL = inductive reactance
$1/\omega C$ = capacitive reactance

Furthermore, for frequencies above 3 kHz, a useful diagnostic for determining the mechanism is the ratio of the rate of change in voltage to the rate of change in current (Van Doren, 1991). For the special cases of sinusoidal signals or resistive loads, the ratio is impedance; otherwise, it is a *pseudo* impedance value. A low value (< 377 Ω) indicates a large change in current and inductive coupling; a high value (> 377 Ω) indicates a large change in voltage and capacitive coupling. At very high frequencies (> 20 MHz), a ratio near 377 Ω indicates electromagnetic coupling.

$$\text{Diagnostic ratio or pseudo impedance} = (\mathrm{d}v/\mathrm{d}t)/(\mathrm{d}i/\mathrm{d}t).$$

Example 6.5.1 An arc welder on the end of a robotic arm generates noise interference in a local embedded controller. The welder produces 120 A at 12 V. Since the unit time difference (Δt or dt) is identical for both voltage and current, the pseudo impedance is the ratio of voltage to current: 12/120 = 0.1 Ω. A low value indicates that inductive coupling is the mechanism of noise interference.

Example 6.5.2 An op-amp circuit is receiving noise interference from a nearby digital switching circuit. The digital circuit switches logic levels between 4.5 V and 1.0 V within 10 ns. Its current changes from 0 to 10 mA within 100 ns. The pseudo impedance is $[(4.5 - 1.0)/10^{-8}]/[10^{-2}/10^{-7}] = 3500\ \Omega$. A high value indicates that capacitive coupling is the mechanism of noise interference.

6.5.1 *Conductive Coupling*

The conductive mechanism for energy coupling requires a connection between the source and receiver that completes a continuous circuit. Fig. 6.9 provides a schematic of the conductive mechanism for energy coupling. Many times these connections are inadvertent and difficult to find; such connections are called *sneak* circuits.

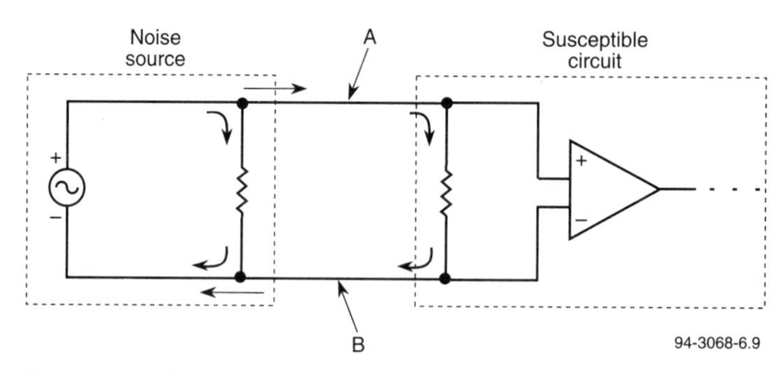

Fig. 6.9 Conductive coupling. If either connection A or B is removed, the conductive noise is eliminated.

A *ground loop* is a complete circuit that allows unwanted currents to flow into the ground (Fig. 6.10). Substantial current in a ground path (as opposed to a return path) can produce voltage differences across the ground resistance and raise the ground potential at the loads. Conversely, significant potentials in the ground can force unwanted current to flow between circuits.

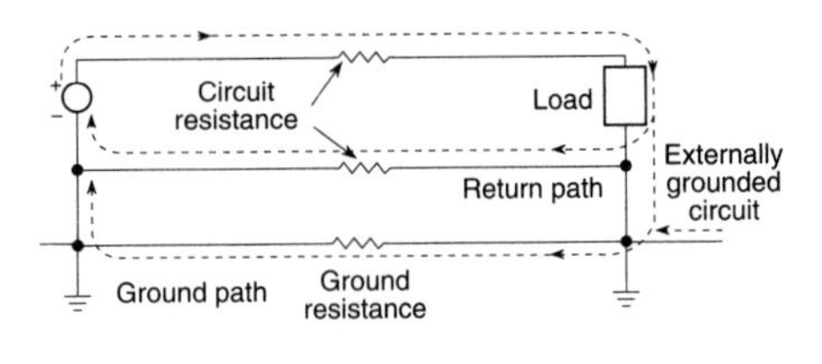

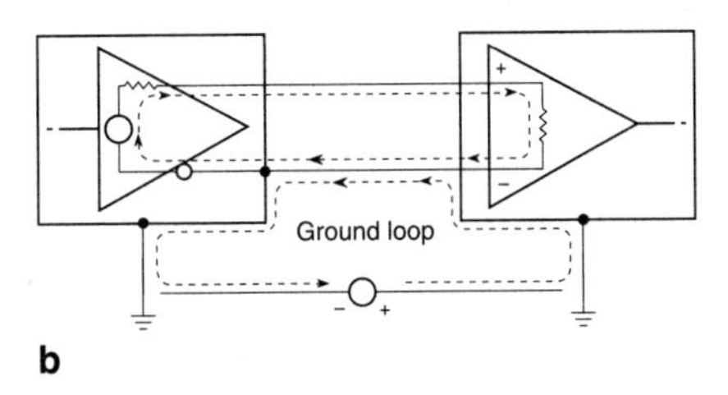

FIG. 6.10 A ground loop. Multiple ground connections can provide multiple return paths that cause significant current flow in the grounding structure, as shown in part (a). Ground potential causes currents in ground loops that unbalance circuits, as shown in part (b).

6.5.2 *Inductive Coupling*

An inductive coupling mechanism requires a current loop that generates a changing magnetic flux. Generally a current transient creates the changing magnetic flux, as follows:

$$\Phi = BA = \mu_0 nIA$$

and

$$v = \mathrm{d}\Phi/\mathrm{d}t = A(\mathrm{d}B/\mathrm{d}t) = A\mu_0 n(\mathrm{d}i/\mathrm{d}t),$$

where Φ = magnetic flux
B = magnetic field
A = loop area
μ_0 = permeability of free space
n = number of turns in the loop
i = current
v = voltage

Clearly, the induced voltage in a magnetically coupled circuit is proportional to the time rate of change of current and loop area. Reducing either the current or the area will reduce the voltage. Since inductance is a function of loop area, reducing the loop area will reduce the inductive reactance of a circuit (see Fig. 6.11).

Remember that current follows the path of lowest impedance, not necessarily lowest resistance. Therefore current will follow the path of minimum inductive reactance; this means that current flow will minimize loop area in a circuit (Fig. 6.12).

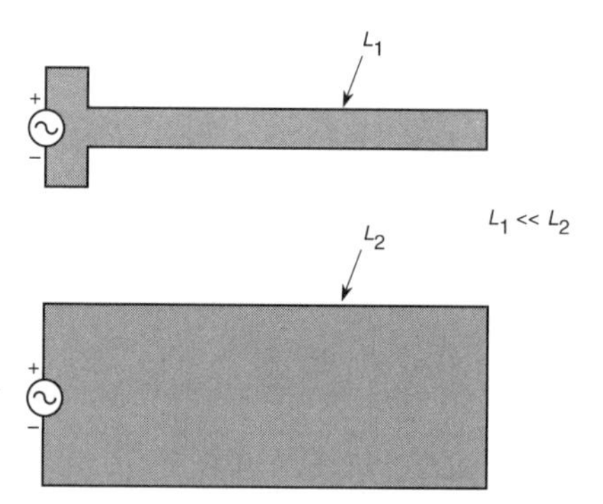

FIG. 6.11 Reducing loop area. Small loops of current have lower self-inductance and thus lower impedance.

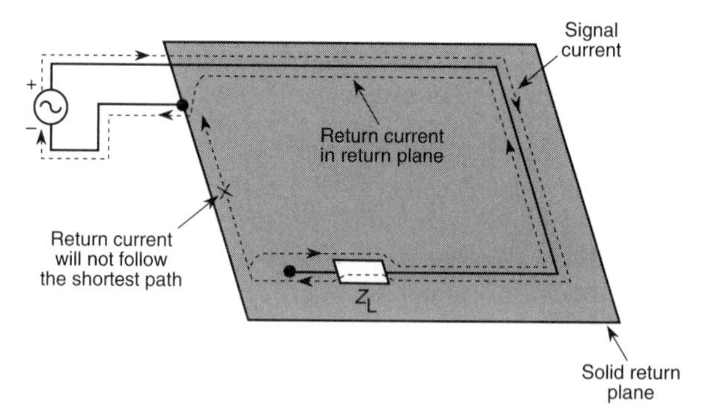

FIG. 6.12 Path of return current. Return current will follow the path of least impedance (in this case, the least self-inductance), not necessarily lowest resistance.

A slot in the ground plane of a circuit board will increase the loop area of a circuit (Fig. 6.13). Avoid such slots because they increase the chance of noise coupling into other circuits.

One of the most common areas for inductive coupling is in cables that connect circuit boards. The long, straight wires encompass significant loop area that provides an inductive reactance. Twisting the pairs of signal and return lines together eliminates the loop area and the mutual inductive coupling between circuits (Fig. 6.14).

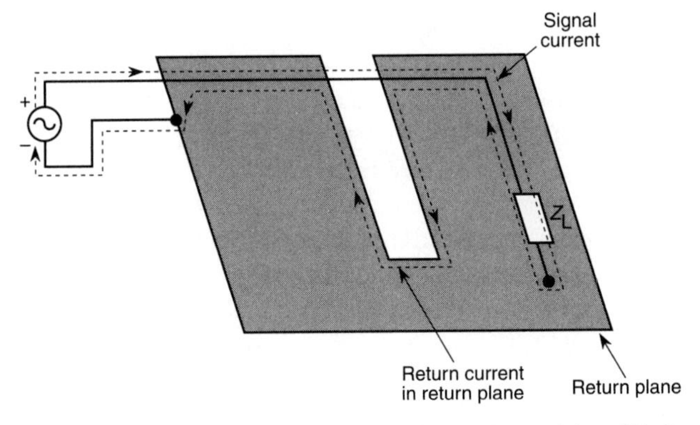

Fig. 6.13 A slot in the return plane increases the current loop area and the self-inductance.

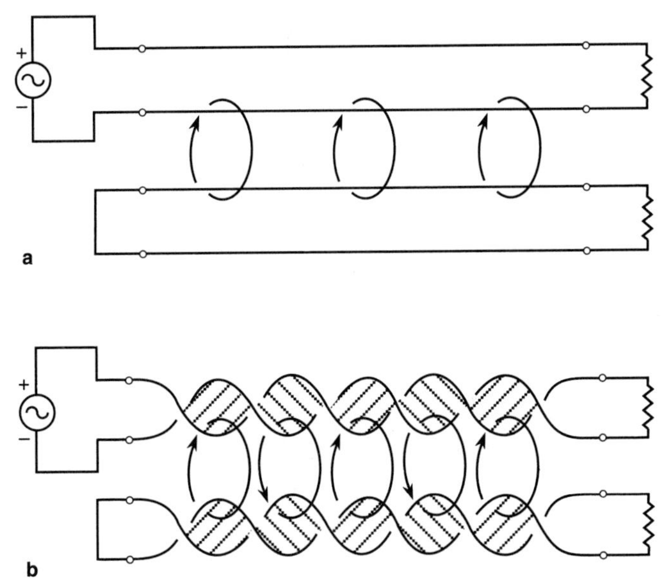

Fig. 6.14 Effect of twisted wire. (a) Straight wires create small loops that can couple magnetically. (b) Twisted wire eliminates the effective loop area of cables and magnetic coupling.

Inductive coupling of noise becomes a factor for frequencies above 3 kHz. Generally, the load impedance is large, while the source impedance is small. Inductive coupling requires a changing current (di/dt) that is large relative to the change in voltage (dv/dt) (Van Doren, 1991):

$$(dv/dt)/(di/dt) << 377\Omega.$$

6.5.3 *Capacitive Coupling*

A capacitive coupling mechanism requires both proximity between circuits and a changing voltage. You can reduce capacitive coupling with separation of conductors and appropriate shielding. Fig. 6.15 illustrates how a properly connected shield can eliminate noise coupled by stray capacitance.

Capacitive coupling of noise becomes a factor for frequencies above 1 kHz. Generally, the total circuit impedance is high; that is, both the source and load impedances are large. Capacitive coupling requires a change in voltage (dv/dt) that is large relative to the change in current (di/dt) (Van Doren, 1991):

$$(dv/dt)/(di/dt) >> 377\Omega.$$

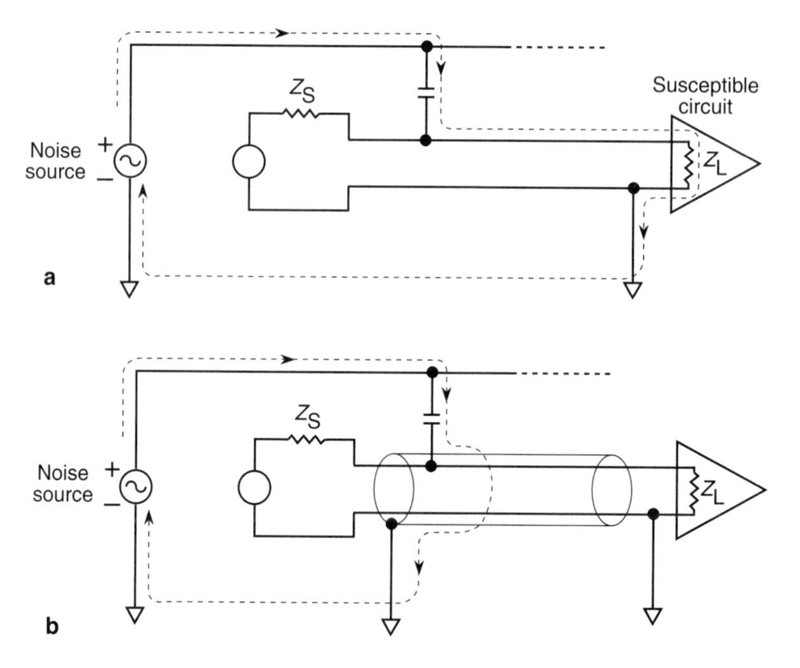

Fig. 6.15 Effect of shielding. (a) Without the shield, stray currents can disrupt susceptible circuits. (b) A properly connected shield can divert capacitively coupled current from susceptible circuits.

6.5.4 *Electromagnetic Coupling*

Electromagnetic, or radiative, coupling becomes a factor only when the frequency of operation exceeds 20 MHz. Below 200 MHz, cables are the primary sources and receivers for electromagnetic coupling; above 200 MHz, PCB traces begin to radiate and couple energy. Signal conductors within circuits must have lengths that are an appreciable fraction of the wavelength to act as antennas. Generally, the length must be longer than 5% of the wavelength ($>\lambda/20$). Radiative or electromagnetic coupling can be diagnosed by a pseudo impedance factor between 100 and 500 Ω (Van Doren, 1991):

$$(\mathrm{d}v/\mathrm{d}t)/(\mathrm{d}i/\mathrm{d}t) \approx 377\,\Omega.$$

6.6 GROUNDING

Grounding provides safety and signal reference. The general principle is to minimize the voltage differential between your instrument and a reference point. Fig. 6.16 diagrams grounding and illustrates two different configurations. Often designers use the return conductors as a signal reference; Fig. 6.17 outlines three configurations for return conductors.

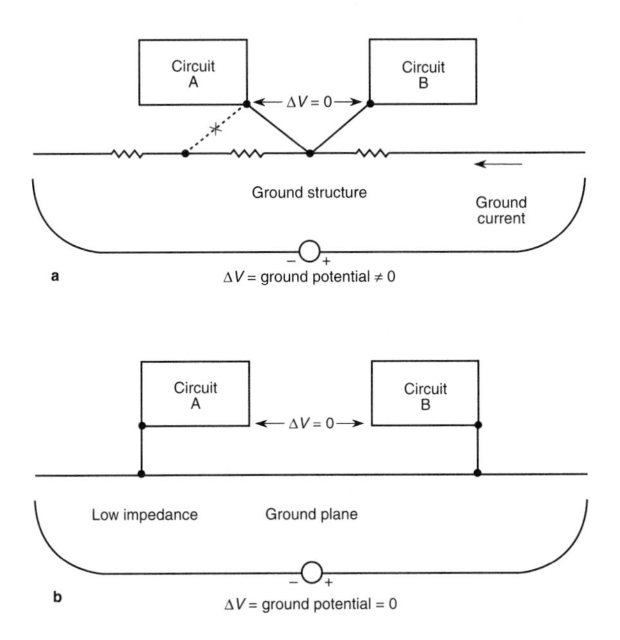

FIG. 6.16 Two grounding configurations. For safety and for signal reference, reduce the potential difference between the ground connections of circuit, instrument, or chassis. (a) Single-point ground. (b) Multipoint ground with low-impedance ground structure.

Safety grounding seeks to reduce the voltage differential between exposed conducting surfaces, while signal referencing seeks to reduce the voltage differential between reference points. Obviously a dynamic tension exists between these concerns: Safety grounding should have many connections between exposed conducting surfaces, and signal referencing should have one connection between reference points at low frequency.

In either case, ground is not the return path for a signal. Both safety and signal grounds nominally conduct no current, whereas a return path routinely conducts current. Unfortunately, return and ground paths are often confused; therefore, be careful when specifying the circuit symbols for each type of ground and return. Table 6.2 outlines the appropriate usage (Van Doren, 1991).

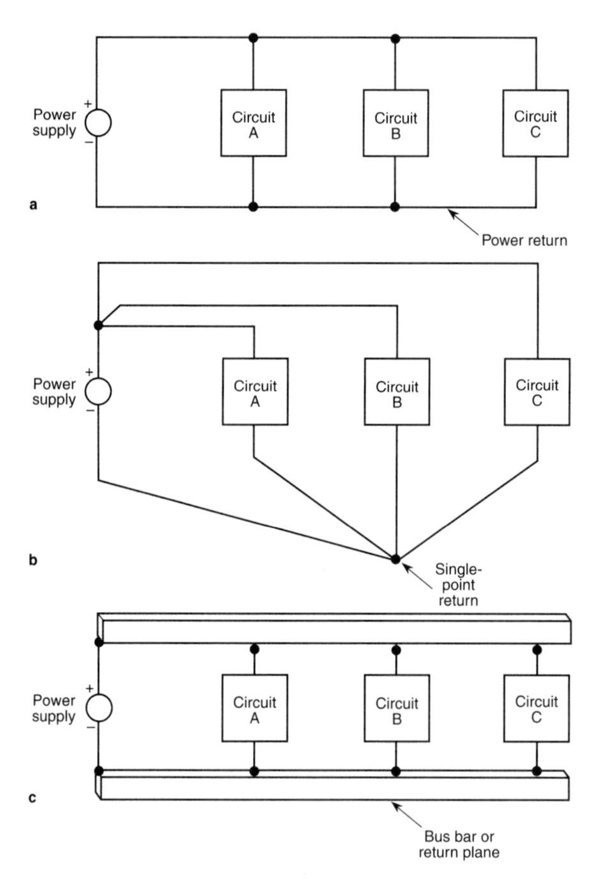

Fig. 6.17 Three return configurations. (a) Series return connection. (b) Single-point return. (c) Multipoint return.

Table 6.2 Ground and return symbols

Symbol	Function	Application
	Safety ground	A connection to an electrical ground structure like building steel or an isolated ground wire
	Signal ground	A connection to a chassis that does not normally conduct current
	Signal return	A conductor that sustains return current for signal or power

6.6.1 *Single-Point Grounding*

In single-point grounding (Fig. 6.18), the separate ground conductors isolate the noise in the return paths of the separate circuits because the single-point reference connection does not complete any ground loops between circuits. Single-point grounding is most appropriate for low-current, low-frequency (< 1 MHz) applications. The ground conductor should be a short strap to reduce high-frequency noise and unsafe voltages.

Single-point grounding has disadvantages. Conductors longer than 5 m (16 ft) are susceptible to high-frequency ground noise. (A braided cable may reduce impedance at high frequencies by increasing the skin effect; that is, current tends to flow along the surface, and braided cable has a large surface area [Williamson, 1986].) Conductors longer than 30 m (100 ft) or those conducting high-fault currents

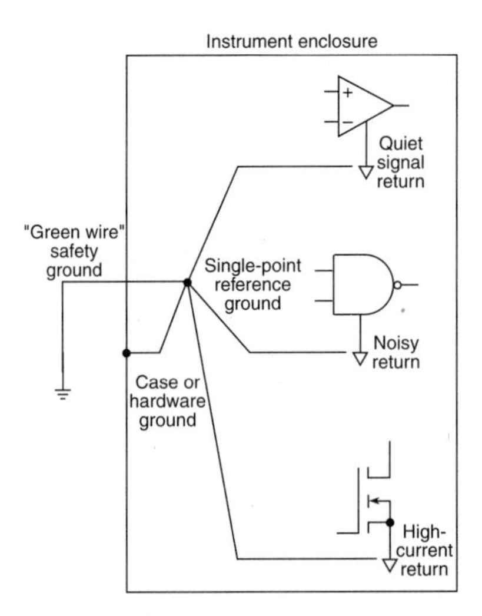

Fig. 6.18 Single-point grounding. Connecting grounds to returns in parallel and then to a single point isolates noise within an instrument. The ground conductors should not carry significant current; that is the function of the return conductors.

are unsafe. The inherent impedance of the conductor will cause large potential differences to exist between the instrument and ground.

The analog-to-digital converter (ADC) is one application that needs a single-point ground for signal reference. Separate references can generate noisy ground loops (discussed below). Fig. 6.19 shows how a single-point ground can reduce noise from the signal reference.

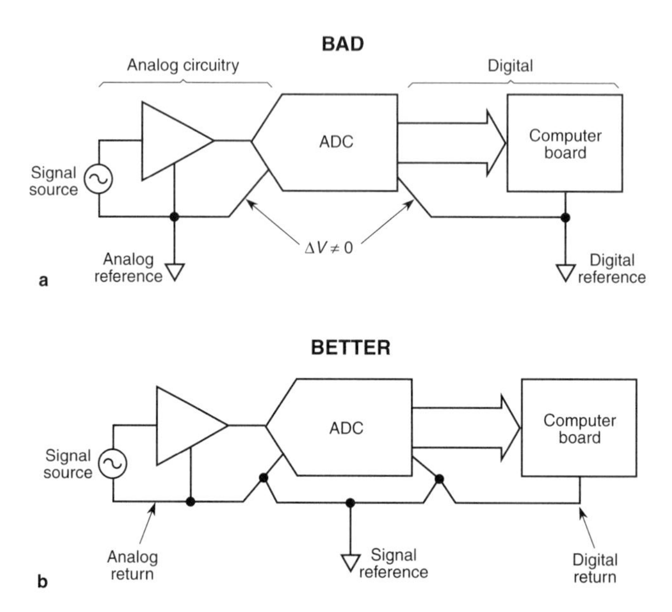

Fig. 6.19 Reducing noise from the signal reference. Tie the analog and digital references together near the ADC. (a) Separate references can lead to large loops of current. (b) A single-point reference eliminates large loops; usually an uninterrupted ground plane suffices.

6.6.2 *Ground Plane or Grid*

A ground plane, within a circuit board, is better for high-frequency (> 100 kHz) operation. Likewise, a ground grid is better for high-frequency or high-fault currents (Fig. 6.20), because it has lower impedance than a single cable.

6.6.3 *Ground Loop*

A ground loop is a complete circuit that comprises a signal path and part of the ground structure. It arises whenever multiple connections to ground are physically separated. External currents in the ground structure generate potential differences between the ground connections and introduce noise in the signal circuit. Fig. 6.21

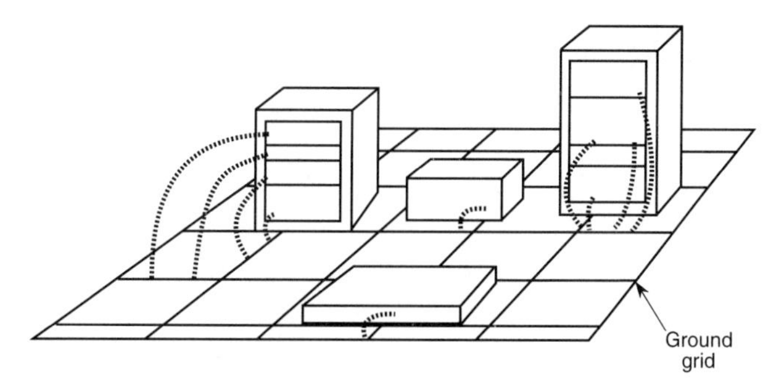

FIG. 6.20 A ground grid. Such grounding is appropriate for high-frequency operation or high-fault currents.

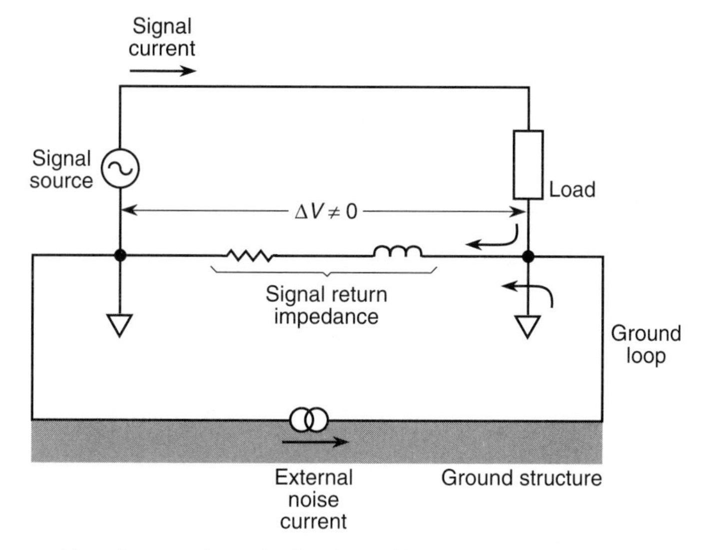

FIG. 6.21 A ground loop. It occurs when a circuit makes multiple connections to the ground conductor or structure.

is a schematic of a ground loop (Bryant, 1991). Generally, the problem arises at low frequencies (< 10 MHz); high frequencies follow the path of minimum impedance that can avoid higher-impedance ground loops.

Ground loops are a particular problem in systems that have low-level signal circuits and multipoint grounds separated by large distances (Ott, 1988). For such systems, either circuit balance or signal isolation can eliminate noise from ground loops (Fig. 6.22). For safety, coordinate the routing of power and signal to reduce noise introduced by the ground structure (Fig. 6.23).

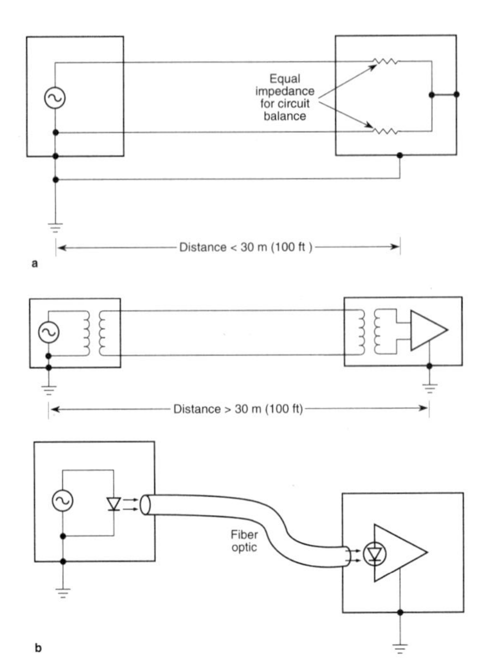

Fig. 6.22 Eliminating ground noise. Circuit balance or isolation removes problems from ground noise. (a) For short distances, a balanced transmission line and single-point grounding reduce noise and safety concerns. (b) For long distances, an isolated signal transmission allows multiple safety grounds while eliminating ground loops.

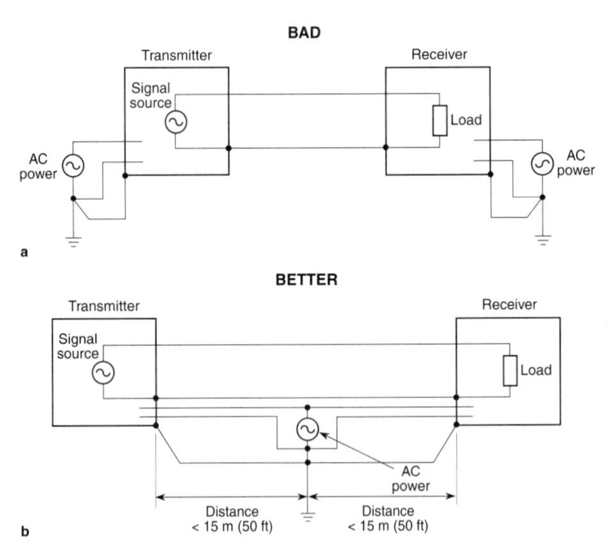

Fig. 6.23 Coordinated routing of power, signal, and safety ground. (a) Separate grounds and power, while safe, introduce noise. (b) For short distances, a coordinated routing of power and ground maintains safety and reduces noise.

6.7 FILTERING

Only filtering reduces conductive noise coupling. A filter can either block or pass energy by three criteria (Van Doren, 1991):

1. Frequency
2. Mode (common or differential)
3. Amplitude (surge suppression)

Most of us are familiar with frequency-selective filters. A low-pass filter passes low-frequency energy and rejects high frequencies, while a high-pass filter passes high-frequency energy and rejects low frequencies. Time-average and time-sync filters are frequency selective as well.

Common-mode noise injects current in the same direction in both the signal and return lines (Fig. 6.24). Fig. 6.25 illustrates a common-mode filter. Differential-mode noise injects current in opposite directions in the signal and return lines.

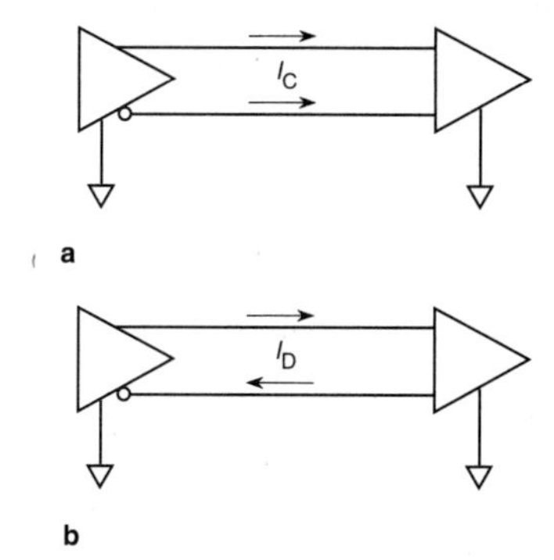

Fig. 6.24 Two modes of operation for noise. (a) Common mode, I_C, is the net current injected in the same direction. (b) Differential mode, I_D, is equal currents injected in opposite directions.

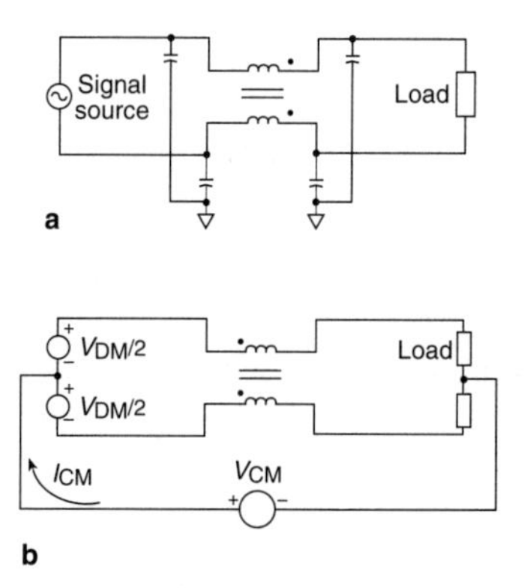

Fig. 6.25 A common-mode filter. The filter either (a) diverts common-mode (CM) noise currents to ground or (b) blocks common-mode currents while passing differential-mode (DM) current.

An amplitude-selective filter generally removes large transients, or *spikes* of noise energy, from a signal line. Surge suppressors that are built into AC power strips are amplitude-selective filters to protect sensitive equipment.

Usually you implement a time-average filter in software to reduce the effect of noise on data within a signal. A time-sync filter prevents interference-sensitive operations from running during a periodic disturbance; for instance, it could suspend circuit function during the periodic transients in a switching power supply.

6.7.1 *Minimize Bandwidth*

A low-pass filter reduces high-frequency emissions and susceptibility for signal applications. Filtering input signals may improve the noise immunity of your circuit.

A digital pulse has a large number of high-frequency components. Any periodic signal may be represented by the sum of sines and cosines of frequencies that are multiples of the fundamental frequency. The sum is called the Fourier series. The coefficients of the Fourier-series expansion represent the magnitudes of the harmonic frequencies. Sharp edges on pulses will have correspondingly large Fourier coefficients. Slowing the rise and fall times of pulse edges will reduce the bandwidth of signals. Fig. 6.26 illustrates the difference in spectral content between a square pulse and a pulse with rounded edges.

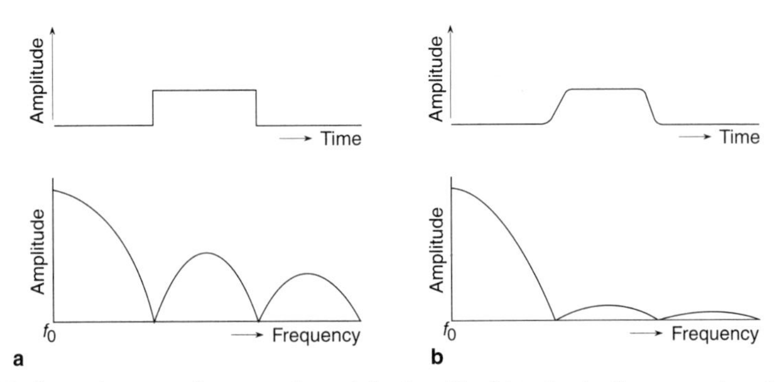

Fig. 6.26 Spectral content of square and rounded pulses. The faster the rise time on a pulse edge, the greater the amplitudes of the harmonics and the greater the potential for noise coupling. (a) Faster edges create greater harmonics. (b) Slower edges reduce the harmonic content.

Fig. 6.27 compares spectral content of several pulse shapes. Filtering your clock signal to reduce the high-frequency harmonics is one area where you may significantly reduce noise interference. But be careful not to violate the minimum slew rate required by the logic circuits.

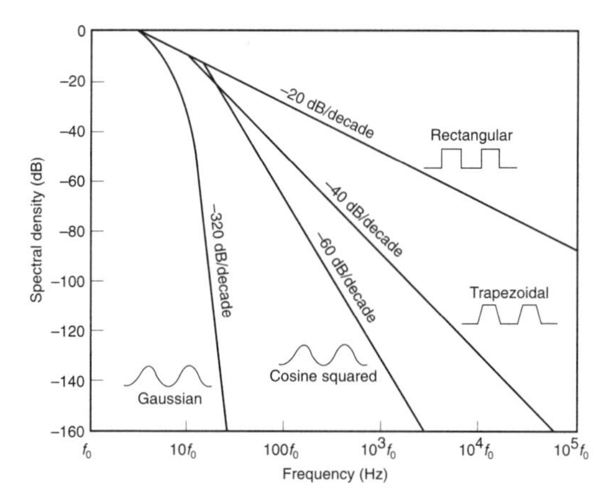

Fig. 6.27 Spectral content of pulse shapes. Pulse shape defines the harmonic content of a signal and its potential for radiating electromagnetic interference. f_0 = fundamental frequency. (Adapted from Brewer, 1991.)

6.7.2 *Ferrite Beads*

Ferrite beads provide one form of filtering based on frequency. A ferrite bead is a magnetically permeable sleeve that fits around a wire. It presents an inductive impedance to signals that attenuates high frequencies. Ferrite beads are best suited to filter low-level signals and low-current power feeds to circuit boards. Fig. 6.28 shows an application for ferrite beads.

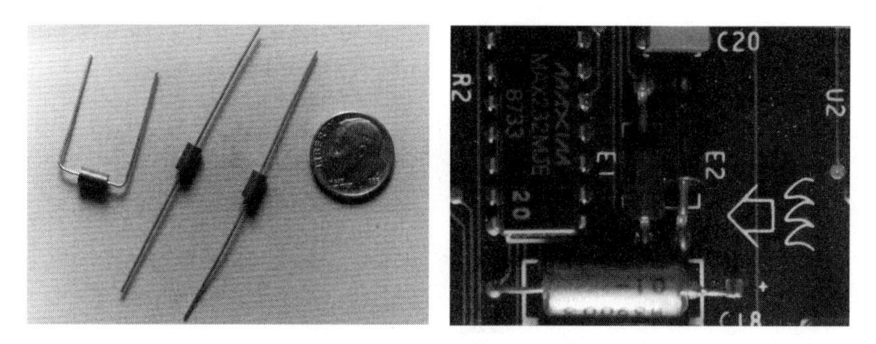

Fig. 6.28 Ferrite beads and an application where they filter RS-232 signals.

6.7.3 *Decoupling Capacitors*

Decoupling capacitors provide filtering based on frequency. They filter and smooth out the spikes in the DC power of ICs. During a logic transition, a momentary short circuit from power to return in a digital device demands a large current transient. A decoupling capacitor can supply the momentary pulse of current, as illustrated in Fig. 6.29, and effectively decouple the switching spike from the power supply.

Inductance in the power supply circuit can accentuate the effect of the switching current transients by producing large voltage spikes. Decoupling capacitors mitigate the effect of inductance by reducing the effective loop area between the power supply circuit and the ICs. In essence, they reduce the impedance of the power supply circuit:

$$Z_0 \approx 0\Omega = \sqrt{R^2 + [\omega L - (1/\omega C)]^2}\ ,$$

where Z_0 = impedance of the power supply circuit
R = resistance of the power supply circuit $\approx 0\ \Omega$
L = inductance of the power supply circuit
C = capacitance of the decoupling capacitor

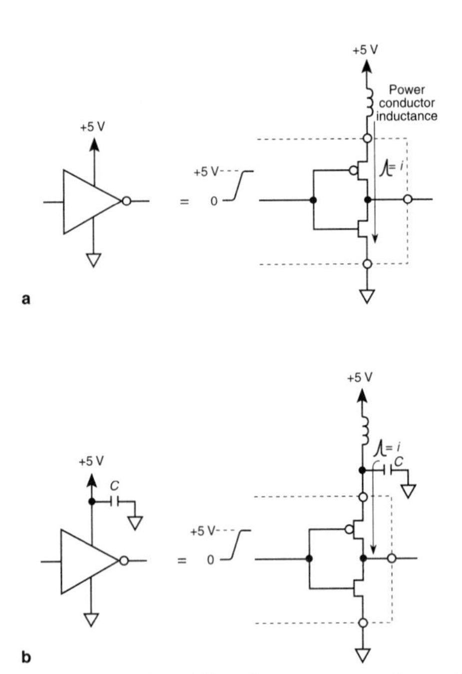

Fig. 6.29 Decoupling, or bypass, capacitors. They filter current transients when transistors switch within a digital logic gate. (a) As input voltage changes logic level, both transistors turn on, momentarily producing a short circuit and drawing a current pulse from the DC power source. (b) A decoupling capacitor can supply the current pulse and reduce the transient propagated through the DC power supply.

But beware of resonance. If you arbitrarily make the decoupling capacitor too large, you will move the resonance frequency of the supply inductance and decoupling capacitor down into the range of operation of your circuit and cause excessive ringing in the supply. Besides, large capacitors have larger parasitic inductances than smaller decoupling capacitors.

6.7.4 *Line Filters, Isolators, and Transient Suppressors*

Line filters and isolators select energy on the basis of mode, while transient suppressors select energy on the basis of amplitude. A common-mode filter for AC power lines (Fig. 6.25) diverts noise to ground, but beware of polluting the signal-reference ground with noise (St. John, 1992). An optoisolator can eliminate common-mode noise by interrupting the conductive path. A differential-mode filter has to separate noise from signal by criteria other than current direction; a low-pass filter is an example of a differential-mode filter that uses frequency as the selection criterion.

Large transients generated by lightning or machinery and conducted by the AC power circuit deserve special attention. Fig. 6.30 demonstrates how machinery switching on or off produces transients through inductive "kick." The opening or closing of switches changes the load current instantaneously and generates a sizable voltage across the line inductance that affects the other loads.

Transient protection can take one of four approaches: filter, crowbar (thyristor), arcing discharge, or voltage clamp (Zener diode or metal-oxide varistor [MOV]). A filter removes the high-frequency components of the energy associated with the sharp edge of a spike. Consequently, the peak of the spike is flattened. A crowbar circuit detects an overvoltage and short-circuits current until the input voltage is cycled off and on again. Arcing discharge occurs across the gap in a gas tube. The initial breakdown of the gas requires a fairly high voltage, but once the arc is established the holding voltage is much lower. A voltage clamp shorts the excess energy to prevent an overvoltage condition. Fig. 6.31 outlines these approaches.

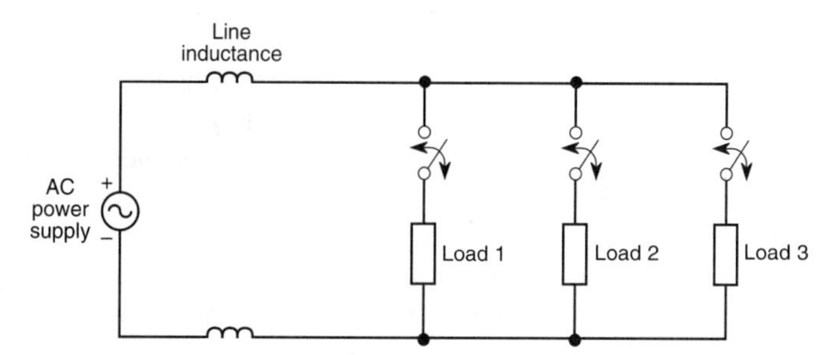

Fig. 6.30 Inductive kick. Switching loads and machinery generate the kick by instantaneously changing the load current.

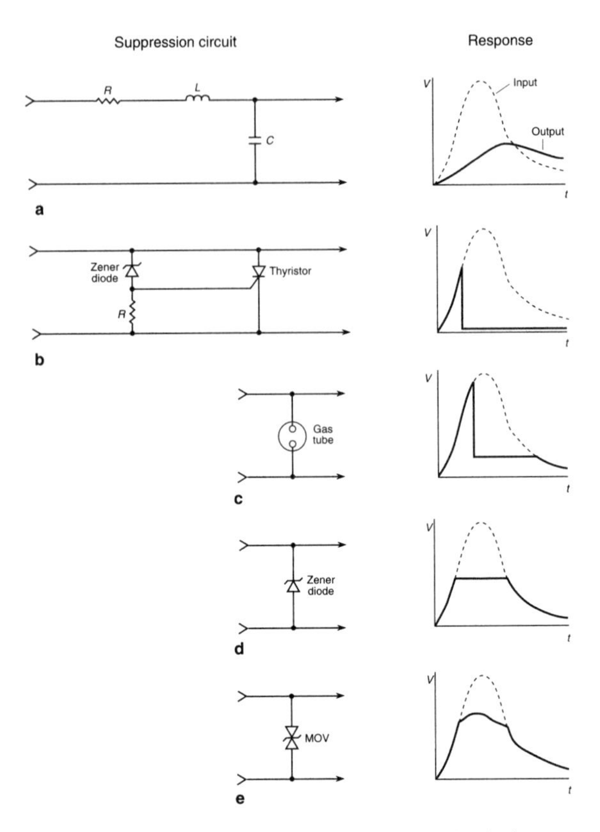

FIG. 6.31 Five ways to suppress transients. (a) Filter. (b) Crowbar circuit shorts output until the input is cycled off and on. (c) Gas tube uses a breakdown phenomenon to short current. (d) Zener diode and (e) MOV are voltage clamps.

Filters and crowbar circuits have very specific applications. Crowbar circuits, for instance, are used in the outputs of power supplies (see Chapter 10). They require more design and assembly and are more expensive than a voltage clamp like a Zener diode or MOV.

Gas tubes are used primarily in telephone circuits to suppress surges caused by lightning. They can sustain very large currents. They cannot protect against electrostatic discharge from human contact or handling and are not useful for overvoltages in AC power lines.

MOVs are low-cost devices with high current capacity. They have a limited life before wearing out, though, and require replacement more often than diodes. Zener diodes are faster MOVs, but they cost more and have a lower current capacity.

6.8 SHIELDING

Shielding either prevents noise energy from coupling between circuits or suppresses it. The energy coupling may be through magnetic flux, electric field, or electromagnetic wave propagation. Because prevention is cheaper and more effective than suppression, I will concentrate on shielding that prevents noise coupling.

6.8.1 *Inductive Shielding*

Inductive shielding is concerned with self-inductance and mutual inductance. It reduces noise coupling by reducing or rerouting magnetic flux. Magnetic noise coupling depends on the loop area and current within both the emitting and receiving circuits, as outlined in Section 6.4.

The most effective inductive shielding minimizes loop area. Separating circuits and reducing the change in current help, while metal, or magnetically permeable, enclosures place a distant third in usefulness. Furthermore, the cost of manufacturing makes magnetically permeable enclosures an even less desirable solution for inductive shielding.

Twisting the signal and return conductors in a cable reduces the mutual inductance and improves the shunt capacitive balance (and common-mode coupling) as already seen in Fig. 6.14. Coaxial cable, or coax, also has minimal loop area and may be preferable for higher frequencies (> 1 MHz) because it provides both good capacitive shielding and a controlled impedance.

Be careful to twist the correct wires together. Always pair signal with return; otherwise, you will not gain any inductive shielding (Fig. 6.32). Fig. 6.33 illustrates the best application of twisted-pair cable; reducing the separation between signal and return lowers both self-inductance and mutual inductance.

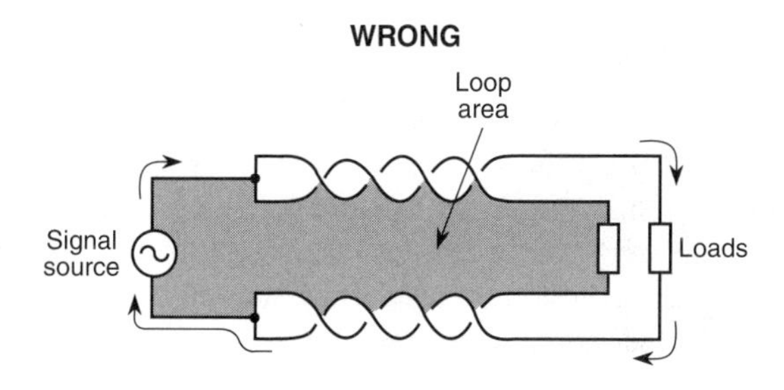

Fig. 6.32 Incorrect application of twisted-pair wire. A large loop area still exists.

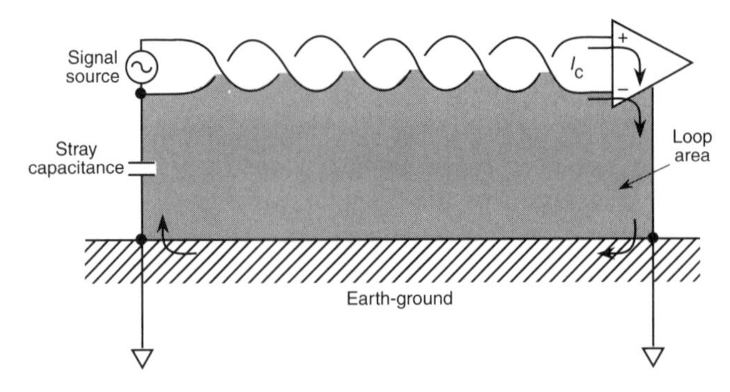

Fig. 6.33 Twisted-pair cable. Twisting the wire and running it close to the ground will reduce the common-mode current, I_C, by reducing the loop area for inductive coupling.

> **Example 6.8.1.1** Lightning damage to a burglar alarm system was caused by inductive coupling because the area of the sensor loops allowed the associated magnetic flux to induce a large transient of current (Doubek, 1991). Twisted-pair cable would have eliminated the possibility of damaging currents that couple magnetically.

Signal routing on circuit boards needs your careful consideration also. Make sure that the return path is always under the signal conductor to minimize loop area. Slots in ground planes can significantly increase inductive coupling of noise by increasing the loop area of the signal path, as already shown in Fig. 6.13.

Enclosures provide magnetic shielding by allowing eddy currents to reflect or absorb interference energy. These enclosures are heavy, expensive, and frequency dependent compared with the inductive shielding described above, but sometimes they are the only solution.

6.8.2 *Capacitive Shielding*

Capacitive shielding reduces noise coupling by reducing or rerouting the electrical charge in an electric field. Capacitive shields shunt to ground charge that is capacitively coupled, as shown in Fig. 6.34 (and previously in Fig. 6.15).

At low frequencies (< 1 MHz), you should connect a capacitive shield at one point if the signal circuit is grounded as in Fig. 6.34d. Multiple connections can form ground loops. Furthermore, you can improve capacitive shielding by reducing the following:

1. Noise voltage and frequency
2. Signal impedance
3. Floating metal surfaces

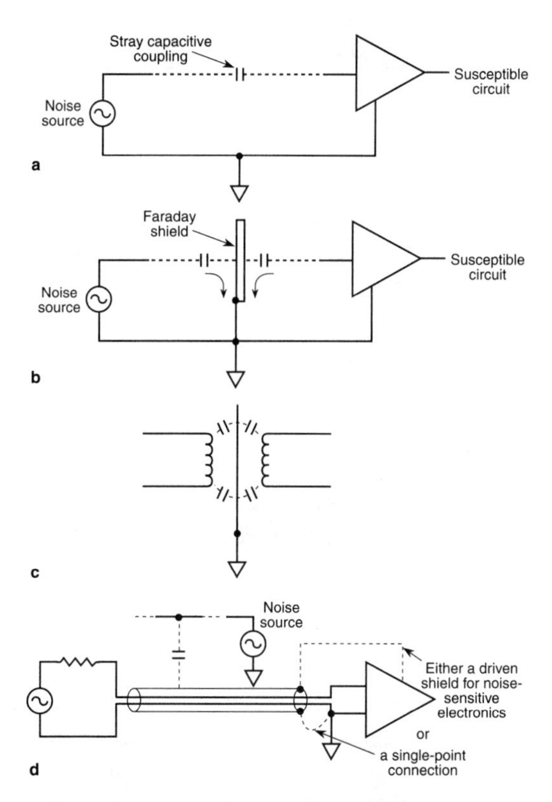

FIG. 6.34 Capacitive shielding. Capacitive coupling provides a path for the injection of noise charges, while an appropriately placed shield prevents the coupling between circuits by shunting charge to ground. (a) Outline of capacitive coupling. (b) Proper placement of a shield to shunt noise charge. (c) A shielded transformer prevents capacitive coupling between windings. (d) A cable shield usually should be connected at only one place to prevent coupling or shunt charge.

Conversely, multiple ground connections are necessary for high frequencies (> 1 MHz). Stray capacitance at the ungrounded end of a shield can complete a ground loop. Therefore, you should ground both ends of a long (relative to wavelength) shield (Ott, 1988).

A metal enclosure can be an effective electrostatic shield, or Faraday cage. It prevents capacitive coupling and need not be grounded if it completely encloses a circuit. Grounding, however, ensures that current from stray capacitance flows to the signal reference ground rather than feeds back and causes crosstalk.

6.8.3 *Electromagnetic Shielding*

Electromagnetic shielding reduces emissions and reception. Emission sources include lightning, discharges, radio and television transmitters, and high-frequency circuits, as outlined in Table 6.3 (Doubek, 1991).

Table 6.3 Range of some possible emission sources

Emitter	Frequency range (MHz)	Comments
AM radio	0.5–1.6	
FM radio	88–108	
VHF television	54–126	
UHF television	470–890	
Citizens band radio	27	Very low power
Radio-control models	27, 49, 72	Very low power
Radar	150–10,000	Some of these can be very powerful sources.
Computers	5–200	Improper cabling can often be the antenna.

Electromagnetic interference (EMI) always begins as conductive (current in wires), becomes radiative, and ends as conductive (fields interact with circuitry), as illustrated in Fig. 6.8. Several techniques can reduce EMI:

1. Reduced bandwidth (longer wavelengths)
2. Good layout and signal routing
3. Shielded enclosures

A shielded enclosure should ideally be a completely closed conducting surface. You can think of an effective enclosure as one that has watertight metallic seams and openings. Openings can leak electromagnetic radiation. Openings include cooling vents, cable penetrations with slots larger than a fraction of a wavelength (> $\lambda/20$), push buttons, and monitor screens (Fig. 6.35).

In the same spirit, cable shields must seal completely around each connector. Fig. 6.36 outlines a good electromagnetic seal for high-frequency applications.

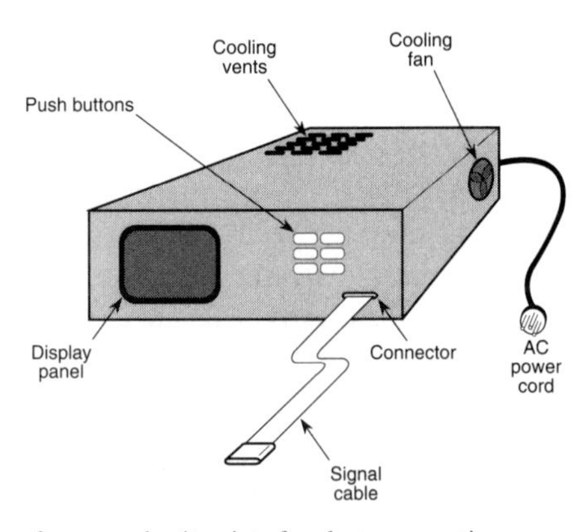

Fig. 6.35 Examples of entry and exit points for electromagnetic energy.

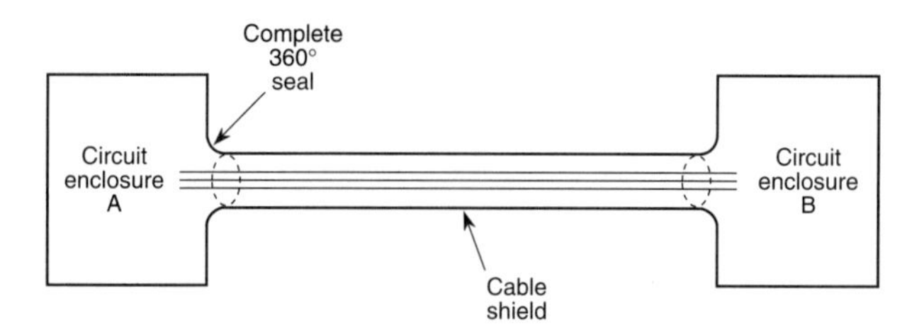

Fig. 6.36 Electromagnetic seal. A cable and enclosure must be completely sealed to prevent leaks of electromagnetic energy, in or out.

6.8.4 *Some Practical Applications*

Undoubtedly, some of you are asking questions like "OK, do I use twisted-pair or coax cable?" or "Do I ground both ends of a shield or not?" The answer is "It depends." Your particular application and frequency of operation will largely determine the necessary shielding.

Twisted-pair cable is usually effective up to 1 MHz. It becomes lossy at higher frequencies. Its advantage over coax is that it is cheaper and mechanically more flexible. Coax, on the other hand, has low loss and less variance in characteristic impedance from DC to very high frequencies (> 200 MHz).

Ott (1988) presents an interesting comparison between cables with various ground and shield connections. Simple twisted-pair cable attenuates noise reasonably well. Ultimately, a single ground connection to both the shield and return line provides the best attenuation of the 50–kHz noise. Fig. 6.37 illustrates the preferred cable schemes for low frequencies.

But as frequency increases, coax cable emerges as the best candidate. A pigtail connection from the shield to ground presents a loop inductance that increases impedance with frequency. Eventually, high frequencies (> 10 MHz) demand a complete 360° seal of the shield at both ends (Fig. 6.36). Fig. 6.38 illustrates the

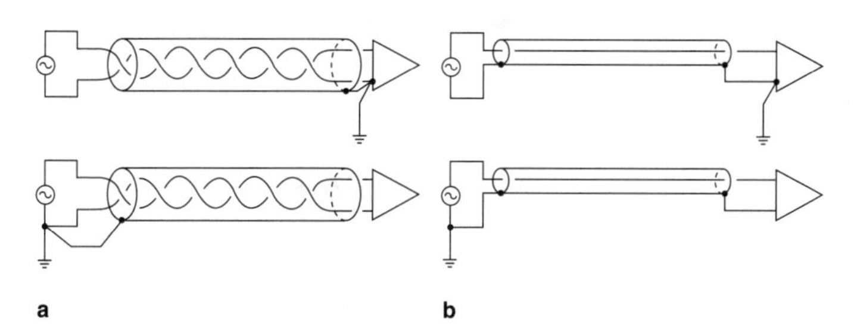

Fig. 6.37 Cable configurations that best attenuate low-frequency noise (< 1 MHz). (a) Shielded, twisted pair. (b) Coaxial cable. (Modified from Ott, 1988.)

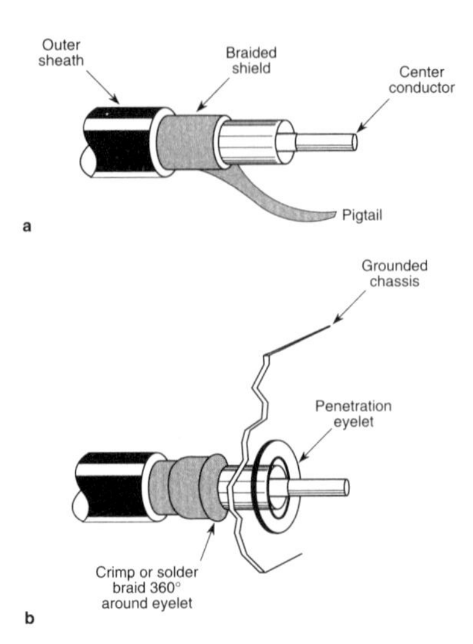

Fig. 6.38 Two termination methods for a coaxial cable shield. (a) The pigtail connection is fine for low-frequency applications. (b) The complete 360° conducting seal is necessary for high frequencies.

difference between a pigtail connection of a shield and a terminated shield with 360° contact.

Ribbon cable is ubiquitous in instrumentation. It is suitable for low-frequency operation (< 1 MHz), and the same rules apply to ribbon cable as to twisted-pair cable. *You should pair each signal with a return conductor or use a return plane for low-level signals or higher frequencies.* Fig. 6.39 illustrates three possible configurations for ribbon cable.

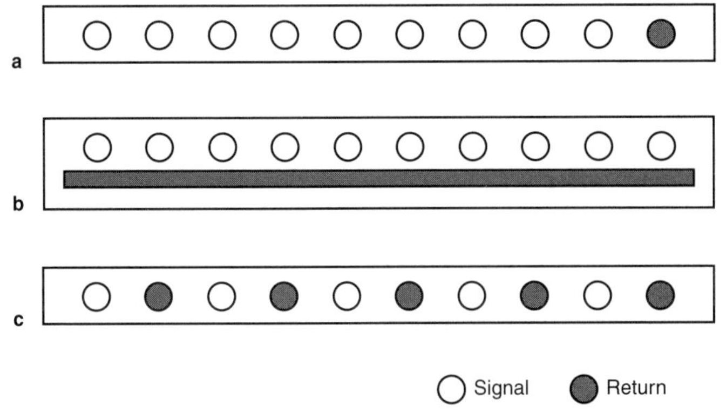

Fig. 6.39 Three possible configurations for ribbon cable. (a) A single return line is good only for low-frequency operation. (b) A return plane supports return currents that minimize the inductive loop area and is okay for high frequencies. (c) Paired signal and return lines allow higher frequencies because they greatly reduced the inductive loop area; twisting each pair is even better.

Enclosures are a singular challenge to seal against electromagnetic radiation and reception. You need to consider all openings, grounding, and the appropriate regulations. Several of the references will help you with the multitude of details (Denny, 1983; Hemming, 1992; IEEE, 1992).

Any opening into the interior of the enclosure is a possible electromagnetic leak. You will have to seal all seams, penetrations, and doors to prevent leaks. Seams must be soldered, welded, or overlapped. Penetrations such as vents and cables need appropriate filters and shields. A honeycomb matrix in vents, for example, acts as a waveguide to filter electromagnetic radiation. Access doors usually leak, particularly at the corners, unless they have good EMI gaskets. Gaskets may be made from a variety of materials, such as phosphor bronze finger stock or copper-beryllium mesh embedded in an elastomer bead.

6.9 PROTECTING AGAINST ELECTROSTATIC DISCHARGE

Electrostatic discharge (ESD) is a discharge at very high voltage and very low current that readily damages sensitive electronics. ESD can range from hundreds to tens of thousands of volts. Any instrument containing integrated circuits is susceptible if not protected.

ESD transfers electrical charge in three stages: pickup, storage, and discharge. Some sources of charge are listed in Table 6.4. Usually, mechanical rubbing between dry, insulated materials transfers the charge from the source to storage. Often the storage medium is a person, who then unwittingly delivers the damaging discharge. Proximity or physical contact discharges the charge from storage. Several conditions, including humidity, speed of the activity, and material, affect the charge transfer (Table 6.5).

The discharge waveform of ESD has a fast rise time and short duration. Fig. 6.40 illustrates a sample waveform for simulating ESD while testing products. You can test your products with several types of ESD simulators for static and dynamic electric fields and dynamic magnetic fields to determine whether they are affected or damaged (Clark and Neill, 1992; Richman, 1991).

You can use several schemes, including grounding, shielding, and transient limiters, to protect circuits from ESD. The methods for grounding and shielding already covered in this chapter work well to control ESD. Sometimes shielding can be too expensive or cumbersome, so the next step is to limit ESD transients that couple through cables and ground connections. Input gates are the most susceptible to damage, so you should use surge limiters on input lines (Fig. 6.41). (Most integrated circuits incorporate reverse-biased diodes connected between the input and either the ground or the power. These diodes divert high-voltage transients from input gates, but they have a limited current capacity.)

Generally you can choose between Zener diodes and MOVs to limit surges. Zener diodes tend to turn on faster, while MOVs are cheaper and handle larger peak current (Demcko, 1991).

Table 6.4 Sources of electrostatic charge

Object or process	Material or activity
Work surfaces	• Waxed, painted, or varnished surfaces • Common vinyl or plastic
Floors	• Sealed concrete • Waxed, finished wood • Common vinyl tile or sheeting
Clothes	• Virgin cotton • Wool • Common synthetic personnel garments • Nonconductive shoes • Common clean room smocks
Chairs	• Finished wood • Vinyl • Fiberglass
Packaging and handling	• Common plastic bags, wraps, envelopes • Common bubble pack and foam • Common plastic trays, tote boxes, vials, part bins
Assembly, cleaning, testing, and repair	• Spray cleaners • Common plastic solder suckers • Solder irons with ungrounded tips • Solvent brushes (synthetic bristles) • Cleaning by fluid or drying by evaporation • Temperature chambers • Cryogenic sprays • Heat guns and blowers • Sandblasting • Electrostatic copiers

Table 6.5 Typical electrostatic voltages

Static generation	Electrostatic voltages	
	10% relative humidity	65% relative humidity
Walking across carpet	35,000	1,500
Walking over vinyl floor	12,000	250
Common plastic bag picked up from bench	20,000	1,200
Work chair padded with polyurethane foam	18,000	1,500

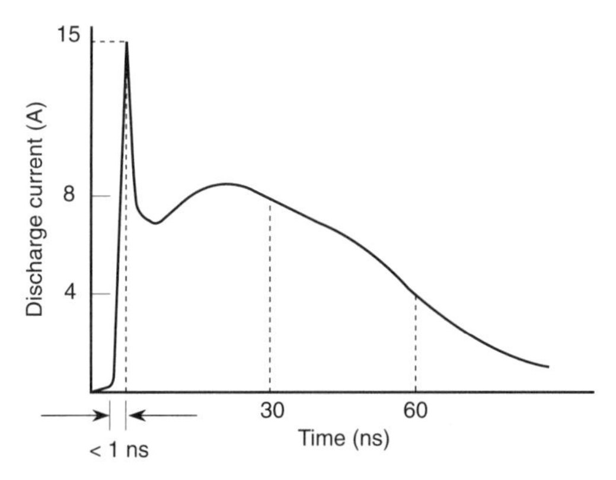

Fig. 6.40 Discharge waveform at 4 kV according to IEC 801-2 regulations. This is one example of several different waveforms for testing protection circuits against ESD.

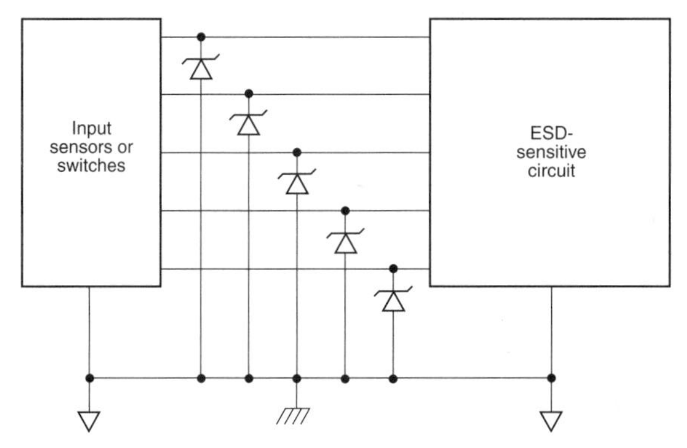

Fig. 6.41 Preventing damage by shunting high-voltage transients away from circuits with Zener diodes or MOVs.

Example 6.9.1 Management in a hotel called in a consultant because the new electronic locks on the room doors were failing. Guests were walking through carpeted hallways to reach their rooms and building up charge on their bodies. When they stuck the plastic keys into the electronic locks on the doors, ESD would sometimes destroy the electronics. The consultant found that the circuitry had no surge protection at all. He then realized that the locks had been designed in Japan, where the humidity is always high and they do not have that particular ESD problem. Had the designers better understood the actual operating environment, they would have been able to avoid the problems from ESD during the initial design.

Beyond surge limiters, prevention is the best medicine. Proper handling of components will reduce the chance of ESD. *In general you want to eliminate the activities and materials that create high static charge.* Control methods include the following:

- Grounding
- Protective handling
- Protective materials
- Humidity

Fig. 6.42 is a checklist to make work areas less prone to ESD. Fig. 6.43 diagrams a workbench that grounds ESD.

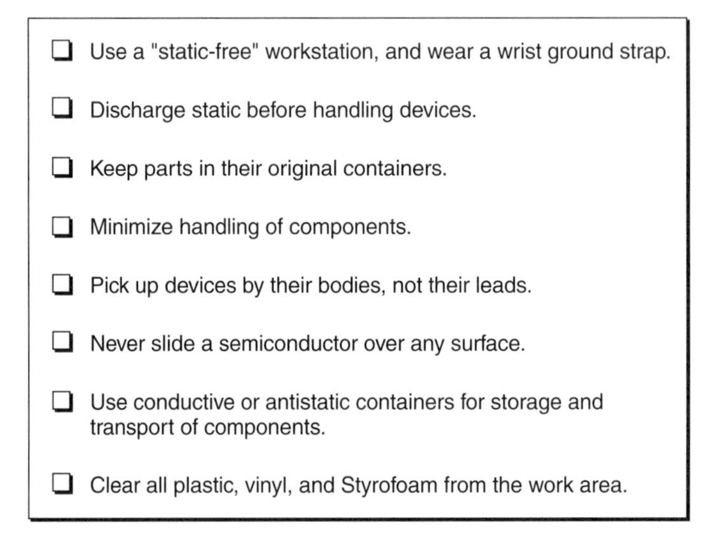

- ❑ Use a "static-free" workstation, and wear a wrist ground strap.
- ❑ Discharge static before handling devices.
- ❑ Keep parts in their original containers.
- ❑ Minimize handling of components.
- ❑ Pick up devices by their bodies, not their leads.
- ❑ Never slide a semiconductor over any surface.
- ❑ Use conductive or antistatic containers for storage and transport of components.
- ❑ Clear all plastic, vinyl, and Styrofoam from the work area.

Fig. 6.42 Checklist for an ESD-safe work area.

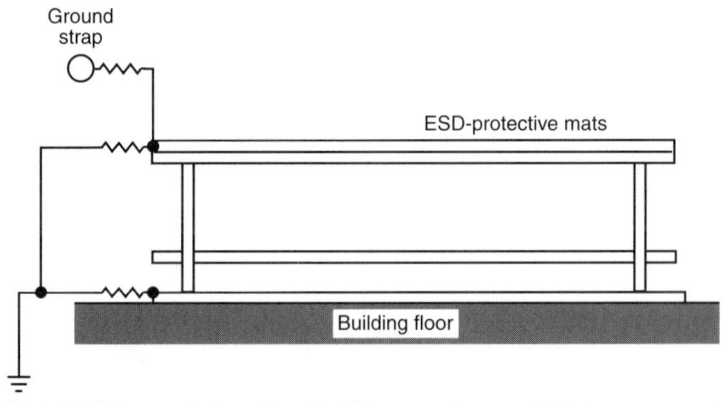

Fig. 6.43 Typical ESD-grounded workbench. This type of setup will help protect static-sensitive components.

6.10 GENERAL RULES FOR DESIGN

When you design and develop a product, you must include grounding and shielding. You may follow these general guidelines: Characterize your system, conform to the appropriate standards, and use good design techniques.

6.10.1 *System Characterization*

Follow these steps to characterize your system design:

1. Establish the following:
   a. Grounding options, both available and needed
   b. Source and load impedances
   c. Frequency bandwidth
2. Determine possible coupling mechanisms and their locations (see Fig. 6.35).
3. Diagram the topology of the circuit paths and reduce the loops:
   a. Ground loops
   b. Inductive loops in signal and power circuits

Several diagnostics can suggest the coupling mechanisms. Table 6.6 lists some clues to the coupling mechanisms.

Table 6.6 Summary of noise-coupling mechanisms and their associated diagnostic clues

Noise-coupling mechanism	Diagnostic clues
Conductive	1. Complete conductive path such as multiple ground connections
	2. Frequency < 10 MHz
Inductive	1. $\frac{dv/dt}{di/dt} << 377\ \Omega$
	2. Ringing on pulse edges
	3. Low source impedance and high load impedance
	4. Laying conductive foil over cable decreases noise level (acts as an eddy current shield)
	5. Large circuit loops
Capacitive	1. $\frac{dv/dt}{di/dt} >> 377\ \Omega$
	2. Rounding of pulse edges
	3. High source impedance and high load impedance
	4. Laying conductive foil over cable increases noise level
Electromagnetic	1. Distance between noise source and susceptible circuit > λ
	2. Frequency > 15MHz

Example 6.10.1.1 A new-model sports car had a disconcerting problem: occasionally the dashboard lights would all illuminate simultaneously. Two service calls later, replacement of a wire harness for the spark plugs solved the problem (Schroeder, 1992). What was the coupling mechanism? The engine compartment and dashboard were separated and have only auto chassis between, so conductive coupling from the shorted cable probably did not explain the entire problem. An ignition wire conducts very high voltage transients at low current—an indication of capacitive coupling. Most likely the transients flowed in the chassis and then coupled through stray capacitance into sensors or cables that led to the dashboard computer to cause the false indications.

Example 6.10.1.2 A motorcycle drove by my house one evening while I was watching television. The picture was interrupted by groups of lines marching across the screen. The interference faded as the motorcycle passed. What was the coupling mechanism? Obviously no physical connection existed between my television and the motorcycle, so the coupling could not be conductive. The distance between the television and the motorcycle was about 15 m (50 ft). The rise time on an arc discharge in a spark plug may be on the order of 1 ns, indicating a maximum frequency of about 250 MHz. Considering the distance and the frequency of reception for VHF television (see Table 6.3) electromagnetic coupling is a good bet.

6.10.2 *Standards*

You will undoubtedly encounter regulations and standards in whatever market your product will compete in. It will have to meet or surpass the limits of emission or susceptibility in both conducted and radiated environments. Table 6.7 lists some of these regulations.

Commercial regulations cover the frequency bandwidth from 9 kHz to 1 GHz. Fig. 6.44 illustrates the limits for radiated emissions.

6.10.3 *Procedure*

Good design techniques for grounding and shielding have a few basic rules:

1. Reduce frequency bandwidth.
2. Balance currents.

Table 6.7 Examples of regulations that you may encounter while developing electronic products

Type	Regulation	Country or countries of jurisdiction
Commercial	CNELEC	Europe
	IEC 801	Europe
	IEC 871	Europe
	FCC	United States
	VCCI	Japan
	VDE	Germany
Military	GAM-EG-13	France
	VG NORM	Germany
	DEF STAN 59-41	United Kingdom
	MIL-STD-461D	United States

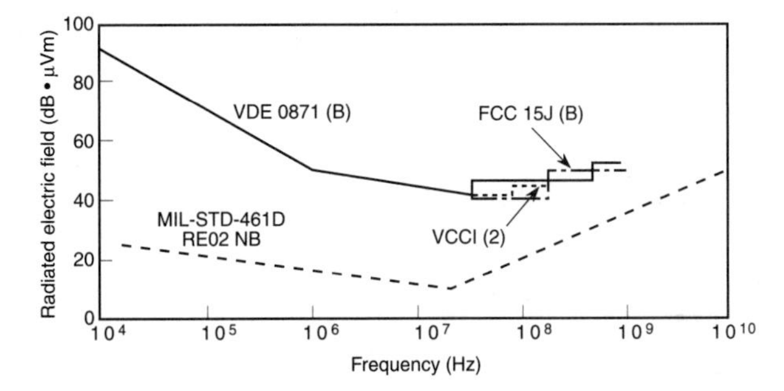

Fig. 6.44 Example of the limits for radiated emissions for several different sets of regulations.

3. Route signals for self-shielding; in particular, circuit boards should have the following:
   a. A return (ground) plane
   b. Short traces
   c. Decoupling capacitors
4. Add shielding only when necessary.

A kind of reciprocity exists for shielding: Anything that reduces emissions usually makes the circuit less susceptible to disruption. If you decrease the area of a circuit to reduce the inductive loop and emission, for instance, then you will also reduce the mutual inductive coupling that could receive noise.

Figs. 6.45 through 6.47 show some instrumentation that will help you determine the best grounding and shielding configurations experimentally. A close-field probe is useful during the breadboard stage to pinpoint circuits that are radiating. During preproduction you can measure conducted EMI with a line

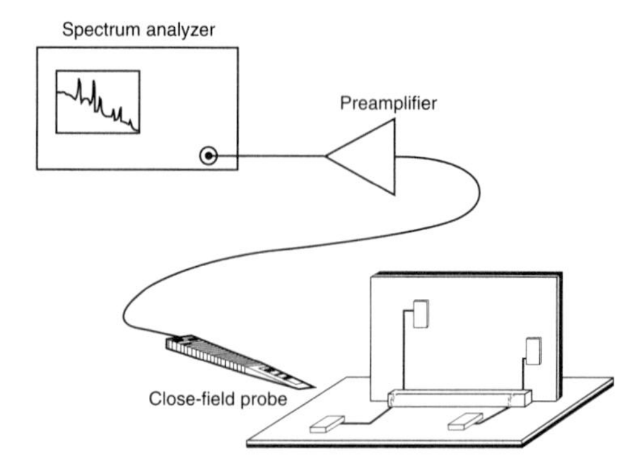

Fig. 6.45 Use of close-field probe to detect individual components, circuits, or shields that are radiating EMI.

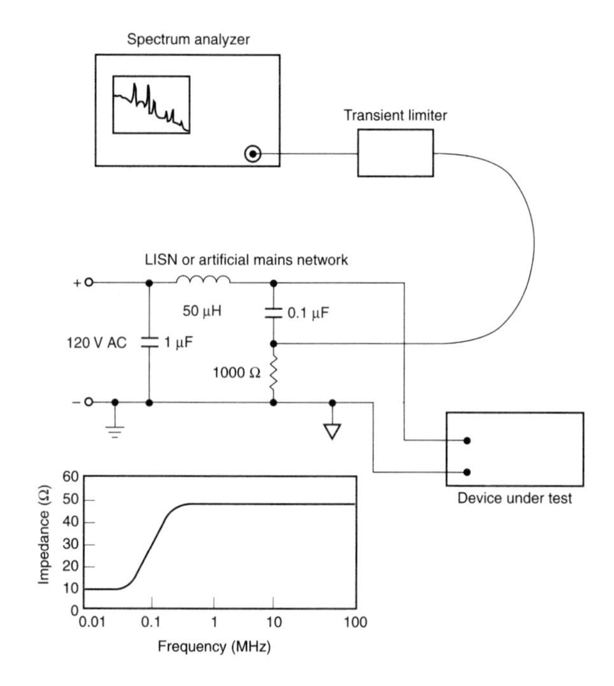

Fig. 6.46 Measuring conducted EMI by powering the device under test through a line impedance stabilization network (LISN). The LISN, or artificial mains network, isolates the test circuit from emissions conducted from the input power mains. The transient limiter presents damage to the spectrum analyzer when the LISN is switched into use.

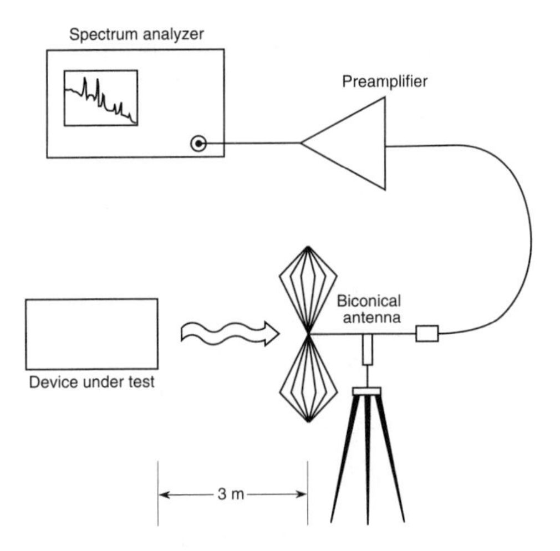

Fig. 6.47 Measuring radiated EMI with an appropriate antenna by positioning the device under test for maximum emissions.

impedance stabilization network (LISN). Finally, you generally measure radiated EMI for the preproduction and production models because so many variables that affect EMI depend on the final physical configuration of the instrument.

6.11 CASE STUDY — EMC DESIGN OF THE HP 54600 SERIES OSCILLOSCOPES*

The EMC design of the HP 54600 Series digitizing oscilloscopes consisted of a combination of circuit board suppression and mechanical design techniques. Since the entire product (including enclosure) was a completely new design, we had an opportunity to design in RFI suppression techniques from the very start of the development. This article describes the design and test methods employed to ensure that the products met international and military EMC standards.

RFI standards for HP products typically include the German FTZ 1046 Class B limit (similar to the U.S.A. FCC Class B limit).† Additionally, we decided to attempt to meet the U.S. MIL-STD-461C RFI requirements as specified in the environmental standard MIL-T-28800D. Meeting these military standards imposed much more stringent design goals than normal. Since one of the primary drivers of the HP 54600 design was cost, early EMC/RFI integration was necessary. By

---

*From Wyatt (1992). © Copyright 1992 Hewlett-Packard Company. Reproduced with permission.

† Starting in 1992, the European EN55011 standard became effective. The HP 54600 Series oscilloscopes meet the Class A limit of this new standard.

considering EMC early in the system design, we hoped to minimize the cost of achieving the multiple goals of both international and military standards. In the end we were forced to offer two of the more costly suppression methods for some of the military requirements as options. We felt that the additional cost involved should not be imposed on the majority of customers, who would not require the additional shielding.

6.11.1 *Enclosure Design*

We knew that older enclosure designs included too many seams for easy RFI suppression. Therefore, we decided to start with fresh ideas. Several concepts were suggested and we finally settled on a two-part molded design, coated on the interior with a 2.5-μm-thick layer of vapor-deposited aluminum. One of the parts is the main cabinet and the other is the front panel. This two-part enclosure concept reduces the number of seams to the minimum required for access to the electronics. The aluminum is applied to all sides of the seam area so that good contact is made between the cabinet and the front panel.

The seam design is a continuous, overlapping "knife-edge," which fits in a slightly zigzag manner between a series of raised bumps and stiffening ribs. These stiffening ribs are spaced about 3 to 6 cm apart all the way around the seam, providing a distributed pressure to ensure a good electrical connection. Molded bumps between the contact fingers act as additional contact points. Fig. 6.48 shows the main seam details. This design provides uniform contact while allowing easy disassembly without the need for special tools. The maximum spacing of about 3.5 cm between contact points yields a theoretical shielding effectiveness of 20 dB at 500 MHz. We performed several tests on both the seam itself and the enclosure as a whole to confirm the shielding performance.

Fig. 6.48 The main seam that joins the front panel to its cabinet is formed by a knife-edge on the cabinet that fits between a series of raised bumps and stiffening ribs on the front panel.

6.11.2 *Enclosure Testing*

Several evaluation techniques were used to assess the RFI performance of the enclosure before building a finished electronic prototype. A number of proof-of-concept tests were performed on a glued-together prototype enclosure. These tests included near-field seam leakage tests using an HP 11940A close-field probe and far-field tests with a harmonic comb generator placed within the enclosure and measured as if it were a finished product.[1]

The first enclosure testing consisted of a measurement of the basic shielding properties of various conductive plating choices. For this test, a pair of HP 11940A close-field probes were used tip to tip, one being the transmitting source and the other the receiver (Fig. 6.49). These probes were connected to an HP 8753B network analyzer and the transfer characteristic of the measurement system (without shielding material) was normalized to zero dB. The shielding samples were then placed between the probe tips and the near-field magnetic shielding effectiveness plot was obtained directly from the screen. This method allowed a quick comparison of several different shielding materials. Since the close-field probes were held by posts on the fixture, it was possible to measure various points of the actual molded parts to verify shielding performance and consistency.

Once molded prototypes were available, the comb generator was mounted within the enclosure. The comb generator simulated a very noisy product by emitting strong harmonics every 5 MHz from 30 to 1000 MHz. With the generator in

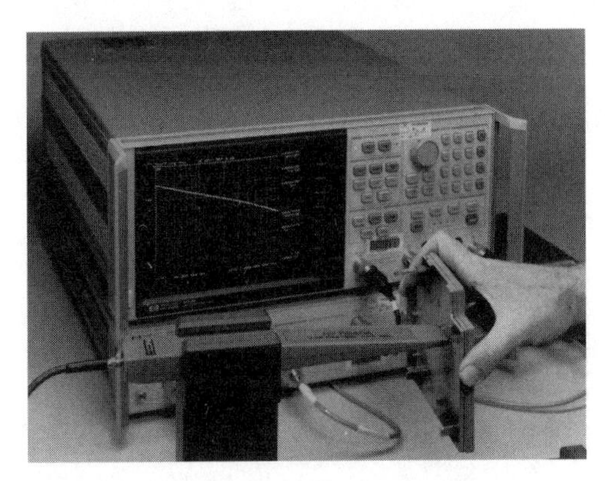

Fig. 6.49 This photo shows an option module cover being measured using the shielding effectiveness fixture and a network analyzer. This yields a continuous plot of shielding effectiveness versus frequency for thin shielding materials. The technique allows a comparison of different shielding techniques.

---

[1] A more detailed description of the harmonic comb generator measurement techniques along with construction details can be found in K. Wyatt and D. Chaney, "RFI Measurements Using a Harmonic Comb Generator," *RF Design Magazine*, January 1991, pp. 53–58.

place, seam emissions were measured with the 11940A close-field probe. A composite emissions characteristic was recorded using a spectrum analyzer and the signal leaks were noted. This emission recording was repeated for each mold trial to track the general progress of the design and to allow improvements to be incorporated into the mold before the electronic prototype was available.

Finally, far-field testing was performed, again using the comb generator as a simulated product. By measuring the difference in signal strength between the generator in the enclosure and the generator without the enclosure, a rough idea of the system shielding effectiveness (including all apertures and seams) was obtained. The video display module, keyboard, and steel deck were mounted within the enclosure to fill the larger apertures and more closely resemble the finished product. The prototype enclosure was tested at a 3-meter distance in an anechoic chamber. First, the harmonic levels of the generator were measured. Then, the generator was placed inside the enclosure and the harmonics were remeasured. For each measurement, the product was rotated in azimuth and the highest harmonic levels were recorded. The difference in readings then indicated the worst-case shielding effectiveness. Fig. 6.50 shows a typical plot of shielding effectiveness versus frequency. This technique exposed weak areas of the total system design before installing the electronics and allowed the mechanical and electronic design efforts to proceed in parallel.

Together, these three evaluation techniques allowed us to determine weak areas of the enclosure design before completion of a prototype oscilloscope. The net result was a good enclosure design early enough in the product cycle so that the costs for RFI reduction were minimized.

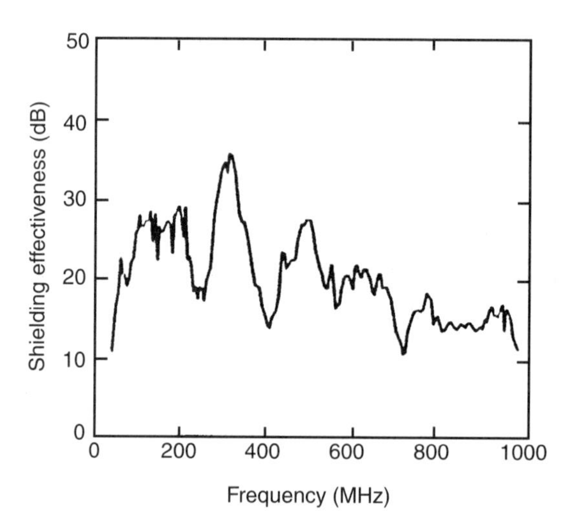

Fig. 6.50 Measured shielding effectiveness of the HP 54600 oscilloscope enclosure including seams and apertures. The harmonic comb generator was used to obtain the plot.

6.11.3 *Option Module Design*

The HP 54600 oscilloscope includes option modules which plug onto the rear of the instrument and provide custom I/O and computer control. These modules are made of molded plastic and coated with the same vacuum metallization process as the main enclosure. The challenge from an EMC point of view was how to connect the module shield and main enclosure shield together well enough to prevent RFI when both coatings were on the inside of their respective moldings.

The option module is connected to the main enclosure by a 50-pin connector and three metal-coated hooks which fit into three openings in the main cabinet. The connector (similar in style to an HP-IB connector) includes a grounded shell extension which overlaps the raised shell of the mating connector. We had originally hoped that these overlapping ground connections and the three hooks would adequately join the module and main cabinet shields together. Unfortunately, the connector half in the option module was soldered directly to the circuit board inside and thus was not grounded well enough to the module shield, so the initial result was very strong emissions once the module was attached to the oscilloscope.

When dealing with enclosure design, it is useful to consider each enclosure (in this case, the module and oscilloscope) as a separately shielded EMC environment. Any untreated penetration of either environment (by a wire, or by a connector in this case) will allow internally generated noise currents to escape, thus reducing the effectiveness of the environment's shield. Since the connector of the module was not directly connected to the module shield, internal noise currents were allowed outside the shield at the connector resulting in an increase in emission level. To fix this, an EMI finger stock strip was staked directly to the module enclosure half and against the metallic coating as shown in Fig. 6.51. The flexible beryllium copper fingers press firmly to the ground shell of the module connector and reduce the emission level by 3 to 8 dB.

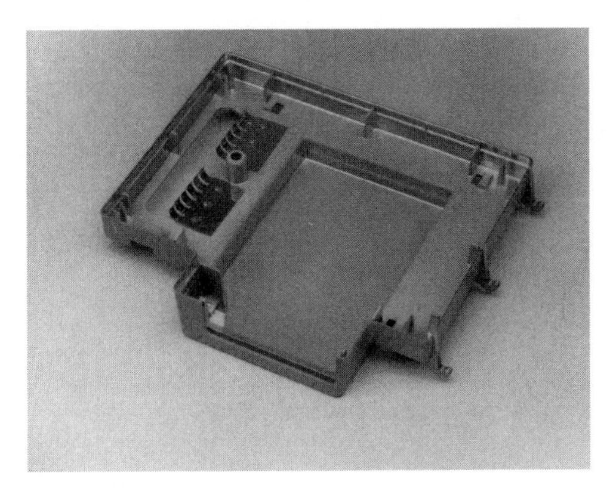

Fig. 6.51 A finger stock strip is used to reduce the radiated emissions from the option module. The fingers contact the module connector ground shell.

6.11.4 *Circuit Suppression Design*

One of the real keys to our radiated emissions success was the source suppression techniques used on the circuit board. Three methods were used: decoupling of the crystal oscillators, proper bypassing of each IC, and RC filtering of clock signals at the primary buffer gate.

It is well-known that crystal oscillators or other high-frequency periodic signals (such as clock lines) are among the primary sources of radiated emissions on circuit boards.[2] Three crystal oscillators are used in the circuit design: 10.73, 40, and 80 MHz. These high frequencies required that close attention be paid to circuit trace layout and board design. All oscillator circuits are well-bypassed and decoupled from the rest of the circuit board with an L filter consisting of a small 77-ohm ferrite bead and a 0.1-μF multilayer ceramic chip capacitor.

All ICs are liberally bypassed from the supply pin to ground with 0.01-μF or 0.1-μF chip capacitors. Each package has one or more as appropriate. The larger ASICs have both bypass capacitors and filter networks to control noise generation.

The last primary RFI reduction technique used is a simple RC filter network on the microprocessor clock buffer/divider (Fig. 6.52). Clock rise times are a major cause of emissions and are often 2 ns or less, depending upon the logic family used. Very often, these fast edges are unnecessary.

If the Fourier series of a trapezoidal waveform is analyzed, it will be noted that the envelope containing the noise current harmonics starts rolling off at a 20-dB/decade rate and then decreases at a 40-dB/decade rate with a break frequency that depends only on the rise (or fall) time of the square wave (Fig. 6.53). By slowing the edges of the clock signal, the harmonic content can be substantially reduced. Radiated clock harmonics have been reduced by 6 to 12 dB merely by adding this simple RC network.

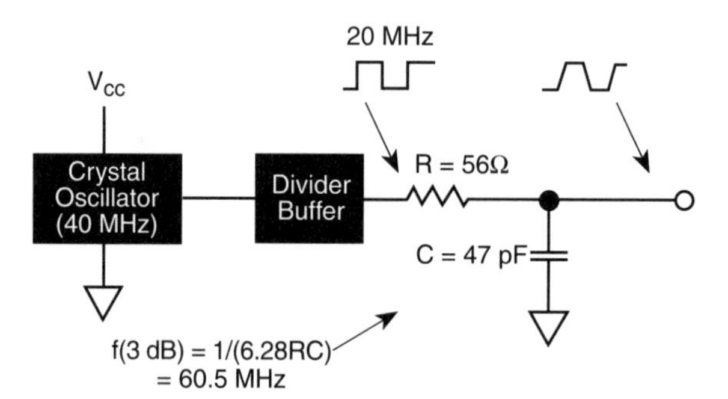

FIG. 6.52 An RC filter is installed following the microprocessor divider/buffer circuit. The filter slows down the rise time of the clock signal and greatly reduces the clock harmonics.

[2] H. Ott, *Noise Reduction Techniques in Electronic Systems*, 2nd edition, Wiley, 1988.

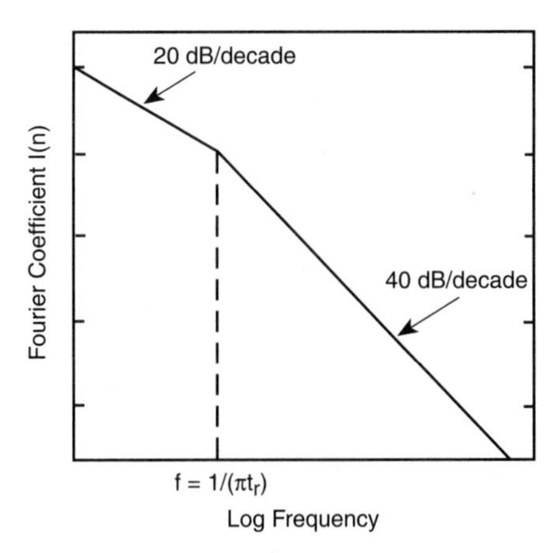

Fig. 6.53 The envelope of the Fourier series of a trapezoidal waveform has a breakpoint that depends purely upon the rise time (t_r). Slowing down the rise time shifts this breakpoint down in frequency and reduces the amplitude of the high-frequency harmonics.

Related to the edge speed problem is the use of fast devices. In one case, a transistor with an f_T of 5 GHz was used to amplify an 80-MHz trigger signal. The rise time for this device was 2 ns and we were measuring a strong harmonic at 800 MHz. By changing to a different device with an f_T of 1 GHz and a rise time of 3.5 ns, the harmonic amplitude was reduced by 6 dB.

6.11.5 *Printed Circuit Board Design*

The circuit board is a six-layer, single-sided, surface mount design. Full ground planes are used to provide both a ground plane and additional isolation between trace layers. All secondary power is isolated with L filters consisting of a multiturn ferrite core and multiple chip capacitors at the board power connector. This prevents circuit board noise currents from traveling out the power supply wiring.

Another suppression technique is component grouping by function and speed. The power input and filtering are located at the rear of the board, well away from other sensitive or noisy connectors. The crystal oscillators are located close to the power filter section and away from sensitive analog inputs. The analog circuitry is located on one side of the board, while the high-speed digital circuitry is located on the other. Plenty of filtering and decoupling is used on all supply traces to prevent noise currents from flowing from noisy areas to sensitive areas of the board. Much of the analog circuitry is contained within a cast shield to prevent interference from external fields.

The circuit board is also attached to the steel deck, which serves three purposes. Besides serving as a structural portion of the system, the deck provides about 10 dB of low-frequency (60 Hz) shielding and also serves as an image plane.[3] An image plane acts to confine external fields generated on the circuit board to the area between the board and the plane. The large metal surface forces any electric (E) and magnetic (H) field lines impinging on its surface to rotate orthogonally. As the image plane is moved close to the board, this forced rotation tends to cancel both the E and H fields that originate on the board. This technique (which is free) further reduces the radiated emission sources within the oscilloscope and helps lower system RFI.

6.11.6 *Power Supply Filtering*

Design constraints on the power supply filter circuit made for another major challenge. The oscilloscope must be able to use any power source worldwide, so it has to operate from 90 to 270 volts and from 44 to 440 Hz, automatically, without user-settable switches. We also had to meet safety leakage current limits (3.5 mA, maximum) at any voltage/frequency combination, and we had to meet both FTZ (Vfg) 1046 Class B (based upon the German VDE 0871 standard) and MIL-STD-461C, CE-03 (U.S. military) conducted emission limits. Finally, because of cost constraints, we could not use shielded power entry modules (PEMs). The entire filter circuit had to be designed with discrete components mounted on the power supply board. To say that this was a challenge would be an understatement.

One major difference between FTZ 1046 and MIL-STD-461 conducted emissions tests is that for FTZ 1046, noise *voltages* are measured, while for MIL-STD-461, noise *currents* are measured. Thus, the filter design for FTZ 1046 needs to look like a low impedance (shunt capacitor to limit the voltage), while the design for MIL-STD-461 needs to look like a high impedance (series inductance to limit the current). Thus, the filter topologies chosen for the two standards tend to be diametrically opposed.

Our filter vendor ultimately proposed a filter that resembles a typical FTZ 1046 topology with series inductors to reduce the MIL-STD-461 noise currents. The circuit is shown in Fig. 6.54.

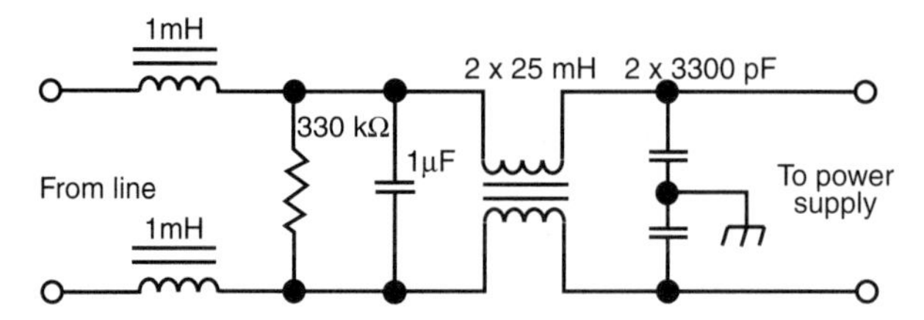

Fig. 6.54 This is the final circuit used for the switching power supply filter. It will allow the HP 54600 Oscilloscope to meet both FTZ 1046 and MIL-STD-461 conducted emission limits.

[3] R. German, H. Ott, and C. Paul, "Effect of an Image Plane on Printed Circuit Board Radiation," *IEEE International Symposium on EMC*, Washington, D.C., August 21–23, 1990, pp. 284–291.

6.11.7 *System Design*

Since the circuit board assembly contains connectors at each end, the usual nuts that attach the oscilloscope probe BNC connectors to the front panel were eliminated to avoid stresses on the enclosure. The BNC connectors are part of a cast shield assembly that attaches directly to the main circuit board and then to the steel deck with a screw. These connectors are simply pushed through matching holes in the molded plastic front panel as the circuit board is installed. This was another EMC design challenge in that we were originally counting on a fairly long path to ensure an adequate ground connection (or cable shield termination) between the BNC ground and the shielded enclosure of the oscilloscope. The result was a higher level of probe cable radiation than we could tolerate.

The solution to this is a thin, nickel-coated, phosphorbronze shim placed between the cast BNC shield assembly and the front panel. This shim includes four connection points (90 degrees apart) which press around the perimeter of each BNC connector. Other tabs press between the deck and front panel. The shim provides a good connection between the circuit board and the enclosure, thereby reducing probe cable radiation.

The other system problem stems from the use of a raster-scanned CRT display. Large magnetic fields (60-Hz vertical and 25-kHz horizontal) are used to drive the electron beam, and these fields radiate from the oscilloscope display. The 60-Hz field can interfere with other nearby CRT displays, and vice versa. Also, other low-frequency fields (from power transformers, typically) can interfere with the oscilloscope display, causing the display to swim or wiggle. Finally, the 25-kHz field radiates strongly out the front of the CRT and can be picked up by the oscilloscope probe tip as far away as one to two feet. If the user measures sensitive circuits directly in front of the CRT, they will tend to pick up this horizontal scanning signal. Passage of the radiated emission and susceptibility portions of MIL-STD-461C requires both of these fields to be considerably reduced. The optional shields previously mentioned were designed to help with both of these problems. While the product meets commercial EMC standards worldwide, some customers may need the additional shielding provided by these options.

Option 001 is a mumetal shield that snaps over the CRT deflection coils. This both reduces the external fields exiting the oscilloscope and, more important, reduces the susceptibility of the display to low-frequency external fields (such as power transformers). Unfortunately, even with the shield in place, we are unable to comply fully with the RE01 (magnetic radiated) or RS02 (magnetic susceptibility) tests of MIL-STD-461C because the fields from these tests can enter through the front of the CRT display and upset the displayed waveform by moving the electron beam. This is a disadvantage of all raster-scanned CRT displays.

Option 002 is a conductive filter shield that is attached in front of the display screen. This shield reduces the E fields at the horizontal scan frequency of 25 kHz exiting the front of the CRT display. This filter screen allows the oscilloscope to meet the RE02 emission requirements of MIL-STD-461C. Additionally, it solves the problem of 25-kHz CRT emissions entering the oscilloscope probe tip.

6.11.8 *Acknowledgments*

I would like to acknowledge the entire HP HP 54600 design team for their assistance in preparing this article and in persevering throughout the EMC design effort. I would also like to thank Dean Chaney for all the EMC testing.

Scope: system or enterprise
Lessons learned:

1. Early consideration of shielding in system design will minimize cost.
2. Combining electronic and mechanical techniques can achieve good shielding.
3. Consider each enclosure as a separately shielded EMC environment.
4. Reduce the number and size of seams for RFI suppression.
5. Treat all penetrations as possible leaks.
6. Electronic suppression includes the following:
   - Capacitive decoupling of ICs
   - Filtering of clock signals
   - Full ground planes
   - Component grouping by function and speed
   - Power supply filtering: at the AC input and on the PCB

6.12 SUMMARY

Grounding and shielding are incredibly important concerns in electronics. Unfortunately, many people treat the subject as a black art. It's not. The individual nature of each problem compounds the mystique, but you can debunk the erroneous impression with thoughtful consideration. Define the possible emission sources, the coupling mechanisms, and the susceptible circuits. Draw a simple schematic of the ground structure. Then use diagnostic clues to eliminate possible explanations for mechanisms.

You must consider grounding and shielding from the beginning of any design. (How many times have you heard that?) Know the standards and regulations that affect your product and its market. If you don't, you will probably be doing an expensive and time-consuming redesign.

6.13 RECOMMENDED READING

Ott, H. W. 1988. *Noise reduction techniques in electronic systems.* 2nd ed. New York: Wiley.
A thorough, readable, well-illustrated, and practical resource for understanding and reducing noise in systems.

A number of good seminars are available. Some of them are given by the following companies:

Diversified Technology, Inc. Rt. 3, Box 2000D Gainesville, VA 22065	(703) 347-0030
The Keenan Corporation The Keenan Building 8609 66th Street North Pinellas Park, FL 34666	(813) 544-2594
Van Doren Company Rt. 6, Box 319 Rolla, MO 65401	(314) 341-4097

6.14 REFERENCES

Brewer, R. W. 1991. Suppress EMI/RFI from the ground up. *Electronic Design,* March 28, pp. 73–86.

Bryant, J. 1991. Ask the applications engineer—10. *Analog Dialogue* 25(2):24.

Clark, O. M., and D. E. Neill. 1992. Electrical-transient immunity: A growing imperative for system design. *Electronic Design,* January 23, pp. 83–98.

Demcko, R. 1991. Miniature multilayer ZnO transient voltage suppressors. *EMC Test & Design,* November–December, pp. 35–37.

Denny, H. W. 1983. *Grounding for the control of EMI.* Gainesville, Va.: Don White Consultants.

Doubek, E. R. 1991. Susceptibility of consumer electronics: A general overview of the immunity problems of consumer electronic products. *EMC Test & Design,* November–December, pp. 31–34.

Hemming, L. H. 1992. *Architectural electromagnetic shielding handbook: A design and specification guide.* Piscataway, N.J.: IEEE Press.

IEEE. 1992. *IEEE recommended practice for powering and grounding sensitive electronic equipment.* IEEE Standard 1100-1992. New York: IEEE Press.

National Fire Protection Association. 1993. *National electrical code handbook.* 6th ed. Quincy, Mass.: National Fire Protection Association.

Ott, H. W. 1988. *Noise reduction techniques in electronic systems.* 2nd ed. New York: Wiley.

Richman, P. 1991. Diagnosing ESD problems in plastic-cased products. *EMC Test & Design,* November–December, pp. 42–44.

Schroeder, D. 1992. Long-term Carrera 4. *Car and Driver* 37(11):153.

St. John, A. N. 1992. Grounds for signal referencing. *IEEE Spectrum,* June, pp. 42–45.

Van Doren, T. 1991. Grounding and shielding electronic systems. Notes from video seminar on the NTU Satellite Network, November 11 and 12.

Williamson, T. 1986. Designing microcontroller systems for electrically noisy environments. *Intel Application Note AP-125,* November, pp. 12–14.

Wyatt, K. D. 1992. EMC design of the HP 54600 series oscilloscopes. *Hewlett-Packard Journal,* February, pp. 41–45.

7

CIRCUIT DESIGN

There is no useful rule without an exception.
—Thomas Fuller

7.1 FROM SYMBOLS TO SUBSTANCE

Funny that I should wait until the seventh chapter in a book about electronic instrumentation before getting into the practical concerns of circuit design! But circuit design is a small portion within the development of successful instruments, and I hope that you are beginning to appreciate the large number of issues that surround that development.

Rather than try to compete with the many good textbooks on electronics, it's far more appropriate for me just to recommend them. This chapter focuses on the common problems found in every project that many textbooks don't cover. It provides a framework for circuit design so that you can see how electronic design fits into the big picture of system development.

Fig. 7.1 illustrates some of the common areas of circuit design. Each section of this chapter covers some portion or aspect of circuit design. The issues range from general concerns for all types of circuits to specific applications. General issues include the selection of technology, reliability, fault tolerance, high-speed design, low power, noise, and error budgets. Two more sections on buses, networks,

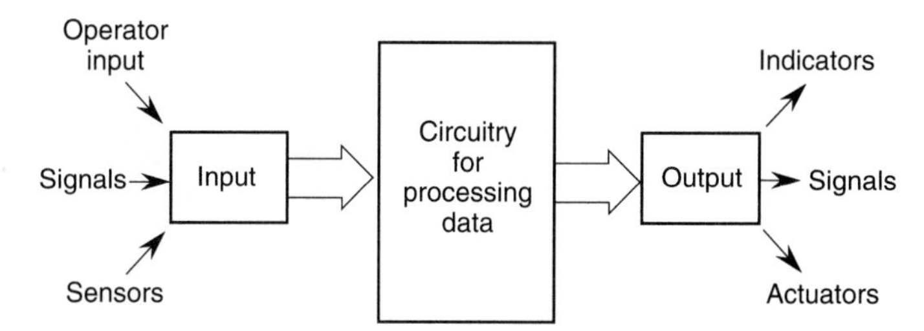

FIG. 7.1 General configuration for the circuits within a system.

and power control cover the middle block of Fig. 7.1, the circuitry for data processing. Finally, two sections are devoted to input and output (I/O). All these issues help engineers convert the intentions of the customer into successful systems.

7.2 CONVERT REQUIREMENTS INTO DESIGN

Establishing requirements is the most difficult part of circuit design. I'll give some tips and thoughts for setting requirements, but your experience is your best guide.

A general-to-specific approach may serve you in establishing requirements (Table 7.1). Start by defining the desired function in broad terms. Then refine the function with operational concerns. Finally, settle on exact regulations and specifications. Setting specifications is one of the most difficult parts of engineering. This is where good judgment and experience are necessary to help you understand the big picture and ask the tough questions that pin down the constraints in a design.

In an ideal, textbook situation in which the specifications are complete, you pick the appropriate technology and begin the circuit design. Unfortunately, the division between solid requirements and design is never sharp because requirements often change late in the effort and spoil the design. Furthermore, the technology interacts with the requirements to mold them. In spite of these concerns, some principles can help bound the design problem, as shown in Fig. 7.2. By knowing the region of operation for your system, you will be able to pick the options

Table 7.1 Some requirements that drive design

System concern	Requirement	Parameters
Function	Response times	s, min, hr
	Data rates	Mbytes/s, kbits/s
	I/O drive	A, V
	Reliability—MTBF	hr
Regulations	FCC	
	UL	
	Military specifications	
Environment	Volume	in^3, m^3
	Weight (mass)	lb, kg
	Vibration	g, m/s^2
	Shock	g, m/s^2
Operation	Bandwidth	Hz
	Resolution	%, mV/LSB
	Speed	ns, μs, ms
	Accuracy	%
	Power consumption	mW
	Noise	$nV/\sqrt{Hz}$, SNR

Note: FCC = Federal Communications Commission; LSB = least significant bit; MTBF = mean time between failures; SNR = signal-to-noise ratio; UL = Underwriters Laboratories.

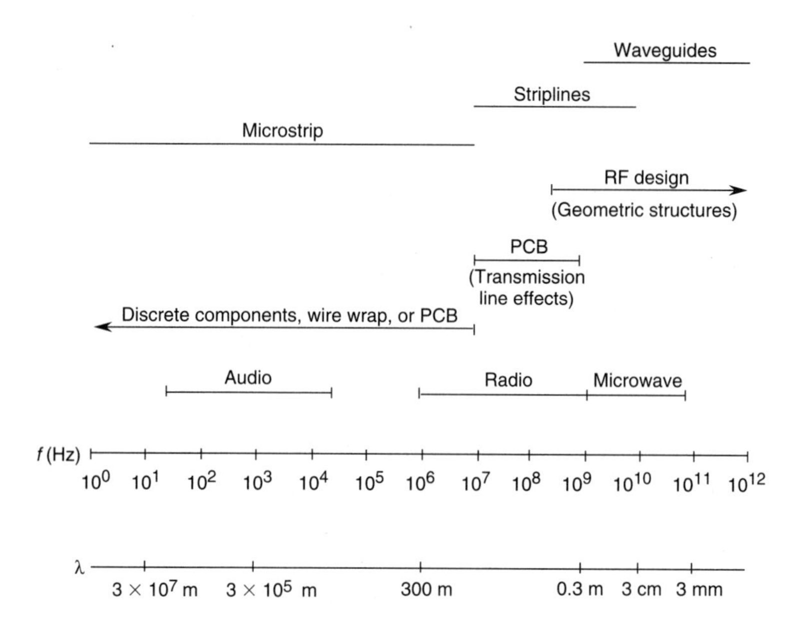

FIG. 7.2 A portion of the electromagnetic spectrum and corresponding technology applications.

available for circuit design. The right choices will reduce parts count, board space, power, and cost and so decrease time to market and increase reliability.

Figs. 7.3 and 7.4 can help you select the appropriate technology and refine the focus of the circuit design. Your time and effort in design increases as the complexity of the function of your system increases. For instance, an ASIC may solve a signal- or data-processing problem optimally in terms of high throughput and low power, but it will cost more than a collection of standard components. Does the development time merit using the ASIC in your product? Will you be able to amortize its cost over the production runs? Will a circuit using many cheap programmable array logic (PAL) chips really be more cost-effective than a single microcontroller that takes less time to implement and program? What is the driving factor—cost, development ease, or time to market? You will have to consider each of these factors and many others before designing the actual circuits.

Consider the factors that go into selecting a processor for an embedded system, for example. First, do you already have experience, software, and development tools for a particular processor? These concerns weigh heavily in choosing a processor and achieving your goals for time to market. Furthermore, you have to specify the performance, number of peripheral functions, memory, and tool support to determine the appropriate processor for your product (Vaglica and Gilmour, 1990).

Performance is related to the architecture of the processor and its processing capacity in bits (4, 8, 16, or 32). Performance is determined by several factors:

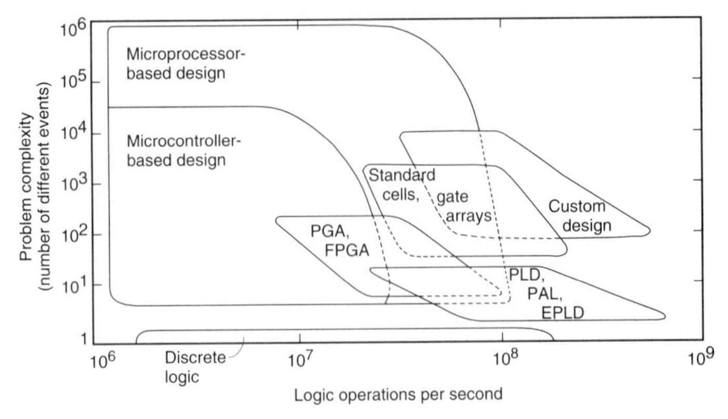

Fig. 7.3 Comparison of the performance of various technologies to the problem complexity handled by the components. (EPLD = electrically programmable logic device; FPGA = field-programmable gate array; PAL = programmable array logic; PGA = programmable gate array; PLD = programmable logic device.)

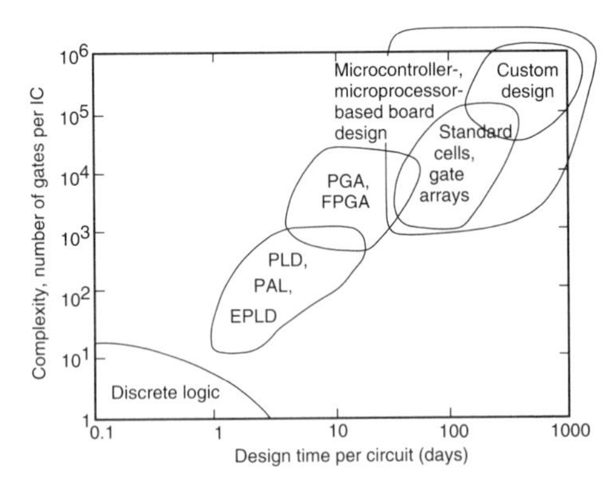

Fig. 7.4 Comparison of the design time required using various technologies versus the complexity of the components.

- Throughput
- Resolution and dynamic range
- Address space and available memory
- Language choice: code size, speed (compilation and actual execution)
- Predominant types of calculations: integer or floating point

Resolutions to 0.5% and dynamic ranges of 1 part in 1000 can easily be achieved by an 8-bit microcontroller. In contrast, resolutions of 1 part per million, dynamic ranges of 10 parts per billion, and high throughput probably will call for a 32-bit microprocessor with direct memory access (DMA) control. The real issue is speed: How much information do you need to process per second? Sure, you could implement 64-bit math on a 4-bit controller, but it would be very slow.

Peripheral functions also drive the selection of a processor. These functions include the following:

- Math coprocessors
- Graphics accelerators
- Interrupt handlers
- Data transfer and communications: DMA, small computer system interface (SCSI), serial I/O ports
- Timers
- Analog-to-digital converters (ADCs)
- Digital-to-analog converters (DACs)
- Power drivers
- Watchdog timing

Some microcontrollers integrate some of these functions onto a single chip to reduce the parts count and increase the reliability. Furthermore, most microcontrollers have memory integrated on the chip; the size of memory can exceed 64 kbytes of ROM and 2 kbytes of RAM.

On the other hand, the architecture of microprocessors does not include memory or peripherals, so you will have to include these components in your design, thereby increasing the parts count and complexity. You can estimate the amount of memory you need by adding up the possible sizes of the following:

1. Data arrays
2. Stack
3. Temporary and permanent variables
4. Compiler overhead
5. I/O buffers

The sum of the estimates will give a minimum size for memory. Always plan for and specify margin in your memory requirements. The possibility of future upgrades and modifications affects the selection of processor, memory, and peripherals. If your system will need more memory to support new algorithms, for instance, will the processor easily accommodate the change? Furthermore, are the components really available, or are they some vendor's dream? Are second sources available?

Power consumption within the processor always plays a role in systems design either for cooling concerns or for sizing a battery. Consequently, power consumption can drive the selection of a processor; some models dissipate considerably less than others or can be powered down for short periods to reduce the dissipation.

Finally, and probably most important, what experience do you have, and what tools are available to support your development? Do hardware emulators exist that will help you debug both circuits and code? What software tools are available to support development on the selected processor? Does the vendor of these tools provide good support? Does the vendor have a good reputation? Development tools that target specific microcontrollers or microprocessors can markedly affect, either for good or bad, your design effort.

7.3 RELIABILITY

Early in any design you will have to specify how long your product will last. What is the minimum acceptable to the customer? What is the longest duration of continued operation that also maintains cost within reasonable limits? What is the reliability of each component, and how do they combine into a total system? Ultimately, what is the reliability of your product? Two factors drive reliability:

1. Complexity: Fewer parts are almost always better.
2. Design margin: You must allow for stressing of components.

You can determine a measure of reliability by two methods: model prediction or prototype tests. Neither provides absolute verification. Models should be used early to update estimates of reliability; however, models are limited and can't predict every outcome. On the other hand, tests of prototypes find many weaknesses and problems, but they are time consuming. Consequently, you will want to use a combination of model prediction and testing to estimate reliability of your product.

You can find standard methods for modeling in handbooks such as MIL-HDBK-217F (United States Department of Defense) or BT-HRD-4 (British Telecom Handbook of Reliability Data). These methods use formulas based on practical experience of failure rates and physical knowledge to relate environmental factors to the reliability of electronic components. The failure rate for a component is generally a base rate that is modified by various factors:

$$\lambda = \lambda_b \pi_e \pi_q \pi_a \ ,$$

where λ = failure rate of the component
λ_b = base failure rate
π_e = environmental factor
π_q = quality factor
π_a = acceleration factor

Reliability of a component is defined as a function of failure rate:

$$R(t) = e^{-\lambda t} \ ,$$

where $R(t)$ = reliability
λ = failure rate
t = time

Reliability of a system is the product of all component reliabilities:

$$R_{\text{system}} = \prod_{i=1}^{n} R_i ,$$

where R_{system} = reliability of the system
R_i = reliability of component i

Once you know these formulas, you can calculate reliability by looking up the appropriate factors for each component and substituting them into the equations.

While these formulas provide some basis for comparing the relative reliability of different designs, they are seldom accurate. Most failure rates relate acceleration factors (such as π_e, π_q, and π_a) to temperature, but none consider the application. You must consider the application and how some of these stresses and susceptibility factors might affect reliability (Brombacher, 1992):

- Corrosion
- Thermal cracks
- Electromigration
- Secondary diffusion
- Ionizing radiation
- Vibration
- High-voltage breakdown
- Aging

Any one of these factors can drastically alter reliability and still not be predicted by standard models. Only care, study, and experience will help you predict reliability more accurately.

Example 7.3.1 A satellite application toggled a bit-slice processor on and off 25 times each second to reduce the overall power consumption. Unfortunately, the power surges at each turn-on accelerated electromigration within certain conductors on the chip die that caused them to open, leading to premature failure of the processor.

7.4 FAULT TOLERANCE

Fault tolerance goes beyond the design and analysis for reliable operation and reduces the possibility of dysfunction or damage from abnormal stresses and failures. It allows a measure of continued operation in the event of a problem. Fault tolerance is primarily a philosophy of system design and architecture. It has three distinct areas: careful design, testable functions, and redundant architectures.

7.4.1 *Careful Design*

You can avoid many failures from abnormal stresses by conservative and careful design. Here are some design techniques that can reduce the probability of failure:

1. Reduce overstress from heat with cooling and lower-dissipation design.
2. Use optoisolation or transformer coupling to stop overvoltage and leakage current.
3. Implement ESD protection (see Chapter 6).
4. Mount for shock (from accidental drops) and vibration.
5. Tie down wires and cables that flex frequently, and use strain relief.
6. Prevent incorrect hookup; use keyed connectors.

All of these techniques fortify a design, but blind application of them leads to costly "gold-plated" or overly conservative design. Analyze the function of your product and its use before picking the appropriate techniques to reduce the danger of abnormal stresses in the particular application.

Example 7.4.1.1 Consider a portable instrument that uses a 9-V battery. It will not need extensive cooling to prevent overstress from heat dissipation, but it may not withstand reversed polarity from its battery. You can either use a mechanically polarized socket in the battery compartment or design a rectifier circuit, such as those in Fig. 7.5, to provide the proper polarity of power.

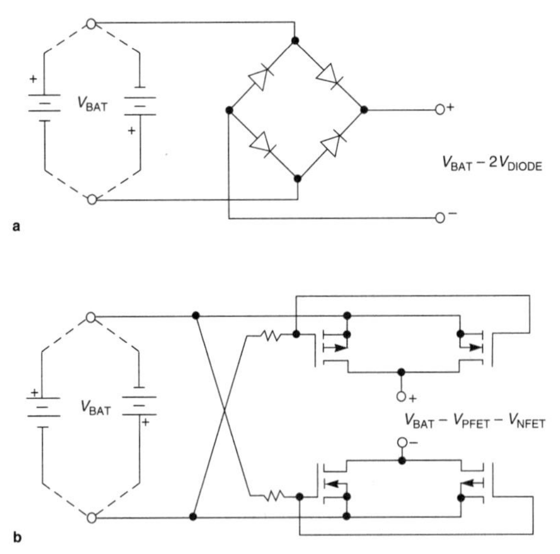

FIG. 7.5 Two schemes for preventing polarity reversal that can destroy circuitry. (a) In a bridge rectifier, the diode voltage drops can be significant. (b) P-channel and N-channel metal-oxide semiconductor field-effect transistors (MOSFETs) provide lower on-resistance and voltage drops for higher efficiency.

7.4.2 *Testable Architecture*

The process of testing and diagnosing failures within a system is another area of fault tolerance. Testing stimulates and records the response of components, boards, and modules; a trained technician or "smart" device makes the corresponding diagnosis.

A testable architecture has two possible configurations. The simpler configuration provides probe points for a technician or instrument to stimulate circuits and record responses. Usually, trained personnel must disassemble the system and remove the circuit for testing. The more complex configuration has dedicated internal circuitry called built-in test (BIT) that tests the system and diagnoses problems without disassembly of the equipment.

Testable architectures are more complex than those that simply incorporate the techniques of careful design. For stimulus and response, you must design access into circuits that tap, break, or invade the normal signal path; for longer chains of connected components, access becomes more critical for isolating problems. Obviously, the more access is needed, the more complex the test configuration becomes. BIT adds considerable complexity and consequently reduces reliability of the overall system. The trade-off for BIT is quicker diagnoses and repair versus higher reliability; you will have to determine the balance of the trade-off in each application.

Regardless of the configuration for testing, an appropriate calibration standard is always necessary when you measure a result. This concern links to maintenance specifications, and you will have to specify the calibration used in the maintenance. Often, high precision and resolution are not needed; the calibration can be as trivial as testing for the presence or absence of a logic level. In any case, you still need to define the appropriate standard.

7.4.3 *Redundant Architectures*

The most complex and fault-tolerant architectures are redundant systems. They use multiple copies of circuitry and software to self-check between functions. These are complex systems that are justified only when downtime for repair and maintenance cannot be tolerated.

A doubly redundant architecture merely indicates a failure in one of the subsystems; this allows for quick repair. A triply redundant architecture uses voting between the outputs of three identical modules to select the correct value. It can have a failure and still operate correctly. Finally, dissimilar redundance compares the output from modules with different software and hardware to select the correct output. It can survive failure and even indicate errors in design if one system is coded correctly and the others are not. Fig. 7.6 illustrates the three architectures.

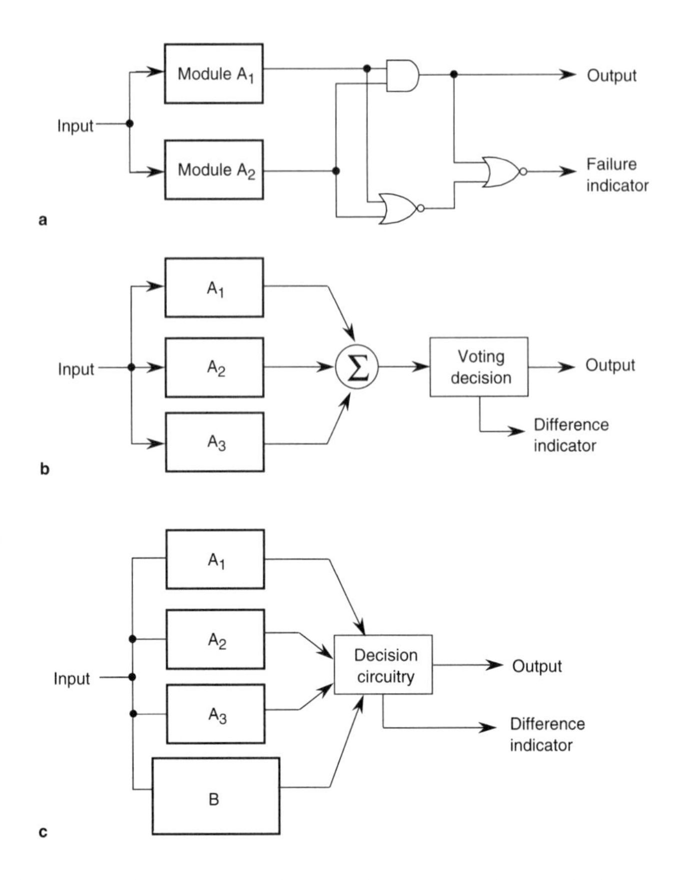

Fig. 7.6 Different configurations for redundant architecture. (a) Doubly redundant architecture can signal failure. (b) Triply redundant architecture operates in spite of failure. (c) Dissimilar redundance operates in spite of failure and indicates improper design or operation.

7.5 HIGH-SPEED DESIGN

Anytime the clock frequency exceeds 1 MHz in a circuit or system, you should begin considering transmission line techniques, because the harmonics generated by the edges of the clock and signal pulses can easily be 20 or 30 times the fundamental. When the clock frequency exceeds 10 MHz, the fundamental wavelength is less than 30 m, and the length of some signal lines in the circuit will be a significant portion of the wavelength of the harmonics.

Two conservative criteria (or rules of thumb) may be used to estimate when transmission line effects begin: circuit dimensions versus signal wavelength and rise time versus propagation delay. First, if circuit dimensions exceed 5% of the minimum wavelength, then the signal path approaches a transmission line:

$$l > \lambda / 20,$$

where l = length of signal path
λ = maximum wavelength of the signal

Second, if the rise time of a signal, t_r, is less than four times the propagation delay of the signal path, t_p, then the signal path approximates a transmission line with a characteristic impedance:

$$t_r < 4t_p,$$

where t_r = rise time of a signal
t_p = propagation delay of the signal path

In either case, you have to design with concerns for transmission lines in mind.

Careful design can avoid problems associated with transmission lines, including bandwidth limitation, decoupling, ground bounce, crosstalk, impedance mismatch, and timing skew or delay. Each of these problems contributes to noise in circuits; Fig. 7.7 compares some sources of noise and the ease of remedies.

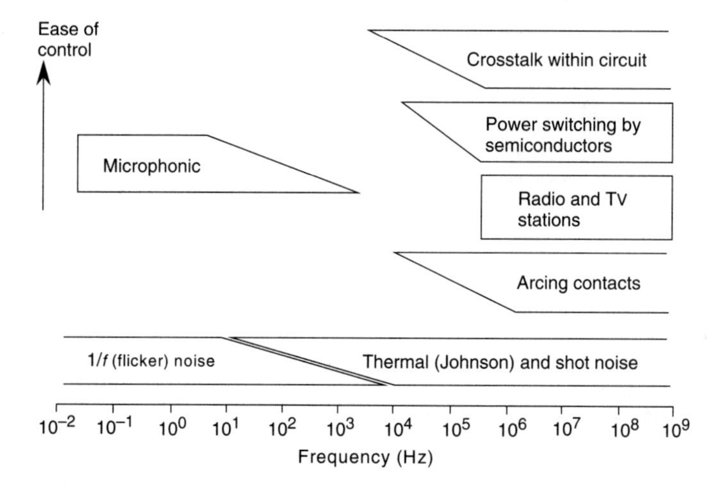

Fig. 7.7 Sources of noise in circuits and comparative ease of design control and elimination from circuit.

7.5.1 *Bandwidth*

Limiting the bandwidth of the signals within a system is the most effective way to reduce noise, EMI, and problems with transmission lines. You may limit the bandwidth two ways: by increasing the rise and fall times of the signal edges or by reducing the clock frequency (usually less effective, since rise and fall times contribute higher-frequency components; the effectiveness also depends on the relationship between the clock rate and rise time). Selecting the appropriate logic family will set the edge rates and the consequent limit on transmission line concerns. Slower edge rates allow longer interconnections between circuits, as shown in Table 7.2. *One criterion for selecting logic according to transmission line effects is a ratio less than 4 between the rise time,* t_r, *and the propagation delay,* t_p ($t_r/t_p < 4$).

Table 7.2 Maximum allowable lengths for interconnection before transmission line effects need attention

Rise time (ns)	Maximum interconnection (cm)	(in.)	Logic families
0.2	0.85	0.34	GaAs
0.5	2.1	0.84	GaAs
1.0	4.3	1.7	ECL 100 K, AS
2.0	8.5	3.4	ECL 10 KH, STTL, FCT
3.0	12.8	5.0	FAST, AC, FACT
5.0	21	8.4	ECL 10 K, TTL
8.0	34	13.4	LSTTL, ALS
10.0	43	16.8	HC
15.0	64	25	
20.0	85	34	
30.0	128	50	
50.0	213	84	15-V CMOS
100.0	427	168	5-V CMOS

Notes: Rise time assumes a linear ramp from 0% to 100%. The length of interconnection is acceptable if the time of propagation is less than the rise time by a factor of 4 or more. Velocity of propagation is assumed to be 17.1 cm/ns (6.72 in./ns).

7.5.2 *Decoupling*

The switching of digital logic causes transients of current on the voltage supply. These current transients will generate voltage spikes or transients on the power supply leads, as shown in Fig. 7.8, through the inductive impedance of the circuit. A decoupling capacitor will reduce the impedance of the power supply circuit by minimizing the inductive loop area. (This is particularly true for two-sided PCBs. Multilayer boards with power and ground planes already have very low impedance and do not need as many decoupling capacitors.) The decoupling capacitor should occupy the shortest possible path, between the voltage and ground pins of the integrated circuit, to minimize the inductive loop (Fig. 7.8c and d).

Select decoupling capacitors that balance capacitance with their equivalent series inductance and resistance (Chapter 12 will introduce these concepts). Incorporating large filter capacitors and ferrite beads may help you reduce noise emissions

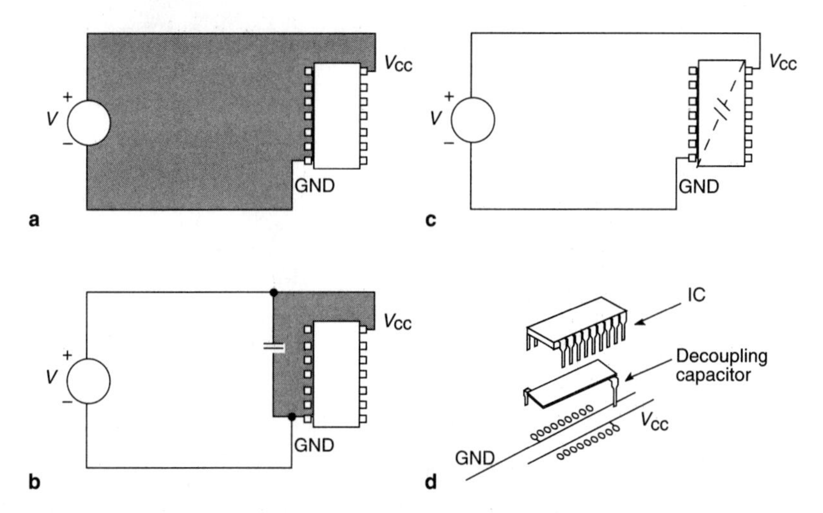

FIG. 7.8 Adding a decoupling capacitator to reduce the power-distribution impedance and consequently noise. (a) Large inductive loop area without decoupling capacitor. (b) Decoupling capacitor reduces the inductive loop. (c) and (d) Closer to the chip is better.

even further; however, you should consult a reference such as the one by Keenan (1985). Here are some general recommendations for decoupling:

1. Use a decoupling capacitor near each chip for two-sided boards. (Fewer are needed on multilayer boards that have power and ground planes.)
2. Use a large filter capacitor at the power entry pins.
3. As a last resort, use a ferrite bead at the power entry point to the circuit board.

7.5.3 *Ground Bounce*

Ground bounce is a voltage surge that couples through the ground leads of a chip into nonswitching outputs and injects glitches onto signal lines. The voltage surge is generated when an output (or simultaneous outputs) switches from high to low as shown in Fig. 7.9. The switching current from discharging the load capacitance may cause a significant voltage to appear across the loop inductance formed, in part, by the ground lead in the chip. A large enough voltage surge on the ground circuit can cause incorrect operation by referencing other outputs above the input thresholds of the driven logic.

Asynchronous signals are the most susceptible to ground bounce. Furthermore, ground bounce is a cumulative problem. The more outputs that switch simultaneously, the greater the magnitude of ground bounce. You can reduce ground bounce by reducing the loop inductance (for example, by removing wire-wrap

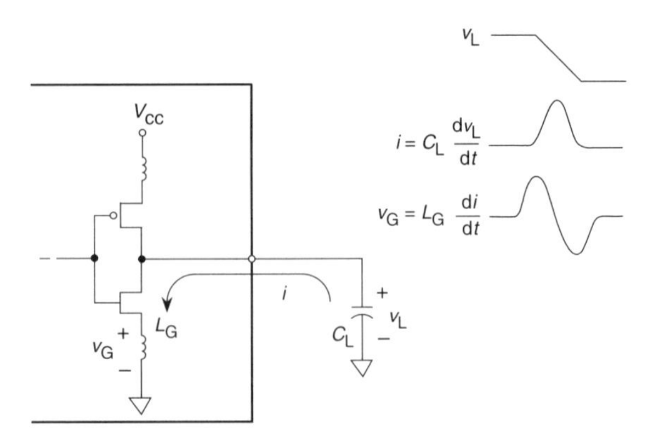

Fig. 7.9 Ground bounce—the noise glitch generated by the inductance of the leads when the load capacitance is discharged.

pins and sockets), by reducing the input gate capacitance, and by choosing logic families that either control the signal transition or have slower fall times.

7.5.4 *Crosstalk*

Crosstalk is the coupling of electromagnetic energy from an active signal to a passive line. The coupling mechanisms are capacitive or inductive (Fig. 7.10). Crosstalk is a function of line spacing, length, characteristic impedance, and signal rise times; decreasing the coupling length and the characteristic impedance and increasing rise times of the signal will reduce crosstalk (Fig. 7.11). These concerns are often best addressed by good layout, as detailed in Chapter 8.

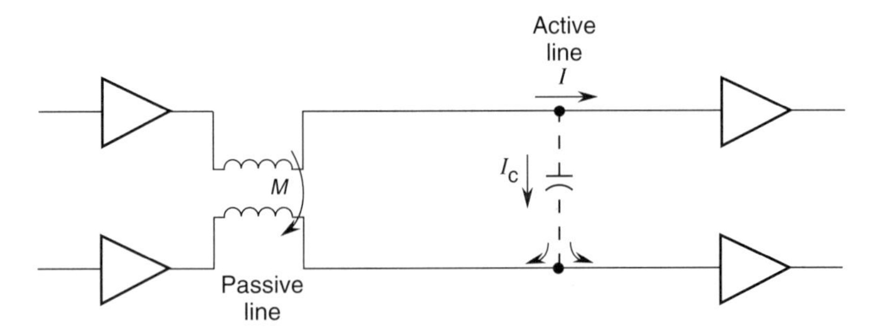

Fig. 7.10 Crosstalk—the inductive and capacitive coupling of signal from an active line to a passive line. (M = mutual inductance; I_c = capacitive current.)

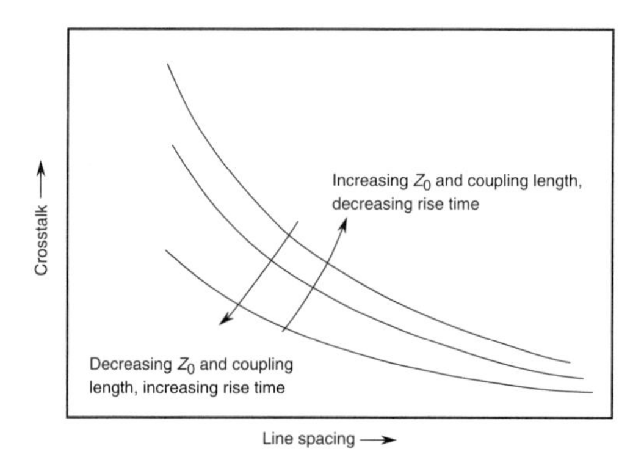

Fig. 7.11 Relationship of crosstalk to line spacing, coupling length, characteristic impedance (Z_0), and signal rise times.

7.5.5 *Impedance Matching*

Impedance matching makes the source and termination impedances equal to the characteristic impedance of the transmission line. Matching will eliminate reflections that cause ringing, undershoot, and overshoot in the signal pulses. Impedance discontinuities occur in two configurations: endpoint and stub.

Endpoint discontinuity means that the ends of the transmission line do not match the characteristic impedance of the transmission line. This problem is easily solved. You can match a source by adding series resistance until the total impedance equals the line impedance. Terminating the other end of a signal line from the driver will complete the impedance matching. The type and value of termination determines how well the signal line is impedance matched. Table 7.3 illustrates some drive configurations and terminations.

Stub discontinuities cause impedance mismatch and signal reflections by connecting multiple circuits to a signal line. Each connection of a stub divides the impedance and splits the power of the signal. You can reduce stub discontinuities by making them very short—even zero with active transceivers to interface the stub to the transmission line. Fig. 7.12 illustrates some networks and their relative goodness for minimizing reflections.

Table 7.4 summarizes guidelines for high-speed design by controlling bandwidth, decoupling, ground bounce, crosstalk, and impedance. (Contrary to some currently held myths, bends and vias in board traces contribute very little change in impedance.)

Table 7.3 Several methods to terminate signal lines by impedance matching, $Z_0 = Z_L$

Terminating configuration	Load impedance, Z_L	Advantages	Disadvantages
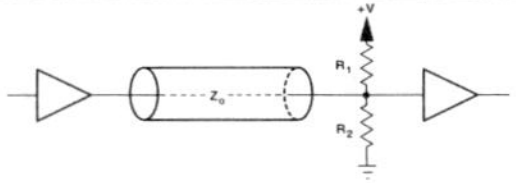	$R_1 \| R_2$, assuming $Z_{receiver} \approx \infty$ otherwise $R_1 \| R_2 \| Z_{receiver}$	Active termination voltage to help a weak driver by speeding transitions. Use with TTL, FAST, and ECL logic.	Constant DC power dissipated through R_1 and R_2. More components required.
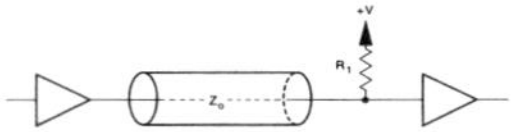	$R_1 \|$ or $R_1 \| Z_{receiver}$	Reduces DC power dissipation.	Additional load will slow weak drivers.
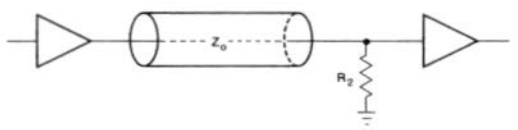	R_2 or $R_2 \| Z_{receiver}$	Reduces DC power dissipation	Additional load will slow weak drivers.
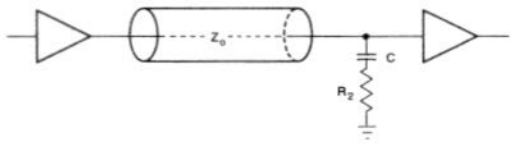	R_2 or $R_2 \| Z_{receiver}$ because $Z_c \approx 0$	Eliminates DC power dissipation. Use with FACT logic.	More components required.
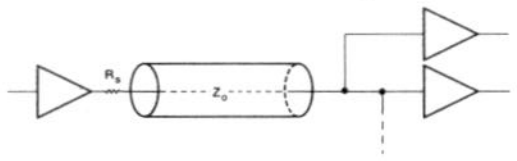	Source impedance $Z_S = R_S + Z_{driver}$	Allows branching at load. Eliminates DC power dissipation. Use with FACT and ECL logic.	
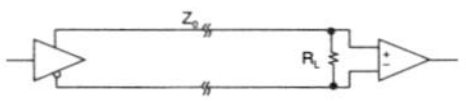	R_1 or $R_L \| Z_{receiver}$	Use with RS-422 and RS-485.	Constant DC power dissipated through R_L.
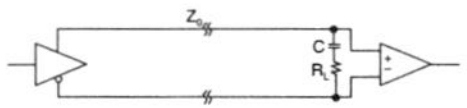	R_L or $R_L \| Z_{receiver}$ because $Z_c \approx 0$	Eliminates DC power dissipation.	Long strings of consecutive 1's or 0's will charge C and cause time jitter in signal transition as C discharges.
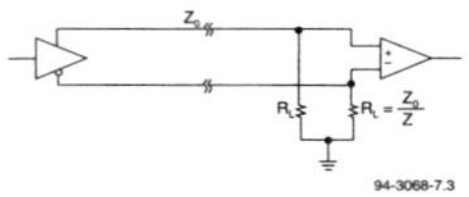	$2R_L$ or $2R_L \| Z_{receiver}$	Reduces common-mode noise.	Constant DC power dissipated through resistors.

Note: Z_0 = characteristic impedance; Z_S = source impedance; Z_c = capacitive impedance.

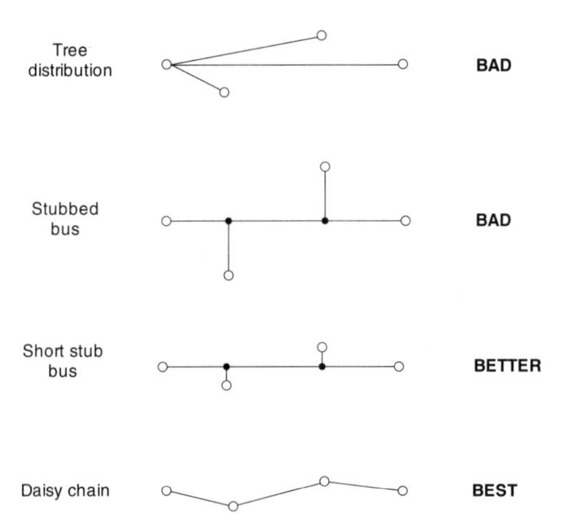

Fig. 7.12 Network geometries. Stubs cause signal reflections on transmission lines that result in ringing, overshoot, and undershoot. Reduce the length of stubs and terminate the ends of the transmission line to eliminate reflections.

Table 7.4 Design guidelines for high-speed circuits

	Reduces			
Guidelines	**Signal reflections**	**Power-distribution noise**	**Ground bounce**	**Crosstalk**
Control the rise time of signal pulses by selecting the appropriate logic family and technology.	√		√	√
Reduce signal frequency.	√			√
Use decoupling capacitors.		√		
Use multilayer printed circuit boards with power and ground planes.	√	√	√	√
Terminate transmission lines where $t_r/t_p < 4$.	√			
Keep stubs short.	√			
Keep signal lines short and perpendicular on adjacent layers. Control spacing between traces.	√			√
Group circuits into separate areas: analog circuits, high-speed logic, and high-current switching.		√		√
Avoid parallel asynchronous lines, wire wrap, and sockets.			√	√

7.5.6 *Timing*

As clock frequency increases, propagation delays, timing skew, and phase jitter can all conspire to render a logic design useless. Differences in propagation delay of the clock signal to different destinations cause the clock to *skew*, or arrive at different times. Furthermore, differences in the propagation delay of the rising edge and of the falling edge within a device change the duty cycle of the clock signal, or *shrink* it. Crosstalk, ground bounce, and propagation delays all contribute to deviation, or phase jitter, in the intervals between a reference clock edge and later edges. Fig. 7.13 illustrates some of these problems.

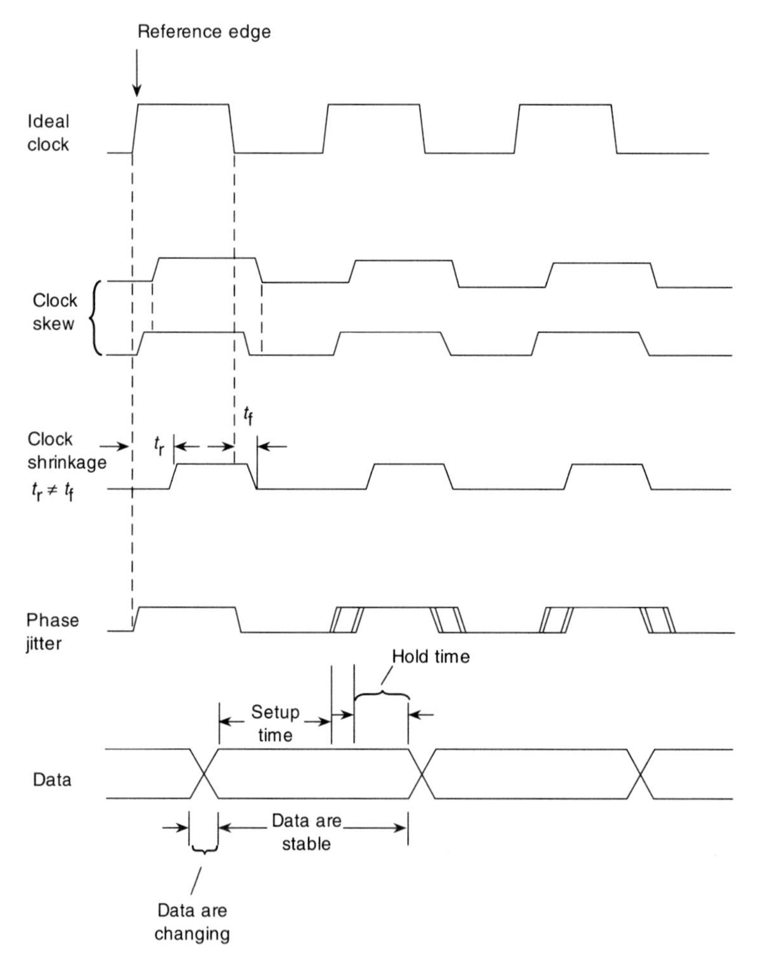

Fig. 7.13 Problems with timing. Adequate margins in the setup and hold times are necessary to account for clock skew, shrinkage, or jitter.

You must ensure adequate setup time and hold time to latch data reliably under these conditions. In general, the following guidelines will determine the margin in the timing of setup and hold (O'Dell, 1990):

1. Account for all components and factors that introduce delay or skew.
2. Get the maximum and minimum values of each factor.
3. Sum the minimum values that increase the timing margin.
4. Subtract from the result of step 3 the maximum values that decrease the timing margin.

The result will be a conservative estimate of the timing margin. The timing margin must be positive to ensure reliability in clocking; a negative margin indicates the possibility of problems.

7.6 LOW-POWER DESIGN

Cellular telephones, TV remote controls, digital multimeters, "floating" instrumentation for isolation, modems, security systems, video camcorders, and laptop computers have one common denominator in design. They are low-power devices. Low power is used for a number of reasons: portability, isolation, battery power, and low heat dissipation. Power is a function of frequency, load capacitance, and voltage; reduction in any of these will reduce the consumption of power by your system. (This really applies only to complementary metal-oxide semiconductor [CMOS] components, the most prevalent in low-power devices, not to certain other logic families such as emitter-coupled logic [ECL].)

$$P = f C V^2 ,$$

where P = power
f = frequency
C = load capacitance
V = DC supply voltage

These seven guidelines in design will minimize power:

1. Lower clock frequency.
2. Lower supply voltage to digital logic.
3. Shut down unused circuits.
4. Put controller into sleep mode when not needed.
5. Terminate all unused inputs. Don't allow any to float. Noise will cause the outputs of gates to transition slowly, causing excess current demand that sometimes destroys the device.
6. Avoid slow signal transitions. Outputs of gates that transition slowly draw excess current, in CMOS and transistor-transistor logic (TTL). (Note: Faster transitions conflict with reducing noise; you will have to strike the balance!)
7. Make normal states use the lowest current; for instance, LEDs should be off.

7.7 NOISE AND ERROR BUDGETS

No system can perfectly reproduce a signal. Physical limits in the components establish noise floors and error baselines beyond which useful information is lost. Three types of errors occur in electronics. Two of them are variations in parameters due to production and environment; the third is noise that is described statistically. Anytime you design a system, you should know these fundamental limits.

Production variations, the first type of error, place an uncertainty on the exact value of a component used in a circuit and therefore can cause uncertainty in the operation of the circuit. Statistical variation in the exact value of a capacitor, for example, prevents the exact calculation of the frequency response for a filter. (Electrolytic capacitors have a +80%/–20% tolerance in value.)

The second type of error comes from environmental influences; for example, temperature can induce drift in the offset of an operational amplifier.

The third type of error is noise within each device. There are four categories of internal noise:

1. Johnson, or thermal, noise places a lower limit on the noise voltage in any detector, amplifier, or signal source. Johnson noise has a flat power spectrum and is "white" Gaussian noise.

$$V_{\text{noise(rms)}} = \sqrt{4kTRW}\,,$$

where k = the Boltzmann constant = 1.38×10^{-23} J/K
T = absolute temperature (K)
R = resistance (Ω)
W = bandwidth (Hz)

Example 7.7.1 A 100-kΩ resistor at room temperature (25°C) has a root-mean-square (rms) noise voltage of 12.8 μV for a 100-kHz bandwidth.

2. Shot noise is the transfer of a quantum of charge. In semiconductors it is the random diffusion of carriers and the generation and recombination of electron-hole pairs. Shot noise is white and Gaussian.

$$I_{\text{noise(rms)}} = \sqrt{2qI_{\text{DC}}W}\,,$$

where q = 1.60×10^{-19} C
I_{DC} = DC current (A)
W = measurement bandwidth (Hz)

Example 7.7.2 A steady current of 10 mA fluctuates 5.7 nA over a bandwidth of 10 kHz due to shot noise.

3. Flicker, $1/f$ or "pink" noise varies with frequency and is ubiquitous in physical systems.

$$V_{\text{noise(RMS)}} = V_f[0.392 + \log_{10}(f_{\text{high}} / f_{\text{low}})],$$

where V_f = noise at a fixed frequency (V/(Hz)$^{1/2}$)
f_{high} = high-frequency corner of bandwidth (Hz)
f_{low} = low-frequency corner of bandwidth (Hz)

The constant of 0.392 adjusts for the noise equivalent bandwidth. The high-frequency corner for the noise equivalent bandwidth is $\pi/2$ times the actual. The low-frequency corner for the noise equivalent bandwidth is $2/\pi$ times the actual. This means that the noise equivalent bandwidth is $\pi^2/4$ (or 2.467) times greater than the specified bandwidth (Fazekas, 1988).

Example 7.7.3 A carbon-composition resistor with a noise of 1 μV/(Hz)$^{1/2}$ at 1 kHz has a $1/f$ noise of 4.1 μV for a frequency band between 10 Hz and 50 kHz.

4. Interference is the coupling of unwanted energy into a device from outside sources. Prime examples are 50-Hz or 60-Hz AC power noise (*buzz*), noise from power equipment, and radio and television transmissions. Even though this type of noise is often the most difficult to diagnose, you have more control over it than other sources. Chapter 6 is devoted to eliminating interference by careful design.

Each electronic component may be modeled as an ideal device with a noise voltage and noise current referenced to its input. Noise or errors accumulate through the root sum square of the component values (assuming that the sources of error are Gaussian) (Cripps, 1989):

$$V_{\text{total}} = (V_1^2 + V_2^2 + \ldots + V_n^2)^{1/2},$$

where V_{total} = total noise (V)
$V_1, V_2, \ldots V_n,$ = individual noise components (V)

Another source of noise is quantization error from analog-to-digital conversion. The quantization noise or error contributes a value equal to the least significant bit (LSB) divided by $(12)^{1/2}$ or 3.464. Finally, the signal-to-noise ratio (SNR) provides a measure of the fidelity of a system.

$$SNR = 10 \log_{10}(V_{signal}^2 / V_{noise}^2),$$

where SNR = signal-to-noise ratio
V_{signal} = full-scale amplitude of the signal (V)
V_{noise} = total noise amplitude of the system (V)

Garrett (1987) goes into great detail on noise, errors, and error budgets and provides several good examples of error budgets. Table 7.5 lists the system errors that you will need to consider and budget for.

Table 7.5 Error budget considerations for a complete system

Section	Component	Description of possible errors
Input	Sensor	Inherent signal noise
	Transmission line	Interference, crosstalk
	Amplifier	Temperature dependence, linearity, CMRR, SNR
	Filter	Variations in component values, bandwidth, pass-band and stop-band ripple, phase linearity
Conversion and processing	Sample-and-hold	Aperture jitter, droop
	Analog-to-digital conversion	Quantization error, differential nonlinearity, integral nonlinearity, aliasing
	Digital filtering	Round-off error in calculations
Output	Digital-to-analog conversion	Quantization error, differential nonlinearity, integral nonlinearity, aliasing
	Level shifting amplifier	Temperature dependence, linearity, CMRR, SNR
	Filter	Variations in component values, bandwidth, pass-band and stop-band ripple, phase linearity

7.8 STANDARD DATA BUSES AND NETWORKS

Many systems have a data bus or network to communicate with other circuit boards or devices. The data bus or network is the backbone of the design and represents a significant portion of the system architecture. Consequently, issues that affect the bus or network affect the entire system.

7.8.1 *Bus Architecture Concerns*

Basic issues in bus architecture are the drive configuration, termination, and signal handshakes. The *Digital Bus Handbook,* edited by Joseph Di Giacomo (1990) provides in-depth coverage of a number of bus architectures.

Drives may have either a single-ended or a differential configuration. A single-ended drive uses one trace or signal line for the transmitted signal and shares the circuit ground for the return signal. A differential drive transmits two signals with reversed polarity on two separate lines; differential drives have greater tolerance for noise than single-ended drives because they reject common-mode noise. Generally, a single-ended drive is satisfactory for signal paths that are short compared with the wavelength of the signal, such as those found in traces on PCBs, or for slower communications, such as RS-232 serial lines. For long cables, a differential drive is far better.

One of the problems associated with driving a bus is that the multiple outputs of transistors connected in parallel present a considerable capacitive load to the drive circuit. Because the capacitive load takes a nonzero time to charge, it slows the transition time of the signal. Fig. 7.14a illustrates an example of capacitive loading on a bus signal line. One solution to the problem of capacitive loading is to isolate the intrinsic capacitance of each transistor from the signal line with a Schottky diode (which has low series capacitance) as shown in Fig. 7.14b. This arrangement

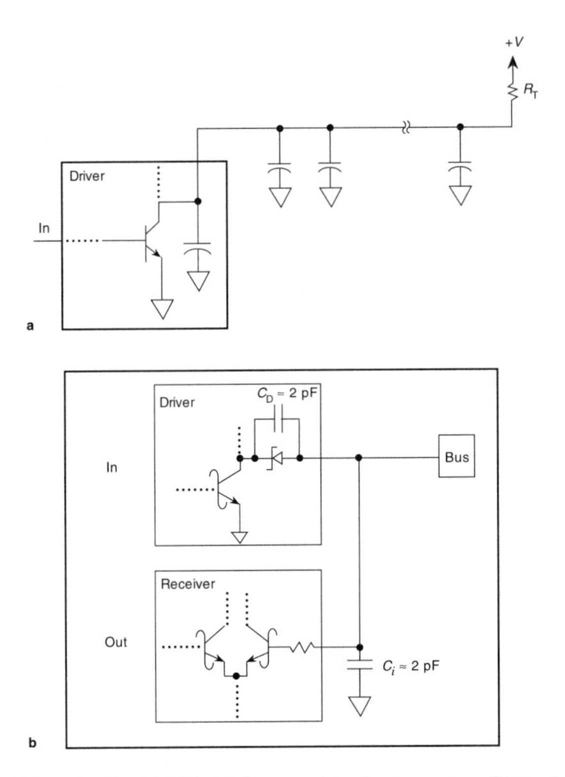

Fig. 7.14 Bus driving circuits. (a) TTL driving capacitive loads on a bus. (b) Back-plane transceiver logic interface circuits isolate capacitive load from the bus.

is called back-plane transceiver logic (BTL) and has been incorporated into the Futurebus+ specification. BTL is much faster than conventional TTL or CMOS drivers because it does not have to wait for signal reflections to settle on the bus, which is a nearly ideal transmission line with matched impedance.

Buses can use either synchronous or asynchronous protocols to transfer data. Synchronous operation shares a clock signal among circuits or boards and transfers data in specific phase to the clock pulses. Synchronous operation is often oriented toward the transmission of blocks of data. Asynchronous operation uses handshake signals to initiate and terminate data transfers; the data tend to be character oriented.

7.8.2 *Serial Communications*

Serial communications transmit and receive 1 bit at a time. Even though serial communications are inherently slower than parallel buses, the serial format minimizes the number of signal conductors and makes it practical to transfer data over long distances. Serial communications make possible a wide variety of schemes to transfer data, such as modems, computer networks, avionics control systems, and distributed manufacturing controls.

Data can flow between devices or systems in three ways: simplex, half duplex, and full duplex (Fig. 7.15). The drive implementation for serial communications may be single-ended or differential, as illustrated in Fig. 7.16.

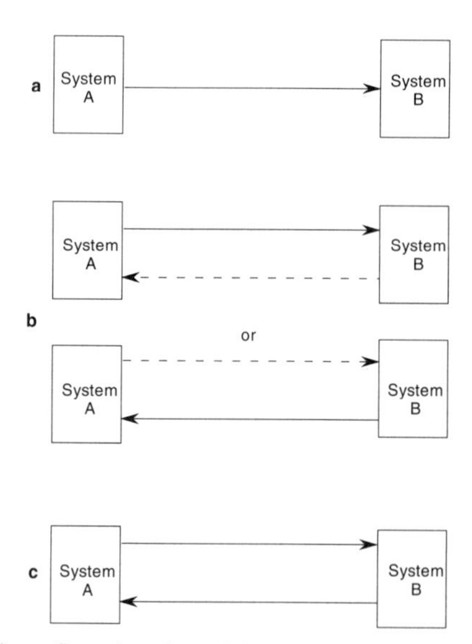

Fig. 7.15 Three basic configurations for serial transmission. (a) In simplex, data flows in only one direction. (b) In half duplex, data flows in both directions but not simultaneously. (c) In full duplex, there is simultaneous two-way communication.

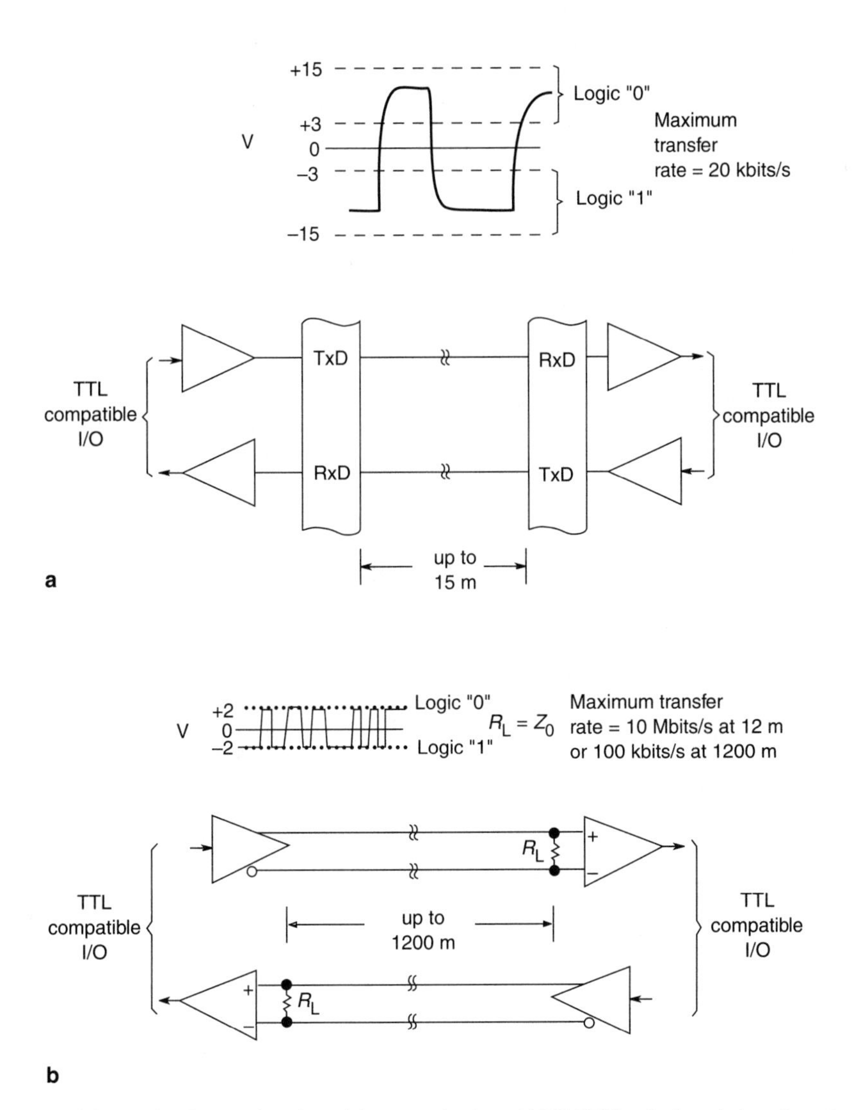

Fig. 7.16 Two implementations for serial communications. (a) RS-232C, a single-ended configuration. (b) RS-422A, a differential configuration.

The transmission protocol may be synchronous or asynchronous. Many computer peripherals use the RS-232 standard and communicate at rates of 1200, 2400, 4800, 9600, 19,200, and 38,400 bits/s. The RS-232 standard is quite old, dating to the 1960s, and has evolved into a number of configurations, as shown in Fig. 7.17. Therefore, understand and clarify the configuration of the communication ports on all the devices and systems before designing a serial link between

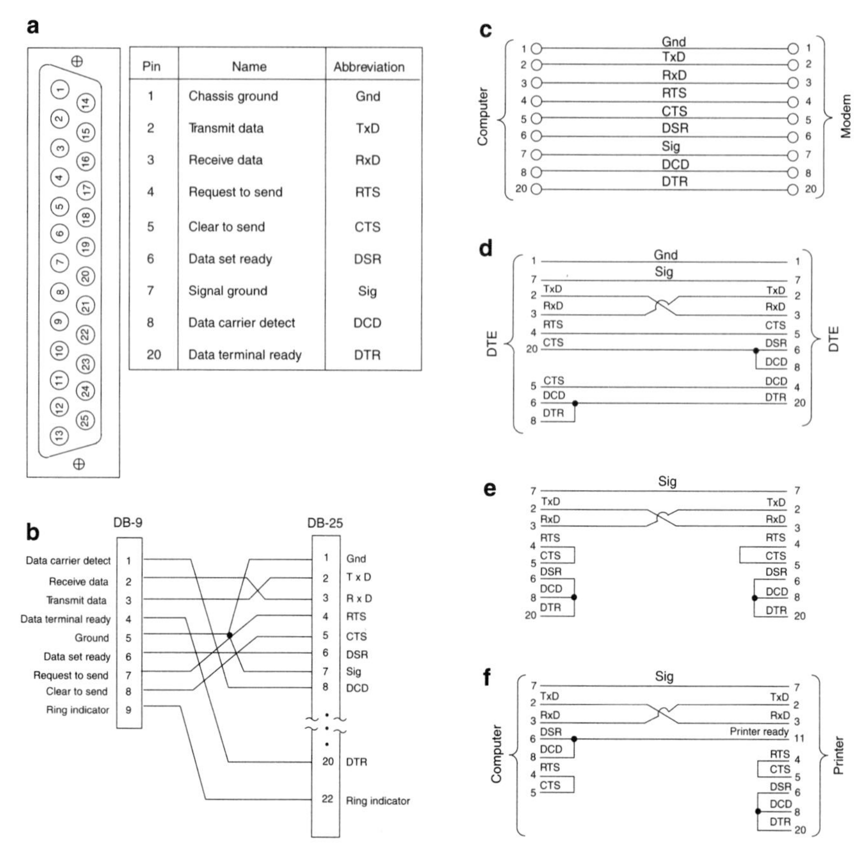

Pin	Name	Abbreviation
1	Chassis ground	Gnd
2	Transmit data	TxD
3	Receive data	RxD
4	Request to send	RTS
5	Clear to send	CTS
6	Data set ready	DSR
7	Signal ground	Sig
8	Data carrier detect	DCD
20	Data terminal ready	DTR

FIG. 7.17 Definition of RS-232 signals and some common configurations. (a) DB-25 connector. (b) An adapter cable from DB-9 to DB-25 connector. (c) Modem cable. (d) Null-modem cable to connect computers together. (e) A minimum-wire null-modem cable. (f) A printer cable.

them. The reference by Putnam (1987) is an invaluable aid when designing in the RS-232 environment.

7.8.3 *Instrumentation and I/O Buses*

Serial data transfer is slower than parallel. Peripheral devices that demand higher rates of data transfer use one of several parallel I/O buses. The architectures of these buses define widths of data paths, transfer rates, protocol, cable lengths, and connector configurations (Table 7.6). IEEE-488 evolved from an instrumentation network developed by Hewlett-Packard in the 1970s; it is used primarily in test and measurement instruments. SCSI generally connects high-speed computer peripherals, such as hard disk drives, to the main processor board. Finally, the

Table 7.6 Comparison of several peripheral I/O buses

Bus	Maximum data transfer (Mbytes/s)	Data width (bits)	Address width (bits)	Number of control lines	Number of peripheral devices	Connector pins	Maximum cable length (m)	Timing	Control
IEEE-488	1	8	—	8	15	24	20	A	Instrumentation networks
SCSI	2 (asynchronous) 5 (synchronous)	8 + parity	—	9	7	50	6 (single-ended) 24 (differential)	A, S	Primary to support computer peripherals like disk drives and CD-ROMs
SCSI-2	10, 40	16, 32	—	9	7	50	6, 24	A, S	
Sbus	100	32	32	18	—	96	—	A, S	I/O coprocessing, supports only three circuit boards
Turbochannel	100	32	32	11	—	96	—	S	I/O address = 27 bits, board size = 11.6 × 14.4 × 3.3 cm.

Note: A = asynchronous; S = synchronous.

Sbus and Turbochannel buses are high-capacity, high-data-rate extensions of workstations to support sophisticated coprocessing systems.

7.8.4 *Back-Plane Buses*

When you move from peripheral devices to processing boards within an enclosure, you need to choose a back-plane bus to connect the circuit boards together. A number of configurations exist that vary in physical size and data transfer capacity. The architectures of back-plane buses define widths of data paths, transfer rates, protocol, and connector configurations (Table 7.7).

The STD 32 and G-64 buses have small form factors and target process control and industrial instrumentation. The EISA, Micro Channel, and NuBus cards serve coprocessing and instrumentation applications that are based on desktop or personal computers. Multibus II and VXIbus systems address sophisticated instrumentation and control problems. VMEbus and Futurebus+ back planes provide a basis for reconfigurable processors and computers. Fig. 7.18 illustrates the differences in sizes of the circuit boards for these back-plane buses.

7.8.5 *Local Area Networks*

Business communications and distributed processing have driven the explosive expansion of computer networks. Entire sections of libraries are devoted to networks; I will introduce the concept of networks only in the context of incorporating them into a system design.

Local area networks (LANs) facilitate communications and support databases between computers. The connections within a LAN affect the performance of the network and define the topology of the LAN; three prominent topologies are ring, bus, and star. The medium for transmission may be one of four types of cable:

1. Unshielded twisted-pair wire
2. Shielded twisted-pair wire
3. Coaxial cable
4. Fiber optic cable

Finally, the layers of communication on a LAN are shown in Fig. 7.19.

LANs have three prominent protocols for data transfer: Ethernet, token ring, and Arcnet. AppleTalk is a higher-level protocol, operating at 230 kbits/s, that can operate on networks with either Ethernet or token ring protocols.

Ethernet is a 10-Mbit/s protocol that is called carrier-sense, multiple-access with collision detection (CSMA/CD). Each node operates in the following manner:

1. Wait for activity on the network to stop.
2. Transmit.

Table 7.7 Comparison of several back-plane bus configurations

Back-plane bus	Maximum data transfer (Mbytes/s)	Data width (bits)	Address width (bits)	Multiplexed data/ address	Timing	Drive type	Connector type	Connector pins	Interrupts	Physical dimensions W × L (cm)	Comments
STD 32	20	8, 16, 32	16, 32	—	A	TTL	CE	108	1	11.4 × 16.5	Wide industrial base; robotics process control, data acquisition
G-64, G-96+	2, 40	8, 16, 32	24	—	A	TTL	DIN	96	6	9.9 × 16.0	Single Eurocard
VMEbus	40	16, 32	16, 32	√	A	TTL	DIN	96	7	23.3 × 16.0	Single or double Eurocard instrumentation computing
VXIbus	1000	8, 16, 32	16, 24, 32	—	A	TTL	DIN	192	7	23.3 × 34.0 or 36.6 × 34.0	Superset of VME instrumentation
Multibus II	40	16, 32	32	√	S	TTL	DIN	96	1	21.6 × 25.7	Instrumentation
Nubus	40	32	32	√	S	TTL	DIN	96	1	10.2 × 30.0	One interrupt per slot
EISA	33	8, 16, 32	16, 24	—	S	TTL	CE	198	11	10.7 × 34.3	PCs
Micro Channel	17	8, 16, 32	16, 24, 32	—	A	TTL	CE	198	11	7.9 × 29.2	IBM PS/2
Futurebus+	400	32, 64	32, 64	√	A	BTL	NA	192	NA	26.5 × 30.0	

Note: NA = not available.

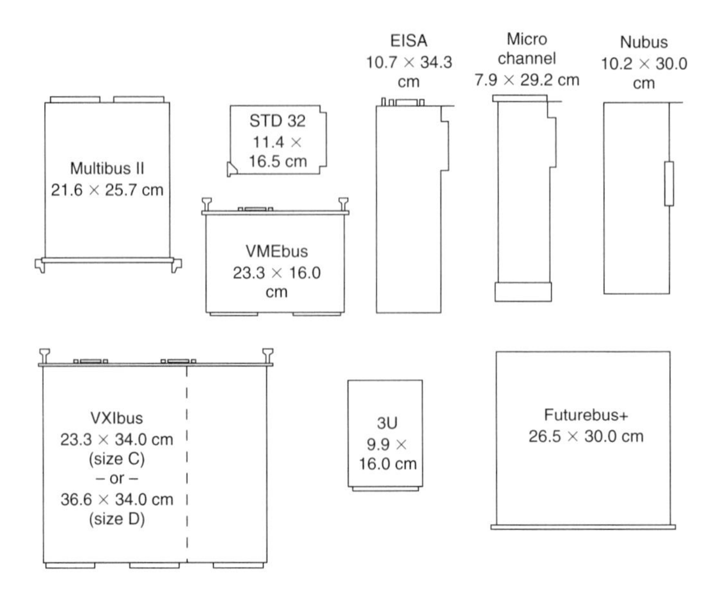

Fig. 7.18 Relative sizes of circuit boards that fit into different back-plane buses.

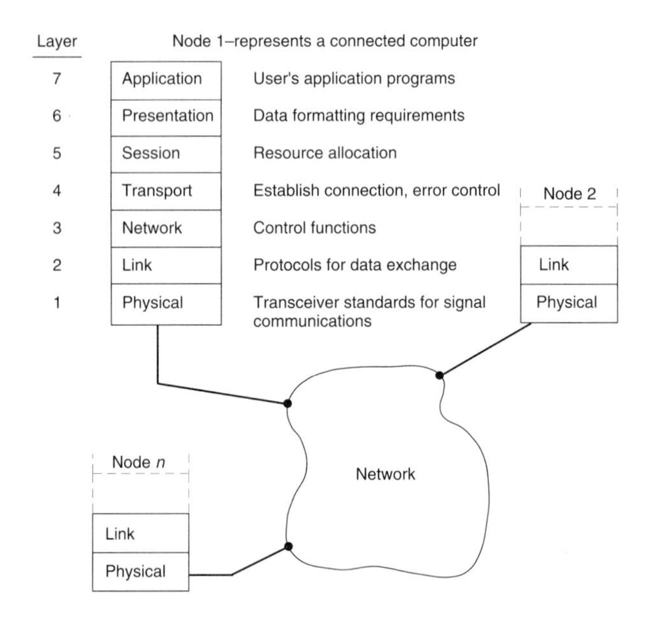

Fig. 7.19 Open system interconnection model for network protocol.

3. Sense if simultaneous transmissions occur.
4. Wait for a random length of time—shortest interval begins transmission.
5. Stop transmission after a prescribed duration.

The actual rate of data transfer will be more like 800 kbits/s because most of the capacity is taken up by software overhead (see Fig. 7.19).

Token ring is a 4- to 16-Mbit/s protocol in which a special message, called a *token*, circulates among nodes. The node with the token gets access to the network, if it needs it, for a limited time.

Arcnet is a 2.5-Mbit/s protocol that polls each node for a message. Each node has a unique address, from 1 to 255, and the lowest address is the controller. Arcnet operates as follows:

1. The controller polls each node for messages.
2. If there is no message, the node stays silent.
3. Otherwise, the node transmits for a prescribed duration.

7.9 RESET AND POWER FAILURE DETECTION

All systems should initialize to a known state whenever power is applied. A reset circuit generates a signal that prevents the generation of unwanted conditions by the system during power application. The reset signal forces the processor to begin execution from a fixed memory location that has code for initializing system operation. The reset signal should also set (or clear) critical output signals to states that do not cause undesirable actions. Reset circuits, therefore, are common to all systems, and you should consider their function early in design.

Reset circuits sense the voltage level of the power supply and generate a reset signal when it falls below a preset value. Upon application of power to a circuit, the reset signal stays active until the voltage of the power supply exceeds the preset value; this action avoids unusual functioning by the system while the supply voltage adjusts to acceptable levels. Fig. 7.20 illustrates several reset circuits.

The simple reset circuit in Fig. 7.20a uses the time constant of the R_1C network to set the desired duration of the reset signal. The Schmitt trigger inverter transforms the exponential charging waveform on the input to a signal transition appropriate for logic gates. The diode, D, allows charge to drain off the capacitor if the DC voltage fails, thereby protecting the inverter from an input voltage higher than its supply voltage. When pressed, the manual push button shorts the charge on the capacitor to ground to generate a reset signal so that a user can initialize the operation of the system even while the supply power is stable. The weakness of this design is that it does not detect failure of the input power.

The reset circuit in Fig. 7.20b provides several additional functions beyond detection of power failure. It has manual reset, it can warn of significant drops in voltage on the unregulated DC power, and it provides a watchdog timer. The watch-

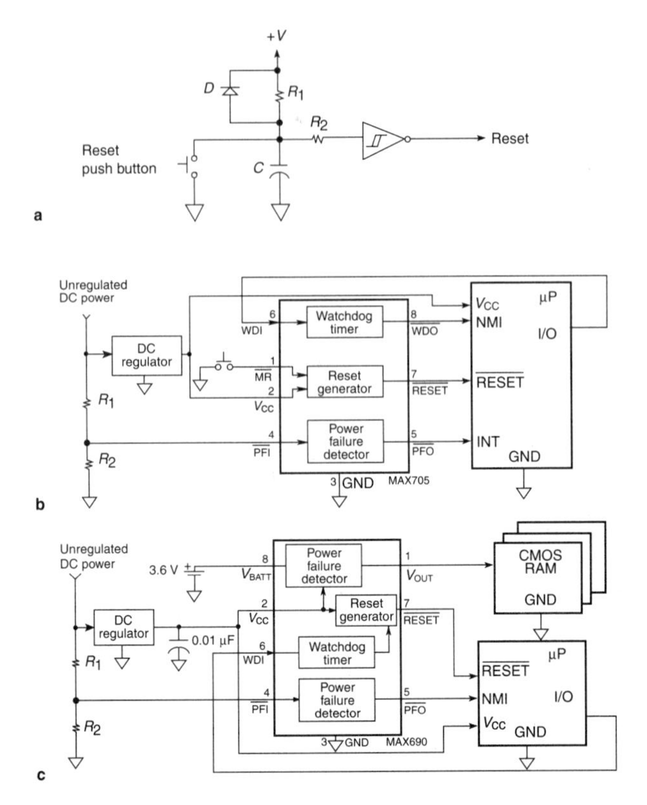

Fig. 7.20 Several circuits for generating reset and detecting power failure. (a) A simple reset circuit. (b) The MAX705 provides a debounced manual reset and a watchdog timer. (c) The MAX690 has battery backup as well as a watchdog timer.

dog timer indicates possible hang-ups in the execution of software within a processor. If the processor fails to signal the watchdog timer within 1.6 s, the watchdog generates an interrupt signal to the processor. The processor then executes recovery routines to resolve the problem.

The reset circuit in Fig. 7.20c switches over to battery backup if power fails and provides a watchdog timer. The resistors, R_1 and R_2, set the lower level of variations on the input voltage that triggers an indication of power failure to the system processor.

Some sophisticated circuits monitor the AC voltage to the power supply and determine when the power will fail during power outages or brownouts (voltage sags). These circuits can provide 10 to 20 ms warning of a power failure over a circuit that senses the DC voltage level of the power supply.

Power supplies and regulators are significant portions of any system, and Chapter 9 covers them in more detail.

7.10 INTERFACE: INPUT

The input to all circuits is some sort of electrical signal. Each signal comes from another circuit, a transducer, or a switch. Most signals need some preprocessing or conversion before the system can assimilate them. For example, switches generate logic transitions that bounce when pressed; that is, there is a series of rapid glitches at the beginning and end of the signal pulse. Generally, you will have to design some circuitry to suppress the glitches produced by bounce. On the other hand, sensors often produce continuously changing analog signals that must be converted to digital logic levels for further processing. Each of these concerns is common to all systems that you will design.

Early in development you will need to define the types of inputs that you expect the system will receive. Once you know the type of input, you can decide on the necessary circuitry to manipulate the input signals.

7.10.1 *Switches*

Switches provide electrical continuity between two or more signal lines. Like a drawbridge controlling traffic, a switch controls charge flow from one line to another. Switches have a variety of configurations: push button, toggle, slide, and rotary (Fig. 7.21).

Charge flow in a switch may simply be through mechanical contact such as a point of contact (Fig. 7.22a) or membrane (Fig. 7.22b), or the charge may flow through a change in capacitance (Fig. 7.22c). Even blocking of a beam of light can be used in a switch (Fig. 7.22d).

The main problem with switches is that the mechanical contacts will resonate or bounce when pressed, causing the current path to open and close many times within milliseconds and leading to high-frequency glitches on the signal lines. These glitches occur at both the beginning and the end of switch activation and, depending on the size and stiffness of the contacts, last between 10 and 50 ms.

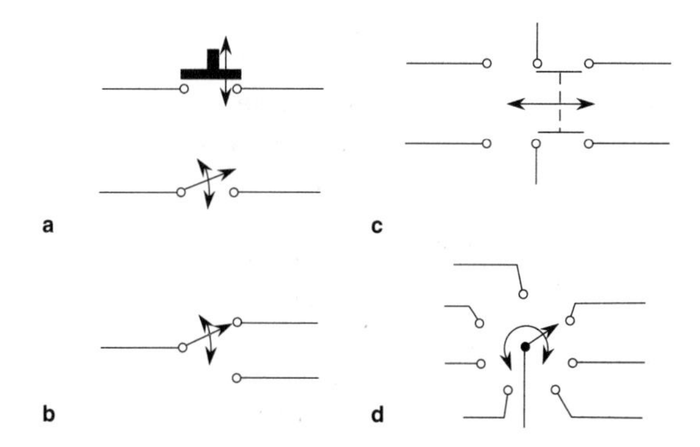

Fig. 7.21 Types of switches. (a) Push button. (b) Toggle. (c) Slide. (d) Rotary.

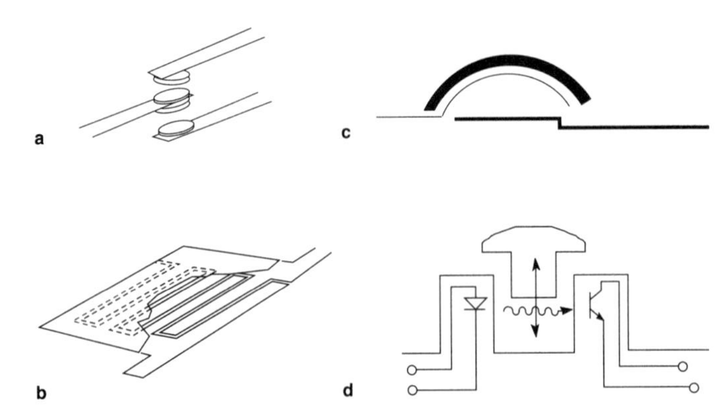

Fig. 7.22 Several ways to make electrical contact: (a) Mechanical contacts. (b) Membrane. (c) Capacitive. (d) Optoelectric.

Fig. 7.23 illustrates some bounce found on the signal lines of switches as well as the circuits to *debounce* or suppress the glitches.

One trade-off that you will want to investigate is when to use a microcontroller to debounce inputs versus when to use discrete logic. This is particularly important for keyboards, in which a large number of switch keys would demand many components for a design using discrete logic. A simple microcontroller can easily sample all the keys rapidly, debounce the inputs, and signal the appropriate input through a serial connection to an embedded processor.

7.10.2 *Sensors*

Sensors transduce signals from the physical domain into electrical signals. Generally the physical signals are properties such as temperature, pressure, force, motion, flow, light, or sound. Table 7.8 lists some sensors and their properties that you may encounter in your designs. Each has its peculiar characteristics, capabilities, and shortcomings.

Sensors are characterized by the nine parameters listed below (Cripps, 1989; Derenzo, 1990; Sheingold, 1981). They help determine the type and amount of signal preprocessing that must be performed before transmitting the input from a sensor to a digital processor.

1. *Threshold:* the minimum value of a measurable quantity
2. *Sensitivity:* the change in output for a unit change in the input
3. *Full range:* the maximum measurable value
4. *Linearity:* the maximum deviation from a straight line between threshold and full range

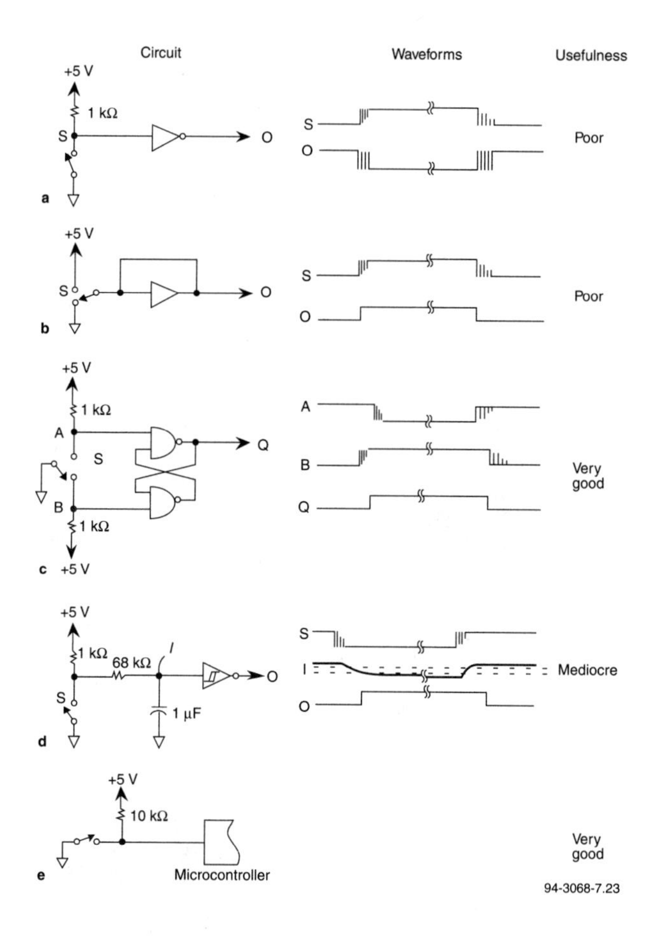

Fig. 7.23 Circuits to remove switch bounce from the input signal. (a) Circuit without debouncing. (b) Active feed around a buffer can debounce a signal. (c) Debounce circuit using an RS flipflop. (d) Debounce circuit using an RC filter and a Schmitt trigger inverter. (e) A software delay can debounce an input signal. (A, B, S = points for switch contacts; O = output signal; Q = latched output signal.)

5. *Accuracy:* the difference between the measured and the true value
6. *Precision:* how well the measurements agree under repeated, identical conditions
7. *Stability:* the ability to maintain the same response and noise level over time
8. *Hysteresis:* how much the measured value depends on previous values
9. *Noise:* input from sources other than the desired source

Table 7.8 A sample of sensor inputs to an electronic instrument

Measured property	Sensor type	Characteristics
Displacement	Strain gauge	Low-level, nonlinear signals
	LVDT	Nonlinear—use feedback to linearize
	Potentiometer	Simple use; mechanical wear limitations
Motion	Hall effect	Mechanical/electrical isolation
	Optoelectronic	Mechanical/electrical isolation
	Tachometer/resolver	Complex mechanics
Acceleration/force	Strain gauge	Low-level signal needs conditioning, nonlinear
	Semiconductor strain gauge	Simple circuit integration; temperature sensitive; good force sensitivity
	LVDT	Low acceleration range; nonlinear
	Piezoelectric	High output impedance; needs charge amplifier conditioning
Temperature	Thermocouples	Low-level signals; poor linearity
	Thermistors	Low cost; high sensitivity; nonlinear
	Resistance temperature detector	Accurate, repeatable, fairly linear, heat dissipation is a concern
	Silicon semiconductor	Simple circuit integration
Light	Phototransistor/photodiode	Submicrosecond response, simple circuit integration
	Photomultiplier	Exquisite sensitivity; needs kilovolt supplies
Pressure	Switch	Simple design and use
	Strain gauge	Low-level signal needs conditioning, nonlinear
	Electromagnetic flowmeter	Needs contact with conductive fluid
	Variable capacitance	Small; nonlinear
Flow	Pitot tube	Nonlinear; uses pressure transducer
	Turbine impeller	Mechanically invasive; simple conditioning
	Matched transistors	Simple circuit integration
	Variable capacitance	Small; nonlinear
Sound	Microphonic	Cheap; simple to use
	Ultrasound	Requires drive circuitry; pressure and temperature sensitivity
Video (light illuminated scenes)	CCD camera	Very high data rates

7.10.3 *Analog Preprocessing*

Most sensors generate an analog electrical signal that must be transformed in some way for further processing. Fig. 7.24 illustrates the components found in many systems that preprocess the analog input. The sample-and-hold (S/H) block and the ADC block are often combined within an IC. They represent important issues for the system and circuit designer and are discussed below.

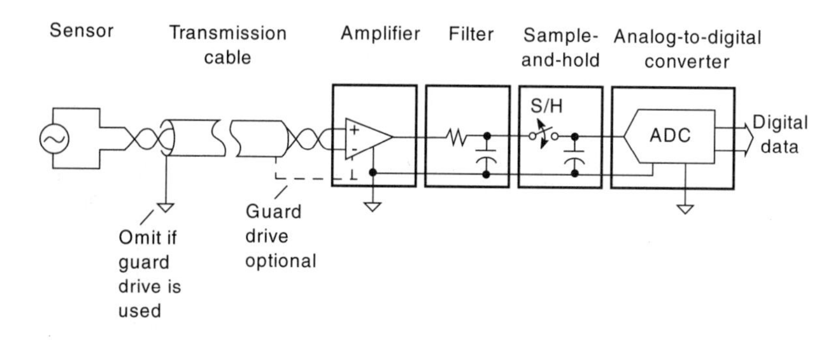

Fig. 7.24 Generalized analog input to a system.

When sensors reside at a distance from the rest of the system, the signals from the sensors must be transmitted to the system. Circuits for signal transmission can use several different configurations (Fig. 7.25). The signal can be an analog voltage (Fig. 7.25a) or an analog current (Fig. 7.25b). A current driver requires only two wires to support data transmission; it derives power from the signal current loop and transmits data by modulating the current in the wires. To increase the immunity to noise, the transmission circuitry can either use frequency modulation (FM) to transmit the signal from a sensor (Fig. 7.25c) or transform the signal into the digital domain with an ADC and transmit a digital signal (Fig. 7.25d). Note that the circuitry in Fig. 7.25c and d combines the two functions of transmission and digital conversion, which may be desirable to reduce component count in some systems.

Sometimes a DC bias must be added or subtracted from the sensor's signal. Fig. 7.26 illustrates two common circuits for adjusting the DC voltage: a general configuration for adding or subtracting voltages and a ground reference that provides a virtual ground midway between 0 and 5 V. The virtual ground allows bipolar inputs to circuits with a single voltage supply, such as a battery-powered instrument.

Some situations require isolating the sensor signal from the system, usually for safety reasons. Isolation cuts the direct current path; that is, it removes the DC bias from the signal. Isolation uses one of three physical principles:

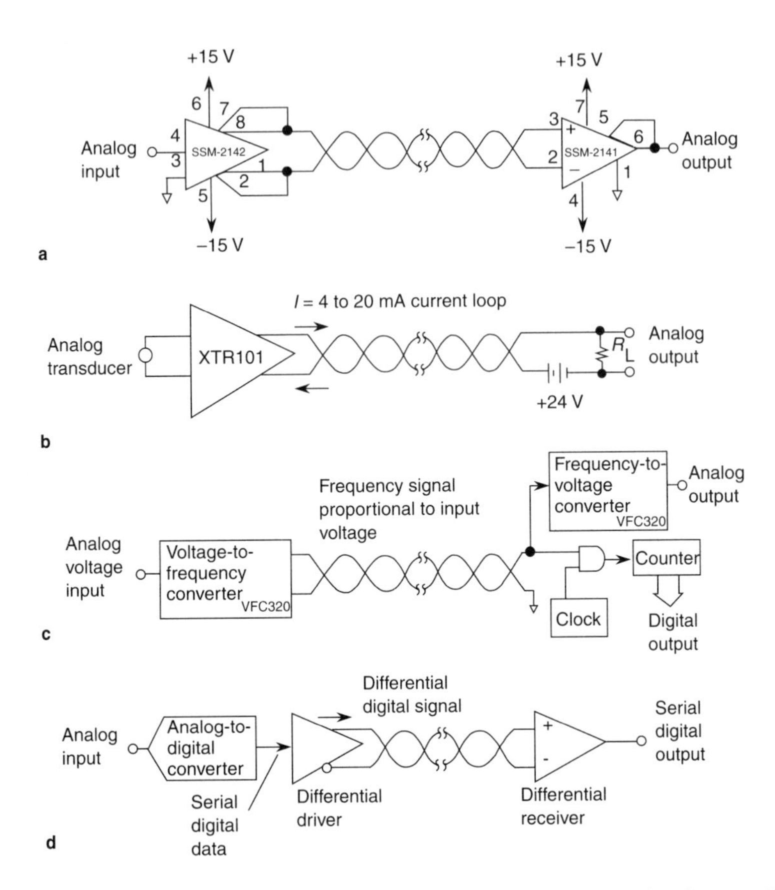

Fig. 7.25 Forms of analog signal transmission. (a) Balanced line driver and receiver for transmitting signals in the audio bandwidth (SSM-2141 and SSM-2142 from Analog Devices). (b) A 4- to 20-mA two-wire transmission system that relies on the magnitude of current, not voltage, to represent the analog value (XRT101 from Burr-Brown Corporation). (c) Voltage-to-frequency conversion and transmission provide a noise-immune signal (VFC320 from Burr-Brown Corporation). (d) Analog-to-digital conversion and digital transmission provide a noise-immune signal but require more complex circuitry.

1. *Optoisolation* has a nonlinear transfer function.
2. *Capacitive isolation* has linear transfer but low current capacity.
3. *Transformer isolation* has linear transfer, but magnetic components can saturate and tend to be bulky.

Optoisolation typically provides 2500 V isolation and is best suited for digital signals or switched inputs (Fig. 7.27a, b, and c). For capacitive or transformer isolation,

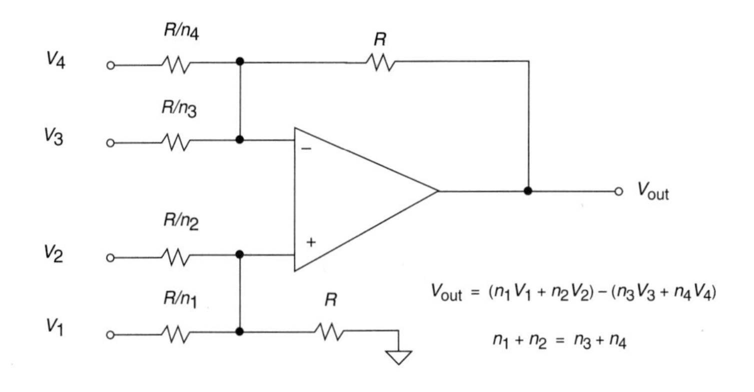

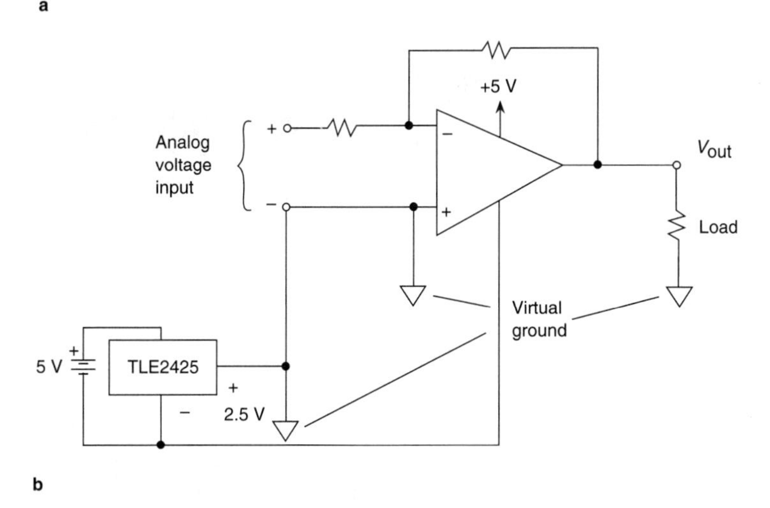

Fig. 7.26 Circuits for adjusting DC voltage. Level converters translate voltage to a range acceptable for analog-to-digital conversion. (a) General circuit to add DC offset. (b) Ground reference IC for circuits with a single voltage supply (TLE2425 from Texas Instruments).

the isolation circuit may be integral to the input amplifier (Fig. 7.27d and e and Fig. 7.28).

One of the most common steps for preprocessing is to use a low-pass or band-pass filter on the analog input from the sensor. Filtering removes noise components that are outside the frequency band of interest. Often a low-pass, or antialias, filter will be used to remove high-frequency signals from the input to an ADC.

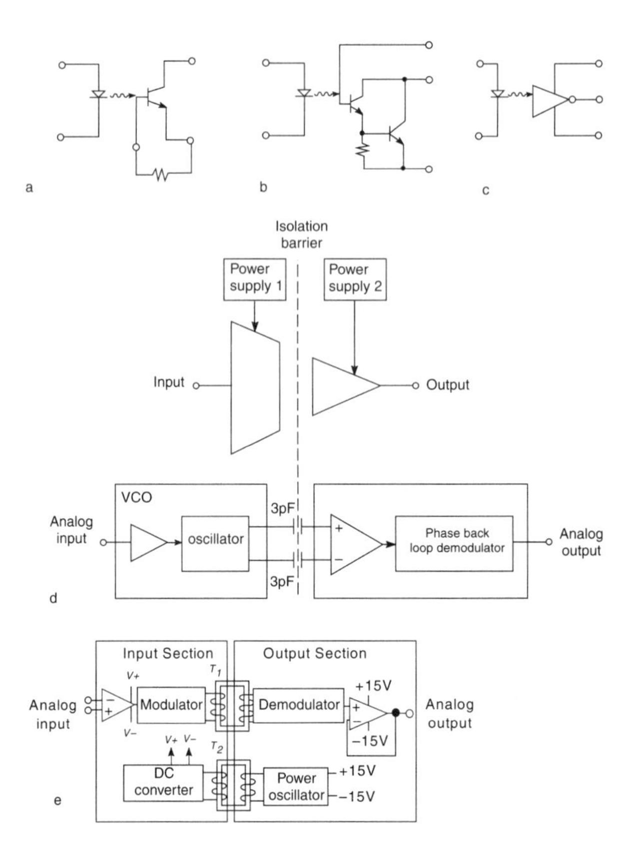

FIG. 7.27 Forms of signal isolation. (a) Optoisolation. (b) Optoisolation using a Darlington pair for increased gain. (c) Optoisolation with logic output. (d) Capacitive coupling for linear transfer and isolation of the signal. (e) Transformer coupling for transfer and isolation of the signal.

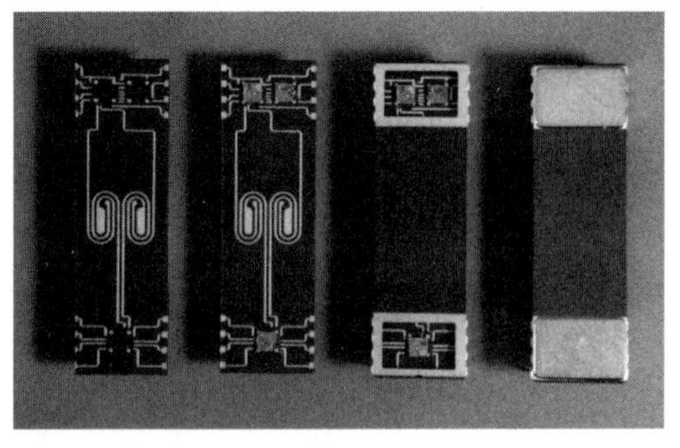

FIG. 7.28 Stages of manufacture in producing a signal amplifier with capacitive isolation (photograph courtesy of Burr-Brown Corporation).

7.10.4 *Analog-to-Digital Conversion*

Analog-to-digital conversion transforms an input signal from the continuous analog domain to the discrete digital domain. The ADC will set the limit on resolution of your system; greater resolution translates into more bits of digital data, which means higher accuracy, wider dynamic range, and slower conversion speed. Table 7.9 outlines and compares the five prominent architectures used in ADCs. Note that the conversion speed is inversely related to the resolution (Fig. 7.29) (Pryce, 1992).

The architectures of ADCs differ significantly. Each configuration has inherent capabilities making it suitable for particular applications (Table 7.9). The general features of each architecture are defined as follows:

1. A *flash converter* uses $2^n - 1$ comparators, one for each possible digital code in an n-bit converter. Each comparator taps a voltage reference from a precision resistor chain and compares the analog input with it. While extraordinarily fast, flash converters are limited by die size to 10 bits, 1023 comparators, and the precision of the resistor chain.
2. A *subranging converter* uses two stages of flash converters to reduce the number of comparators significantly. Because they take two or more clock cycles to perform a conversion, subranging converters are slower but less power hungry than flash converters, which need only one clock cycle per conversion.
3. A *successive-approximation converter* counts a reference signal generated by an internal DAC, and a comparator selects the count or digital code that matches the analog input. These converters need an S/H circuit to present a stationary analog value to the comparator for each conversion.
4. A *delta-sigma converter* uses oversampling, noise shaping, and digital filtering to generate digital codes. The analog components produce a high-frequency, single-bit data stream, with the analog signal represented by the average value of the frequency. The signal, in essence, is low frequency and the quantization noise of the comparator is shaped into a high-frequency region that the digital filter removes.
5. An *integrating converter* integrates the input signal during the sample period and then discharges the integrator during the measurement period. The length of time to discharge represents the amplitude of the input. Noise on the input is integrated out to give the converter an intrinsic ability to reject noise.

Several parameters characterize ADCs (see Chapter 12 for more detail).

1. *Missing codes:* specific digital values or codes that aren't produced at the output regardless of the value of the analog input. An ADC should have no missing codes at the desired resolution. You might use an ADC with more advertised resolution than you need to eliminate missing codes.
2. *Differential nonlinearity:* Differences in the step width between each digital

Table 7.9 Comparison between various configurations of ADCs

Configuration	Conversion speed (samples/s)	Resolution (bits)	Relative power consumption	Requires S/H?	Multiplex input?	Applications and comments
Flash converter 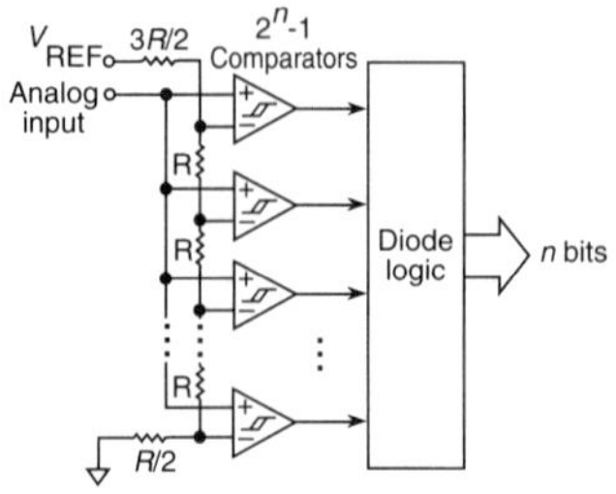	1M–500M	4–10	Highest	No, not for all applic-ations	Yes	• Communications • Radar • Digital oscilloscopes
Subranging converter 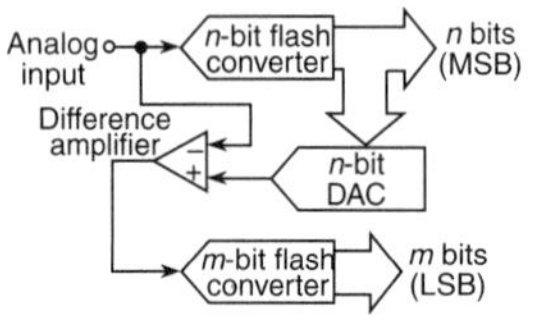	100k–40M	8–16	Medium	Yes	Yes	• Video image capture
Successive-approximation converter 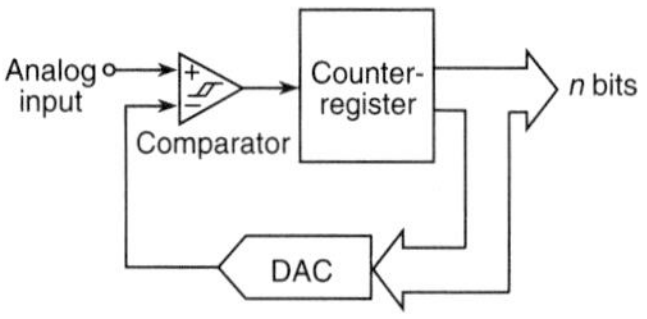	1–1000k	8–16	Medium to low	Yes	Yes	• Widest number of applications • Low cost
Delta-sigma converter 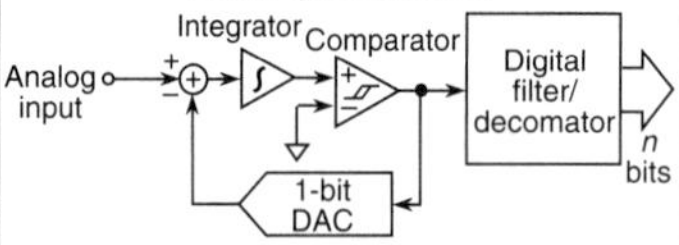	10–100k	16–20	Medium to low	No	No	• Digital audio • Instrumentation*
Dual-slope integrating converter 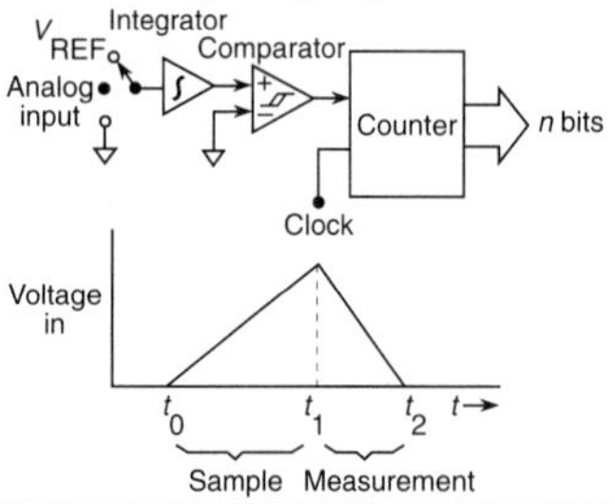	2–200	16–24	Low	No	No	• Digital voltmeters • Low frequency transducers – Pressure – Temperature

*Advantages: few precision components, simple linear phase digital filter.
Disadvantages: filter delays prevent multiplexing and loop stability, poor step response.

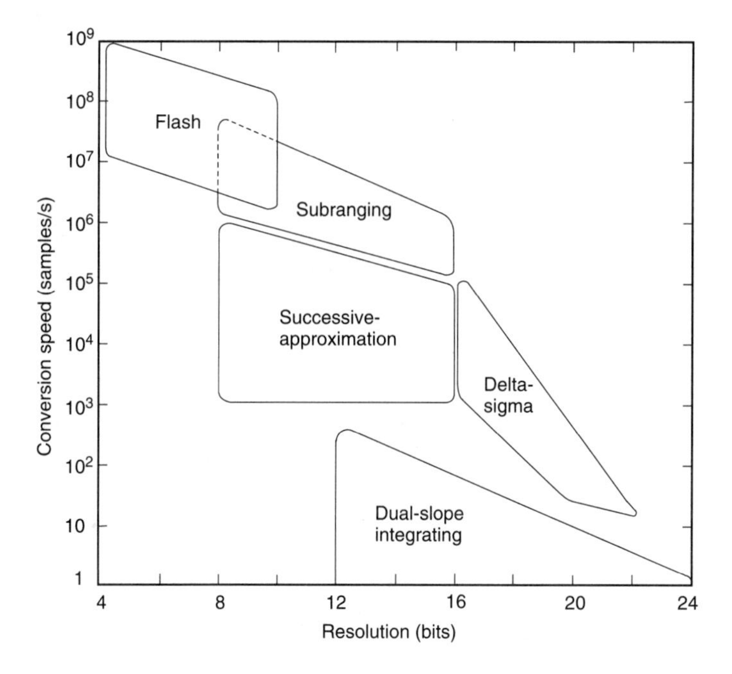

FIG. 7.29 Common regions of operation for various configurations of ADCs.

code. They vary from −1 LSB to greater than +1 LSB. A missing code is a differential nonlinearity of −1 LSB; strive for differential nonlinearity of less than ± 0.5 LSB in your designs.

3. *Integral nonlinearity:* the deviation from a linear transfer function. It is defined by the noise and SNR.
4. *Aperture jitter:* the uncertainty in the settling time of the S/H circuitry. This is a particular concern for high-speed ADCs.

These parameters relate to the fabrication and layout of the ADC circuitry; consequently, a monolithic or hybrid integrated circuit containing all the necessary components will almost always solve your design problem more efficiently than you could if you built up the circuit yourself.

7.11 INTERFACE: OUTPUT

Output circuitry completes the transfer of data and information through a system, from sensors and input circuitry, through internal processing, to the output stages. Outputs indicate conditions, actuate devices, and pass signals to other systems.

The different types of indicators and actuators used in outputs are not numerous. You will typically encounter only LEDs, lamps, and liquid crystal display (LCD) panels for display indicators, with an occasional gas-discharge display thrown in for good measure. (Computer screen monitors are also widely used as display indicators, but you will probably not design one in favor of buying it off the shelf. They are system design projects in themselves.) To design actuators, you will probably use only coils, relays, solenoids, and motors. They come in a variety of forms and sizes, however. Finally, you may have to generate analog signals with a DAC to pass on to other devices. That's really it for indicators and actuators.

As with the input interface, you will have to define early in development the types of outputs you expect the system to drive. Once you know the type of output, you can decide on the necessary circuitry to generate the output signals. Again, you should use monolithic or hybrid solutions for motor drives or DACs rather than build up the circuit yourself.

7.11.1 *LEDs*

Light-emitting diodes are small, low-voltage, low-power components that provide visual indication. They are simple to implement and can be found on circuit boards and front panels. The forward voltage drop across an LED may range from 1 to 2 V. Generally, you must put a current-limiting resistor in series with the LED to drive it from digital logic, as shown in Fig. 7.30. The value of a current limiting resistor ranges from 100 to 2000 Ω. Some models of LEDs have the resistor built in, thus reducing parts count and assembly time.

You can readily find LED displays of alphanumeric characters that are easy to use because they have integrated data latches and LED drivers. Beware! These components are power hogs.

7.11.2 *LCDs*

Liquid crystal displays are low-power panels that are suited for battery-powered applications. LCDs can provide large panels of characters and custom shapes and symbols; many laptop computers use them for their screens. While LCDs require AC drives, these generally are built into the LCD module and the module interfaces easily to digital logic.

7.11.3 *Lamps*

Incandescent lights in electronic systems? You've got to be joking! That's right, you'll find incandescent lamps in many applications that require a high level of luminance: turn indicators in automobiles, process controls for manufacturing, flashlights and torches, and warning lamps, for example.

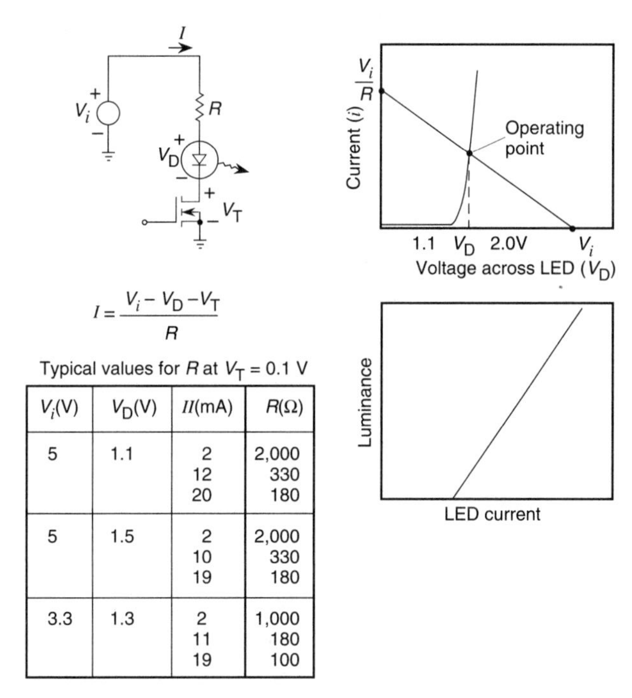

Typical values for R at $V_T = 0.1$ V

V_i(V)	V_D(V)	I(mA)	R(Ω)
5	1.1	2 12 20	2,000 330 180
5	1.5	2 10 19	2,000 330 180
3.3	1.3	2 11 19	1,000 180 100

Fig. 7.30 Current-limiting resistors. A load resistor limits current through an LED to extend its operating life and prevent a short circuit.

Incandescent lamps have several notable characteristics. They convert electrical energy to light by heating a filament to high temperature. The conversion process is inefficient, and lamps represent a significant demand on power. In addition, the filament has low resistance without power applied; turning on the lamp draws a nonlinear, transient current spike to heat the filament and raise its resistance. That spike, called *inrush current*, can be as much as 10 to 15 times the nominal current of a hot filament as shown in Fig. 7.31.

Example 7.11.3.1 Beware! Turning on several lamps simultaneously can generate a momentary but significant current drain on the power supply. I designed a system once that turned on 13 lamps simultaneously during a self-test. The first time I tried it, the transient caused by the inrush current forced the power supply into foldback current limiting and shut it down. I had to add a large power resistor in the power line to limit the inrush current and prevent the power supply from shutting down. Crude, but it works.

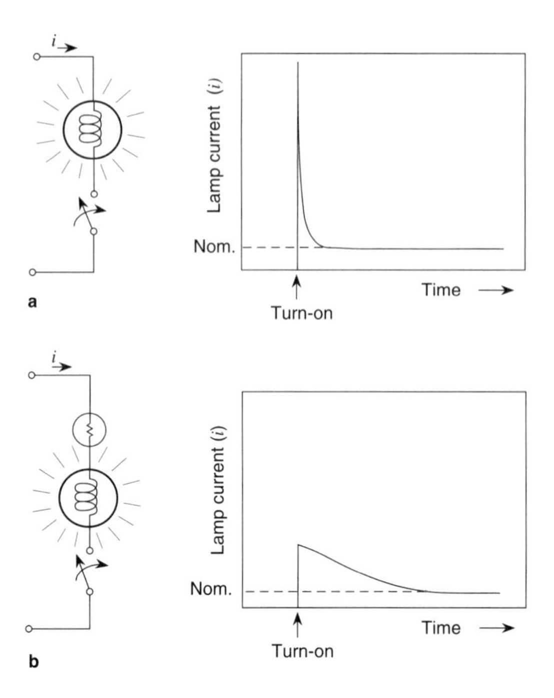

Fig. 7.31 Lamp filaments. They have very low resistance when off (cold) and high resistance when on (hot). (a) A typical turn-on transient. (b) Reducing the turn-on transient also causes the filament to take longer to warm up.

Limiting the inrush current increases lamp longevity, but it also increases the time to turn on a lamp. Warm-up duration for some filaments is about 100 ms; if circuitry limits the inrush current, then warm-up can take 1 s or more (Valentine, 1990).

Beyond inrush current and lamp longevity, you will need to consider the switching components and configuration for controlling lamps. Fig. 7.32 illustrates some of the possible circuits for controlling a lamp. Low-side switching (Fig. 7.32a, b, and c) is simpler and requires fewer components than high-side switching (Fig. 7.32d through g). Table 7.10 gives some of the concerns for these two types of circuits.

Note that both configurations in Fig. 7.32 use metal-oxide semiconductor field-effect transistors (MOSFETs) exclusively instead of bipolar transistors. MOSFETs have higher transconductances that survive short circuits long enough to blow a fuse, whereas bipolar transistors can pull out of saturation during a short circuit and burn up before a fuse can blow (Valentine, 1990).

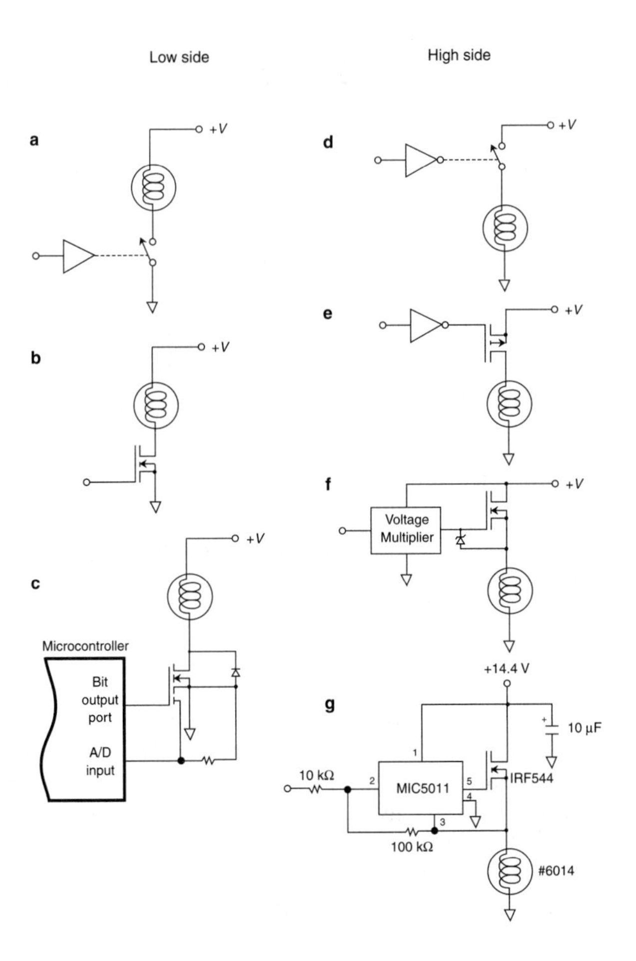

Fig. 7.32 Low-side versus high-side switching in lamp circuits. (a) General configuration for low-side switching. (b) Simple application of N-channel MOSFET to lamp control. (c) Microcontroller during MOSFET with current sense can control lamp and determine whether the lamp is open or the socket shorted. (d) General configuration for high-side switching. (e) Application using a P-channel MOSFET to control a lamp. (f) A voltage multiplier can raise the bias level of the gate above the source to allow use of an N-channel MOSFET in high-side switching. (g) A practical implementation for high-side switching with N-channel MOSFET.

7.11.4 *Relays and Solenoids*

Both relays and solenoids use electromagnetic coils to control mechanical armatures. Relays can switch high current or voltage with small control currents. Solenoids generate either a straight-line motion or a rotation in the armature to cause displacement in space.

Table 7.10 Comparison of switching configurations that control lamps

Switching mode	Advantage	Disadvantage
Low side	Generally simplest and cheapest circuit to build	1. The lamp is always connected to the supply voltage. An accidental short across the bulb's socket would stress the power voltage. 2. The constant voltage potential at the socket will drive a leakage current in moist environments (such as an automobile turn indicator) and accelerate corrosion.
High side	Isolates load from supply; reduces chance of accidental shorts	1. P-channel MOSFETs are more expensive. 2. N-channel MOSFETS require extra circuitry to multiply gate voltage.

The coils in relays and solenoids are inductors that present a special concern to the designer. Remember that a changing current through an inductor develops a proportional voltage:

$$V = L(\mathrm{d}i/\mathrm{d}t)\,.$$

Consequently, current in a coil, or inductor, can't change instantaneously. If it changes rapidly, it induces a large voltage across the coil. For relays and solenoids, that large voltage is countered by a nearly equal voltage across the interrupting switch or transistor (Fig. 7.33). This results in either arcing across the contacts in a mechanical switch or breakdown in a transistor. You can suppress this inductive kick with a conductive, dissipative circuit around the coil as shown in Fig. 7.33c. The dissipating circuit around a coil is called a *snubber* or *transient suppression network.*

For AC circuits, solid-state relays can avoid the inductive kick by using zero-crossing switching. A circuit detects when the AC *voltage* approaches zero before turning on the transistors and detects when the AC *current* approaches zero before turning off the transistors.

Exceeding the breakdown voltage of the transistor is another concern for the designer. A simple Zener diode, connected as shown in Fig. 7.34, can provide secondary protection.

7.11.5 *Motors*

Motors also use electromagnetic coils to control mechanical armatures, but the rotating magnetic field presents other challenges to the designer. First, you have the choice between DC motors and stepper motors. (Control for AC induction motors and synchronous motors is a design field to itself, and I'm not even going to attempt it here. Besides, most electronic design for appliances and instrumentation

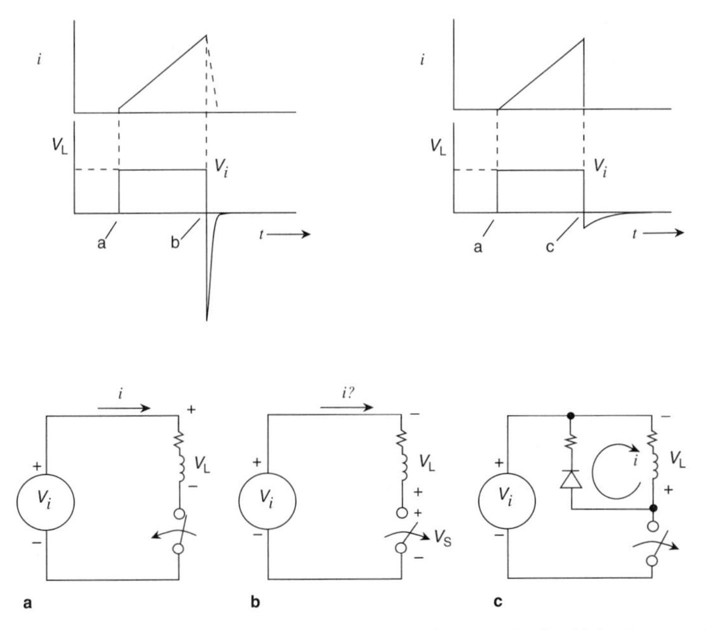

Fig. 7.33 Inductive kick. Coils, in relays and solenoids, give an inductive kick whenever the current is interrupted that forces a large voltage across the interrupting switch. The induced voltage across the switch causes arcing in mechanical contacts and breakdown in transistors. (a) A closed circuit that establishes current in coil. (b) Interrupting the current with a switch induces a very large voltage across the switch, $V_S = V_i + V_L$. (c) A diode and resistor in parallel with the coil will dissipate the inductive energy and minimize the voltage across the switch.

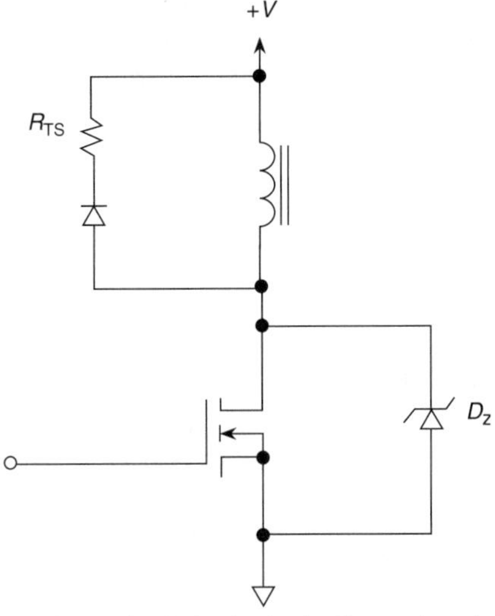

Fig. 7.34 A general scheme for driving inductive loads. Choose the value of R_{TS} to set the settling time of the inductive current; the time is proportional to R_{TS}. Choose the reverse voltage of D_Z below the breakdown voltage of the transistor.

will use DC motors or steppers.) DC motors typically are cheaper and smaller than stepper motors, but they require commutation and may demand more complex control. Fans, automobile door locks and electric windows, audiocassette players, and many kitchen appliances use DC motors. In contrast, the speed and position of stepper motors are more easily controlled. The discrete, precise movements of stepper motors drive mechanisms found in printers, video camcorders, and photocopiers.

The rotor within a DC motor supports a rotating magnetic field. Commutation reverses the current polarity within the field coils to reverse the magnetic flux and drive the rotor. Brush commutation is a mechanical means of reversing the current. Brushless DC motors use electronic means to reverse the current. DC motors with brush commutation tend to be cheaper, while brushless DC motors are used in high-reliability, high-performance applications.

Second, after you have selected a motor, you need to design a drive circuit for it. Several different configurations of drive circuits are available.

The H-bridge circuit exercises complete control over the rotation, both speed and direction, of a DC motor (Fig. 7.35a). A crossover current occurs, however, when one transistor turns off and the other turns on in this configuration (Fig. 7.35b). Controlling the inputs to each transistor separately eliminates the crossover current at the expense of more complex control circuitry (Fig. 7.35c).

Pulse-width modulation of the gate signals controls the speed of a DC motor. Increasing the duty cycle increases speed, while decreasing the duty cycle reduces the speed of rotation.

The H-bridge drive circuits in Fig. 7.35 control DC motors with brush commutation. Brushless DC motors require more complex control circuits than DC motors with brushes. Brushless DC motors have either Hall-effect or optical sensors to encode the shaft position and commutation logic to generate the correct drive polarity for the motor. Fig. 7.36 illustrates just one example of a drive circuit for a brushless DC motor.

Note that the chip in Fig. 7.36 incorporates a charge pump to provide gate drive for N-channel MOSFETs in the high side of the motor drive. While P-channel MOSFETs might reduce the parts count, they require up to four times the die area within the chip to achieve an equivalent conductance to N-channel MOSFETs and therefore are more expensive. N-channel MOSFETs, however, require gate voltages above the drain potential to turn on; therefore, they require some sort of dynamic gate drive that increases the gate voltage above the drain.

Stepper motors come in two configurations: bipolar and unipolar windings. Bipolar-wound steppers have one winding per phase, and the current reverses in the winding to reverse the stator flux. Unipolar-wound steppers have two coils per phase, and the coils alternate to reverse the stator flux. Generally, bipolar stepper motors require twice as many transistors in their drive circuitry as unipolar stepper motors do, and they require careful timing to avoid short circuits through the transistors. Unipolar stepper motors, while simpler to control and cheaper, have less

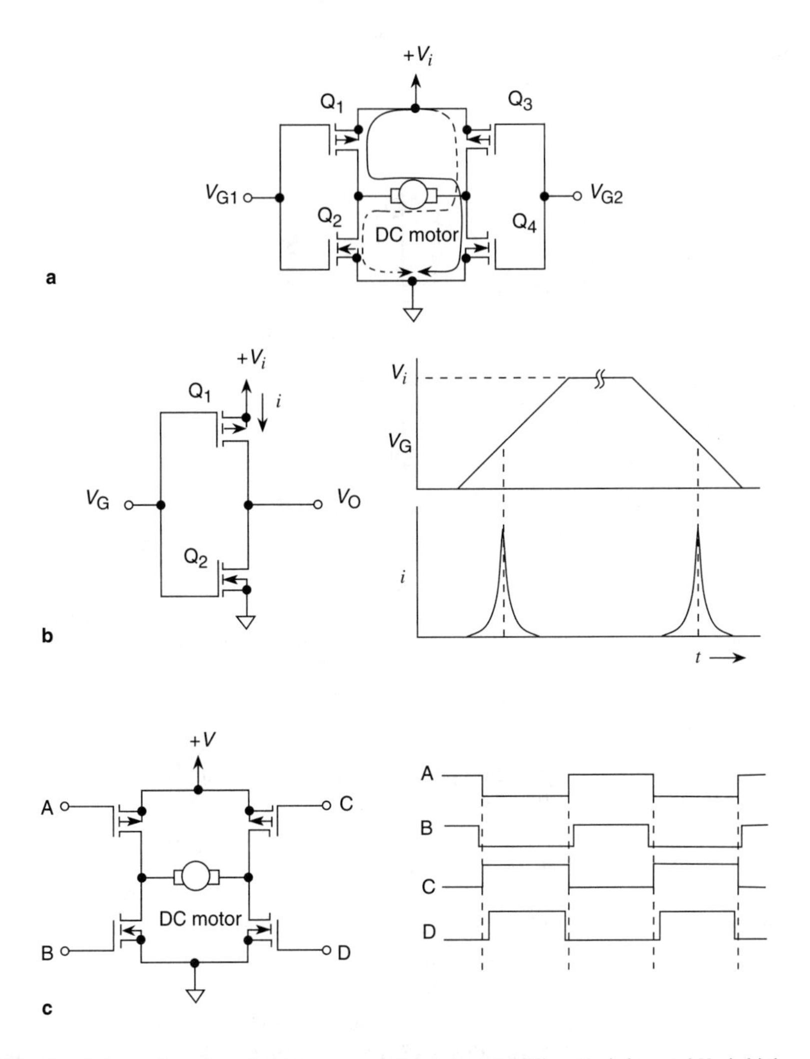

Fig. 7.35 H-bridge configurations for driving small DC motors. (a) When V_{G1} is low and V_{G2} is high, Q_1 and Q_4 are on and the current follows the solid line; the motor rotates in one direction. When V_{G1} is high and V_{G2} is low, Q_2 and Q_3 are on and the current follows the dashed line; the motor rotates in the other direction. (b) The crossover current through Q_1 and Q_2 occurs when one transistor turns off and the other turns on. Crossover current is the short-circuit current through both transistors when they are partially on (or off). (c) Proper communication or phasing will reduce the crossover current.

torque at low step rates (Cerato and Scurati, 1990; Pryce, 1989). Fig. 7.37 illustrates two examples of control circuits for stepper motors.

The frequency of the phase change controls the speed of a stepper motor. Increasing the frequency of the phase change increases the speed, while decreas-

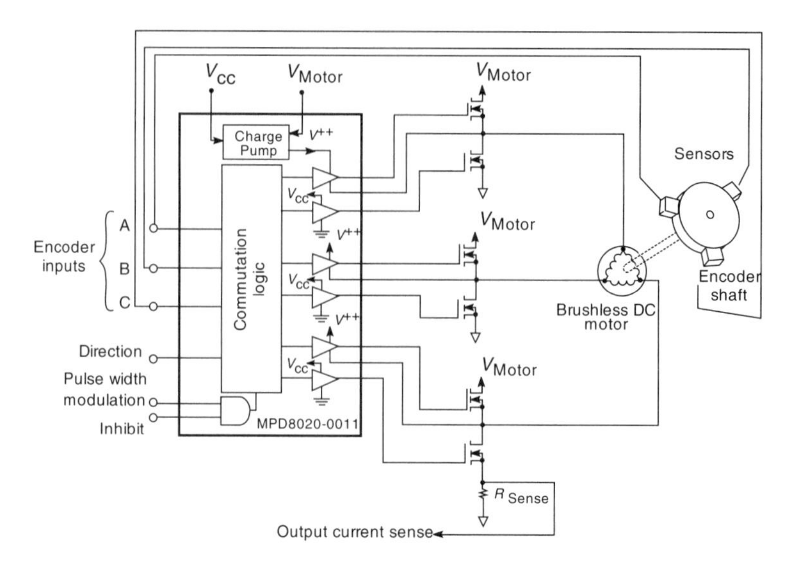

FIG. 7.36 The MPD 8020-0011 from Micrel, one example of many driver chips available for controlling brushless DC motors. It uses feedback from Hall-effect or optical encoders to develop the appropriate drive signals for a three-phase motor. It also provides short-circuit protection. The charge pump generates suitable voltage to drive the N-channel MOSFETs on the high side of the circuit.

ing the frequency reduces the speed of rotation. The duty cycle, typically 50%, remains constant regardless of the frequency.

7.11.6 *Digital-to-Analog Conversion*

Beyond actuators and indicators, your system may have to transmit signals to other devices, instruments, or systems. If the signals are analog, you must convert the digital codes of your system to analog signals. Fig. 7.38 shows the two approaches available for DAC; either use a monolithic component, or build it from discrete components yourself. Monolithic DACs, like that in Fig. 7.38a, have built-in references, timing, and current-to-voltage conversion. If your application can tolerate some ripple in the analog signal and it needs to be cheap, then you can use the RC filter as a leaky integrator to produce an average DC signal. Consider the trade-off between incorporating a monolithic component and designing the crude DAC implementation such as found in Fig. 7.38b, which requires computing and timing resources from your system plus your effort to perfect the circuitry.

Monolithic solutions generally use current steering to convert digital codes to analog (Fig. 7.39a). An R-2R ladder network implements the current steering by dividing the current among the resistive elements according to the significance of each representative bit. Two architectures are available: the voltage output DAC and the multiplying DAC.

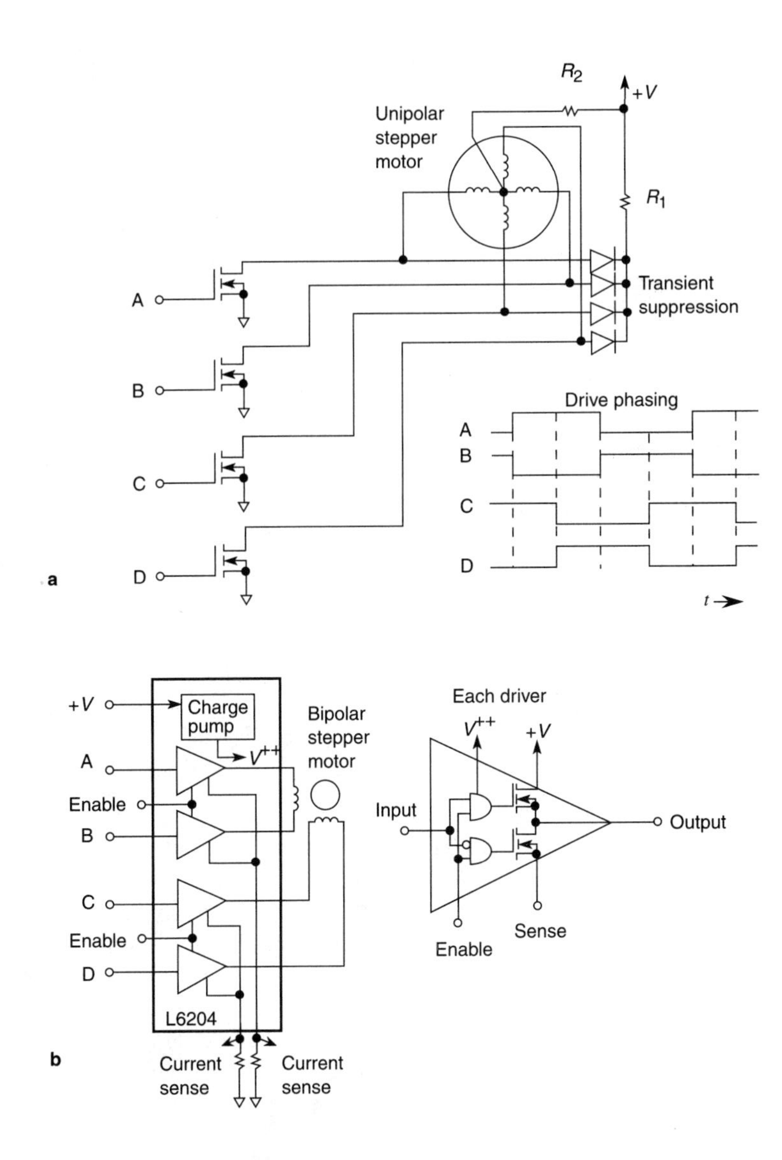

Fig. 7.37 Two drivers for stepper motors. (a) Unipolar stepper motor. (b) Bipolar stepper motor using an L6204 dual full-bridge driver from SGS-Thomson.

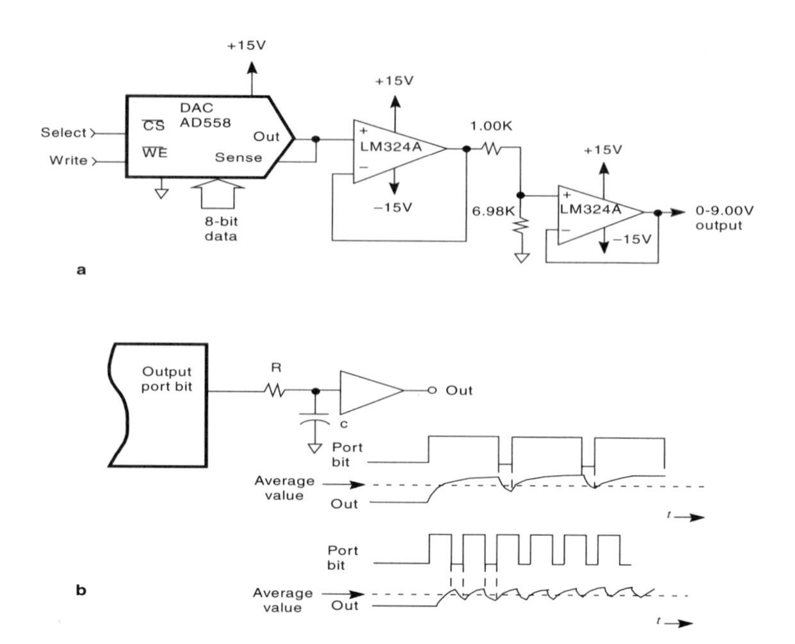

Fig. 7.38 Two possible configurations for digital-to-analog conversion and output. (a) An accurate, multichip solution. (b) A software-driven output that converts pulse-width-modulated signal to an average analog signal.

The voltage DAC (Fig. 7.39b) requires only a single-supply voltage to operate and outputs a positive signal from a positive reference. The multiplying, or current, DAC (Fig. 7.39c) requires a positive and negative supply, inverts the polarity of the output signal from that of the reference, and usually requires an external buffer. The reference signal in a multiplying DAC, however, may be another analog signal that the digital input attenuates (multiplies) according to its code. This provides "two-quadrant" multiplying as shown by the small graph in Fig. 7.39c. You can find more information on DAC architecture and design in the *Analog-Digital Conversion Handbook,* edited by D. H. Sheingold (1986).

7.11.7 *Analog Drive*

You may need to adjust the DC offset of the analog signal produced by your system. Again, Fig. 7.26 provides a schematic of the general configuration for level shifting. Fig. 7.38a used two op-amps to translate a 0- to 10.3-V signal to a range between 0 and 9.00 V, for example. In addition, you may need to filter the output signal to shape the frequency bandwidth. Finally, terminate unused op-amps in dual or quad packages by grounding the positive input and connecting the negative input to the output.

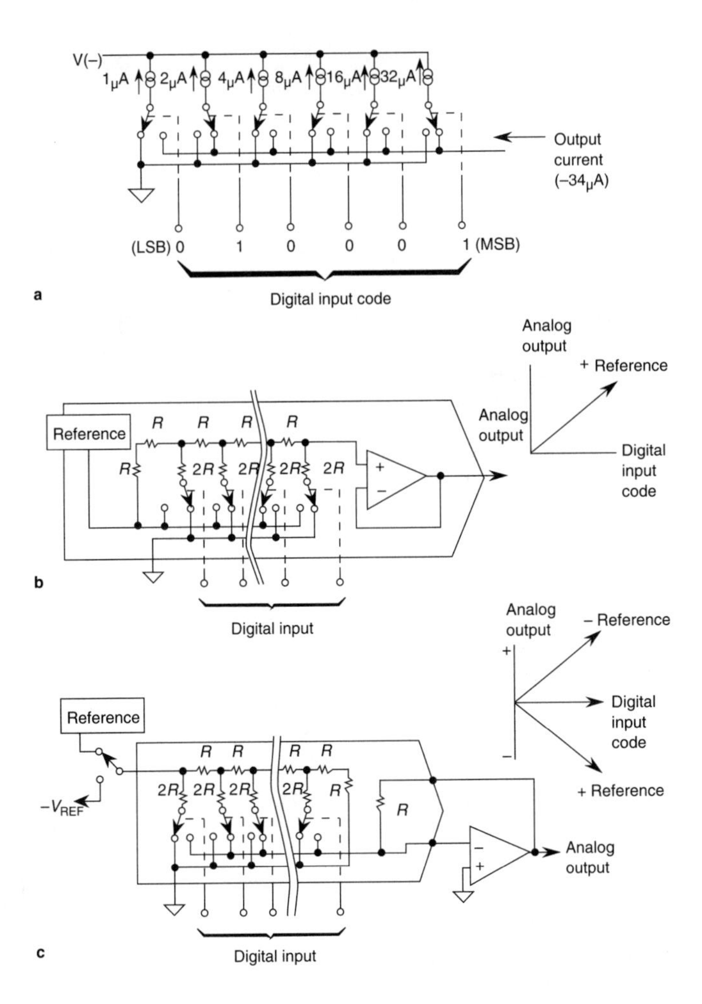

FIG. 7.39 Architectures for DACs. (a) Example of current steering to generate an analog current, 22 hex (or 34 decimal) produces 34 μA of current. (b) Voltage output DAC. (c) Multiplying or current output DAC with an external op-amp buffer.

7.12 BREADBOARDS, EVALUATION BOARDS, AND PROTOTYPES

Simulations and models are limited, and often you will need to explore actual device and circuit operations to verify circuit design. As explained in Chapter 1, you should experiment to refine your circuit designs. Breadboards, evaluation boards, and prototypes are progressively more sophisticated methods to investigate circuit operation (Fig. 7.40).

A breadboard facilitates quick connection of components to check operation. They are limited in operation by crosstalk and noise as frequency increases beyond

FIG. 7.40 Examples of a breadboard, an evaluation board, and a wire-wrapped prototype board.

5 or 10 MHz. The pins tend to corrode with time, and the connections become unreliable. In spite of these concerns, breadboards can expedite your experiments with circuits.

Evaluation boards, from a chip's vendor, feature a particular component, such as an ADC or a microcontroller, and provide enough circuit support to test it thoroughly. Additionally, they can connect to other circuits to emulate your final system. Evaluation boards for microcontrollers and microprocessors balance the system cost with the development time for a custom design. They offer debugged hardware that allows early software prototyping and development. An evaluation board has the advantages of both the vendor's experience for implementing the particular component and the vendor's development tools, such as programming languages and libraries (Sperry, 1993).

Embedded personal computers (PCs) can reduce the development time even further if the production quantities are small and time is critical. They eliminate circuit design and production, have all the advantages of evaluation kits, and have all the circuits and resources to start software development immediately. On the other hand, they are more expensive in large quantities than a custom design and may not perfectly fit your application.

Finally, you can use facilities, like that shown in Fig. 7.41, for quick turnaround of PCBs to build prototype circuits. A prototype allows you to study the circuit board and design in a nearly final state. For high-frequency circuits, this may be the only way that you can experiment with the design.

Fig. 7.41 Equipment for producing prototype circuit boards quickly (photograph courtesy of Direct Imaging, Inc.).

7.13 SUMMARY

This chapter is not a comprehensive guide to electronic design; rather it is an overview of common problems and concerns found in circuit design. Its purpose is to raise your awareness of the problems, introduce methods for attacking them, and indicate their relationship to the design of the total system. You have to do the rest.

Issues that affect the overall design are reliability, fault tolerance, high speed, low power, and noise and error budgets. Specific issues in circuit design are data buses and networks, reset and power failure detection, input, and output.

7.14 RECOMMENDED READING

Di Giacomo, J., ed. 1990. *Digital bus handbook.* New York: McGraw-Hill.
Good coverage of a number of buses for commercially available circuit boards.

Frederiksen, T. M. 1988. *Intuitive operational amplifiers: From basics to useful applications.* New York: McGraw-Hill.
If you do any analog design, get this book! It's tremendous. The material is readable and very useful.

Garrett, P. H. 1987. *Computer interface engineering for real-time systems: A model-based approach.* Englewood Cliffs, N.J.: Prentice-Hall.
Broad treatment of errors and error budgets in data acquisition and signal processing.

Horowitz, P., and Hill, W. 1991. *The art of electronics.* 2nd ed. Cambridge: Cambridge University Press.
One of the finest books for good, basic electronic design. It covers both analog and digital circuits. The material is written and presented excellently.

Johnson, H. W., and M. Graham. 1993. *High Speed Digital Design: A Handbook of Black Magic*. Englewood Cliffs, N. J.: PTR Prentice-Hall.
If you design digital circuits, then run, don't walk, to your nearest bookstore and get this book. It contains some of the best practical tips I have ever found on design and measurement, and it is well written.

Sheingold, D. H., ed. 1986. *Analog-digital conversion handbook*. 3rd ed. Englewood Cliffs, N.J.: Prentice-Hall.
An in-depth primer on ADCs and DACs.

7.15 REFERENCES

Brombacher, A. C. 1992. *Reliability by design: CAE techniques for electronic components and systems*. New York: Wiley.

Cerato, S. and M. Scurati. 1990. Design stepper motor drives with smart bridge circuits. *Electronic Design*, April 12, pp. 107–112.

Cripps, M. 1989. *Computer interfacing: Connection to the real world*. London: Edward Arnold.

Derenzo, S. E. 1990. *Interfacing: A laboratory approach using the microcomputer for instrumentation, data analysis, and control*. Englewood Cliffs, N.J.: Prentice-Hall.

Di Giacomo, J., ed. 1990. *Digital bus handbook*. New York: McGraw-Hill.

Fazekas, P. 1988. A systematic approach facilitates noise analysis. *EDN*, May 12, pp. 153–162.

Garrett, P. H. 1987. *Computer interface engineering for real-time systems: A model-based approach*. Englewood Cliffs, N.J.: Prentice-Hall.

Keenan, R. K. 1985. *Decoupling and layout of digital printed circuits*. Pinellas Park, Fla.: TKC.

O'Dell, G. M. 1990. Timing analysis forestalls failures in digital circuits. *EDN*, May 24, pp. 157–160.

Pryce, D. 1989. Dedicated circuits suit a wide variety of tasks. *EDN*, September 4, pp. 59–76.

Pryce, D. 1992. Speed-resolution tradeoff key to choosing ADCs. *EDN*, August 6, pp. 39–49.

Putnam, B. W. 1987. *RS-232 simplified: Everything you need to know about connecting, interfacing, and troubleshooting peripheral devices*. Englewood Cliffs, N.J.: Prentice-Hall.

Sheingold, D. H. ed., 1981. *Transducer interfacing handbook: A guide to analog signal conditioning*. Norwood, Mass.: Analog Devices.

Sheingold, D. H., ed. 1986. *Analog-digital conversion handbook*. 3rd ed. Englewood Cliffs, N.J.: Prentice-Hall.

Sperry, T. 1993. A single-board computer update. *Embedded Systems Programming*, April, pp. 55–60.

Vaglica, J. J., and P. S. Gilmour. 1990. How to select a microcontroller. *IEEE Spectrum*, November, pp. 106–109.

Valentine, R. 1990. Don't underestimate transistor-based lamp-driver design. *EDN*, June 7, pp. 119–124.

8

CIRCUIT LAYOUT

Good order is the foundation of all things.
—Edmund Burke

8.1 MUNDANE BUT NECESSARY

Circuit boards help distill concepts for electronic circuits into real systems. Though often overlooked, circuit boards are a necessary step in the complex process of developing a functional product. Packaging, component layout, and the routing of signal traces affect the operation of circuits. Poorly selected placement, packages, and routing will increase noise, susceptibility, and EMI in electronic circuits. Consequently, the design and manufacture of circuit boards need serious consideration early in development (along with everything else).

Concomitant with early planning, you should foster good communication with the manufacturer to eliminate errors and delays from misunderstanding or insufficient specifications. Consider this example of how one company formerly did business: "We used to design and lay out a circuit board, then send it to manufacturing who would change the board to accommodate their pick-and-place machines and soldering equipment. This, of course, would change the design and maybe violate some electrical rules, so the electrical engineer would rework everything, and so on (Donlin, 1993, p. 142)."

This chapter outlines procedures and good practices for preparing circuit boards. I will introduce the design and construction of printed circuit boards. Fig. 8.1 gives one view of how these different circuit technologies, such as ICs, MCMs, and PCBs, relate.

8.2 CIRCUIT BOARDS

Circuit boards combine electronic components and connectors into a functional system through electrical connections and mechanical support. They have different configurations: wire wrap, stitch weld, PCB, chip on board, and MCM. Fig. 8.2

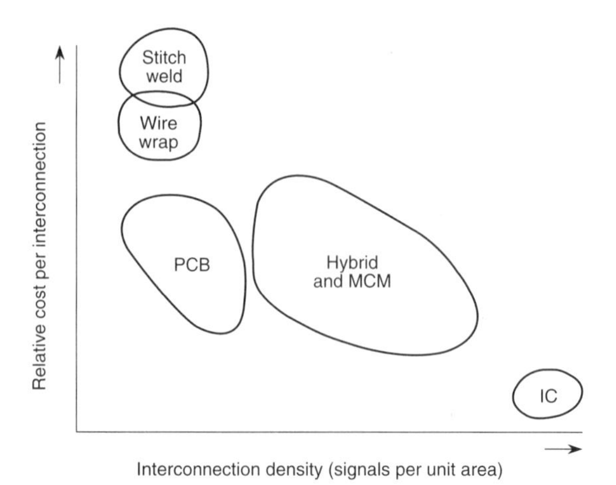

Fig. 8.1 Technologies available for connecting components and circuits.

Fig. 8.2 A sampling of different types of circuit boards: wire wrap, printed circuit, hybrid, flexible printed circuit, and stitch weld.

illustrates several types of circuit boards. This chapter will focus on PCBs because they are used frequently in mass-produced electronics.

Wire wrap is generally suitable for prototype development because you can easily change the connections; circuit modifications and corrections are straightforward. The downside is that large circuit boards require extensive effort because every wired connection must be wrapped around a pin; consequently, wire wrap is much less useful for production quantities. Furthermore, wire-wrapped circuit boards are limited in operation to less than 5 or 10 MHz, above which the loop inductances in the wired connections distort signals.

Stitch weld connects components with point-to-point wiring on circuit boards much like wire wrap. However, the stitch-weld posts are shorter and the wire is welded to the pins, not wrapped; therefore the loop inductances are lower and allow much higher-speed operation (100 MHz). Stitch weld resists vibration and shock better than wire wrap and has been used in spacecraft. But stitch weld is more expensive and requires a special welding station.

PCBs have etched and plated connections that are well suited for both manufacturing and rugged environments. PCBs make automated placement and soldering of components possible, and they control impedances more effectively than wire wrap.

MCMs achieve higher levels of circuit density by bonding the bare die of ICs onto a substrate. The compact packaging of MCMs improves signal speeds and reduces load capacitance. Historically, MCMs and hybrids have been expensive to fabricate, but recent advances in materials and processing are now making MCMs cost-effective in some products.

8.2.1 *PCBs*

You will probably incorporate PCBs into most of your electronic products for reasons of cost, manufacturing ease, and reliability. Understanding their construction will help you grasp the majority of interconnection techniques used in circuit boards.

PCBs contain layers of insulating material and copper conductors with plated holes, called *vias*, to interconnect the conductors. Here are some common terms that are used in their construction (DeSantis, 1993):

- *Laminate:* layers of glass cloth impregnated with resin such as epoxy or polyimide pressed together under heat to form an insulating base for circuit boards
- *Prepreg:* glass cloth or fabric saturated with epoxy resin and partially cured
- *Copper-clad laminate:* laminate with copper foil bonded to one or both sides, sometimes known as *core* in multilayer circuit boards; also called *clad*
- *Copper foil:* thin layer of copper sheet that bonds to laminate; classified according to weight per unit area, commonly in two weights:
  - 1-oz copper = 0.31 kg/m^2 (1 oz/ft^2) = 36 μm (0.0014 in.) thick
  - 2-oz copper = 0.61 kg/m^2 (2 oz/ft^2) = 71 μm (0.0028 in.) thick
- *Copper plating:* deposited layer of copper that coats vias and signal traces and forms a conducting path between sides of a circuit board
- *Etching:* removal of unwanted copper foil from the core by electrochemical reaction

- *Photoresist:* a thin photosensitive polymer that supports photographic patterns of the signal traces and pads for etching
- *Solder mask:* dark-green coating, called *resist*, that covers the entire board except the solder pads to prevent solder bridges between traces and to resist moisture and scratches; usually applied with silk screen

The signal conductors can reside on different sides or layers of PCBs, as shown in Fig. 8.3.

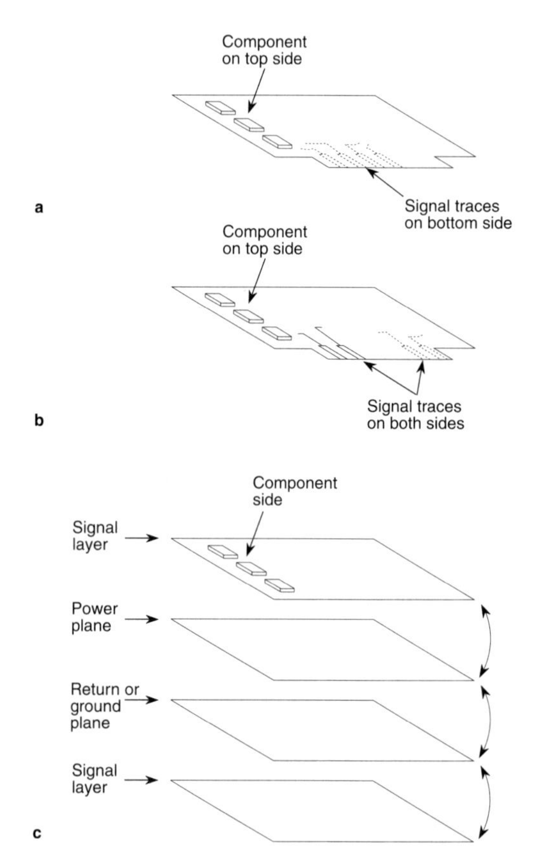

Fig. 8.3 Types of PCBs. (a) Single-sided PCBs have components and jumpers on one side, signal traces on the other. (b) Double-sided PCBs have signal traces on both sides. (c) Multilayer PCBs have four, six, eight, or more conductive layers laminated together.

8.2.2 *Single-Sided PCB*

The single-sided PCB is still used in products like audio amplifiers because it is cheap and can operate satisfactorily at low frequencies (typically less than 25 kHz). Cost is low because production of the PCB has few steps; only one side needs etching, and plating is not necessary. The circuit board does not use plated-through holes, since the signal traces are on only one side; instead, signals cross over wire jumpers that loop from one side to the other.

8.2.3 *Double-Sided PCB*

A double-sided PCB has signal traces on both sides of the circuit board and plated-through vias. Double-sided PCBs can pack components a little more densely than single-sided PCBs and can support higher-frequency operation *if laid out very carefully*. Fig. 8.4 illustrates a cross section of a double-sided PCB.

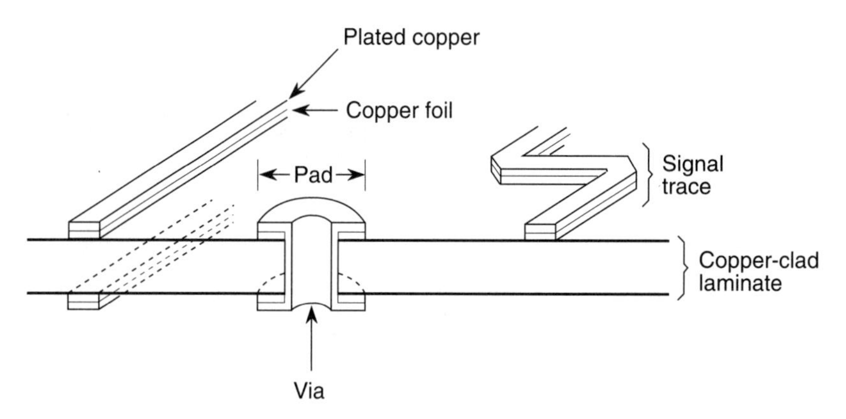

Fig. 8.4 Double-sided PCB, with signal traces on both sides of the board. Vias provide electrical connections to both sides.

8.2.4 *Multilayer PCB*

A multilayer PCB is a stack of alternating layers of copper-clad laminate, or core, and prepreg (Fig. 8.5). Multilayer PCBs support very dense circuit connections because they can have 20 or 30 conducting layers laminated together. Multilayer PCBs control impedance much more tightly than double-sided PCBs and are absolutely necessary for high-frequency circuits.

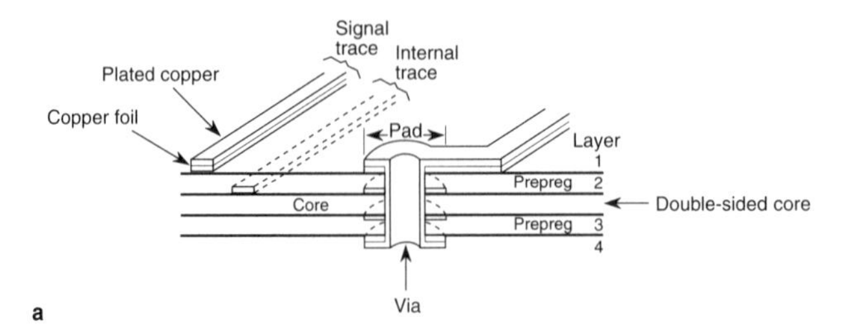

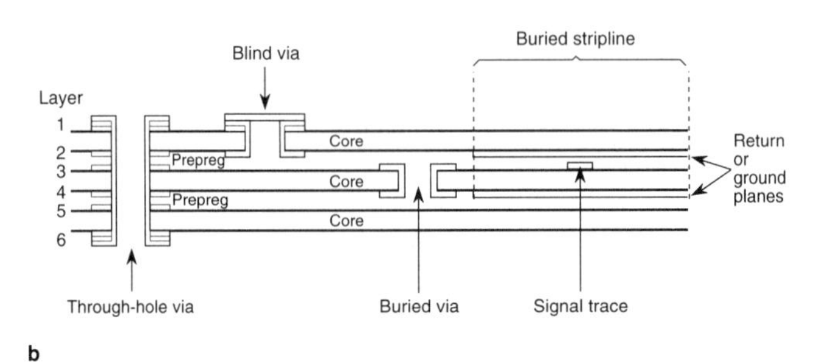

Fig. 8.5 Multilayer PCB, with alternating layers of core and prepreg. (a) Prepreg layers have traces on one side, while core may be double-sided. (b) Through-hole vias connect all layers in the PCB, while blind and buried vias connect only two layers. A stripline is a controlled impedance trace between two ground planes.

Through-hole, buried, and blind vias connect signal paths on different layers together. Through-hole vias penetrate all layers and can connect signals on each layer. Buried vias connect traces on two sides of an internal layer. Blind vias are exposed on one external layer and connect traces on the two sides of that layer.

8.2.5 *Assembly and Materials*

The sequence for producing a multilayer PCB is roughly as follows (Davis, 1993):

1. Send the Gerber database of the PCB to the manufacturer via modem, tape, or diskette. (The *Gerber database* is a digital file that describes the locations of each part, via, and signal trace on the PCB.)
2. Cut core material, together with prepreg, from sheets to a standard size.
3. Apply photoresist to copper-clad laminate (steps 3 through 9 are for inner layers of the PCB).
4. Place *negative*-image artwork of the signal traces on the photoresist. The clear areas on the film represent the conductors.
5. Expose the artwork and photoresist to ultraviolet light. The light fuses the polymer fibers within the photoresist.

6. Wash the clad to remove the unfused photoresist. This operation exposes the unwanted copper for later removal. The fused photoresist protects the copper that represents the future signal traces.
7. Etch the exposed copper from the clad.
8. Wash the fused photoresist from the clad.
9. Apply a coating to the remaining copper to prevent oxidation.
10. Repeat steps 3 through 9 for each inner-layer core.
11. Bake the inner-layer cores to remove moisture.
12. Stack the cores with prepreg between each clad. The prepreg provides insulation and a dielectric medium.
13. Laminate the cores into a PCB with heat, pressure, and vacuum.
14. Sandwich the PCB with aluminum or fiberboard to prevent splintering from drilling.
15. Drill all the holes.
16. Plate the internal walls of the holes with copper. (Holes must be greater than a minimum diameter to allow free flow of the plating bath.)
17. Apply photoresist to copper-clad laminate (steps 17 through 25 are for the outer layers of the PCB).
18. Place the *positive*-image artwork of the signal traces on the photoresist. Now the dark areas on the film represent the conductors.
19. Expose the artwork and photoresist to ultraviolet light. The light fuses the polymer fibers within the photoresist.
20. Wash the clad to remove the unfused photoresist. This operation exposes the copper that represents the future signal traces.
21. Plate the exposed clad with additional copper.
22. Plate the exposed copper with tin. (Tin acts as a resist to the chemicals that wash the photoresist and etch the unwanted copper.)
23. Wash the fused photoresist from the clad.
24. Etch the exposed copper from the clad.
25. Remove the tin from the copper traces and pads.
26. Apply the solder mask.
27. Mark the PCB with an epoxy-based ink to outline and designate components.
28. Electroplate gold onto any contact fingers with an underplating of nickel.
29. Coat the exposed traces and pads with solder.
30. Cut the PCB to its final shape.

Variations on this theme are used for producing different types of circuit boards.

The material in a PCB affects circuit performance and mechanical reliability. Laminates with a low dielectric constant result in faster signal propagation. A laminate whose thermal coefficient of expansion (TCE) matches that of the component package will reduce the probability of fracture in the solder joints. As always, trade-offs exist between materials. Teflon, for instance, supports the highest propagation speed for signals, but its trace and via density is lower than that of epoxy or polyimide, and it has a large TCE. Table 8.1 lists characteristics of some PCB laminates.

Table 8.1 Properties of some PCBs (Gheewala and MacMillan, 1984; Reynolds, 1984)

Material	Dielectric constant (ϵ_R)	Propagation speed (cm/ns)	Loss factor (ψ)	TCE (ppm/K)	Maximum number of conductor layers	Comments
Cofired ceramic	10.0	9.5	0.002	7	300	Good thermal match to ceramic component packages, high cost, brittle
Epoxy-fiberglass	5.0	12.0	0.020	16	30	Lower cost, light weight, poor thermal conductivity
Polyimide-fiberglass	4.0	15.0	0.010	4	20	Absorbed moisture increases dielectric losses
Teflon	2.2	20.0	0.002	50	10	Lowest dielectric constant provides highest signal speed; trace and via density lower than other materials'

8.2.6 *Soldering*

Solder is a tin-lead alloy that has good conductivity and a low melting point compared with other metals. It bonds well to copper and other metals used in component leads. Solder attaches the leads of components to the PCB in several different configurations, as shown in Fig. 8.6.

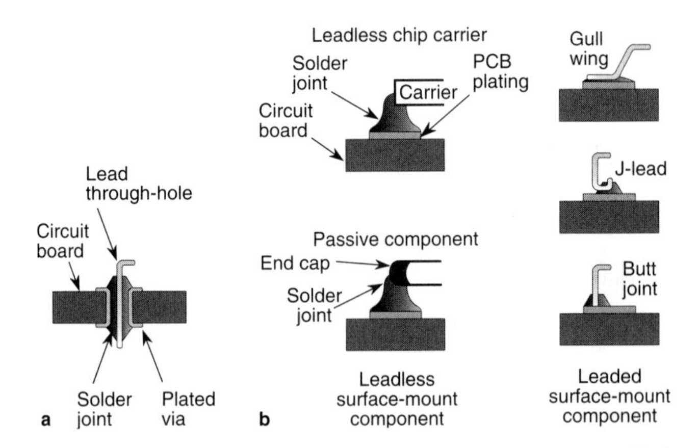

Fig. 8.6 Component soldering. (a) Through-hole soldering. (b) Surface-mount soldering.

Manufacturing equipment can solder all components to a circuit board in a few seconds. There are two primary forms of soldering: wave soldering and vapor-phase reflow. Wave soldering coats the bottom of a PCB in a bath of molten solder and attaches all DIPs and PGAs to the circuit board simultaneously. Vapor-phase reflow melts a solder paste on each pad and solders surface-mount components to PCBs.

8.2.7 *Assembly Costs*

Cost depends on the volume of production, processes used, components selected, and equipment. The more PCBs made per production run, the lower the unit cost of the circuit board. Cost is directly related to the labor during assembly; reducing labor will reduce cost. Specifying the components for ease of assembly by both machine and manual labor will reduce cost. Pick-and-place equipment that has both efficient setup and feeders that can handle the component packages needed will reduce production time and cost. The choice of equipment will constrain the specifications of clearances, substrate thicknesses, maximum number of components, minimum spacing between components, and the sequence of assembly.

Even the design of the PCB directly affects the cost of production. Factors that affect cost include the following (Blankenhorn, 1993):

1. Number of layers: Fewer layers can be cheaper to produce than more, but other considerations, like those listed below, can cause the reverse to be true.
2. Size: A smaller board with more layers may be less expensive than a larger board with fewer layers.
3. Number of boards in a system: A larger board may reduce the number of connectors, the complexity of interconnection, labor, packaging, and maintenance compared with multiple smaller boards.
4. Vias: Smaller holes may cost more, but they may save enough space to reduce the number of layers needed.
5. Pads: Smaller size will allow more traces between adjacent holes. Making pads teardrop shaped will reduce breakout and increase production yield.
6. Plating: Eliminating tin-lead plating reduces warpage from hot-air leveling, but bare copper needs a spray coating to prevent oxidation and so requires extra care in handling.
7. Testing: Electrical tests for open circuits and short circuits are worthwhile, even for prototypes, because the small signal tracks are difficult to see, trouble-shoot, and repair.

8.3 COMPONENT PLACEMENT

The placement of components affects circuit operation, manufacturing ease, and the probability of design errors. Improper layout can degrade operation or even prevent a circuit from working. Furthermore, poor layout can make manufacture of the circuit boards costly and difficult. Thoughtless placement of components complicates the design of the PCB and will increase the chance of wrong connections.

You should group circuits according to their characteristics to maintain the correct operation of each circuit. In general, follow these rules:

1. Group high-current circuits near the connector to isolate stray currents and near the edge of the PCB to remove heat.
2. Group high-frequency circuits near the connector to reduce path length, crosstalk, and noise.
3. Group low-power and low-frequency circuits away from high-current and high-frequency circuits.
4. Group analog circuits separately from digital logic.

Grouping components and circuits appropriately will reduce crosstalk and noise and will dissipate heat efficiently. Fig. 8.7 illustrates grouping of components according to these guidelines.

Careful placement of the components will make production of circuit boards easier and less error prone. Determine the location and direction of the components so that a pick-and-place machine can easily assemble a circuit board without manual intervention. (Logical and consistent orientation helps even if assembly is manual.) Allow plenty of clearance around mechanical supports; it is embarrassing and costly to find a trace shorted to an enclosure post or a component jammed against the card cage. If possible, leave room around large and complex components for sockets on the prototype boards to speed testing and development.

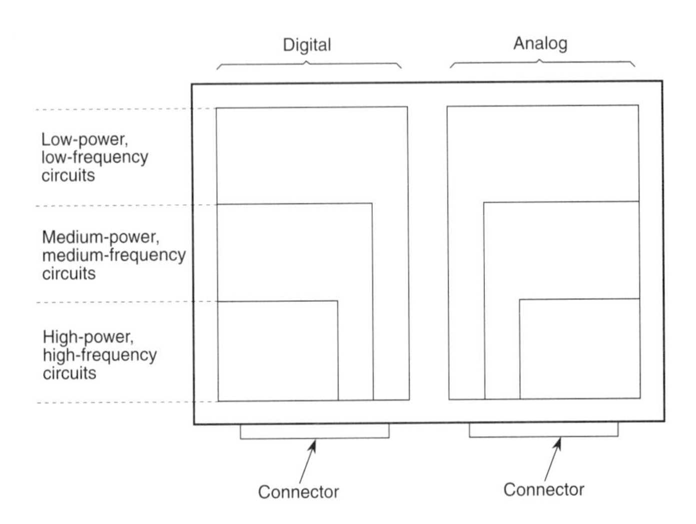

Fig. 8.7 Grouping of components. Arrange high-power and high-frequency circuits near the connector and away from the low-power circuits.

8.4 ROUTING SIGNAL TRACES

You must know the basic principles of signal propagation and circuit layout to design effective systems. Even though computer-aided design (CAD) can lay out PCBs, you still need to understand the reasons for each layout decision. Trace density (lines per unit area) determines the number of signal layers, the size of the PCB, and cost. But problems—such as false triggering of logic, setup and hold violations, and transmission delays—arise from poor layout. Bad techniques cause these problems by forming capacitive and inductive parasitics with stubs, vias, IC pins, multiple loads, and traces (Arrington and Brown, 1989).

8.4.1 *Trace Density*

As you squeeze signal traces together on a board, you can space components closer and reduce the size of the circuit board. With 0.13-mm (0.005-in.) width and 0.13-mm (0.005-in.) spacing, three traces will fit between two leads on a DIP, since leads have 2.5-mm (0.10-in.) centers and the surrounding pads are about 1.5-mm (0.06-in.) in diameter. With 0.2-mm (0.008-in.) width and 0.2-mm (0.008-in.) spacing, two traces will fit between two leads on a DIP, and with larger widths and spacings, only one trace. SMT leads have smaller pads on 1.3-mm (0.05-in.) centers (some SMTs have even smaller centers); these spacings restrict the number of traces between pads.

Smaller boards, allowed by higher trace densities, provide flexibility in packaging your product and may reduce the cost of materials. The trade-off, however, is greater cost and difficulty in producing the denser circuit boards. Furthermore, higher trace densities may degrade signal integrity.

8.4.2 *Common Impedance*

You should minimize the number of circuits that share the same return path. Voltage drops (caused by current switching) on the ground line (return path) increase system noise. Fig. 8.8 illustrates an example of how common impedance can introduce error into a circuit.

You can reduce the voltage drops, and hence the noise, by lowering the effective impedance. An unbroken return plane is the best way. Choosing the right logic family and using decoupling capacitors will help by reducing the magnitude of the current pulses (see Section 7.5). Finally, if you have the freedom, choose IC packaging and bond wires with low impedance so that the ground pins will cause the lowest voltage shift for the specified current. The lead frame and bond wires inside an IC can be primary contributors to inductance.

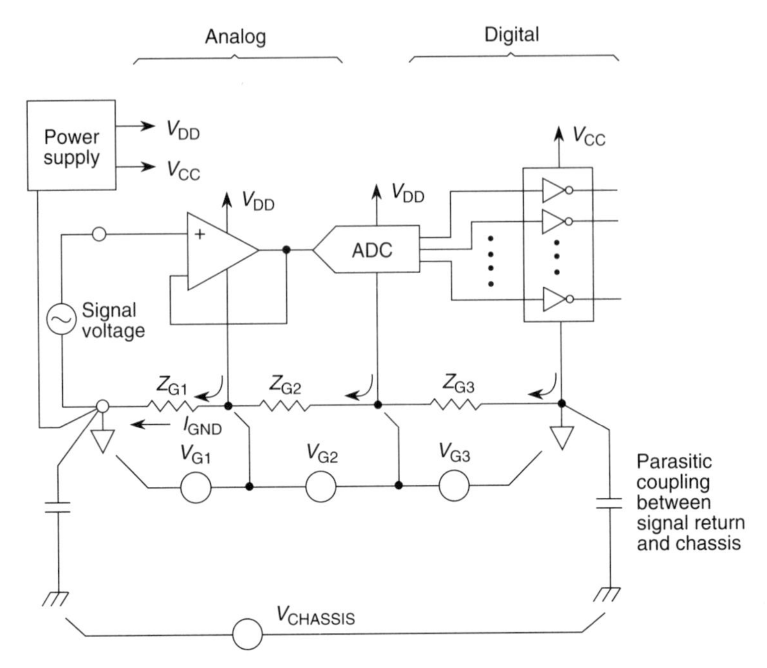

Fig. 8.8 Common impedance. Common-impedance paths cause components to reside at different ground potentials from one another. Larger common impedance will cause larger ground potentials (V_{G1}, V_{G2}, V_{G3}) and coupling to the chassis ($V_{CHASSIS}$); these will bias the signal voltage and introduce error.

8.4.3 *Distribute Signal and Return Carefully*

Address the issue of return paths early in design. Long return (ground) paths can shift the ground potential excessively, decrease noise margins, and cause false switching. Remember that an equal return current mirrors every current in a signal line; at higher frequencies the current tries to return in the plane directly below the signal trace. If the return is longer than the signal, then the current has a high inductance path that can cause noise spikes in the ground system (Cutler, 1986). Fig. 8.9 illustrates an example of a return path for current that is long compared with the signal path.

Large loops of current have high inductance or impedance, and radiated noise is often proportional to return path impedance and loop area. Section 8.6 also covers return paths in cables and connectors.

8.4.4 *Transmission Line Concerns*

Signal conductors are never ideal transmission lines. Furthermore, as frequency goes up and PCB trace widths and spacings narrow, analysis and modeling become even more challenging.

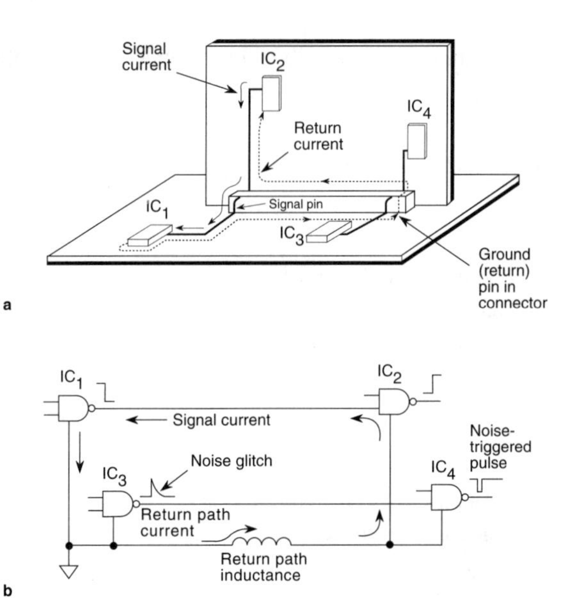

Fig. 8.9 Effect of ground pin placement. Poor placement of ground pins causes long return paths and large loop inductance. (a) The return current follows the path of lowest impedance in the ground under the signal trace, but it makes a long excursion to its connector pin. (b) Schematic representation of a long return path. The voltage across the inductance can cause the output of IC_3 to "glitch" high enough to trigger IC_4.

First of all, characteristic impedance, Z_0, depends on frequency. Classical analysis uses a single frequency, so Z_0 is a fixed value. Conversely, square waves or pulses have multiple harmonics, and higher frequencies attenuate more than lower frequencies. This is called *dispersion*—signals at different frequencies propagate at different speeds.

Second, propagation delay of signals can corrupt circuit operation. Propagation delay depends on both interconnection length and signal velocity. You can reduce delay by shortening interconnection length, and you can increase signal velocity by using materials with a low dielectric constant, ε_r, in the PCB.

Finally, line resistance, skin effects, and dielectric losses can degrade signals and introduce delay and error into circuit operation. (These problems really only become important for gigahertz frequencies and trace lengths greater than 10 cm [4 in.].) These effects degrade signal amplitude as follows (Gheewala and MacMillan, 1984):

$$V_{out} = (e^{-\alpha l})V_{in} ,$$

where α = loss factor
l = length (cm)

For loss due to line resistance, the factor is calculated as follows:

$$\alpha_R = \Delta R / 2Z_0 \, .$$

The skin loss factor is represented by the following equation:

$$\alpha_{\text{skin}} = (\pi/WZ_0)\sqrt{(f/\sigma)} \, ,$$

where σ = conductivity (Ω^{-1} cm^{-1})
f = frequency (GHz)
W = conductor width (cm)

Dielectric loss uses the following factor:

$$\alpha_{\text{dielectric}} = (\pi\psi/c) f \, ,$$

where ψ = dielectric loss (Ω^{-1} cm^{-1})
c = velocity of light (cm/s)
f = frequency (GHz)

8.4.5 *Trace Impedance and Matching*

Impedance of the signal conductors directly affects circuit operation. A circuit with lower characteristic impedance, Z_0, radiates less and is less susceptible to interference than a circuit with higher impedance. Impedance mismatches lead to reflections that can both delay switching and trigger logic falsely. Fig. 8.10 illustrates the effect of impedance mismatch.

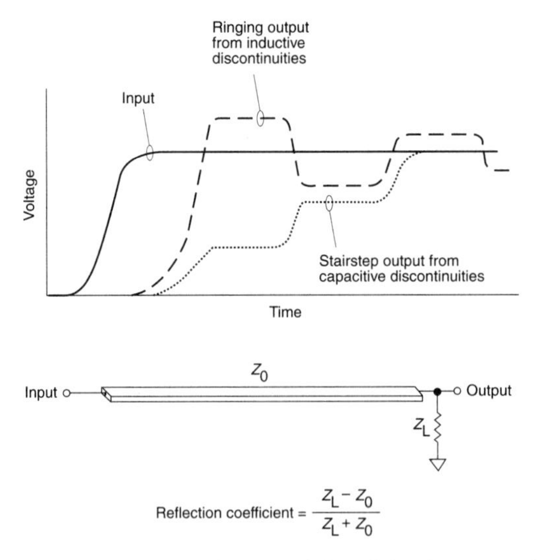

Fig. 8.10 Impedance mismatch and resulting degradation of signals along a printed circuit trace.

You can do several things with PCBs to improve impedance matching and minimize the resulting reflections. Tailor both the dielectric constant of the PCB materials and the spacing between signal traces and the return plane to match impedance and reduce radiation and susceptibility. Keep wire bonds and leads short in IC packages, and keep the packages as small as possible to reduce inductive impedance (Beresford, 1989). Terminate the signal trace appropriately, as shown in Fig. 8.11 (also see Section 7.5). (You can reduce the number of vias, pins, and pads because each one represents a small impedance mismatch. But their contribution to mismatch is usually minimal, and the routing from fewer vias becomes a nightmare.)

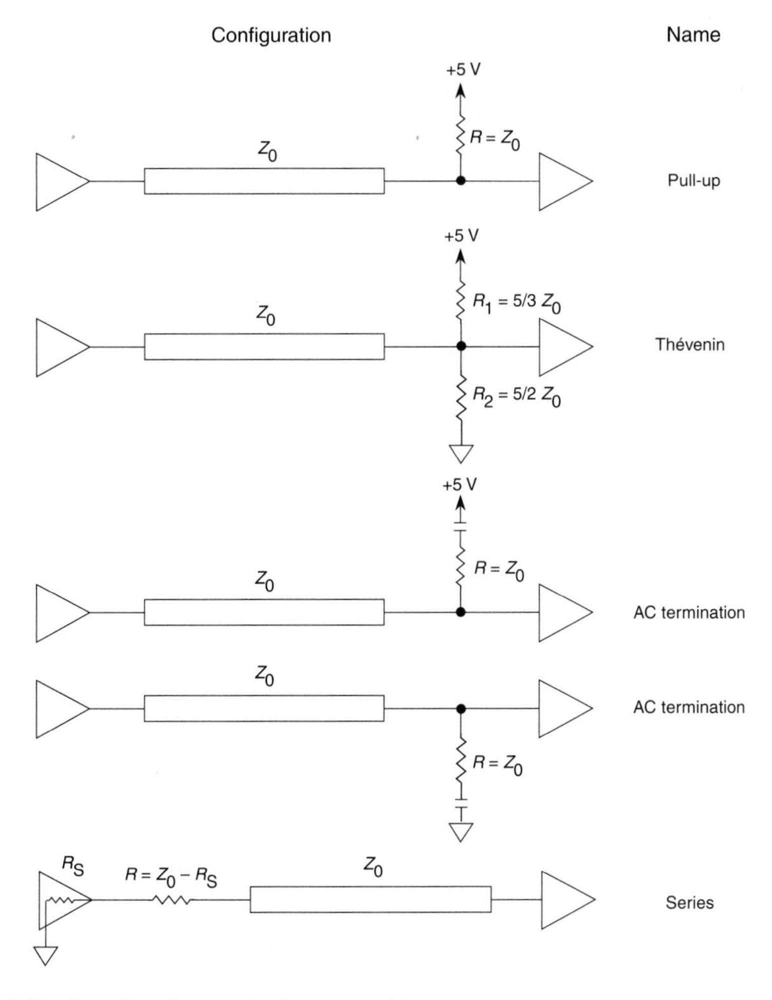

FIG. 8.11 A number of ways to terminate a signal trace.

You need to know the characteristic impedance, Z_0, of signal traces when you match impedances. Figs. 8.12 and 8.13 give the Z_0 of two widely used configurations for signal conductors. Ideally the propagation delay, t_p, depends only on the dielectric constant, ε_r, and is independent of line width and spacing.

Multiple loads have stubs, nonuniform impedances, and mismatches that compromise noise margins and can cause inadvertent switching. Careful layout of the signal path can improve the situation by eliminating stubs, as shown in Fig. 8.14.

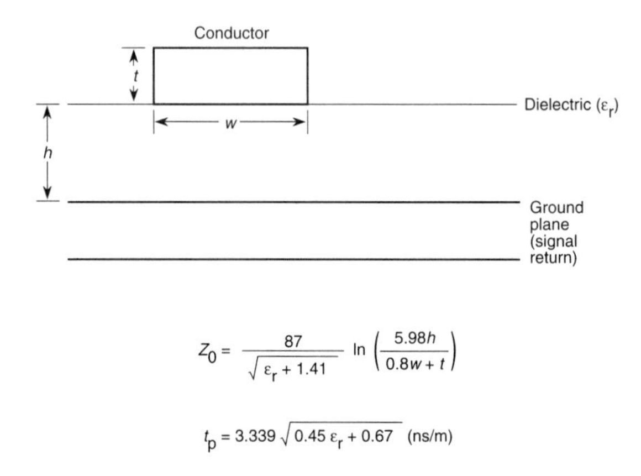

FIG. 8.12 Calculation for characteristic impedance of a microstripline.

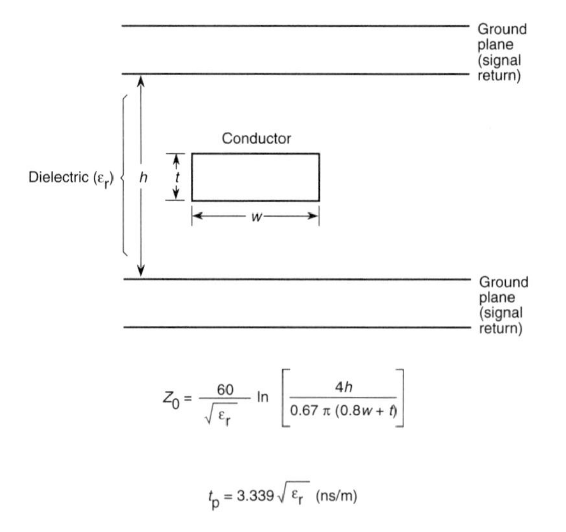

FIG. 8.13 Calculation for the characteristic impedance of a stripline.

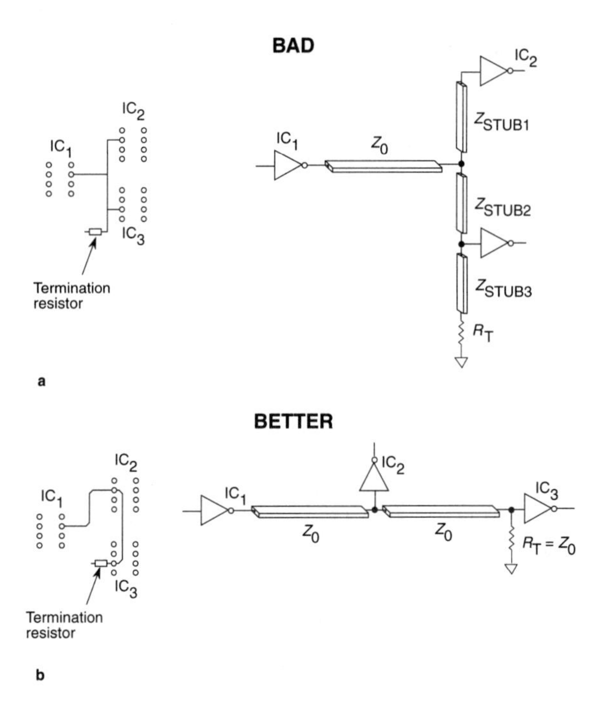

Fig. 8.14 Using a single serpentine trace to connect multiple inputs and eliminate stubs that cause reflections in the signal. (a) Poor layout degrades signal edges at the inputs to IC_2 and IC_3. (b) Using both a serpentine trace and a termination resistor near the last input eliminates stubs and preserves signal integrity.

8.4.6 *Avoiding Crosstalk*

To reduce crosstalk, follow a few simple guidelines when routing signal traces on a PCB (Beresford, 1989; Cypress Semiconductor Corp., 1993):

1. Don't run parallel traces for long distances—particularly asynchronous signals.
2. Increase separation between conductors.
3. Shield clock lines with guard strips.
4. Reduce magnetic coupling by reducing the loop area of circuits.
5. Sandwich signal lines between return planes to reduce crosstalk.
6. Isolate the clock, chip-select, chip-enable, read, and write lines (because crosstalk occurs in synchronous systems on the pulse edges when data are sampled).

Obviously, these guidelines can run counter to high trace density. You will have to balance noise, EMI, and trace density.

8.5 GROUNDS, RETURNS, AND SHIELDS

Grounding and return paths are often the most important concerns in laying out a circuit board. A proper ground and return scheme will shield and suppress most EMI in electronics and reduce errors caused by noise. Good design is straightforward.

8.5.1 *Grounding*

Grounding provides a reference point for signals. The signal reference should be a single point that is as close as possible to the power entry to the PCB. A single-point reference eliminates the problem of common impedance introducing noise, as detailed in Section 8.4. A ground plane connected to the single-point reference will also reduce common impedance.

Be sure to separate the analog and digital circuits so that current pulses from the digital circuits will not corrupt sensitive analog circuits. (Some people advocate separating the ground plane in two: one portion devoted to analog circuits, the other to digital. I'm not sure that this is necessary, since return currents flow under the signal traces and should never interfere with other circuits.) If you do separate the analog and digital ground planes, then the ADC and DAC components should straddle the split between grounds: the analog side of the component over analog ground, and the digital side over digital ground. Furthermore, use separate power supplies and connect their ground leads to the single-point reference.

8.5.2 *Distribute Power and Return Carefully*

You want low impedance and minimum voltage drop within the power distribution of the PCB for optimum performance of the circuits.

If you have a double-sided PCB, lay out the power and return together and symmetrically, or in a grid as suggested by Johnson and Graham (1993). Reduce the inductive loop area between the power and return traces (Fig. 8.15). In most cases, though, you are still better off going to a multilayer PCB with power and return (or ground) planes. They minimize the loop area and provide the lowest impedance.

Example 8.5.2.1 As a young designer, I implemented the power and ground on one circuit board in a system using the interleaved comb arrangement shown in Fig. 8.15a. The circuitry was slow, old-style CMOS running with a 100-kHz clock. The circuitry on the double-sided PCB generated so much interference that the system failed miserably in testing for conformance to MIL-STD-461C. I had the PCB redesigned for four layers: two signal layers, one power plane, and one return plane. Then the system easily conformed to MIL-STD-461C with the circuitry running on the new, multilayer PCB. Clearly, low clock frequency and slow CMOS logic do not reduce EMI nearly as well as a full return (ground) plane does.

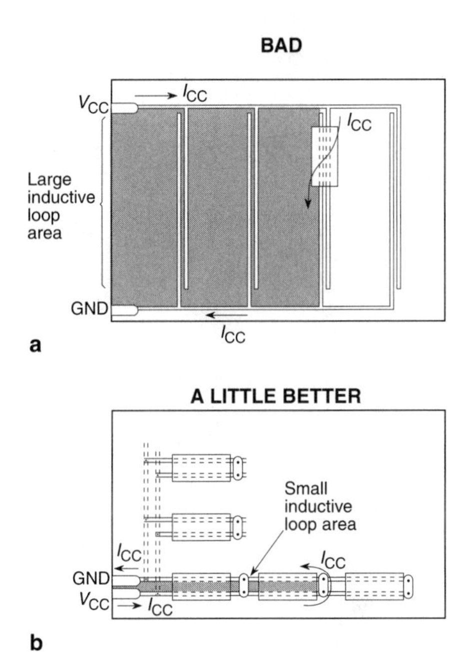

FIG. 8.15 Reducing inductive loop area. Orientation and location of the supply rails are important. (a) Interleaved combs arrangement can create large loops. (b) Running power and return together reduces the loop area. Johnson and Graham (1993) suggest a grid arrangement for the power and return rails to further reduce the loop area of signal currents, but return and power planes are by far the best solution.

Provide separate ground pins for I/O circuits and core circuits in an IC package if you are building hybrids, MCMs, or custom chips. Decoupling each circuit will reduce interference between circuits from switching transients on the power and ground pins.

Use 1-oz copper in the power and return planes for currents less than 15 A, use 2-oz copper for currents less than 30 A, and use multiple planes for currents beyond 30 A. These rough guidelines should help keep voltage drop less than 2% in the distribution circuits.

8.5.3 *Shielding*

A return (or ground) plane is the most effective shield for any circuit. Power and return planes provide circuit paths with the lowest impedance, which reduces radiation, noise, and crosstalk. Minimizing the spacing between power and return will minimize impedance (and therefore radiation and susceptibility) (White, 1982):

$$Z_0 = (120\pi/\sqrt{\epsilon_r}) \cdot (h/d),$$

where: h = separation of planes
d = smaller dimension of a two-dimensional plane
ϵ_r = dielectric constant of substrate board relative to air

Furthermore, adding decoupling capacitors near the chips will reduce impedance, as shown in Fig. 8.16. Conversely, breaking up the power and return planes will restrict current flow and increase impedance and voltage drops. In general, these guidelines will ensure an effective shield (White, 1982):

1. Use power and return planes with minimum separation.
2. Place decoupling capacitors near (or in) IC packages.
3. Don't disrupt the power and return planes with slots or traces.
4. Route digital traces over digital return (if they must cross into the analog region, keep them short and mostly static, such as gain and multiplexer [mux] controls for an ADC).
5. Route analog traces over analog return.
6. Fill the regions between analog traces with copper foil and connect to ground.

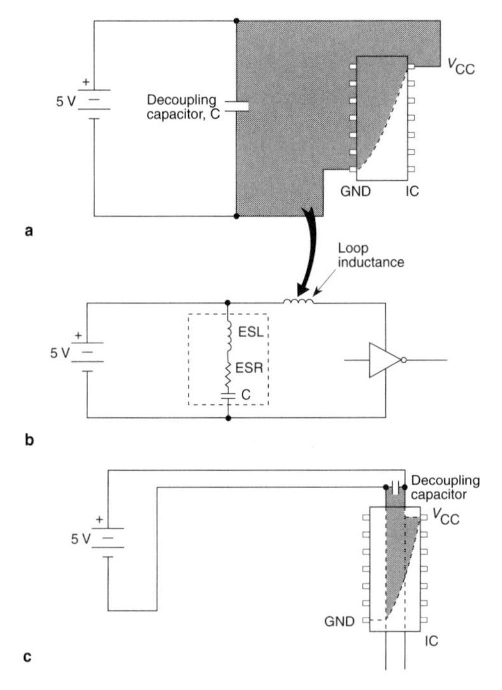

Fig. 8.16 Power supply decoupling. (a) Location of the decoupling capacitor determines the loop area and hence the inductance. (b) Equivalent circuit. (c) Putting the decoupling capacitor near the chip and running power and return together reduces the loop area and inductance.

You can add shielding in other ways too. Mumetal and steel enclosures provide some measure of shielding against magnetic fields. Metal enclosures and cable braiding shield against capacitive coupling of electrostatic fields (see Chapter 6). But the most effective shielding is proper layout of your circuits that eliminates large loops and common impedance.

8.6 CONNECTORS AND CABLES

Thus far I have covered how to lay out signal traces on a PCB, but circuits don't operate in isolation from the outside world. Often signals have to enter and exit a PCB through connectors and wires. Both mechanical and electrical issues surround connectors and cables.

Connectors are the mechanical and electrical interface between a cable and a circuit board. They have two pieces that fit together and hold through friction. The metal-to-metal contact provides the electrical path. The friction fit must be enough to make some "gastight" contacts but not too tight to prevent disengagement when unplugging the connector. Fig. 8.17 illustrates a sample of different configurations for connectors.

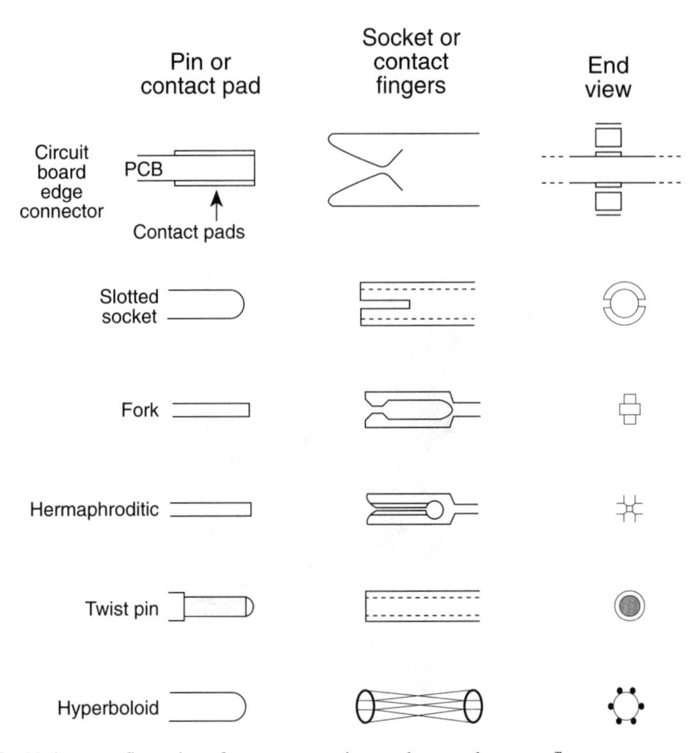

Fig. 8.17 Various configurations for connector pins, sockets, and contact fingers.

Shape or keying *polarizes* a connector so that it cannot be plugged in the wrong way (Fig. 8.18).

Strain relief between a connector and a cable will reduce flexing and consequently failure from mechanical fatigue (see Chapter 5 for more on fatigue, tie-down, and cable routing). Fig. 8.19 shows how loops can relieve strain on cables and how tie-downs can reduce flexing.

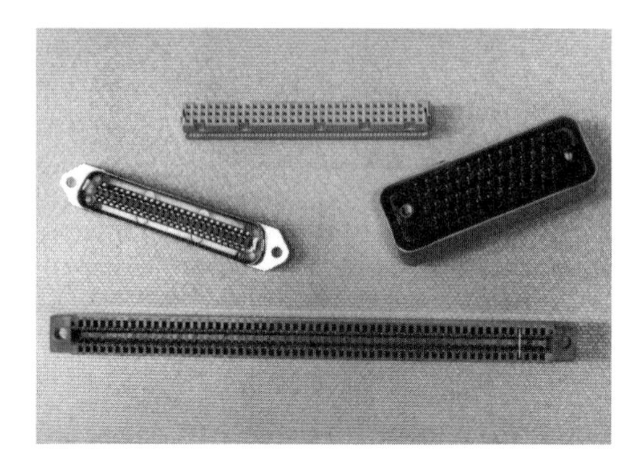

Fig. 8.18 Several ways to key or shape connectors to prevent incorrect connection.

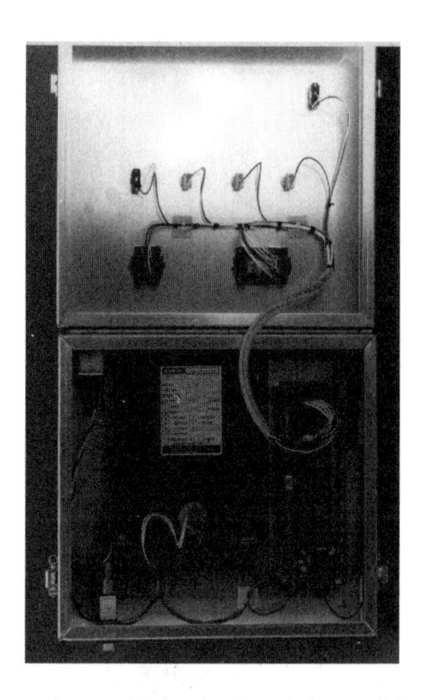

Fig. 8.19 Ways to relieve strain on cables and reduce fatigue and breaking—loops in wires, cable lacing, and tie-downs.

Cables and connectors can cause many electromagnetic problems. Every signal current has an equal return current; if the return line is longer than the signal line, then the current has a high-inductance path that causes a noise spike in the ground system. Long return (or ground) paths can cause excess ground shift, decreased noise margins, and inappropriate switching. Therefore, reduce the inductive loops by interspersing ground lines with signal lines. A ratio of 4 or 6 signal pins for each ground pin will reduce return loops and inductance, and twisting each signal line with a paired return line is the best way to eliminate noise problems due to magnetic induction. A shield around the cable that is correctly terminated to ground will eliminate noise problems due to capacitive coupling. Fig. 8.20 illustrates ground pin assignment in connectors for optimum performance.

Here are some guidelines for connectors and cables:

1. Preassign connector ground pins.
2. Distribute and intersperse grounds (return paths).
3. Place clock next to a ground line.
4. Minimize I/O.

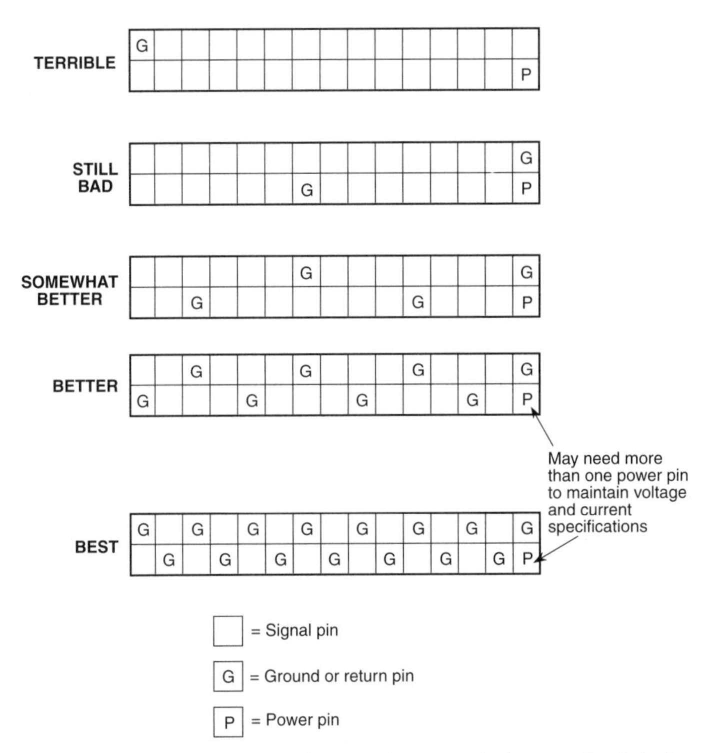

Fig. 8.20 Ground pin assignment. Interspersing ground or return pins between signal pins in a connector will reduce the loop inductance in signal cables.

5. Use long rise and fall times to reduce high-frequency harmonics.
6. Keep current to less than 1 A per connector pin; otherwise use multiple pins or special pins with large current capacity.

Sometimes systems need boards that can be connected while power is on. Hot insertion of circuit boards requires careful design of the connectors so that power application to the circuit has an appropriate sequence and prevents latch-up of ICs. If you are using edge card connectors, you can stagger the contact fingers to ensure the appropriate sequence of connections when the board is inserted into the system (Figs. 8.21 and 8.22) (Barnes, 1989).

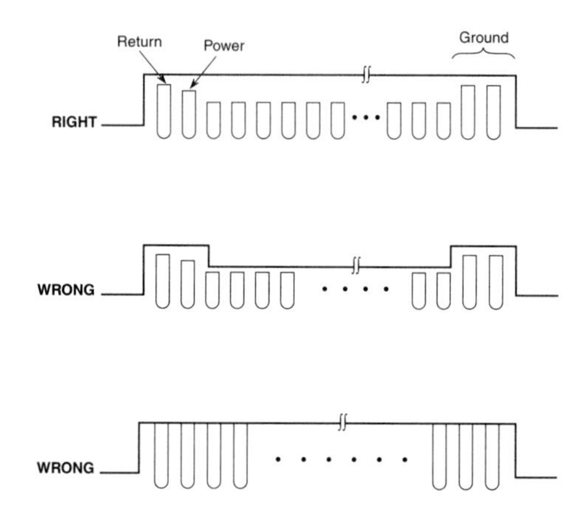

Fig. 8.21 Staggering the depth of finger contacts to ensure the correct sequence of connections between connector and PCB. Don't stagger the PCB to follow the finger length; doing so removes the timing margin during insertion.

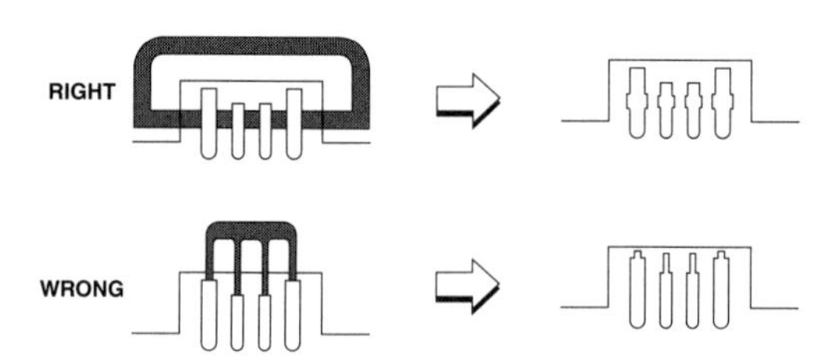

Fig. 8.22 Plating contact fingers. If you gold-plate staggered contact fingers, don't leave nipples (from the shorting bar used during plating) on the end of the fingers.

8.7 DESIGN FOR MANUFACTURE

Good design facilitates production. The key to design for manufacture is good communication between the designer and the manufacturer. Circuit boards, done right the first time, require complete specifications and a well-defined review process. Moreover, certain design practices will help ensure quality PCBs (Clark, 1989):

- Drill selection
- Thermal relief for soldering
- Solder mask
- Trace tolerances
- Overetch or underetch
- Aspect ratio and hourglass effect in plating
- Test coupon

8.7.1 *Good Design Specifications*

Cherf and Mason (1993, p. 68) write, "Lack of correct information in the preproduction phase is the number one cause of lengthened production cycles and increased cost to the final customer." The manufacturer needs all the required documentation in a readable and consistent format. The design should be free of unnecessary complexity, such as unneeded pads, traces, and spacings. You should perform a design rule check to eliminate unnecessary complexity before turning the design over to the manufacturer. Correctly label the layers of the PCB. When changes occur, and they will, establish the responsibility and authority to make changes between you and the vendor. Finally, visit the facilities of PCB vendors to understand the manufacturing process better. Fig. 8.23 is a checklist for producing PCBs (Tolhurst, 1993).

- ❑ Deliverables list: computer file (Gerber), forms, reviews, checklist.
- ❑ Overall specification: nominal crosstalk, impedance, layer stack-up.
- ❑ Mechanical data: components, fasteners, stiffeners.
- ❑ Placement data and instructions: communicate with the PCB designer and use good notes to give reasons for circuit operation.
- ❑ Routing data and instructions: communicate with the draftsman and use good notes to give reasons for circuit operation.
- ❑ Silk-screen guidelines: component outlines, mark pin 1.

FIG. 8.23 Checklist for PCB design specifications.

8.7.2 *PCB Configurations*

Mechanical issues dominate good design practices for PCBs. Clearances, tolerances, panel shape, and coordinate references all affect the manufacturability. Furthermore, proper tolerances and assembly ensure mechanical interchangeability.

Appropriate clearances in the artwork prevent traces from shorting to the edge of the chassis or to washers. Don't extend the return and power planes beyond what is necessary for circuit operation; leave clearance around holes for rivets or bolts. Specify tolerances that allow for positions and diameters of holes when drilling and for width and spacing when etching traces. Allow 1 to 2 cm (0.5 to 1 in.) of copper surface on the perimeter of the PCB panel to clamp the electrode when plating copper. Provide a 1.3-mm (0.05-in.) gap between the plating perimeter and the edge of the actual circuit to avoid overplating and shorting (Taylor, 1993).

Use absolute coordinates for dimensions to reduce the frequency of errors. A drilled hole in the lower left corner of the PCB will serve as a reference. If the PCB is circular, use the center as origin.

Avoid long U- or L-shaped boards; they are difficult to wave-solder. Likewise, small circuit boards should be grouped onto panels of standard stock (Fig. 8.24) to ease handling by the equipment.

Popular, standard sizes of stock for copper-clad laminate are as follows (Jodoin, 1993):

12 × 16 in.	16 × 18 in.	18 × 24 in.
12 × 18 in.	16 × 24 in.	20 × 24 in.
		21 × 24 in.

Mandating the enclosure size without regard to manufacture of the PCB will compromise manufacturability. Therefore, consider package size and shape along with manufacturability of the PCB early in development to reduce conflicts, errors, and time.

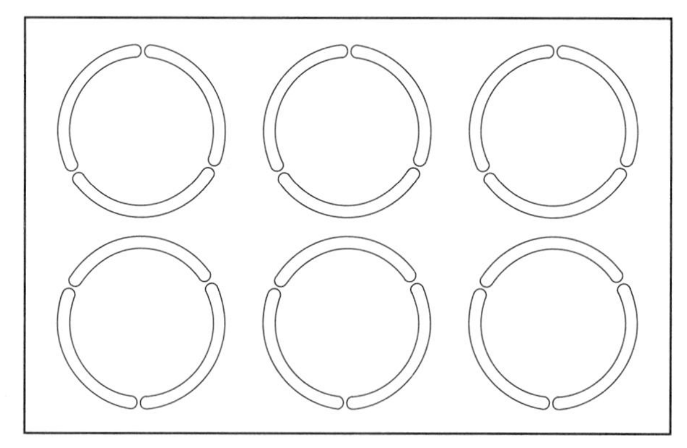

Fig. 8.24 Grouping (*panelizing*) small or circular circuit boards onto standard-sized stock for manufacturing and soldering PCBs.

You should round internal corners of a PCB to relieve stress and reduce cost (Fig. 8.25). Square internal corners can be used, but they are more expensive to manufacture and more likely to fracture. The radius of corners or holes should be 1.6-mm (0.062-in.) or greater.

For edge connectors, chamfer the edge of PCBs to spread contact fingers (Fig. 8.26). Chamfers reduce chipping of the board by connectors and ease insertion.

Don't use a nut and bolt to make electrical contact between the case of a component and the PCB. Compression is not reliable; it can shear the plating in a via and cause an open or intermittent connection. Use a solder lug and short jumper to complete the connection between the case of a component and the PCB (Fig. 8.27). Don't plate fixture holes for bolts; they can short power and ground planes.

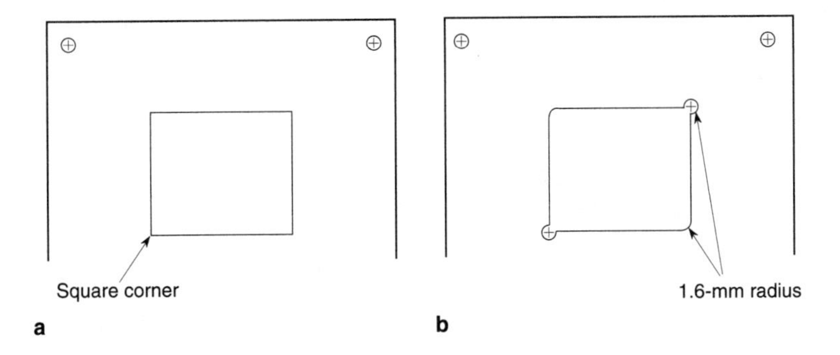

Fig. 8.25 Rounding internal corners to reduce stress in the PCB and cost in manufacture. (a) Avoid square corners. (b) Use either rounded corners or holes at the corners.

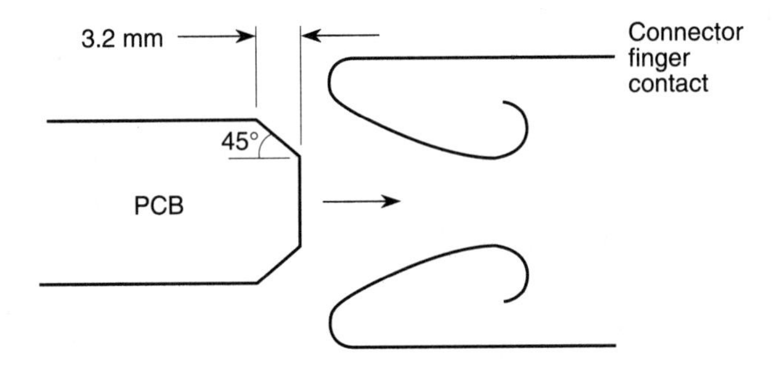

Fig. 8.26 Chamfered leading edge of a PCB that inserts into a connector will reduce chipping of the PCB.

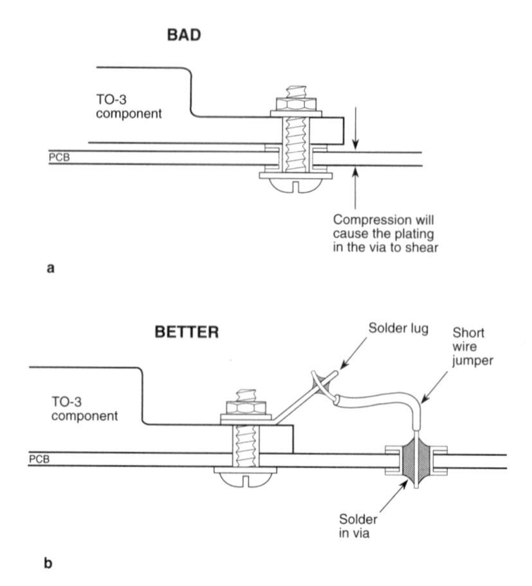

Fig. 8.27 Completing an electrical connection. (a) Don't rely on compression. (b) Use a solder lug and a wire jumper to make the connection.

8.7.3 *Assembly and Inspection*

Assembly of circuit boards involves placing the components, soldering them in place, and inspecting the solder joints. You can control the quality of PCB assembly through procedures and inspections. Good procedures build in the quality; inspections measure the results. Each phase—design, manufacturing, and test—affects the ultimate yield, cost, and reliability of a circuit board.

The test coupon that accompanies a PCB allows inspection for quality of manufacture. Slicing the coupon and examining it under a microscope reveals the distribution of copper plating and solder in the vias. Even distribution of copper plating and solder within specifications indicates proper manufacture of the PCB. You can also verify the characteristic impedance of the circuit board by measuring the electrical properties of a coupon.

For large production runs, pick-and-place equipment *stuffs* or places components in the correct locations on the PCB. Manual stuffing of the components is appropriate for prototypes or small production runs, but defect rates for hand insertion are three times those for machine insertion. Correcting wrong insertions of components raises manufacturing cost (Donlin, 1993).

Soldering can be manual or automatic, such as wave solder or vapor-phase reflow. In either case, you should inspect the solder welds by one of four methods (Reynolds, 1984):

1. Tip the board and view with the eye.
2. Remove some SMTs and inspect solder pads (this is sampling for process control).
3. X-ray joints (expensive and cumbersome).
4. Laser-inspect joints (heat solder and measure with infrared detector).

Another tip for inspection is to provide an area on the PCB for a handhold to grip the board and avoid contact or damage to the components during inspection.

8.8 TESTING AND MAINTENANCE

While you design for manufacture, you should plan for testing and maintenance of circuit boards. Testing and maintenance have five levels: component, circuit, board, module, and instrument. Testing and maintenance encompass a variety of concerns, from repair access to potential problems. Testing and maintenance do not predict problems, but they can shorten the time to repair.

8.8.1 *Testing*

Testing has several different configurations and fixtures. Fixtures provide the necessary power, inputs, and outputs to exercise parts, from components to modules, independent from the system. Even with fixtures, accessibility constrains the effort to test circuit boards. The tests themselves may either thoroughly check every conceivable condition and state or perform a functional test that covers some portion of operation.

Component fixtures support tests of individual ICs (Chapter 12 describes some of these tests). A component fixture holds the IC in a socket, supplies power, and provides interface circuitry and signals to test parameters of the IC's operation. Board fixtures allow testing of complete circuits to ensure functionality. A board fixture clamps the board, supplies power, and provides a probe to sample signals from test points on the PCB. You should avoid placing too many test points in one area because the probe can exert too much pressure and break the PCB (Donlin, 1993).

Circuit layout is more than placing ICs conveniently on a PCB; it defines accessibility and the ease of testing. Good layout and design will accommodate testing at various levels during production by making parts of the circuit accessible (Bostak et al., 1993):

1. Partition the system, both functionally and physically, so that faults can be detected independently.
2. Make circuits testable without the front panel.
3. Provide test outputs compatible with standard test equipment.
4. Eliminate warm-up problems or delays to allow rapid testing.

Testing takes several forms: scan, functional tests, and BIT. Scan testing stimulates points within a chain of circuitry, records the results, and compares them with expected or calibrated data. The circuitry that implements scan testing simplifies testing at the expense of more complexity and real estate on the IC or circuit board. Fig. 8.28 diagrams two configurations for scan testing. Functional tests exercise a subset of all the operations that the circuit may generate. Functional tests require less specialized circuitry but do not check all possible failure modes. BIT usually dedicates some circuitry for testing the remainder of the system; it can implement scan or functional tests. BIT speeds diagnostics at the cost of complexity and lower overall reliability.

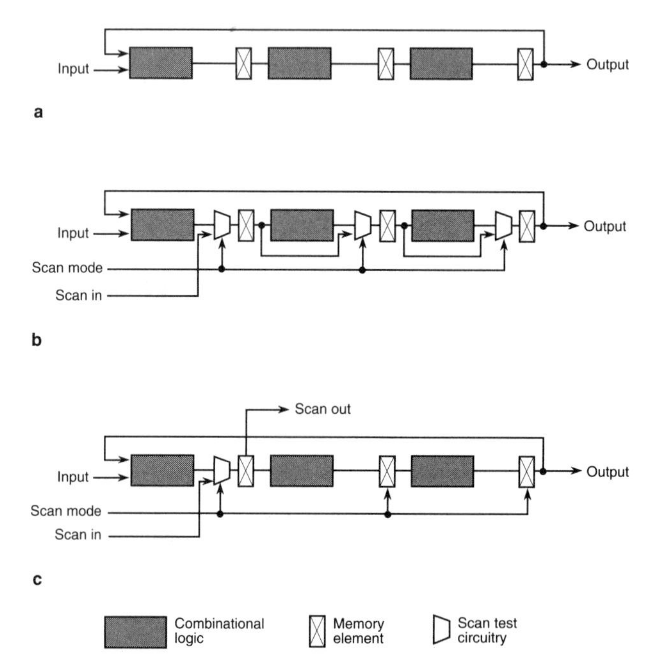

FIG. 8.28 Scan testing. (a) Boundary scan speeds testing in long chains of circuitry. (b) Scan circuitry can inject test signals to exercise each section of circuitry and avoid memory delays, but it consumes real estate on the chip. (c) Partial-scan logic strikes a balance between testability and complex test circuitry.

8.8.2 *Maintenance*

When you define a concept and set requirements for a product, you must decide how much maintenance it will incur throughout its life. Some appliances, for instance, are throwaways, because repair would cost more than purchasing a new one; therefore, maintenance is not a consideration. Others, like control processors for a traffic light system, require extensive planning, support, and maintenance to run smoothly for years.

Concerns in the maintenance of circuit boards are access and ease, support equipment, and instruction and repair manuals. Design for maintenance should make circuit boards easy to locate, disconnect, and remove. Card locks should swivel smoothly, and channels should resist twisting and binding the PCB. You must identify the appropriate support equipment, whether custom or general purpose, in the repair manuals and explain how to connect and use the equipment.

Many maintenance problems are mechanical. Plan for flex fatigue in cables, corroded connectors, and dirty contacts in potentiometers and switches—or just eliminate them from the design. Physical abuse—such as dropping, reversed power polarity, and static discharge—accentuates the need for conservative design and maintenance.

8.9 SUMMARY

This chapter illuminates some of the steps to building operational circuits. While many of the principles seem low tech, your designs probably won't function properly without faithful adherence to good practices.

The layout of components and routing of signals require careful consideration during design; otherwise the circuit will generate EMI. Shielding usually requires a good return plane. Cables must have return paths that match the signal lines.

You must have clear and regular communication with the manufacturer of the PCB to root out problems and design a board that can be built. Maintenance requires access and clear instruction. All this begins with the concept, of course. Plan early, and plan well.

8.10 RECOMMENDED READING

Clark, R. H. 1989. *Printed circuit engineering: Optimizing for manufacturability.* New York: Van Nostrand Reinhold.
Thorough coverage of PCB manufacture.

Cypress Semiconductor Corp. 1993. System design considerations when using Cypress CMOS circuits. In *Cypress semiconductor applications handbook,* pp. 1-1–1-32. San Jose, Calif.: Cypress Semiconductor Corp.
A good article about transmission line concerns and signal routing.

Johnson, H. W. and M. Graham. 1993. *High-Speed Digital Design: A Handbook of Black Magic*. Englewood Cliffs, N. J.: PTR Prentice Hall.
Excellent suggestions for layout and circuit design.

8.11 REFERENCES

Arrington, B., and M. Brown. 1989. Resolving transmission line effects in PCB designs. *High Performance Systems,* December, pp. 85–88.

Barnes, B. 1989. Insert boards into a live system without any hitches. *Electronic Design,* May 11, pp. 74–83.

Beresford, R. 1989. How to tame high speed design. *High Performance Systems,* September, pp. 78–83.

Blankenhorn, J. C. 1993. A rose by any other name. *Printed Circuit Design,* August, pp. 15–17.

Bostak, C. J., C. S. Kolseth, and K. B. Smith. 1993. Concurrent signal generator engineering and manufacturing. *Hewlett-Packard Journal,* April, p. 33.

Cherf, R., and D. Mason. 1993. Crucial DFM considerations. *ASIC & EDA,* April, pp. 68–72.

Clark, R. H. 1989. *Printed circuit engineering: Optimizing for manufacturability.* New York: Van Nostrand Reinhold.

Cutler, R. D. 1986. Ground pin optimization in high-speed digital systems. *VLSI Systems Design,* March, pp. 46–48.

Cypress Semiconductor Corp. 1993. System design considerations when using Cypress CMOS circuits. In *Cypress semiconductor applications handbook.* pp. 1-1–1-32. San Jose, Calif.: Cypress Semiconductor Corp.

Davis, B. 1993. A tour through the board manufacturing process. *Printed Circuit Design,* August, pp. 11–14.

DeSantis, J. A. 1993. An overview of PCB constructions. *Printed Circuit Design,* April, pp. 48–53.

Donlin, M. 1993. A tale of two companies. *Computer Design,* April, pp. 142–145.

Gheewala, T., and D. MacMillan. 1984. High-speed GaAs logic systems require special packaging. *EDN,* May 17, pp. 135–142.

Jodoin, C. J. 1993. Designer's viewpoint. *Printed Circuit Design,* May, pp. 25–26.

Johnson, H. W., and M. Graham. 1993. *High-Speed Digital Design: A Handbook of Black Magic*. Englewood Cliffs, N. J.: PTR Prentice Hall.

Reynolds, R. A. 1984. Clad-metal-core PC boards enhance chip-carrier viability. *EDN,* August 23, pp. 211–215.

Taylor, C. 1993. PCB design: A mechanical viewpoint. *Printed Circuit Design,* February, pp. 57–60.

Tolhurst, B. 1993. Front end. *Printed Circuit Design,* May, pp. 51–52.

White, D. R. J. 1982. EMI control in the design of printed circuit boards. *EMC Technology,* January, pp. 74–83.

9

POWER

. . . the pains of power are real, its pleasures imaginary.
—Charles Caleb Colton

9.1 SOURCES AND REQUIREMENTS

Electronic devices and equipment require electrical power. Sources provide that power with either alternating current (AC) or direct current (DC). Generally AC has a frequency between 47 and 440 Hz and an rms voltage between 85 and 270 V. Most home appliances in the United States operate from nominal values of 120 VAC at 60 Hz; air conditioners and electric clothes dryers use 240 VAC. In Britain the appliances operate from 240-VAC, 50-Hz power, while Europe and a good portion of the rest of the world operate from 220-VAC, 50-Hz power. In contrast, aircraft avionics use either 28-VDC or 28-VAC, 400-Hz power (though some systems are beginning to use 270 VDC). Numerous sources supply DC power. DC voltages range from 1.2 V to hundreds and even thousands of volts for diverse applications from wristwatches to motor drives.

You must understand the characteristics and peculiarities of the electric power sources before you can complete the successful design of an instrument. AC line power, for instance, is subject to dropouts, spikes, brownouts, and blackouts caused by lightning strikes, heavy machinery, and summertime loads. Battery-powered sources of DC power drop in voltage as a function of time and current and cannot sustain heavy current demands. Beyond the electrical characteristics of the power sources, electromagnetic emission standards and physical configurations for power sources and converters vary from nation to nation. Fig. 9.1 illustrates some AC line plugs used in different countries around the world. Clearly, you must investigate the power and specify the characteristics of that source power in the requirements.

Most likely your equipment will not be able to use the source power directly. While the source often is 120-VAC, 60-Hz or 240-VAC, 50-Hz power, many instruments require low DC voltages to operate circuits (Table 9.1). A power conversion device or circuit is needed to convert the source power to the desired form and

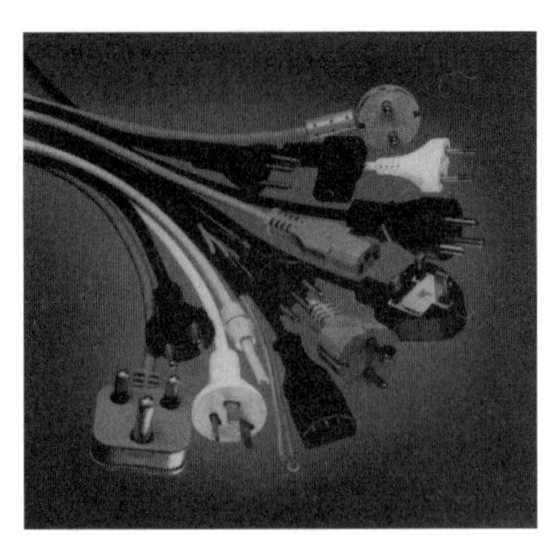

Fig. 9.1 Selection of line plugs from around the world (photograph courtesy of Electri-Cord Manufacturing Co., Inc.).

Table 9.1 Some standard DC voltages used in instrumentation

Voltage (V)	Application circuitry
+3, +3.3	Low-power logic: CMOS and bipolar/CMOS
+5	TTL, NMOS, CMOS logic
±12, ±15	Analog signal conditioning, CMOS logic
−5.2	ECL logic

level. Again, you must understand and specify the electrical environment within your instrument before selecting an appropriate power converter.

Besides maintaining voltage levels, power conversion devices must provide sufficient current, filter noise, respond to load variations, and start up and shut down predictably. These concerns derive from the circuit requirements within your device or instrument. In addition, the power converter must not return power transients (spikes) or noise to the power source.

Power conversion must be specified in the initial concept development and not, as it often is, left as a last-minute detail. The power converter, simply by its presence, will affect the operation of your circuitry. Thermal dissipation, efficiency, electromagnetic emissions, size, and weight contributed by the power converter all factor into the design of the instrument. Early treatment will prevent an overweight, costly device that has been specified too broadly and suffers from poor design and shortsightedness. Understanding and specifying the source power, the load environment, and the characteristics of the power converter will bound ignorance and optimize your converter design.

9.2 OUTLINE FOR POWER DESIGN

The first step is to understand the concepts and definitions of power conversion within your product. Next you need to define the source of power and know its actual condition. Only then can you begin to define the necessary power design. This chapter follows the same general outline by presenting the basic definitions of power conversion first; then presenting some more detailed issues of power conversion, distribution, and line conditioning; and finally presenting details of electromagnetic interference, reliability, and alternative sources of power.

9.3 BUY VERSUS BUILD

Incorporating a power converter into a device or an instrument often raises the question of buying the converter, designing and building it in house, or having a custom converter designed for your application. Generally the answer reduces to final cost. You must consider resident expertise, production quantity, manufacturing cost, nonrecurring engineering (NRE) expense, and facilities for test, burn-in, and qualification. The cost of research and development or NRE can be amortized over a large production run. Materials, components, labor, and inventory all contribute to the unit cost.

One possibility is to use stand-alone power supplies. These are wall-plug units or separate modules such as those used with laptop computers. They have the advantage of economy, since production and testing are already included in their purchase price and they relieve you from the burden of designing and testing a custom power converter. They also have either Underwriters Laboratories or some sort of governmental approval—you won't have to get it yourself.

Possibly even more important to your considerations is the availability of technical expertise to design and test power converter designs. Power converter design is challenging and requires extensive knowledge and testing. For small production runs, the cost of developing the facilities and expertise is usually prohibitive. Otherwise, if your application demands a custom power supply, specialty companies and consultants can provide valuable help in designing and testing your power converter.

9.4 POWER CONVERSION CHOICES

Many devices and instruments require some form of regulated DC power. Fig. 9.2 samples the wide variety of power converters available. This chapter will introduce the basic concepts for power converters that supply DC power to circuits from AC or DC sources. (Power converters have traditionally been called *power supplies*. I will use the terms interchangeably to avoid both confusion and semantic pickiness.)

Power converters or power supplies contain three or four general stages to convert power (Fig. 9.3). First, the converter transforms AC power to some inter-

Fig. 9.2 Variety of power supplies available. Note the different sizes and power capacities (photograph courtesy of Astec Standard Power).

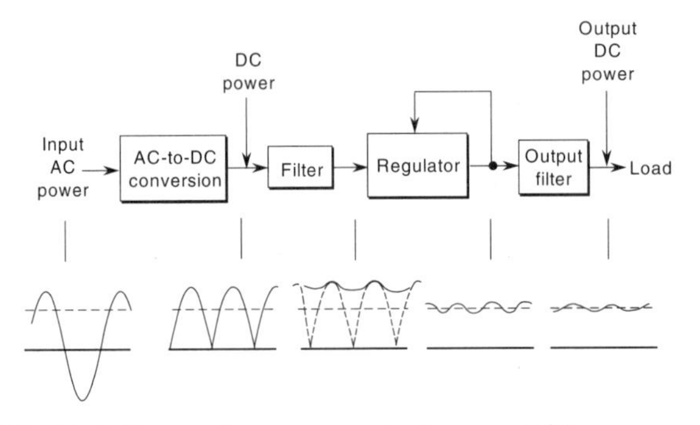

Fig. 9.3 Simplified block diagram of a power converter (power supply).

mediate form of DC power. Then it filters the DC power to remove gross variations in the voltage and supplies the filtered DC power to the regulator. The regulation stage uses feedback to control the output voltage level at the load. Finally, the output filter smooths the variations in the load voltage.

Three common types of power converters use this generalized scheme: ferroresonant, linear, and switching (Table 9.2). Ferroresonant converters have good efficiency and are simple and reliable. Linear converters have excellent regulation and tend to be quiet over the electromagnetic spectrum, but they have poor efficiency of conversion. Switching converters, or switchers, while complex, demonstrate high efficiency and energy density that allow small package designs.

Table 9.2 Comparison between the general characteristics of power supplies

Power supply type	Source input			Output regulation				Power		Size		Reliability (MTBF)	Cost
	Voltage	Frequency	Power factor	Line regulation	Transient response	Current limiting	Over-voltage limiting	Efficiency	Density	Volume	Weight		
Ferroresonant	Insensitive ±15% variation	Tuned, requires strict frequency control	Excellent	Fair	Poor	Inherent	Inherent	Very good 80%	Poor	Large	Very heavy	Excellent	Moderate
Linear	Tolerant ±10% variation	±10%	Good	Excellent	Excellent	Must be designed into circuit	Must be designed into circuit	Poor ≤ 40%	Poor ≤ 0.12 W/cm^3 (2 W/in.^{3})	Bulky	Heavy	Good	Low to moderate
Switching	Can accept wide variation ≥ 100%	≥ 100%	Poor, correction circuit must be added	Very good	Good	Must be designed into circuit	Must be designed into circuit	Excellent > 90%	Excellent > 3 W/cm^3 (50 W/in.^{3})	Very small	Very light	Fair	High

9.4.1 *Ferroresonance*

A ferroresonant power converter incorporates a constant-voltage transformer that regulates voltage before the rectifier that converts AC power to DC. Ferroresonant converters differ in this respect from the basic scheme presented in Fig. 9.3. The constant-voltage transformer operates the secondary winding in saturation. Since saturation maintains a fairly constant flux density within the transformer core, the output voltage remains constant by tracking the flux density regardless of the input voltage. The transformer core consists of two magnetic circuits separated by a magnetic shunt. The magnetic shunt provides a measure of decoupling that prevents saturation in the primary winding core. Fig. 9.4a illustrates a constant-voltage transformer. The tank circuit generates a large reactive current in the secondary winding that causes the saturation in the core. The tank circuit maintains a resonance, tuned to the input source frequency, between the capacitor and the ferromagnetics of the core; hence the name *ferroresonance*.

The constant-voltage transformer outputs a clipped AC waveform that simplifies rectification and filtering. The clipped waveform produces less ripple in the output voltage after filtering than a rectified sine wave. Fig. 9.4b illustrates a simplified block diagram of a ferroresonant power converter.

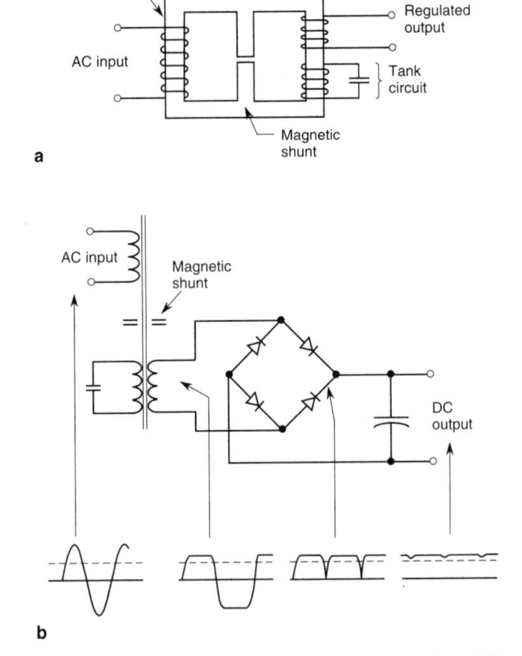

Fig. 9.4 Basic components and design of a ferroresonant power supply. (a) Transformer and magnetic circuits. (b) Simplified schematic and attendant waveforms.

Ferroresonant power converters have several favorable properties—specifically, current limiting, line regulation, and efficiency—that result from the transformer operation. Inherent current limiting occurs when excessive load current reduces the mutual flux between the primary and secondary windings, the quality factor, Q, of the tank circuit drops, and the output voltage on the secondary windings falls. Overvoltage circuitry is unnecessary because of the inherent line regulation. Ferroresonant converters approach 80% efficiency in power conversion, since they have no dissipative elements. In addition, their few components result in a simple design that ensures very reliable operation.

Ferroresonant power converters have significant limitations, however. The source frequency must be stable and very near the tank circuit's resonant frequency; otherwise the output voltage will vary. The load regulation of ferroresonant converters has long response times to transient changes in load. Finally, the converters are bulky and very heavy.

9.4.2 *Linear*

A linear power converter follows the general scheme presented in Fig. 9.2. The AC-to-DC conversion stage includes a transformer and rectifier. The filters use capacitors to reduce the voltage ripple by removing the high-frequency components. The regulation stage, or regulator, uses a pass transistor as a variable resistor to control the voltage across the transistor and flatten the voltage peaks. The regulator incorporates a voltage reference and monitors the output DC voltage to complete the feedback loop with the pass transistor. Fig. 9.5 illustrates these stages of a linear power converter and some components within the regulator.

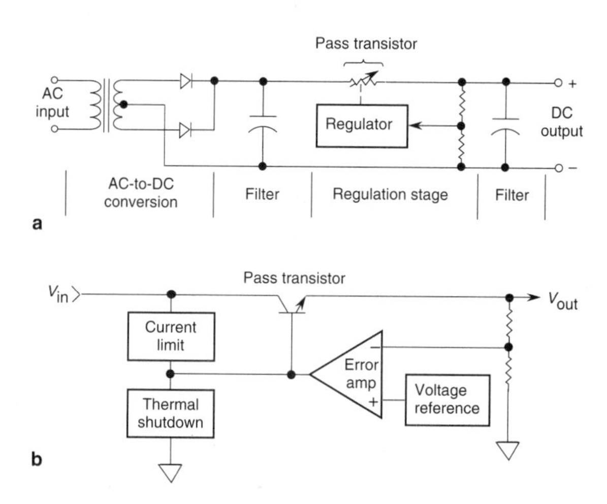

Fig. 9.5 Basic components in a linear power supply. (a) Outline of a linear power supply. (b) Basic regulator design.

The feedback within a linear regulator gives outstanding load regulation and very clean output (low ripple and noise). Linear regulators can respond within microseconds to changes in the load. The transformer provides isolation from the AC input; combined with the regulator, it provides excellent line regulation. In other words, input variations from the AC source do not transfer to the DC output and its load.

Linear power converters tend to be fairly simple. You can scatter regulators about a variety of circuits to provide excellent local regulation. Also, linear power converters may be easily adjusted to change the output DC voltage. Most low-power laboratory supplies are linear designs.

Linear power converters have some drawbacks, particularly in efficiency and weight. Efficiency in power transfer is inherently low, usually less than 50%. The voltage drop required across the pass transistor is the primary contributor to the lower efficiency. Some voltage must be dropped across the pass transistor, so the transformer must supply correspondingly higher voltage to the regulator. The difference between the voltage from the transformer and the DC output from the regulator is called *headroom*. Regulators typically require headroom of 2 or 3 V. Low-dropout regulators have headroom as low as 0.2 V. When you specify a linear regulator, be sure to include headroom for situations with low input voltage and a full load on the output.

Another disadvantage of linear power converters is their relative weight and bulk. For AC sources operating at 50 or 60 Hz, the transformer must be large to transform the power efficiently.

9.4.3 *Switching*

In contrast to the size of linear devices, the high operating frequencies of switching power converters significantly reduce the size and weight of the magnetic components. Switching power converters have a variety of circuit topologies, some of which depart significantly from the general scheme in Fig. 9.3. The basic design of switching power converters uses transistors as nondissipative switches to channel high-frequency currents (20 kHz to 1 MHz) through small inductors and transformers. Consequently, switching power converters can generate DC output voltages either lower or higher than the input voltage.

Changing current through an inductor develops a proportional voltage:

$$V = L(\mathrm{d}i/\mathrm{d}t).$$

By controlling the change in current, you can generate the desired voltage. Likewise, a changing voltage across a capacitor develops a proportional current:

$$I = C(\mathrm{d}v/\mathrm{d}t).$$

Two different circuit topologies, forward mode and boost mode, can provide you with some insight into switching power conversion.

The forward-mode, step-down, or buck regulator switches between two states. In one state, the switch is closed, and the input source charges the energy storage elements and drives the load. In the other state, the switch is open, and current from the energy storage elements drives the load. Fig. 9.6 outlines the operation of a forward-mode regulator.

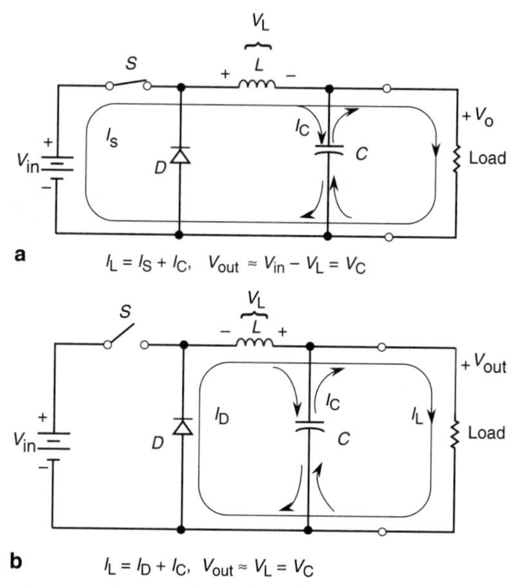

Fig. 9.6 Forward-mode or step-down regulation for switching power conversion. (a) Switch closed: Input source charges *L* and drives the load; *C* provides filtering by charging and discharging. (b) Switch open: Storage energy in *L* drives the load, and *C* provides filtering.

While the switch is closed, the input source, V_{in}, drives current, I_s, through the inductor, *L*, and into the load. The energy storage capacity of the capacitor, *C*, provides filtering of the output voltage. The diode, *D*, is reverse biased.

When the switch opens, current cannot stop flowing in the inductor instantly. The voltage across the inductor reverses, and the diode begins conducting to maintain the current through the inductor. The collapsing magnetic field within the inductor replaces the current no longer supplied by the input source to the load. The capacitor continues its filtering function.

Current continually flows through both the inductor and the load from state to state as the switch cycles between open and closed. The amount of energy supplied to the load is proportional to the duty cycle of the switch. A forward-mode regulator can only step down the output voltage from that of the input. An approximate relationship for output voltage is

$$V_{out} \approx V_{in}\,(\text{duty cycle})\ .$$

Conversely, the boost regulator increases the output voltage over the input by switching between two states to convert power (Fig. 9.7).

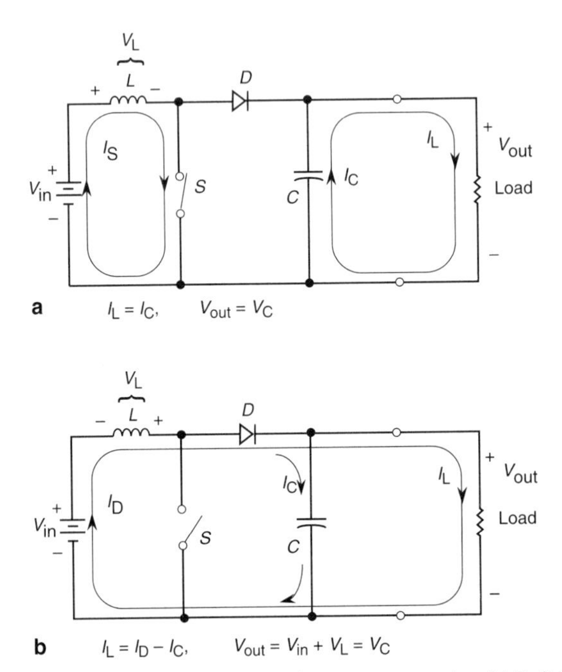

Fig. 9.7 Boost-mode or step-up regulation for switching power conversion. (a) Switch closed: Input source stores energy in L while C drives current through the load. (b) Switch open: Voltage reverses on L, adding to the input voltage; current charges C and drives the load.

While the switch, S, is closed, current flows through the inductor, L, storing energy in its magnetic field. During this state the capacitor, C, supplies current to the load; the reverse-biased diode, D, prevents the current from shorting through the closed switch.

When the switch opens, current flow in the inductor cannot change instantly. The voltage across the inductor reverses, drives current through the load, and charges the capacitor.

A boost, or step-up, regulator increases the output voltage over the source input voltage. Again, the duty cycle of the switching action controls the value of output voltage. The approximate relationship for output voltage is

$$V_{\text{out}} \approx V_{\text{in}} + V_{\text{L}} \approx \frac{V_{\text{in}}(T_{\text{off}} + T_{\text{on}})}{T_{\text{off}}}; T_{\text{on}} \leq T_{\text{off}},$$

where T_{on} = time while switch is closed and the inductor is charging
T_{off} = time while switch is open and the inductor is discharging

The duty cycle for T_{on} usually varies between 0% and 50%.

Several other circuit topologies are used in addition to the two basic regulators just described. These circuits switch the input to the transformer to achieve regulation. Four configurations are commonly used: flyback mode, forward mode, half bridge, and full bridge (Fig. 9.8). Each configuration has advantages and limitations (Table 9.3).

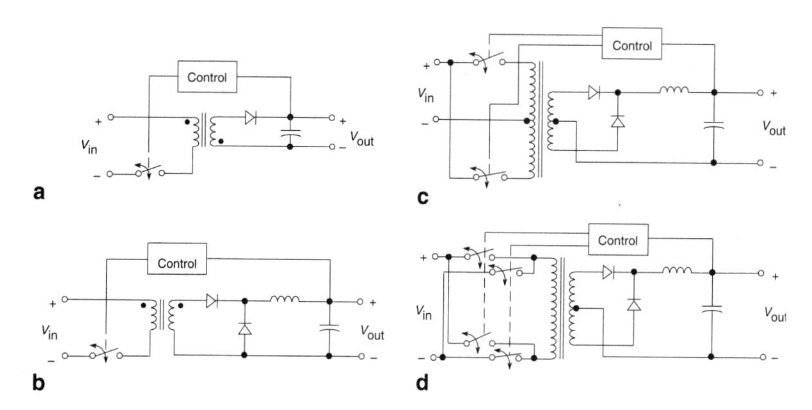

Fig. 9.8 Various circuit configurations for switching regulators. (a) Flyback-mode regulation. (b) Forward-mode regulation. (c) Half-bridge regulation. (d) Full-bridge regulation.

Table 9.3 Comparison of various configurations for switching regulators

Converter mode	Cost	Component count or complexity	Regulation	Noise and ripple	Power output (W)
Flyback	Lowest	Very low	Poor	High	< 100
Forward	Low	Low	Fair	Fair	< 100
Half bridge	Medium	Medium	Good	Low	> 100
Full bridge	High	High	Good	Low	> 250

Switching power converters, while more complex than linear or ferroresonant converters, have a number of advantages over them. Switching converters can have extraordinary power density, exceeding 3 W/cm^3 (50 W/in.^{3}), and very high efficiency—some converters are more than 95% efficient. These two factors combined provide for very compact packages. Switching converters can also be designed to accept a wide range of source voltage and frequency. Therefore, you can readily design a switching power converter as a universal power supply that can use a wide variety of voltages and frequencies from international sources of power.

Switching power converters have certain limitations when compared with linear and ferroresonant converters. They do not regulate load and line as accurately or precisely as linear converters. The high frequency and sharp transitions of the transistors' switching generate significant EMI, both conducted and radiated, that requires filtering and suppression. The switching action can cause complex current-voltage relationships that can force power factor correction in the source power. Finally, the large number of components in switching converters increases manufacturing cost and reduces reliability.

9.5 DEFINITIONS AND SPECIFICATIONS

Good systems engineering requires an understanding of definitions and appropriate specifications. You can design or specify a power converter that is less expensive, lighter, smaller, less complex, and more reliable if you understand the needs and the environment and do not specify overly tight tolerances on the parameters. Some of the more common definitions follow.

9.5.1 *Line Regulation*

Line regulation is the maximum variation in output voltage for input voltage that varies between the specified limits. The variation may be expressed as a percentage across a band or as a percentage on either side of nominal. Linear converters typically exhibit 0.01% line regulation across band or ±0.005% around the nominal, switchers exhibit 0.2% line regulation across band or ±0.1%, and ferroresonant converters may have 1.5% line regulation.

9.5.2 *Load Regulation*

Load regulation is the maximum variation in output voltage for step changes in load. The variation may be expressed as a percentage change while load changes from no load to full load. Linear converters typically exhibit 0.01% load regulation, while switchers exhibit about 0.2% load regulation. Closely associated with load regulation are the transient response parameters overshoot, undershoot, and settling time (Fig. 9.9).

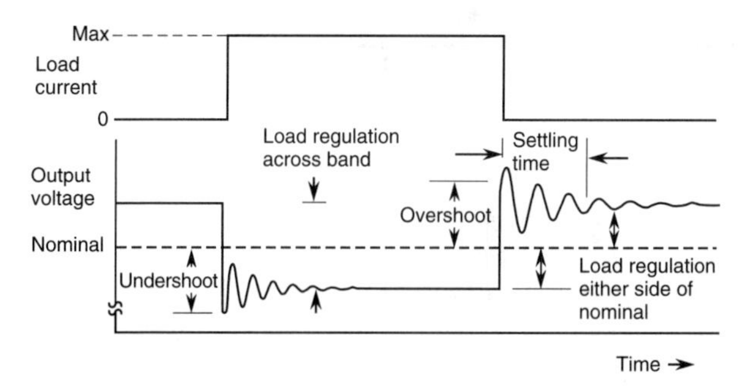

Fig. 9.9 Load regulation and associated parameters.

9.5.3 *Holdup Time*

Holdup time is the time that output voltage remains within tolerance after the input fails and input voltage is zero.

9.5.4 *Noise and Ripple*

Noise and ripple are the periodic variations from the nominal output voltage and are described by a single value. You should specify this value in peak-to-peak terms. The switching action within switching converters produces sharp transients that have large peak-to-peak values but fairly small rms values. In contrast, linear and ferroresonant converters are very quiet, typically producing ripple less than 0.01% nominal.

9.5.5 *Periodic and Random Deviation*

Periodic and random deviation (PARD) is the sum of all the noise and ripple components. Its definition is identical to the definition for noise and ripple.

9.5.6 *Overvoltage Crowbar*

The *overvoltage crowbar* shorts the output to ground in the event of a failure within the converter that causes the output voltage to rise to between 120% and 140% of nominal. The term *crowbar* derives from the analogous situation in which a crowbar is placed across the terminals of a car battery. Linear and switching converters must have overvoltage circuits added, as shown in Fig. 9.10. Ferroresonant converters have inherent overvoltage protection.

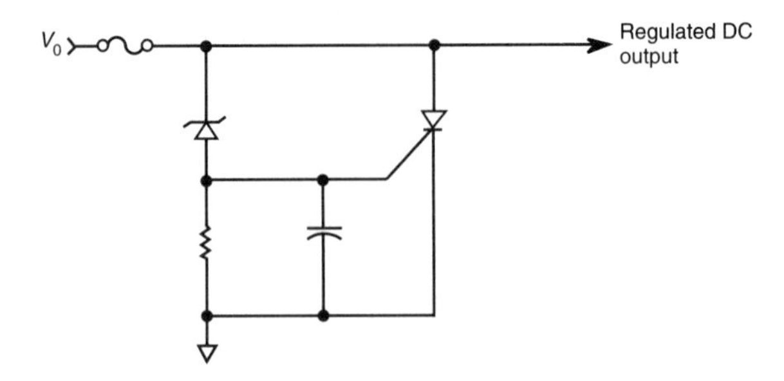

FIG. 9.10 A typical overvoltage crowbar circuit.

9.5.7 *Foldback Current Limiting*

Foldback current limiting reduces both the output voltage and the current should the load develop a short circuit. The overcurrent threshold is typically set at 120% of the rated maximum current. Linear and switching converters must have circuits that implement foldback current limiting added (Fig. 9.11). Ferroresonant converters have inherent overcurrent protection.

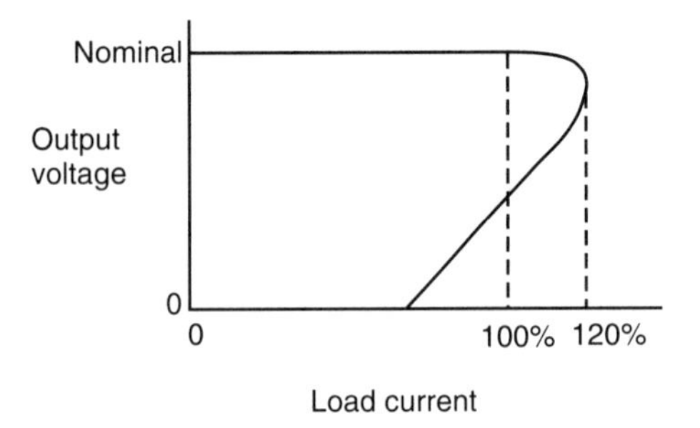

FIG. 9.11 Foldback current limiting.

9.5.8 *Turn-On Overshoot and Turn-Off Undershoot*

Turn-on overshoot and *turn-off undershoot* specify the peak ringing at the output when power is applied or removed. Often specifications will not allow overshoot or undershoot.

9.5.9 *Turn-On Delay*

Turn-on delay marks the elapsed time between the application of input power and the time when output voltages settle at nominal values.

9.5.10 *Thermal Regulation*

Thermal regulation is described by the temperature coefficient, which is the maximum change in output voltage per degree Celsius (%/°C).

9.5.11 *Stability*

Stability is the maximum change allowed in output voltage after warm-up for a specified period. You should also specify the time for warm-up, or the time to reach thermal equilibrium.

9.5.12 *Remote Sense*

Remote sense compensates for voltage drop in the cables that distribute power. Remote sense regulates voltage at the load rather than at the power converter terminals, as shown in Fig. 9.12. You will specify the amount of voltage drop that must be compensated by remote sense. Typically, 200 to 500 mV suffices.

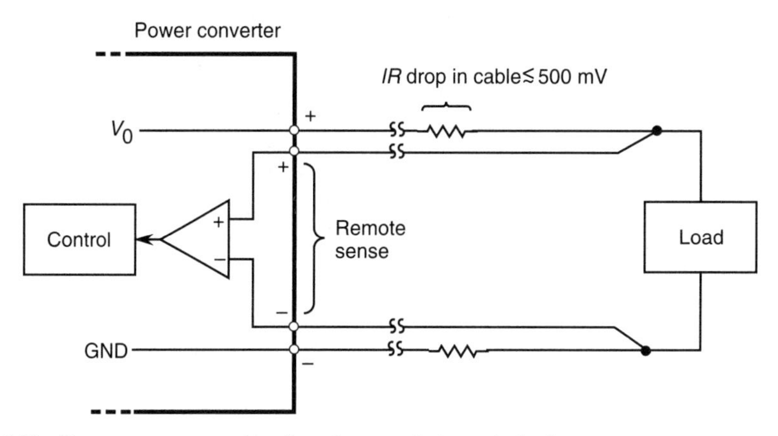

Fig. 9.12 Remote sense connections for voltage regulation at the load.

9.5.13 *Inrush Current*

Inrush current is the surge of current into a power converter that occurs when you apply source power. (Charging of the filter capacitors contributes the most to inrush.) You should specify both the peak current and its duration when defining the inrush current. Fig. 9.13 illustrates two possible methods of limiting the inrush current.

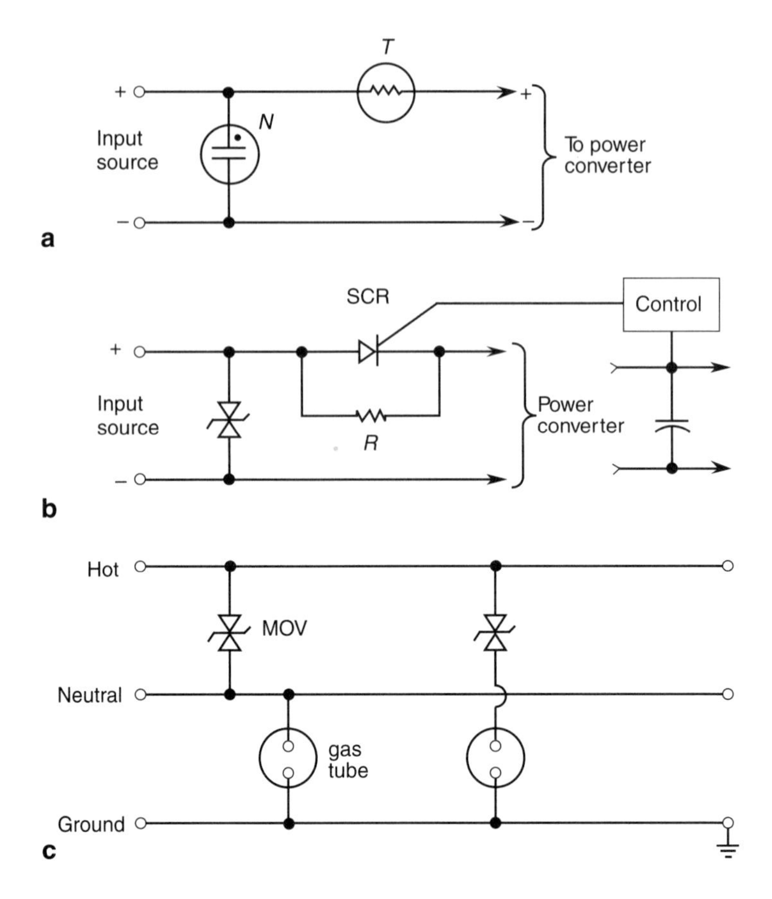

Fig. 9.13 Several schemes for limiting inrush current and protecting against input voltage spikes. (a) The neon bulb, *N*, shunts high-voltage transients, while the thermistor, *T*, limits inrush current. (b) An MOV or a silicon junction device shunts high-voltage transients, while the resistor limits inrush current; the SCR turns on after a preset level is reached and short-circuits the resistor. (c) Surge protection on the AC power input to a device destined for the international market.

9.5.14 *Input Voltage Transients*

Voltage spikes generated by the electric motors in heavy machinery can damage power converters. You should specify the peak and duration of voltage transients that your power converter can accept. Fig. 9.13 also shows how you can protect power converters from input voltage transients.

With these definitions, you can develop reasonable specifications for your power converter. Table 9.4 lists the general specifications for defining a power converter.

Table 9.4 Specification checklist for power supplies

General specification	Parameter	Units	Comments
Input	Voltage range	V	E.g., 90–130 VAC; 80–270 VAC
	Frequency	Hz	E.g., DC, 50, 60, 47–440 Hz
	Inrush current	A, ms	Peak current and duration
	Voltage transients	V, μs	Peak voltage and current
	Fusing	A	
	Power factor	%	
Output	Voltage	V, ±%	Nominal plus range of variation
	Stability	% V, hr	Percentage variation for specified period
	Current		
	—Average	A	
	—Peak	A	
	—Duty cycle	%	
	Turn-on delay	ms	
Regulation	Line regulation	%	
	Load regulation	%	For no load to full load
	—Settling time	ms	
	—Overshoot	%	Percentage of nominal
	—Undershoot	%	Percentage of nominal
	Holdup time	ms	
	Thermal	%/°C	
Protection	Overvoltage	%	120% to 140% of nominal
	Overcurrent	A	Usually 120% of maximum
	Thermal	°C	Circuitry shutdown temperature
Noise	Ripple	% or mV	
	Bandwidth	Hz	0–50 MHz
	Common mode	%	Percentage of nominal
Power conversion	Efficiency	%	Eff = $(P_{out}/P_{in}) \times 100\%$
	Thermal dissipation	W	$P_{diss} = P_{in} - P_{out}$
Interface control	Remote sense	mV	E.g., maximum = 200 or 500 mV
	Logic turn-on	TTL level	
	Power-fail detection	%, ms	Duration before output falls below specified percentage of nominal
	Voltage adjustments	%	Percentage of nominal
Environment	Storage temperature	°C	E.g., –55°C to +125°C
	Operating temperature	°C	E.g., 0°C to 71°C
	Shock	*g*, Hz	
	Vibration	*g*, Hz, hr	
	Cooling	W	Conduction, convection, forced air, liquid cooling
Reliability	MTBF	hr	E.g., 100,000 hr
	Burn-in	hr, °C	Cycle of temperature over time
Mechanical	Volume	cm^3	
	Weight	kg	
	Connectors		Give types
	AC ground		Specify
Additional military specifications or ruggedized concerns	Humidity	%	E.g., 5% to 80%, noncondensing
	Altitude	m	Vacuum requires more insulation
	Salt spray	—	MIL-STD 810
	Standards	—	Safety, VL, IEC, VDE, EMI, MIL-STD-461C
	Configuration control	—	E.g., mounting holes do not change without notification
	Process control	—	E.g., no capacitors with PCBs

9.6 POWER DISTRIBUTION

Transmission and distribution of power within an instrument and between modules require careful attention. You must incorporate the general scheme for power distribution into the conceptual development and then clearly specify the power distribution in the requirements. Two configurations exist for providing power: a centralized supply and a distributed power system (Fig. 9.14). The centralized supply converts, regulates, and transmits power from a single location. The distributed power system transmits unregulated power to each subsystem, and a local converter then regulates the power within each subsystem.

Several factors will influence your selection of the appropriate configuration for power distribution. A small instrument with low, well-defined power demands and a fixed architecture with limited expansion favors a centralized supply. A large

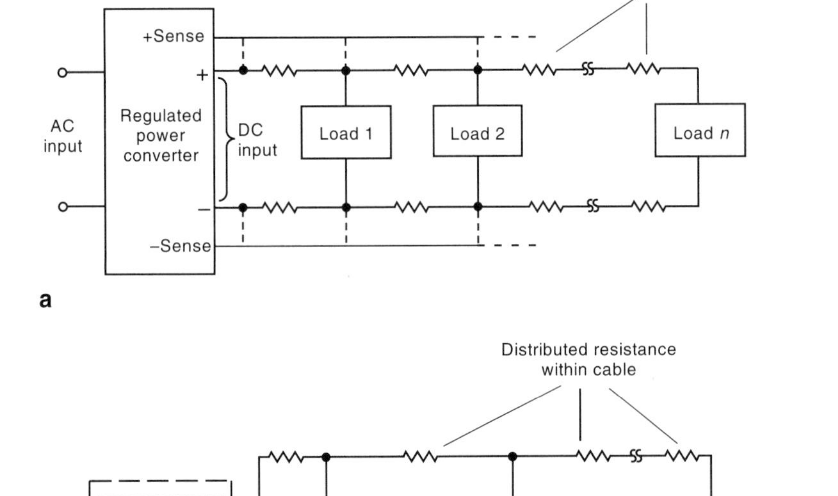

Fig. 9.14 Two schemes for distributing power within an instrument. (a) Centralized power supply. (b) Distributed power supply.

rack of equipment with multiple modules that are isolated from one another, large expansion capability, and wide variations in demand will have a distributed power system. The trade-offs between centralized power and distributed power are trade-offs between simplicity, on the one hand, and weight, redundancy, fault tolerance, and local regulation, on the other (Ormond, 1992; Questad, 1991).

Example 9.6.1 Consider the power losses and voltage variations in an example system that has logic circuitry requiring 200 W at 5 VDC ± 5%. A centralized supply must provide 40 A of current while the power and return cables must each have less than 6.25 mΩ of resistance:

$$P = 200 \text{ W}$$

$$V_{nom} = 5 \text{ VDC}, V_{hi} = 5.25 \text{ V}, V_{lo} = 4.75 \text{ V}$$

$$I = \frac{P}{V_{nom}} = 40 \text{ A}$$

$$R_{cable} \leq \frac{\Delta V_{cable}}{I} = \frac{V_{hi} - V_{lo}}{I} = \frac{5.25 - 4.75}{40} = 0.0125\,\Omega.$$

A distributed system needs to provide only 1 A of current at 250 VDC and uses local DC-DC converters to step down and regulate the voltage to 5 VDC. With converter efficiencies of 80%, the power and return cables may each have as much as 25 Ω of resistance:

$$P = 200 \text{ W}$$

$$V_{nom} = 250 \text{ VDC}, V_{lo} = 200 \text{ V}$$

$$\eta = 80\%$$

$$I = \frac{P}{\eta V_{nom}} = 1 \text{ A}$$

$$R_{cable} \leq \frac{\Delta V_{cable}}{I} = \frac{V_{nom} - V_{lo}}{I} = \frac{250 - 200}{1} = 50\,\Omega.$$

This example demonstrates that higher transmission voltage and lower current in the distributed system will reduce line losses, relax tolerances on voltage variation, and allow lighter cables.

For large systems, regulation is much more precise and accurate with a distributed system but difficult, if not impossible, with a centralized system. In the scheme shown in Fig. 9.14, what location would you choose for the remote sense to minimize voltage variations for a centralized supply? While you could optimize the regulation for one of the loads, you cannot assume that the other loads will behave identically.

A distributed power system often uses DC-DC converters for local regulation (Fig. 9.15). DC-DC converters generally incorporate overvoltage and overcurrent protection and allow a wide range of input voltages. These features provide isolation and fault tolerance within a system; failure in one subsystem will not bring down the entire power system.

On balance, distributed power supplies are complex and are not necessary for many devices and instruments. Furthermore, you must exercise care with power sequencing and limit faults to one subassembly without disturbing others. Table 9.5 summarizes some of the differences between centralized and distributed power systems.

Regardless of the power distribution scheme, you will need to consider EMI. If you minimize the inductive loop area between the power (+) and return (–) conductors, you will reduce the received and transmitted noise significantly. Therefore, power bus bars and cables should run very close together, eliminating wide separations between these conductors (Fig. 9.16).

You may use single-chip regulators and converters to regulate low-power circuits. These components incorporate one of three conversion techniques: linear, switching, or charge pump. Linear regulators are the quietest devices with the best regulation; however, they are less efficient than switching regulators. Switchers may step down or step up voltage with high efficiency, but they require bulky inductors. Charge-pump converters are efficient and don't need inductors; instead they use capacitors to double the voltage of the output. Charge-pump converters can generate only very low current and tend to be fairly noisy. Figs. 9.17 through 9.20 illustrate some regulator circuits.

Fig. 9.15 DC-DC converter components for distributed power systems (photograph courtesy of Vicor Corporation).

Table 9.5 Comparison between centralized and distributed power systems

Parameter	Centralized	Distributed
Complexity and weight	Lower for small devices, significantly higher for large systems	High for small devices, significantly lower for large systems
Fault tolerance and isolation	Difficult to achieve, requires redundant supplies	Inherent
Regulation and remote sense	Difficult to achieve for large systems	Simple and inherent
Protection (overvoltage, overcurrent)	Global, affects all subsystems	Local, isolates subsystems
IR drops and losses	Small for low-power devices, significant problem for high-power applications	Insignificant to device or system
Power conductors	Small for low-power devices and instruments, large and heavy for high-power applications	Very small
Expansion capability	Simple for low-power devices and instruments; capacity must be built into converter, resulting in large and heavy supplies, in large systems	May be as simple and as straightforward as plugging in another subsystem module; capacity must be in the source

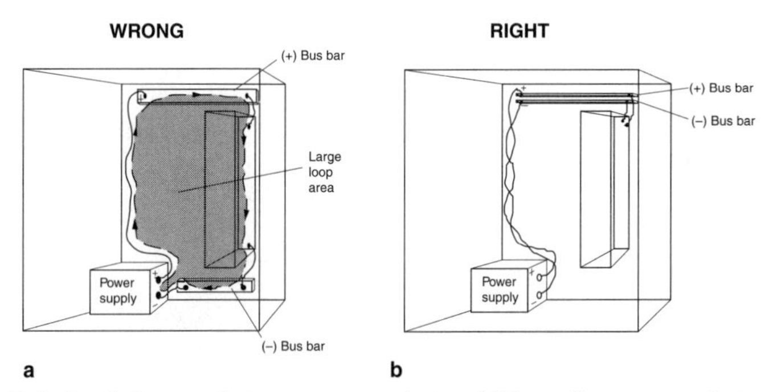

Fig. 9.16 Reducing the loop area between power conductors. (a) Separating power conductors produces a large inductive loop. (b) Twisting the cable together and running the bus bars side by side minimizes or eliminates the inductive loop.

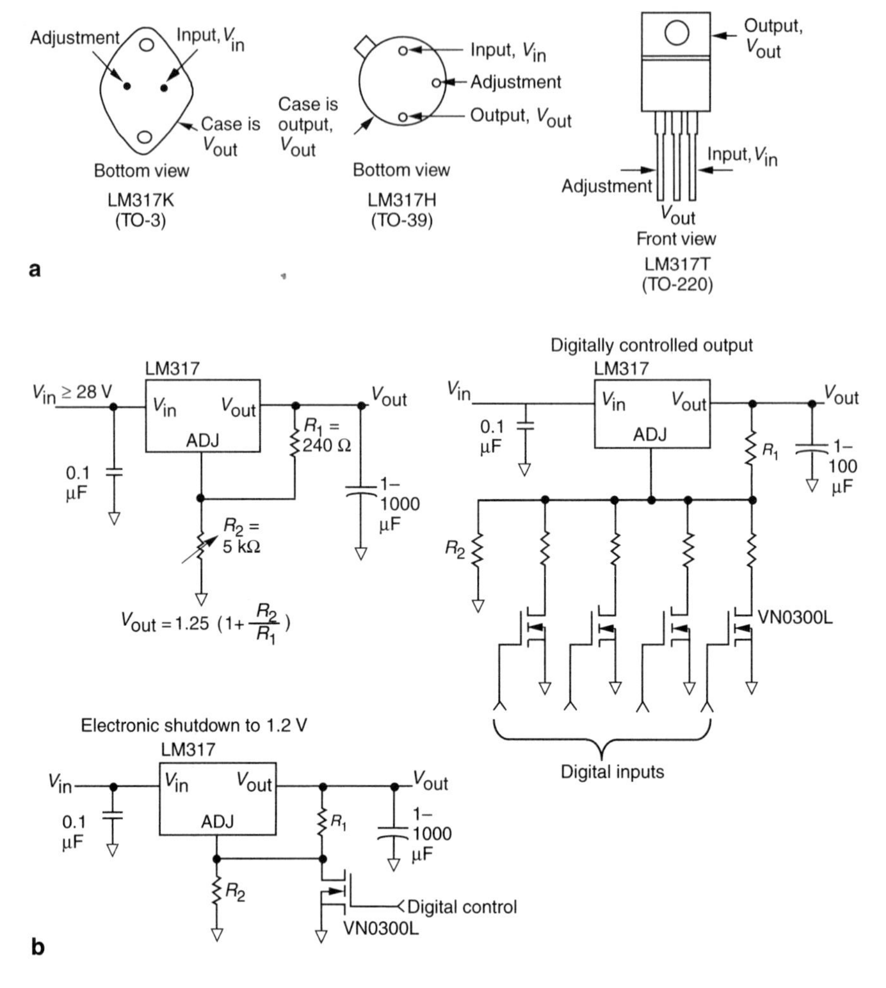

Fig. 9.17 An adjustable linear regulator and sample circuits. (a) Some available packages for a regulator with 1.2–25 V adjustable output. (b) Possible applications.

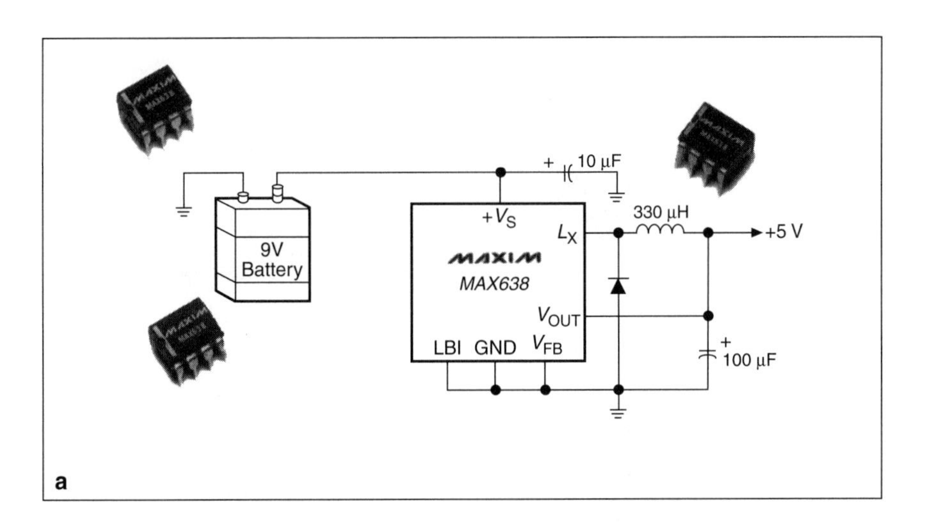

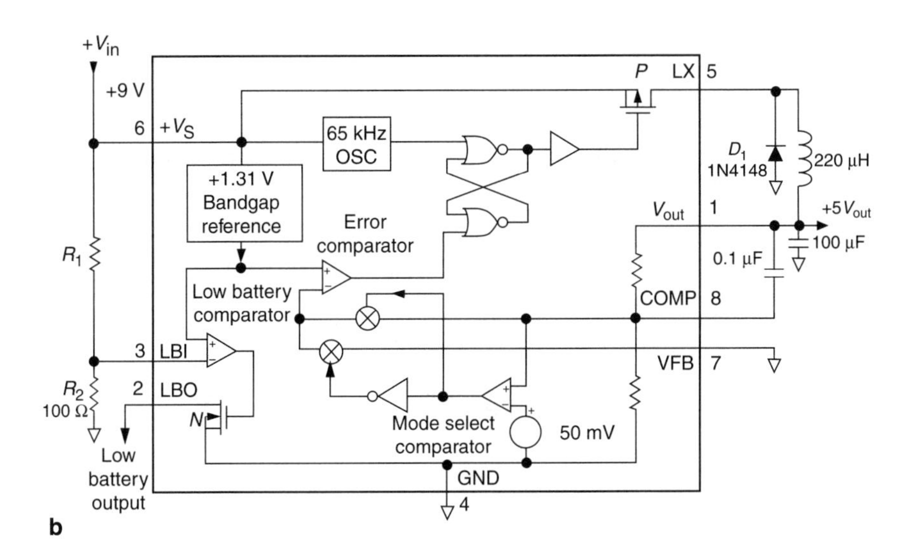

Fig. 9.18 A step-down regulator circuit for converting 9 VDC to 5 VDC. (a) MAX 638 in an 8-pin DIP (photograph courtesy of Maxim Integrated Products, Inc.). (b) MAX 638 block diagram and typical circuit from Maxim Integrated Products, Inc.

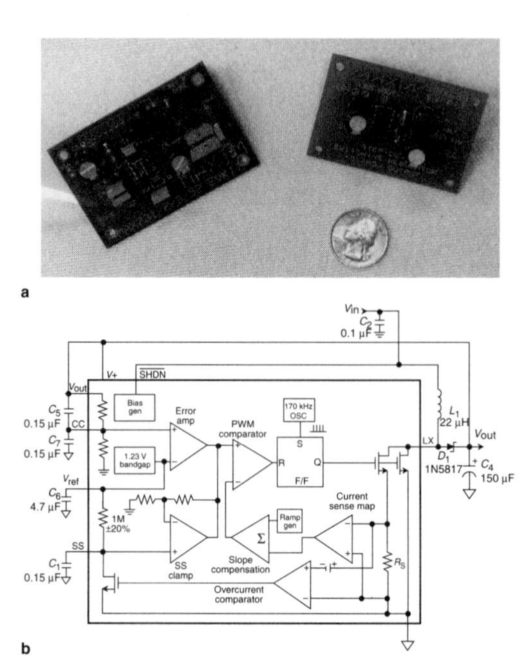

FIG. 9.19 Two different switching regulators. (a) Evaluation boards for the MAX 639 (a step-down regulator) and the MAX 731 (a step-up regulator). (b) MAX 731 block diagram and a typical step-up regulator circuit for converting 2.5 to 5.25 VDC to 5 VDC (with permission from Maxim Integrated Products, Inc.).

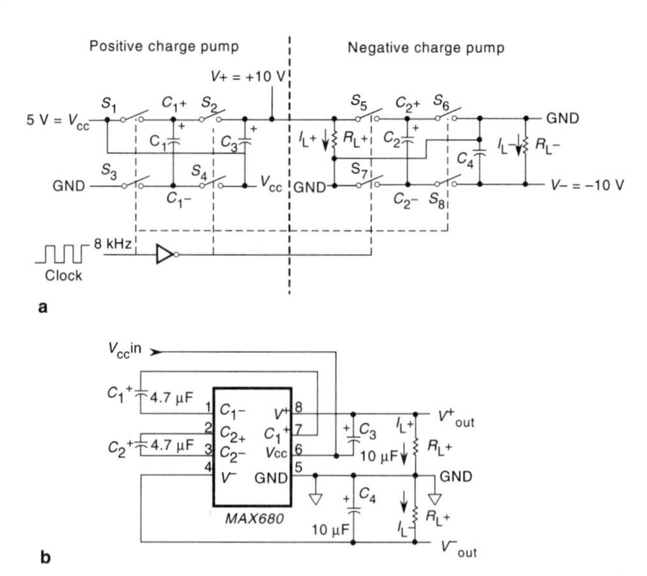

FIG. 9.20 A charge-pump converter to develop ±10 V from +5 VDC. (a) MAX 680 schematic outline. (b) Typical circuit from Maxim Integrated Products, Inc.

9.7 LINE CONDITIONING

AC line power is not always clean and reliable; significant departures from an idealized sine wave occur frequently in all sources. Switching of heavy machinery, construction accidents, transformer failure, air-conditioning loads during the summer, and lightning strikes all contribute to variations in AC power. Voltage variations such as dropouts, brownouts, overvoltage, and spike transients can devastate sensitive electronics or, at the very least, interrupt operation (Fig. 9.21).

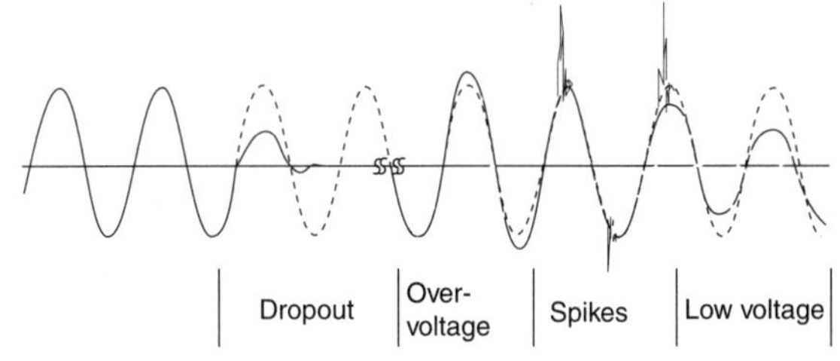

Fig. 9.21 Variations in AC source voltage.

9.7.1 *Uninterruptable Power Supplies*

You have a variety of options for isolating your instrument from these variations. Depending on the operation, you may use filtering or transient suppression to clean up the AC source power, or you may incorporate a UPS to ride out the power failures. A UPS includes battery backup, power filtering, and a DC-to-AC inverter to isolate your equipment from the AC power source (Fig. 9.22). The UPS normally draws power from the AC input and trickle-charges the battery. When AC power fails, the power control unit switches in the battery to drive the DC-to-AC inverter.

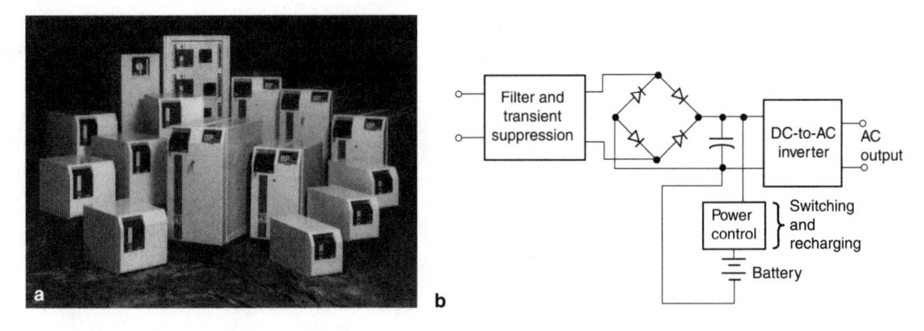

Fig. 9.22 Uninterruptible power supplies. (a) A line of commercial UPSs (photograph courtesy of Best Products, Inc.). (b) One possible UPS configuration.

9.7.2 *Power Factor Correction*

While AC power sources inject disturbances into the input of a power converter, power converters also return disturbances back into the AC source. These disturbances take the form of noise currents with harmonics of the fundamental frequency. The noise harmonics originate from the charging of the input filter capacitors at each voltage peak. Since charging occurs only at the peaks, the current waveform is decidedly nonsinusoidal (Pryce, 1991). Harmonics of the fundamental in the current waveform contribute no effective power (Roark and Pechi, 1991). Fig. 9.23 demonstrates the generation of the current waveform and shows how harmonics contribute no power.

No power is contributed by harmonics of the line current:

$$P = \int_0^{2\pi} V \sin\ \omega t \cdot I_n \sin(n\omega t + \phi) \cdot \mathrm{d}\omega t = 0\ ,$$

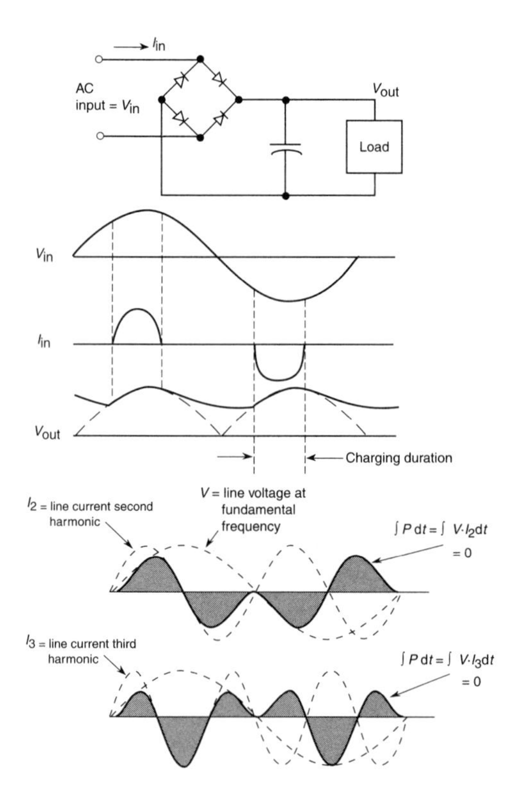

Fig. 9.23 Generation of a nonsinusoidal current waveform and harmonics that contribute no power.

where V = peak voltage
I_n = peak current of nth harmonic
$n > 1$

The power actually delivered to the converter is the product of the fundamental component of the line voltage and the line current and is called the real power, P_r. The sum of all harmonic components of the line current multiplied by the line voltage is the apparent power, P_a. Power factor is the ratio between the real and apparent power and defines how effectively the power converter uses the source power:

$$P_f = \frac{P_r}{P_a} = \frac{VI_1}{\sum_{i=1}^{40} VI_i} = \text{power factor,}$$

where P_r = real power
P_a = apparent power
V = line voltage
I_i = harmonic line current
i = harmonic number

Power factor is a complex interplay of converter efficiency, line voltage, and load. For some converters, power factor may fall with either increasing line voltage or decreasing load. In addition, a converter with a high power factor and low efficiency may exceed the rating for the line current of the AC source (Roark and Pechi, 1991).

Your instrument and its power converter may have to meet standards for power factor such as the International Electrotechnical Commission's IEC 555-2 standard for equipment sold in Europe. If your equipment is built to U.S. military standards, then it must meet MIL-STD-1399. You have several avenues to correct power factor: Buy a converter with power factor correction built in, incorporate a front-end module to correct power factor, or design and build your own circuitry for power factor correction (Hagiwara, 1991; Maliniak, 1992).

9.8 ELECTROMAGNETIC INTERFERENCE

Power converters generate, transmit, and receive significant amounts of EMI or noise. The principles for grounding and shielding discussed in Chapter 6 apply here to power converters. Noise generated or received by power converters is usually magnetic in origin, though some mechanisms are occasionally electrostatic. Noise generally conducts via the power lines. Filtering the AC source at the converter will eliminate differential-mode and common-mode noise from the input conductors generated by switching power converters. Fig. 9.24 illustrates an input filter. Noise

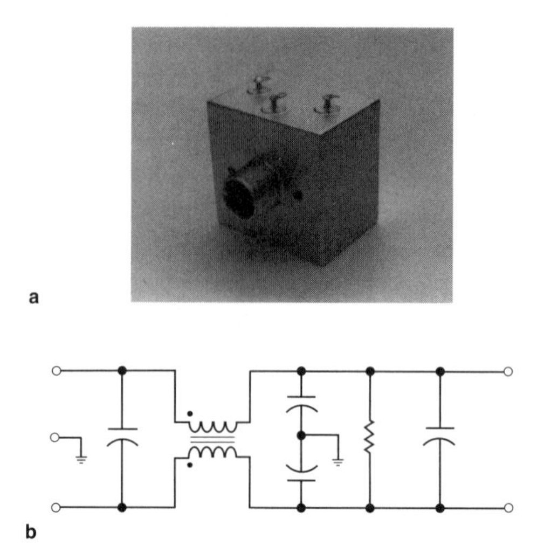

FIG. 9.24 One configuration for an in-line filter to attenuate noise between the AC source and a switching power converter. (a) A commercial line filter (photograph courtesy of Filter Networks, Inc.). (b) Schematic of a line filter.

from the output terminals caused by common-mode current can be eliminated by using an electrostatic shield in the transformer, selecting an appropriate chassis ground connection, and balancing impedances in the leads (Maier, 1991; Questad, 1990).

Reducing the separation between power conductors will reduce the inductive loop and any magnetic coupling to noise sources. Placing Mumetal enclosures around susceptible or emissive circuits is a final measure to attenuate magnetic coupling. Hnatek (1989) discusses electromagnetic compatibility for power converters.

You will encounter safety and EMI regulations when designing and developing an instrument and its power converter. Pertinent standards, depending on the application and the international market, are listed in Table 9.6.

Table 9.6 Electromagnetic compatibility regulations and some applications

Regulation	Application
FCC	Commercial devices and home appliances in United States
VDE 0871B IEC 161 CSA SEV	Commercial devices and home appliances in Europe and Canada
MIL-STD-461C	Military equipment
TEMPEST	High-security military and intelligence operations

9.9 RELIABILITY

You can improve reliability within power converters only when you understand the sources of stress that cause failure such as overvoltage transients or spikes, excessive current pulses, and heat. Clamping diodes, snubbers, transient suppressors, and line filters all reduce voltage and current stresses on the components, particularly switching transistors. Failure from thermal cycling is common in linear pass transistors. Improving the efficiency of power transfer reduces power dissipation and heat generation. The resulting lower temperatures reduce the thermal stress on the components and improve reliability.

Linear power converters tend to dissipate more heat than switchers because of their inherently lower efficiency. You must design and manage thermal transfer carefully to improve reliability, especially around the series pass transistor, where most of the heat is dissipated.

Switching power converters experience greater voltage stresses; clamping circuits and snubbers reduce these stresses. There are three sources of heat within switching converters:

1. The output diodes
2. Core losses within transformers and inductors
3. Voltage drops due to the resistance of the windings within transformers and inductors

Schottky diodes, with a lower forward voltage drop, will reduce the heat generation of the output diodes (Brown, 1990; Shepard, 1984).

You may use mean time between failures (MTBF) as a figure of merit in estimating reliability. MTBF comprises failure rates of components and gives a composite picture of the ultimate failure rate of the device:

$$\mathrm{MTBF} = \frac{1}{\sum_{i=1}^{n} \lambda_i}$$

where λ_i = failure rate of the ith component
n = number of components

One source of information about failure rates, λ, for components is MIL-HDBK-217F. Environmental factors, voltage stresses, and construction parameters will alter the failure rate. In fact, the demonstrated MTBF will often be significantly greater than the calculated MTBF. Components added to power converters to reduce stresses and heat generation lower the calculated MTBF but often improve the actual reliability.

During the late 1970s and early 1980s, the U.S. Navy faced a crisis from failed electronics. Approximately 20% of the failures were in the power converters.

In response to these failures, the Navy/Industry Power Supply Ad Hoc Committee published NAVMAT P-4855-1A to suggest ways to improve reliability (Military-power-supply failures, 1984; Newhart, 1986). NAVMAT P-4855-1A promotes environmental tests to screen power converters. The environmental tests comprise parametric screening, vibration, and 48 hours of thermal cycling. During each thermal cycle, the supply is cycled on and off in a random sequence. At the same time, a full load is maintained at the converter's output. Fig. 9.25 illustrates a thermal cycle. These environmental tests "shake out" the early failures much as burn-in does, as discussed in Chapter 13. Even though these tests originated for military systems, they may be applied to commercial products to increase reliability.

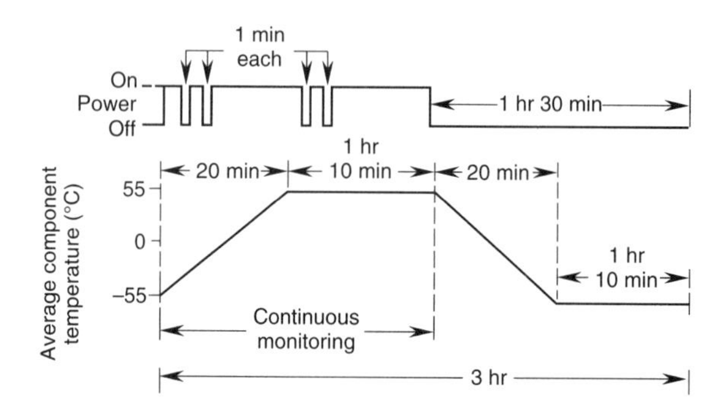

FIG. 9.25 Environmental-stress screening from NAVMAT P-4855-1A to shake out early failures by thermal cycling of fully loaded power supply (not to scale).

9.10 BATTERIES

Many applications require batteries as their primary source of power. Handheld tools, watches, calculators, small appliances and toys, cassette tape and CD players, and laptop computers, to name a few, all employ batteries as their power source (Fig. 9.26). Batteries also figure as important temporary elements for energy storage in the power systems of automobiles, aircraft, and satellites. You must understand the particular characteristics and limitations of batteries before incorporating them into a system design.

Battery power systems have distinct advantages and equally significant limitations and disadvantages. Batteries make power available to remote or portable applications, where power lines cannot reach. They provide excellent isolation and "no-noise" power to very sensitive circuits. On the other hand, batteries generally are heavy and bulky. They have a limited energy capacity and can deliver power for only limited durations. Finally, environmental concerns can hamper their disposal.

Fig. 9.26 A sample of devices and equipment that use batteries.

Batteries are classified as either primary (nonrechargeable) or secondary (rechargeable). Batteries have a wide variety of chemical compositions; each has characteristic properties, including energy density, discharge service life, shelf life, and leakage. Energy density is measured in watt-hours per kilogram or watt-hours per liter. Battery voltage varies as a function of remaining charge and load. Fig. 9.27 illustrates a generalized discharge curve for a battery. Batteries have a finite storage life before they experience self-discharge; temperature, leakage, and internal chemistry determine the usable shelf life. Secondary batteries must be recharged carefully using specific charging regimens such as constant current, taper current, constant voltage, or trickle charge; they also have a finite cycle life—that is, number of times they can discharge and recharge. Table 9.7 summarizes some types of batteries and their characteristics.

You will have several concerns when specifying and designing battery-powered systems. Be sure that you understand the discharge rates used when manufacturers define the discharge curves for their products. Discharge rates affect the actual life of the battery. Some manufacturers specify a discharge rate that gives

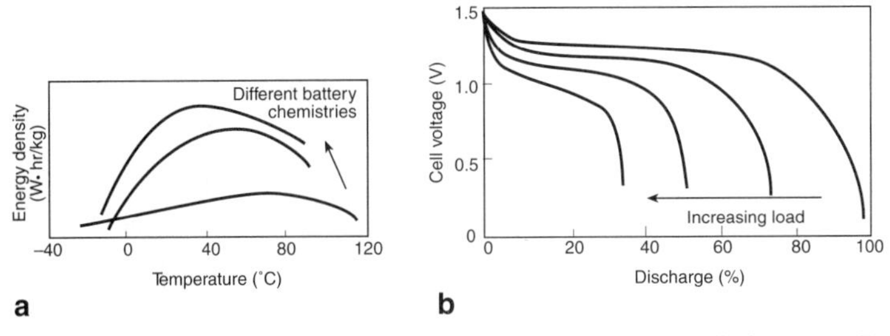

Fig. 9.27 Generalized characteristics of batteries for energy density and discharge. (a) Generalized relationships for energy density to temperature. (b) Generalized discharge for various loads.

Table 9.7 Comparison of several battery types and characteristics (Crompton, 1990; Quinnel, 1991)

Battery type	Average on-load cell voltage (V)	Energy density (W·hr/kg)	Operating temperature (°C)	Shelf life: 80% capacity at 20°C (months)	Relative cost	Charging method, number of cycles
Primary						
Carbon-zinc	1.2	55–77	−7 to 54	18	Lowest	—
Alkaline manganese	1.25	66–99	−29 to 54	18	Very low	—
Mercuric oxide	1.25	99–123	0 to 54	24	Medium	—
Silver oxide	1.5	110–126	0 to 54	18	Expensive	—
Lithium–sulfur dioxide	2.75	260–330	−40 to 60	60	Expensive	—
Zinc-air	1.2	155–330	−29 to 52	48	Expensive	—
Secondary				50% charge retention at 27°C		
Sealed lead-acid	1.5–2.0	18–33	−40 to 60	18	Medium	Taper current, 100–150
Nickel-cadmium	1.0–1.25	18–44	−40 to 70	2	High	Constant current, 100–1500
Silver-zinc	1.3–1.55	55–209	−40 to 74	12	Very expensive	Constant current or volts, 10–200
Nickel hydride	1.2–1.3	56–67	−18 to 27	1	Expensive	Constant current, 300–500

the capacity of the battery discharged within 1 hour, while others specify the capacity discharged over 10 hours. Better yet, describe your load conditions and get battery performance accordingly to make reasonable comparisons between candidate batteries. Because temperature significantly affects energy density, you will need to specify enough capacity to ensure adequate power over the range of the operating temperatures.

Example 9.10.1 Why do you think batteries in household fire alarms always seem to fail in the early morning? Because the house temperature drops, causing voltage to drop even more in the discharged battery. The alarm's low-battery detection circuitry senses the low voltage and activates the alarm to alert you for a battery replacement.

Operational concerns include recharging profiles and rates, discharge rates and recharging cycles, and even battery insertion and polarity reversal. Some batteries require constant-current recharging; silver-zinc batteries can withstand constant-voltage recharging. Lead-acid batteries can undergo 150 full charge-and-discharge cycles, but they will tolerate significantly more cycles if discharge takes only a fraction of the rated capacity (Quinnel, 1991). You may want to incorporate trickle charging circuitry into your design to maintain charge on the battery. Consider battery polarity: Can users destroy or damage your instrument if they reverse battery connections? To avoid confusion, you could design a universal connection, like that shown in Fig. 7.5, that employs a full-bridge rectifier for presenting the correct polarity to the circuitry regardless of the battery orientation. Table 9.8 provides a checklist for selecting batteries.

Table 9.8 Checklist for selecting a battery

General specification	Parameter	Units	Comments
Physical	Volume	cm^3 or $in.^3$	
	Weight	kg	
Electrical	Load		
	Voltage discharge	V	
	Current profile	A	
	Duty cycle	%	
	End-point voltage	V	
	Energy density	W·hr/kg	Plot of energy density vs. temperature
	Temperature	°C	
	Cycle life	—	Number of discharge-charge cycles
	Charge retention	months	% remaining capacity at specified temperature
	Recharging	—	Method: constant current, taper current, constant voltage, or trickle
Environmental	Storage temperature	°C	
	Operating temperature	°C	
	Shock and vibration	g	
	High-altitude use	m	
	Humidity	%	
Special concerns	Shape	—	
	Shelf life	months	Retaining 80% capacity
	Terminal types	—	
	Disposal	—	Nickel hydride may replace nickel-cadmium because of concerns about heavy-metal waste
	Transportation		Lithium is explosive in large quantities

9.11 OTHER POWER SOURCES

A whole world of different, unusual, and even exotic power sources exists beyond the power sources already mentioned. Special applications may demand one of these sources. Each has its advantages and its idiosyncrasies. Some of these sources include stand-alone generators, solar cells, thermionic generators, and fuel cells.

Automobiles have alternators that provide loosely regulated DC power. The electrical environment within an automobile power system is riddled with high-voltage transients and low-voltage dropouts. If you design electronic equipment for an automobile, you must thoroughly understand the variations within the power system.

Many devices—such as calculators, remote radios, and satellites—employ solar cells as primary power sources (Hu and White, 1983). In the future, more applications may incorporate thermionic converters (nuclear reactors on satellites currently do) and fuel cells (Appleby and Foulkes, 1989; Russell, 1967). Each source generates characteristic voltages with specific current and load capacities. You must know intimately the characteristics of the source, its environment, and its load.

9.12 CASE STUDIES

I have helped design power subsystems for several types of projects. The two described in this section were completed for the Johns Hopkins University Applied Physics Laboratory. One device was a handheld medical instrument; the other was a large, distributed system for data acquisition in a submarine. Each project had its unique concerns.

9.12.1 *Handheld Tester*

An implantable medical device called a neurostimulator is used to treat patients with excruciating lower back pain (see the case studies in Chapters 2 and 4). The neurostimulator works by stimulating the spinal cord with electrical pulses that effectively jam the pain signals traveling up the spinal cord (Fowler and North, 1991). A neurostimulator arrives sealed in a sterile package. Originally, a physician could not know whether the device was functional without surgically implanting it and asking the patient whether it reduced or stopped the pain. I designed a tester that evaluates a device before surgery and visually indicates its operation, thereby assuring the physician that the neurostimulator is functioning before implantation.

I designed the tester to be small, light, and simple to use. Two rows of eight low-power LEDs indicate the functional operation of a neurostimulator. Testing a device takes only about 30 s. Generally, a physician needs only one implantable device at a time, so the tester would be used intermittently, allowing it to be switched off between tests to conserve its battery.

I planned for a single 9-V battery to provide power to the tester. Carbon-zinc batteries can provide approximately 250 mA·hr at 9.5 V, while alkaline batteries provide 500 mA·hr at 9.4 V. The circuitry and LEDs draw a maximum of 40 mA at 5 V, worst case, so a power converter must transform the 9-V battery power to a regulated 5-V supply.

The design effort for the power converter focused on deciding between a linear and a switching power converter. One concern was the life of the battery; the tester should run continuously for several hours in a worst-case scenario. The linear approach for the power converter used a low-dropout regulator like the MAX 667 (Fig. 9.28). In the worst case, a linear regulator provides a constant 40-mA current as the battery voltage drops over its lifetime from 9 to 5.2 V.

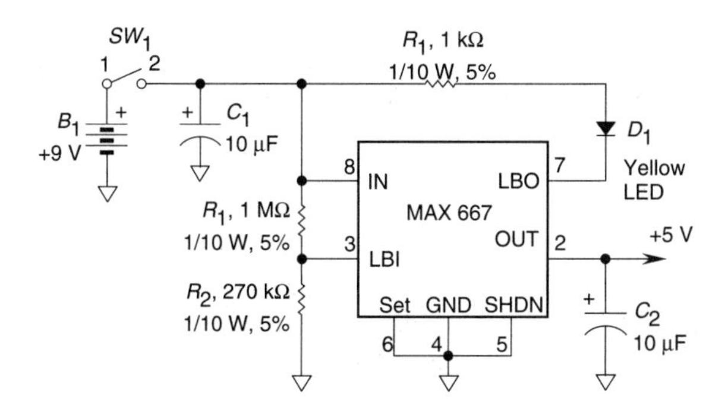

FIG. 9.28 Linear regulator circuit for the handheld tester.

The switching power converter could use any one of a number of step-down regulators. Conversion efficiency was assumed to be a generic 85 percent, even while the battery voltage dropped from 9 to 5.4 V. The battery current is a function of battery voltage:

$$\frac{P_c}{V_{batt}} = \frac{P_{batt}}{\eta V_{batt}} = I_{batt},$$

where P_c = power dissipated by the circuit (40 mA at 5 V = 200 mW)
P_{batt} = power supplied by the 9-V battery
η = efficiency = 85%
V_{batt} = battery voltage
I_{batt} = current produced by the battery

Fig. 9.29 graphs the current provided by the battery to the switching power converter as a function of battery voltage.

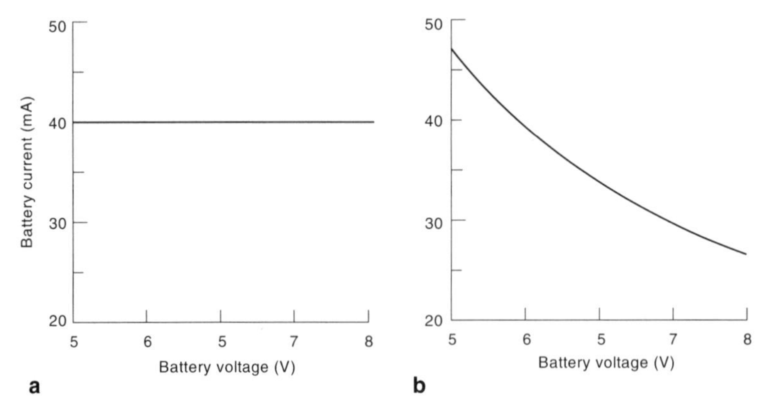

FIG. 9.29 Battery current as a function of battery voltage for (a) a linear regulator and (b) a step-down converter.

I compared battery longevity by assuming a constant 40-mA current with a linear regulator and an average current of 30 mA with a switcher. Table 9.9 compares the battery life for the two converters.

Table 9.9 Comparison of battery life between two different converter circuits

	Battery operational lifetime (hr)	
Battery type	**Linear regulator**	**Step-down converter**
Carbon-zinc	6.25	8.3
Alkaline	12.5	16.7

Both configurations provided sufficient operational lifetime for the battery. The distinguishing feature was that the linear regulator, a MAX 667, provided a low-battery indicator, whereas most switching power converters did not. Resistors R_1 and R_2 in Fig. 9.28 set the low battery voltage at 6.12 V to warn of impending battery failure.

Scope: ad hoc
Duration: 2 weeks
Number of people involved: 1 engineer, 1 technician
Level of effort: full time
Lessons learned: Both converters are simple and provide adequate battery life. The linear regulator is slightly smaller and cheaper to implement, while the switcher is more efficient. However, the linear regulator's low-battery detection tilted the selection in favor of a linear power converter.

9.12.2 *Data Acquisition System*

I helped design a data acquisition system that collected input from sensors distributed around the hull of a submarine (Fig. 9.30). The system supplied electrical power to the sensors from an adjustable DC power supply. The power system presented a significant design challenge.

In operation, the sensors detected light pulses and digitized the representative analog values. The resolution was 14 bits, and the analog range was 0 to +10 V. This resolution corresponded to 610 μV. Each sensor had an analog circuit board and a digital circuit board. The power available to each sensor ranged from 18 to 36 VDC. Fig. 9.31 outlines the circuitry within the sensor.

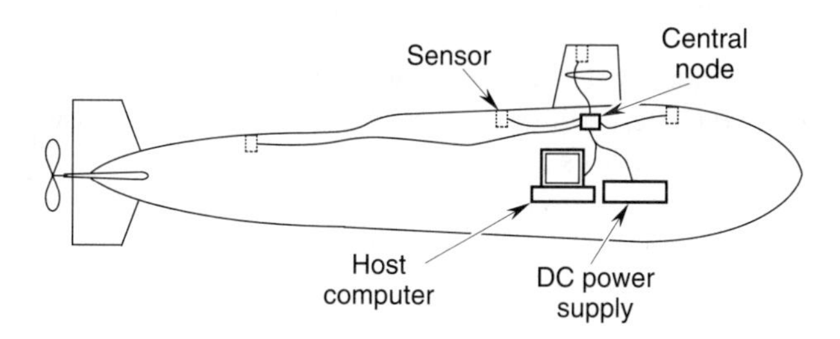

Fig. 9.30 Distributed data collection system.

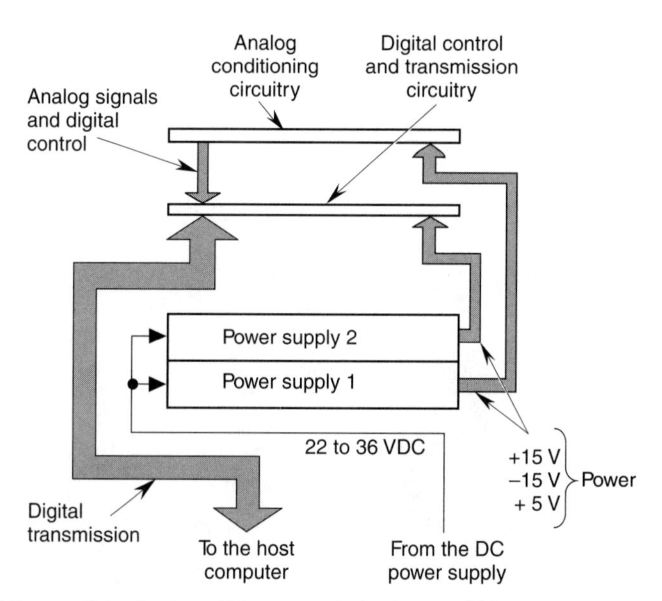

Fig. 9.31 Diagram of the circuitry within a sensor in the data acquisition system.

The requirements specified DC voltages for supplying power over the long cables to the sensors, supposedly to reduce crosstalk and interference in the data lines (a trivial concern because the data transmission was digital with large noise margins and good rejection of common-mode noise and well shielded but we, the design engineers, had little voice in the matter). The cables were restricted to 22-gauge wire (approximately 16 Ω/1000 ft, or 52 Ω/km). The power distribution system had a star topology (Fig. 9.32).

The design decisions came down to two issues: (1) which type of power converter—linear regulator or switcher—would be used in the sensors and (2) what cable arrangement would satisfy the power distribution.

Linear regulation could provide the lowest noise and ripple and would be easy to implement. At 610-ΩV resolution, noise and ripple in the power supply were important considerations. However, the inefficiency in power conversion would increase power loss and voltage drop in the cables.

Switching conversion would be much more efficient and reduce voltage drops in the cables. The converters would need extra filtering to reduce noise and ripple on the outputs, as well as EMI suppression on the input.

The number of power wires dedicated to distribution and the attendant line losses drove the selection of DC-DC converters. Linear regulation would have required at least three wires: one or more return wires, one wire for +15-V and +5-V power, and one wire for −15-V power. For the worst-case voltage variation of +36-V input to a sensor, linear regulation would be only 33% efficient and dissipate nearly 22 W in the power conversion for each sensor. DC-DC converters need only two wires for input, were about 75% efficient, and dissipated only 3.6 W per sensor (Table 9.10).

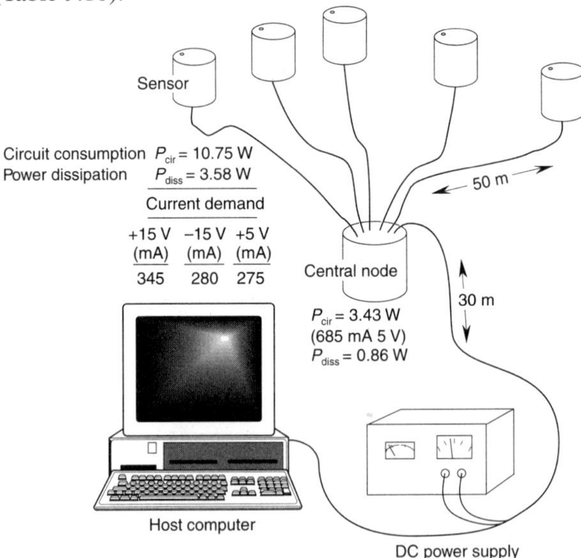

FIG. 9.32 Power and data routed through a star-topology network.

Table 9.10 Power loss per sensor for different input voltages and converter types

Converter type	Voltage tap from converter circuit	Power loss at 18 V (W)	Power loss at 36 V (W)
Linear	+15	(18 – 15)(0.345) = 1.035	(36 – 15)(0.345) = 7.245
	–15	(–18 – (–15))(–0.280) = 0.84	(–36 – (–15))(–0.280) = 5.88
	+5	(18 – 5)(0.275) = 3.575	(36 – 5)(0.275) = 8.575
Total loss		5.45	21.65
DC-DC converter		$P_{diss} = (1 - \eta)P_{cir}/\eta$ $\eta = 85\%$	$\eta = 75\%$
	+15	15 · (0.345) = 5.175	
	–15	–15 · (–0.280) = 4.2	Power consumed by the
	+5	5 · (0.275) = 1.375	circuit, P_{cir} = 10.75
Total loss, P_{diss}		1.90	3.58

The calculation of power dissipated by the DC-DC converters was derived as follows:

$$P_{in} = P_{cir} + P_{diss} ,$$

where P_{in} = input power ($V_i I_i$)
P_{cir} = power consumed by the circuitry
P_{diss} = power dissipated in conversion

Since $P_{cir} = \eta P_{in}$, where η = efficiency (between 0 and 1), and $P_{diss} = (1-\eta)P_{in}$,

$$P_{diss} = \frac{(1-\eta)P_{cir}}{\eta} .$$

The efficiency of DC-DC converters varied as a function of input voltage, as shown in Fig. 9.33 for one model.

Next, the power distribution required attention. The calculation of current and voltage drops uses the following derivation:

$$P_{in} = V_{in} I_{in} .$$

Current input to a sensor was

$$I_{in} = \frac{P_{in}}{V_{in}} = \frac{P_{cir}}{\eta V_{in}} = \frac{\Sigma P}{\eta V_{in}} ,$$

where $\Sigma P = (+15 I_{+15}) + (-15 I_{-15}) + (+5 I_{+5})$.

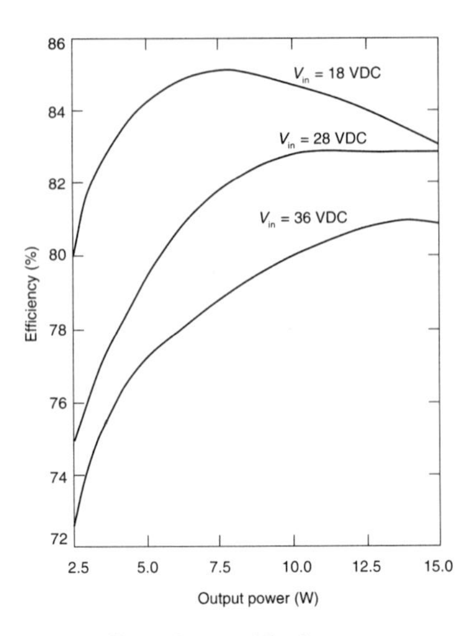

Fig. 9.33 Typical efficiency versus line voltage and load.

Total current distributed to all sensors was

$$I_{\text{sen}} = NI_{\text{in}} ,$$

where I_{sen} = total current distributed
N = number of sensors

Current required by the central node was

$$I_{\text{cn}} = \frac{(+5I)}{\eta V_{\text{cn}}} ,$$

where V_{cn} = input voltage to the central node

Total current sourced by the DC power supply was

$$I_{\text{tot}} = I_{\text{sen}} + I_{\text{cn}} .$$

Voltage drop along a cable was

$$V_{\text{drop}} = IrL ,$$

where I = current in the cable
r = resistivity or resistance per unit length
L = length

Total drop for the send and return wire pair was

$$V = 2V_{drop} .$$

Using this derivation, I developed a computer simulation to study how converter efficiencies affected the power required from the DC power supply and voltage drops in the cable as shown in Table 9.11.

The interaction between converter efficiencies, resistive line losses, and power consumption was nonlinear and complex. Experimental results are shown in Fig. 9.34 for five sensors with power consumed and cable lengths as were depicted in Fig. 9.32.

Table 9.11 Voltage drops and power required as a function of DC-DC converter efficiency (as shown in Fig. 9.33) for data acquisition system depicted in Fig. 9.32

Central node voltage	η (%)	Power Supply Voltage (V)	Power Supply Current (A)	Sensor Voltage (V)	Sensor Current (A)	Cable drops (V) To central node	Cable drops (V) To sensor
24							
	65	38.3	4.45	19.5	0.85	14.3	4.47
	70	37.0	4.06	19.9	0.77	13.0	4.07
	75	35.9	3.73	20.3	0.71	11.9	3.73
	80	35.0	3.45	20.5	0.65	11.0	3.45
	85	34.3	3.21	20.8	0.61	10.3	3.21
	90	33.6	3.00	21.0	0.57	9.6	3.00
28							
	65	39.4	3.57	24.4	0.68	11.4	3.58
	70	38.5	3.28	24.7	0.62	10.5	3.28
	75	37.7	3.03	25.0	0.57	9.7	3.03
	80	37.0	2.82	25.2	0.53	9.0	2.82
	85	36.4	2.64	25.4	0.50	8.4	2.63
	90	35.9	2.48	25.5	0.47	7.9	2.47
32							
	65	41.7	3.02	29.0	0.57	9.7	3.01
	70	40.9	2.78	29.2	0.53	8.9	2.77
	75	40.3	2.58	29.4	0.49	8.3	2.57
	80	39.7	2.40	29.6	0.45	7.7	2.40
	85	39.2	2.25	29.8	0.43	7.2	2.22
	90	38.8	2.12	29.9	0.40	6.8	2.11

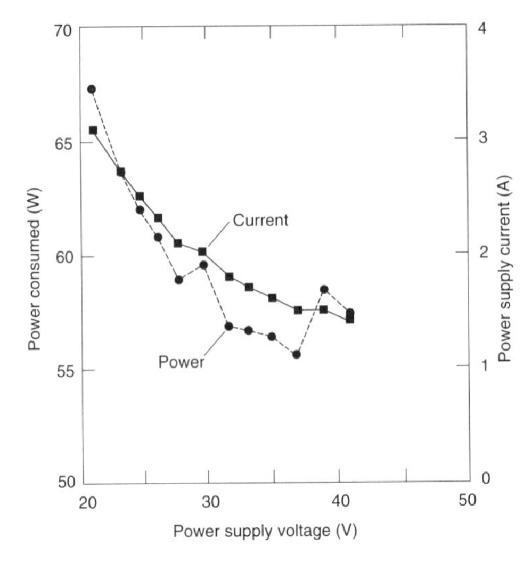

Fig. 9.34 Power consumption is a nonlinear function of voltage variation in the distributed power supply.

Scope: process or project
Duration: 10 months
Number of people involved: 1 program manager, 1 project scientist, 1 project engineer, 7 engineers, 3 technicians, shop personnel for fabrication

Level of effort: full time
Lessons learned: Distributed power systems are complex. You must consider the interactions between converter efficiencies, resistive line losses, and power consumption:

1. Linear regulation has the lowest noise and ripple, but it requires at least three heavy gauge wires to conduct the necessary current.
2. DC-DC converters need only two wires for input and are efficient.

Early involvement with the design engineers could have provided alternative sources of power and a simpler design. Unfortunately, management mandated low-voltage DC power, thereby restricting the choices for power conversion.

9.13 SUMMARY

You must begin considering an instrument's power system during the conceptual stage of development. Too often project design incorporates the power converter too late, and the result is an expensive module that is bulky, heavy, or too tightly specified (and therefore, too costly). You must research, study, and define the capabilities and environment of the power source, the power converter, and the load. Finally, you will have to test the power converter to ensure that it meets applicable regulations.

9.14 RECOMMENDED READING

Brown, M. 1990. *Practical switching power supply design*. San Diego: Academic Press.
A good introductory walk-through of switching power converters.

Hnatek, E. R. 1989. *Design of solid-state power supplies*. 3rd ed. New York: Van Nostrand Reinhold.
A thorough treatment of power converters; treats EMI concerns well.

Pressman, A. I. 1991. *Switching power supply design*. New York: McGraw-Hill.
A thorough and complete treatment of switching power converters.

Shepard, J. D. 1984. *Power supplies*. Reston, Va.: Reston Publishing.
A basic, readable introduction into the definition, design, and test of common power supplies.

9.15 REFERENCES

Appleby, A. J. and F. R. Foulkes. 1989. *Fuel cell handbook*. New York: Van Nostrand Reinhold.

Brown, M. 1990. *Practical switching power supply design*. San Diego: Academic Press.

Crompton, T. R. 1990. *Battery reference book*. London: Butterworths.

Fowler, K. R., and R. B. North. 1991. Computer-optimized neurostimulation. *Johns Hopkins APL Technical Digest* 12(2):192–197.

Hagiwara, A. 1991. IEC 555-2 poses new challenges for power supply users. *Electronic Products*, March, pp. 25–27.

Hnatek, E. R. 1989. *Design of solid-state power supplies*. 3rd ed. New York: Van Nostrand Reinhold.

Hu, C., and R. M. White. 1983. *Solar cells: From basics to advanced systems*. New York: McGraw-Hill.

Maier, C. 1991. Taking the mystery out of switching-power-supply noise. *Electronic Design*, September 26, pp. 83–92.

Maliniak, D. 1992. DC supply keeps line power factor near unity. *Electronic Design*, May 28, pp. 33–37.

Military-power-supply failures give rise to unofficial MIL standard. 1984. *EDN*, January 26, pp. 49–57.

Newhart, M. 1986. Military power supplies look to escape customization trap. *Electronic Design*, November 20, pp. 29–34.

Ormond, T. 1992. Distributed power schemes put power where you need it. *EDN*, July 6, pp. 158–164.

Pryce, D. 1991. Specialized ICs correct power factor in switching supplies. *EDN*, July 4, pp. 106–114.

Questad, P. 1990. Designing with a distributed-power architecture. *Electronic Design*, April 12, pp. 95–105.

Questad, P. 1991. Distributed power systems. *Electronic Products*, March, pp. 41–43.

Quinnel, R. A. 1991. The business of finding the best battery. *EDN*, December 5, pp. 162–166.

Roark, D., and L. Pechi. 1991. A primer on power factor correction. *Electronic Products*, March, pp. 29–31.

Russell, C. R. 1967. *Elements of energy conversion*. Oxford: Pergamon.

Shepard, J. D. 1984. *Power supplies*. Reston, Va.: Reston Publishing.

10

COOLING

In most equipment, the cooling design is as important as the electronic design.
—Allan Scott (1974, p. 1)

10.1 HEAT TRANSFER

All electronics generate heat. No device transforms or transmits electrical energy with 100% efficiency; the loss in energy between the input and output of an electronic device manifests itself as thermal energy dissipation, or heat. Generally, the transfer of thermal energy within electronics is described in terms of power and is measured in watts. The heat generated and dissipated is the difference between input power and output power (Fig. 10.1).

Heat has two deleterious effects on electronics. It reduces reliability by inducing thermal stress within components, which eventually leads to failure, and it alters circuit operation in temperature-sensitive devices. Therefore, you must control the heat from electronics to reduce failure in the components and drift in temperature-dependent operations.

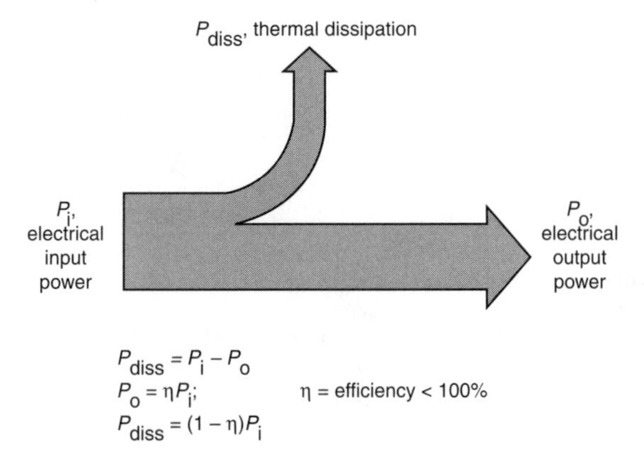

Fig. 10.1 Thermal dissipation within electronics resulting from inefficient power conversion.

Thermal energy flows from hotter objects to cooler objects (or, more specifically, from regions of higher temperature to regions of lower temperature). A component that dissipates heat must be at a higher temperature than its immediate surroundings. The circuit board must be warmer than the internal environment of the instrument, which must be warmer than the ambient environment outside the instrument. Consequently, temperature will increase in a component until it is hot enough for the heat flow to overcome the temperature drops in the thermal path to the lower temperature of the ambient reservoir. If the thermal path is insufficient and the temperature drops are too great, the dissipating component will not transfer enough heat, will exceed its physical limits, and will fail. Proper heat transfer maintains acceptable temperatures within the instrument and its components.

Sources of heat are regions of higher temperature. These sources can be inside or outside an instrument. Heat generated within an instrument originates from power dissipated by individual components; the dissipated heat then flows from the inside to the outside of the instrument. Conversely, heat can also flow into the instrument from hotter outside sources. A satellite in full sunlight, for example, must reflect the intense solar radiation to prevent high temperatures within its circuitry. Likewise, the electronics module within the engine compartment of an automobile must cope with both the heat from the engine and the heat dissipated by its internal components. By mapping the regions of high and low temperature, you can begin planning to manage heat transfer within and through your instrument.

10.2 APPROACH TO THERMAL MANAGEMENT

Knowledge of the application, the environment, and the possible mechanisms of cooling is necessary before you can begin the design of the instrument. A friend of mine worked on a project where, he said, "I had to decide if I could get 70 watts [of heat] out of a one cubic foot box using only natural convection. A decision at this level affects the whole architecture of the design" (Steve Zeise, letter, July 25, 1994).

After specifying the application, the environment, and possible mechanisms, you can then use the appropriate tools to determine the cooling design. You need to manage heat transfer at four levels within the design of your instrument:

1. Enclosure
2. Module
3. Board
4. Component

This chapter introduces the basic mechanisms and calculations of cooling and then presents successively more complex levels of cooling design. Finally, Section 10.12 gives some rules of design for cooling systems.

10.3 MECHANISMS FOR COOLING

Heat transfer occurs through three mechanisms (Scott, 1974):

1. Conduction
2. Convection
3. Radiation

Conduction is the transfer of thermal energy through molecular vibration and motion within solid materials. *Convection* is the transfer of thermal energy through molecular motion and fluid flow in liquids and gases; it is a mass transport phenomenon. Heat transfer occurs by *radiation* when the energy of molecular vibration converts to electromagnetic energy in the infrared spectrum. Fig. 10.2 illustrates these three mechanisms of heat transfer.

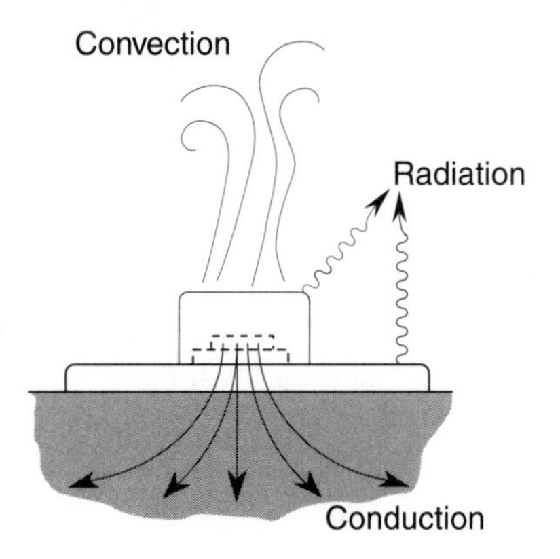

Fig. 10.2 Mechanisms of thermal energy transfer.

You can employ structures and mechanisms that use conduction, convection, and radiation to control heat transfer. Mechanisms to control and manage thermal energy transfer can be either passive or active. Such methods depend on the temperature differential between hot and cool regions to establish thermal paths for heat dissipation. Passive methods do not use external energy sources to transfer heat but rely on conduction and convection to dissipate heat. In special circumstances, radiation from the warm components provides cooling. Four mechanisms—cold plate, cooling fins, heat pipe, and evaporation—constitute the passive methods. The cold plate relies on conduction to remove heat from warm components. Cooling fins (Fig. 10.3), remove heat through convection and, secondarily, through radiation. The heat pipe transfers thermal energy through conduction and internal convection. Evaporation is used in specific situations to transfer large amounts of thermal energy through boiling a liquid that produces an exhaust gas.

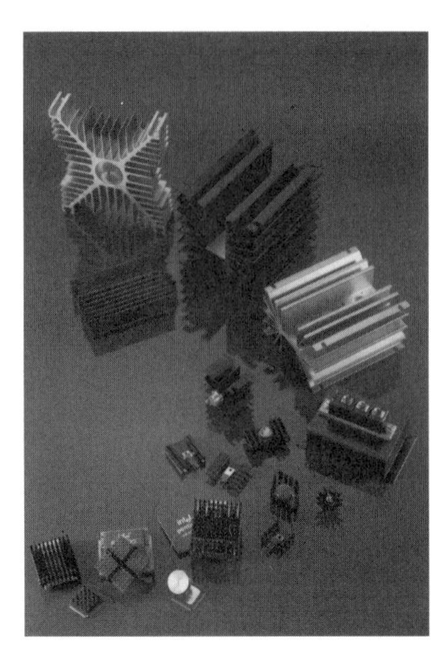

FIG. 10.3 Cooling fins in a variety of shapes. They provide cooling through convection (photograph courtesy of Wakefield Engineering).

Active mechanisms enhance heat transfer by using external power to provide effective thermal paths. Active mechanisms often employ forced convection to cool warm components. Fans such as those shown in Fig. 10.4 are the most common and familiar mechanism for producing forced air convection. Liquid cooling is becoming more prevalent in equipment and components that have high power densities and dissipation. Refrigeration can transfer even more heat by using either a thermodynamic cycle or the Peltier effect in a thermoelectric cooler.

FIG. 10.4 Fans for improving cooling through forced convection. Note the wide variety of shapes, sizes, and capacities for filling many different applications (photograph courtesy of EBM Industries, Inc.).

10.4 OPERATING RANGE

Once you have determined the environment and sources of heat, you must establish the allowable temperatures. Most likely you will design your equipment to operate in one of three temperature ranges:

1. Commercial: 0°C to 70°C (32°F to 158°F)
2. Industrial: –40°C to 85°C (–40°F to 185°F)
3. Military: –55°C to 125°C (–67°F to 257°F)

To achieve the desired operating range, you must account for the total heat transfer within the instrument, make allowances for equipment upgrades that demand more power, and consider the capabilities and capacity of the cooling equipment. The total heat transfer includes the dissipation from each component and the heat from outside sources:

$$P_{\text{total heat}} = P_{\text{in}} + \sum_{j=1}^{n} P_{\text{diss}\,j}\,,$$

where P_{in} = input thermal energy
$P_{\text{diss}\,j}$ = heat dissipated from component j
n = number of components

These calculations should allow for inaccuracies in the assumptions and specifications and for circuit growth. A designer must plan, for example, to manage the additional heat load contributed by add-in circuit boards in a desktop computer. Finally, the capacity of the cooling equipment determines the effectiveness of heat transfer and the operating temperatures. The next three sections describe some basic calculations for evaluating heat transfer.

10.5 BASIC THERMAL CALCULATIONS

10.5.1 *Conduction*

Heat transfer occurs across a temperature gradient. While temperature differential drives heat flow through a material, the inherent thermal conductivity of the material and its physical dimensions determine the amount of heat transferred. Heat transfer is proportional to the area of the thermal path, the thermal conductivity (Table 10.1), and the temperature differential and inversely proportional to the length of the thermal path (Fig. 10.5):

$$Q = \frac{kA\,\Delta T}{L}\,,$$

Table 10.1 Thermal conductivity for some representative materials at 100°C (Scott, 1974, pp. 18–19)

Material	Thermal conductivity, k (W/M·°C)
Air	0.026
Epoxy	0.2–0.8
Stainless steel	16–26
Low-carbon steel	46
Iron	67
Nickel	90
Brass	115
Pure silicon	146
Aluminum	171–193
Gold	298
Copper	394
Silver	417
Diamond	630

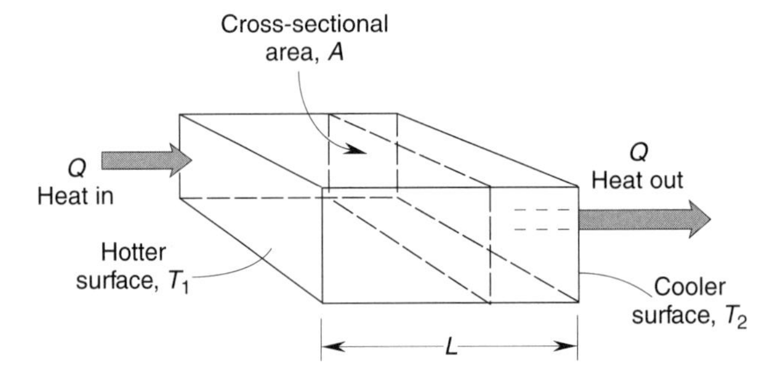

Fig. 10.5 Geometry of heat transfer.

where Q = heat transferred (W)
k = thermal conductivity (W/m·°C)
A = cross-sectional area (m^2)
L = length (m)
ΔT = $T_1 - T_2$ = temperature differential (°C)

A useful concept in heat transfer is thermal resistance. The thermal resistance of an object limits the heat flow through the object for a given temperature differential. Thermal resistance is directly proportional to length and inversely proportional to thermal conductivity and area. It is measured in degrees Celsius per watt.

$$\theta = \frac{L}{kA} = \frac{\Delta T}{Q},$$

where θ = thermal resistance (°C/W)
k = thermal conductivity (W/m·°C)
A = cross-sectional area (m^2)
L = length (m).

Clearly, greater thermal resistance will reduce heat flow. Thermal insulators have large thermal resistances. For a given heat flow, a large thermal resistance maintains a large temperature differential between the hot input and the cooler output.

The thermal resistances of different materials stacked together sum to give the overall thermal resistance, as shown in Fig. 10.6:

$$\theta_{\text{total}} = \sum_{i=1}^{n} \theta_i = \sum_{i=1}^{n} \frac{\Delta T_i}{Q} = \frac{1}{Q} \sum_{i=1}^{n} \Delta T_i = \frac{T_1 - T_n}{Q},$$

where θ_{total} = overall thermal resistance (°C/W)
θ_i = thermal resistance for material i (°C/W)
ΔT_i = temperature differential across material i (°C)
Q = heat flow (W)
i = 1 (source) to n (ambient)

Therefore, if you know the temperature at the source, T_1; the ambient temperature, T_n; and the power dissipated, Q, you can calculate the required total resistance.

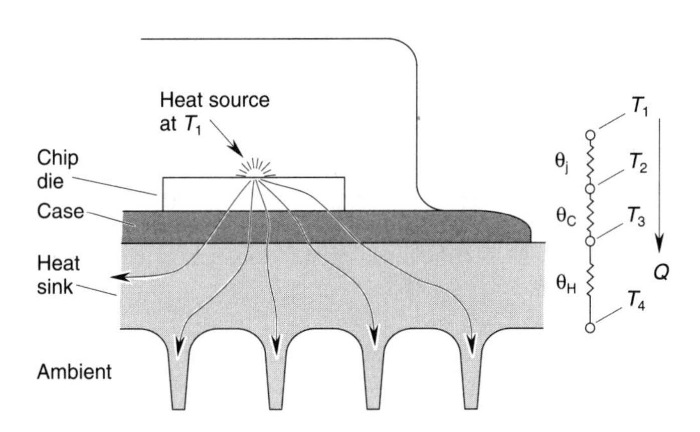

$$\theta_{\text{total}} = \theta_j + \theta_C + \theta_H = \frac{1}{Q}[(T_1 - T_2) + (T_2 - T_3) + (T_3 - T_4)] = \frac{T_1 - T_4}{Q}$$

$$Q = \frac{T_1 - T_4}{\theta_{\text{total}}}$$

Fig. 10.6 Calculation of thermal resistance for a combination of materials. Total thermal resistance from the heat source to the ambient reservoir is the sum of the individual resistances.

Conversely, knowing the total thermal resistance and the power, you can calculate the temperature of the source above the ambient temperature:

$$T_1 = Q\,\theta_{\text{total}} + T_n\ ,$$

where T_1 = temperature at the source (°C)
T_n = ambient temperature (°C)
θ_{total} = overall thermal resistance (°C/W)
Q = heat flow (W)

10.5.2 *Convection*

Convection improves the heat transfer from electronic circuits and modules and their heat sinks. The heat transferred is proportional to the area of the heat sink, the differential between the surface and ambient temperatures, and a convection coefficient:

$$Q = \alpha A\ \Delta T_s\ ,$$

where Q = heat flow (W)
α = convection coefficient ($W/m^2 \cdot °C$)
A = area (m^2)
ΔT_s = temperature differential (°C)

The convection coefficient describes the effectiveness of the heat transfer and includes factors for physical orientation, flow rates, and flow characteristics.

Convection is either natural or forced (Fig. 10.7). Natural convection relies on localized heating of the air (or surrounding fluid) to alter its density and buoyancy, causing it to rise and draw cooler air (or fluid) into the area around the warm component. Forced convection mechanically propels air (or fluid) past the heat sink to increase the transfer of heat. Evaporation or boiling of a fluid has the greatest capacity for cooling because the phase change from liquid to gas absorbs large amounts of energy, expediting the removal of thermal energy.

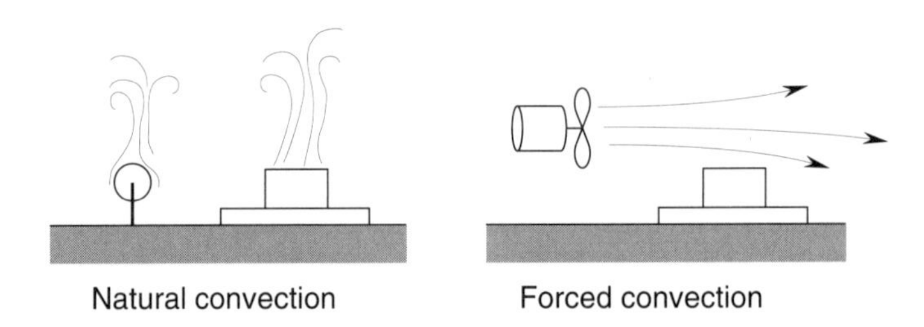

Fig. 10.7 Convection cooling of electronics.

For natural convection, the convection coefficient is a function of the surface temperature and orientation of the heat sink. In forced convection, the convection coefficient is a function of the linear velocity and turbulence of the air over the heat sink. In either case, Table 10.2 also shows that the convection coefficient also depends on the length, shape, and spacing of fins in the fluid flow direction.

Table 10.2 Typical values for the convection coefficient (Incropera and DeWitt, 1990)

Convection process	Convection coefficient, α (W/m² · °C)
Still air	5–25
Forced air	25–250
Forced liquids	50–20,000
Evaporation	2500–100,000

10.5.3 *Radiation*

Radiation generally contributes the least to cooling. The heat transfer follows the Stefan-Boltzmann law and depends on the fourth power of the absolute temperatures:

$$Q = \epsilon\, \sigma\, A(T_s^4 - T_\infty^4),$$

where Q = heat flow (W)
ϵ = emissivity of the solid surface
σ = Stefan-Boltzmann constant, 5.67×10^{-8} (W/m²·K⁴)
A = area (m²)
T_s = temperature of the emitting surface (K)
T_∞ = temperature of the ambient environment (K)
$T_s > T_\infty$

The emissivity is a function of the color, texture, and shape of the heat sink; some typical values are given in Table 10.3. The area is the unobstructed view from the heat sink; hence, fin spacing and orientation are critical.

Table 10.3 Some typical values for emissivity (Incropera and DeWitt, 1990)

Material	Emissivity, ϵ
Metals, highly polished	0.02–0.08
Metals, unpolished	0.10–0.45
Metals, oxidized	0.25–0.70
Ceramics	0.35–0.80
Carbon, graphites	0.75–0.95
Anodized finishes	0.90–0.98

10.5.4 *Further Considerations*

Cooling of electronic equipment often uses a combination of conduction, convection, and radiation. The total heat transferred is the sum of the individual processes:

$$\begin{aligned} Q_{\text{total}} &= Q_{\text{conduction}} + Q_{\text{convection}} + Q_{\text{radiation}} \\ &= \frac{\Delta T_{\text{c}}}{\theta} + \alpha A\, \Delta T_{\text{s}} + \epsilon \sigma A (T_{\text{s}}^4 - T_{\infty}^4). \end{aligned}$$

You must understand the processes and their thermal paths to assign the energy transfer appropriately and to calculate an accurate result.

The calculations presented here are for steady-state situations. If your instrument cycles power on and off, you should study the transient response of the temperature, an example of which is Fig. 10.8. High-temperature spikes can induce severe transients of thermal stress within components and lead to failure.

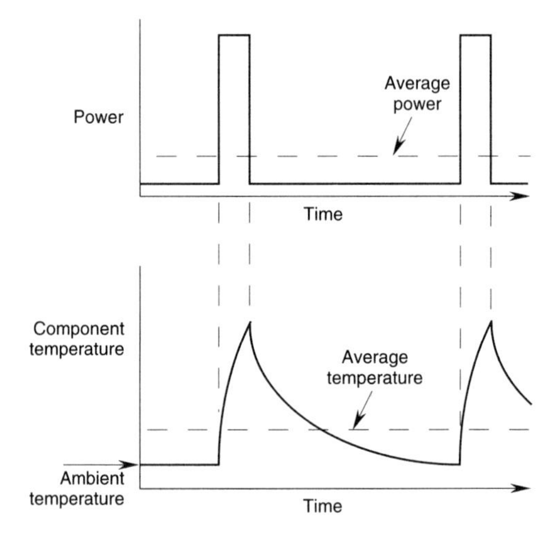

FIG. 10.8 Effect of cycling power on and off. Power cycles in circuits can cause high temperatures and induce thermal stress in components.

10.6 COOLING CHOICES

After establishing the environment, sources, and operating temperatures, you are ready to select the method of cooling. Generally, you may follow this procedure:

1. Fix the desired junction temperature, T_{1}.
2. Calculate the thermal resistance from junction to case without a heat sink, $\theta_{total} = \theta_{1} + \theta_{C}$.
3. Determine the power dissipated, Q.
4. Calculate the surface temperature, $T_{s} = T_{1} + Q\,\theta_{total}$.
5. Determine the available surface area, A.
6. Select the form of cooling from Fig. 10.9 by taking the ratio of power dissipated to surface area at the operating temperature, Q/A at T_{s}.

If your circuitry is a module, you can shorten the procedure. Given the surface area, desired surface temperature, and power dissipation, you can select the appropriate cooling mechanism.

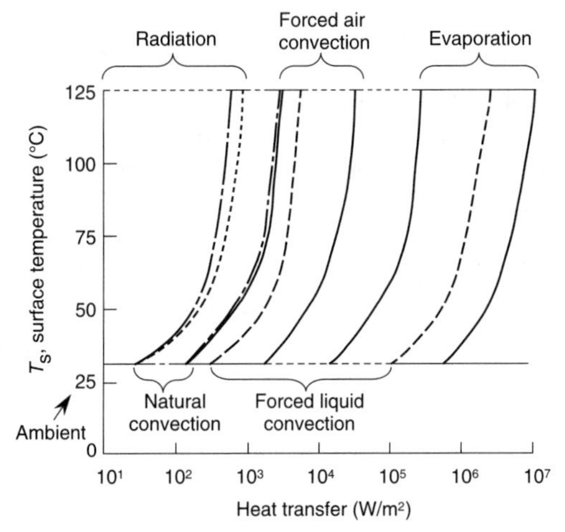

Fig. 10.9 Regimens of cooling for different processes.

10.7 HEAT SINK SELECTION

All circuits transfer heat from the electronic components to an intermediate material by conduction and then use radiation, natural convection, forced convection, or evaporation to transfer heat from the intermediate material to the cooler ambient environment. Sometimes, the intermediate material is a finned heat sink.

You calculate heat transfer through a heat sink by defining three thermal resistances. Their values add in series: from the source, to the semiconductor junctions, to the ambient surroundings (see Fig. 10.6). The resistances are as follows:

1. Junction resistance, θ_j, from the semiconductor junctions to the component case
2. Case resistance, θ_C, from the component case to the heat sink
3. Heat sink resistance, θ_H, from the heat sink to the ambient environment

The junction and case resistances depend on the materials and manufacture of each electronic component and are supplied by the manufacturer of the component. Table 10.4 gives some typical values of combined junction and case resistances. For some components, such as the pin grid array (PGA), 8-pin DIP, and TO-92 packages, the leads contribute significant thermal paths through the case.

The choice of component packages, even their orientation on the circuit board, contributes to heat transfer and affects the operating temperature of the circuitry. A less expensive package with a greater thermal resistance may require a larger, more

Table 10.4 Some typical values for junction and case resistance

	Thermal resistance (°C/W)	
Component type	**Junction to case**	**Junction to ambient (without heat sink)**
PGA	10–20	20–35
Leadless chip carrier	20–40	40–100
8-pin DIP		
Plastic	50–80	140–170
Ceramic	40–70	120–150
16-pin DIP		
Plastic	35–50	65–80
Ceramic	30–45	80–100
40-pin DIP		
Plastic	25–40	50–70
Ceramic	15–25	45–60
TO-92	40	160
TO-39	20	170
TO-220	4	50
TO-3	2.5	35

expensive heat sink than a costlier package with a lower thermal resistance that uses a smaller, less expensive heat sink. The orientation of a component affects its ability to transfer heat to the copper clad within the PCB. A diode or resistor conducts most of its heat through the leads. However, even leads have finite thermal resistance, and these values set an upper limit on the power dissipated into the circuit board. DIPs and surface-mount packages cannot dissipate more than 2 W of power through their leads into a circuit board (Simpson, 1991).

The material interface between the component and the heat sink is the main conduction path for heat, and it must have direct contact between surfaces. A microscopic view (Fig. 10.10) demonstrates that surfaces are never perfectly smooth, so the cross-sectional area of the actual conduction path is drastically reduced. A thermal grease compound or thermal gasket can fill the voids, thereby increasing surface contact, reducing thermal resistance, and improving thermal conduction. Clamping the two surfaces together further improves heat transfer.

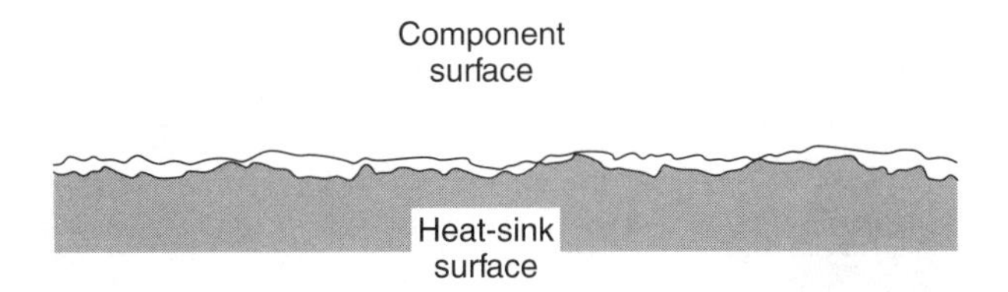

Fig. 10.10 Enlarged view of interface between component and heat sink. Real surfaces are not perfectly flat and present significant resistance to heat conduction across their interface.

Once you have determined a suitable temperature range and conduction path, your next step is to choose or design a heat sink. Since junction and case resistances are set by the component manufacturer, you control the thermal resistance of the conduction path by designing, selecting, and mounting an appropriate heat sink. The thermal resistance depends on the amount of power dissipated, surface dimensions, fin spacing, and physical orientation. Designing a custom heat sink requires techniques far beyond the scope of this text; they can be found in texts such as the one by Incropera and DeWitt (1990). However, heat sinks in diverse configurations are readily available (Table 10.5).

Orientation and physical dimensions of the heat sink are important considerations. A vertical surface or fin transfers the most heat through natural convection, while a horizontal surface facing down transfers the least. Hot air rises, drawing in cool air from the bottom. A vertical surface facilitates convection airflow. A horizontal surface facing down forces the warmed air to diffuse sideways before rising and drawing in cooler air to absorb heat. Increasing the length of the convection

Table 10.5 Example heat sinks and their characteristics (photographs courtesy of Wakefield Engineering)

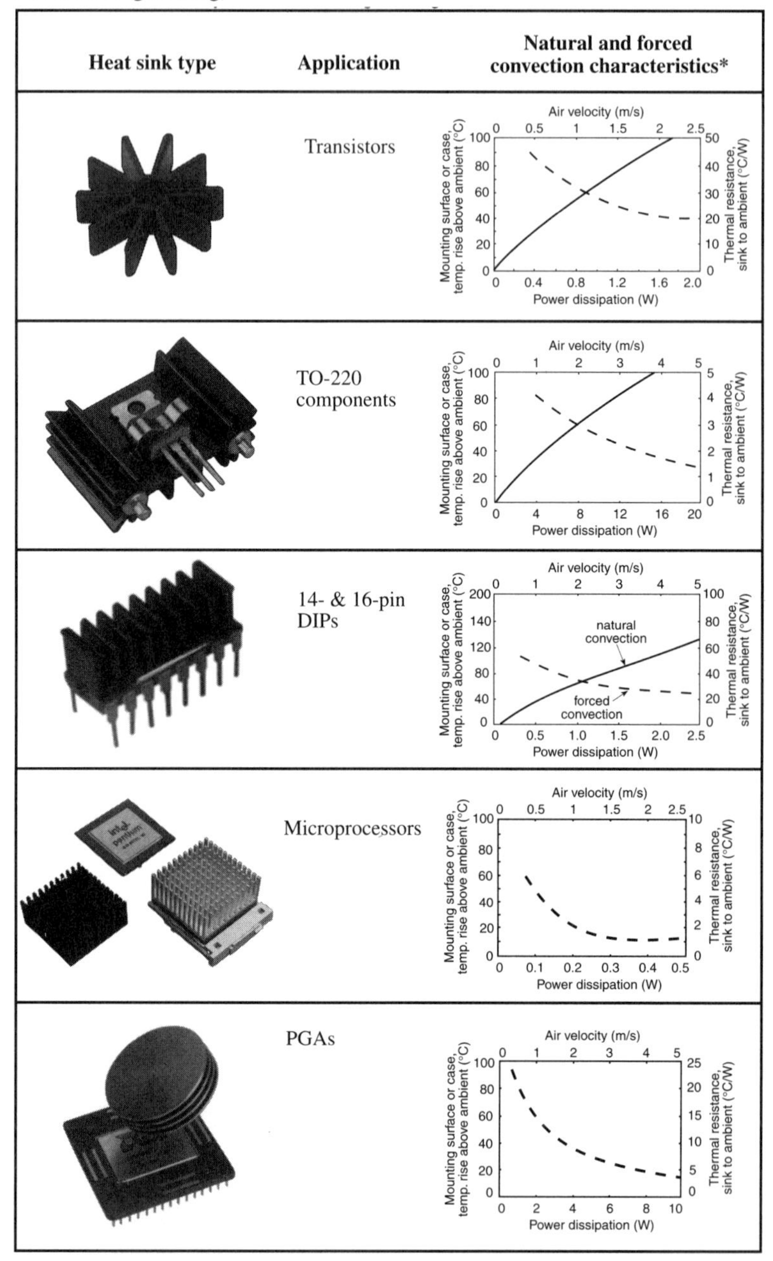

Heat sink type	Application	Natural and forced convection characteristics*
	Transistors	
	TO-220 components	
	14- & 16-pin DIPs	
	Microprocessors	
	PGAs	

Table 10.5 Example heat sinks and their characteristics (photographs courtesy of Wakefield Engineering) (continued)

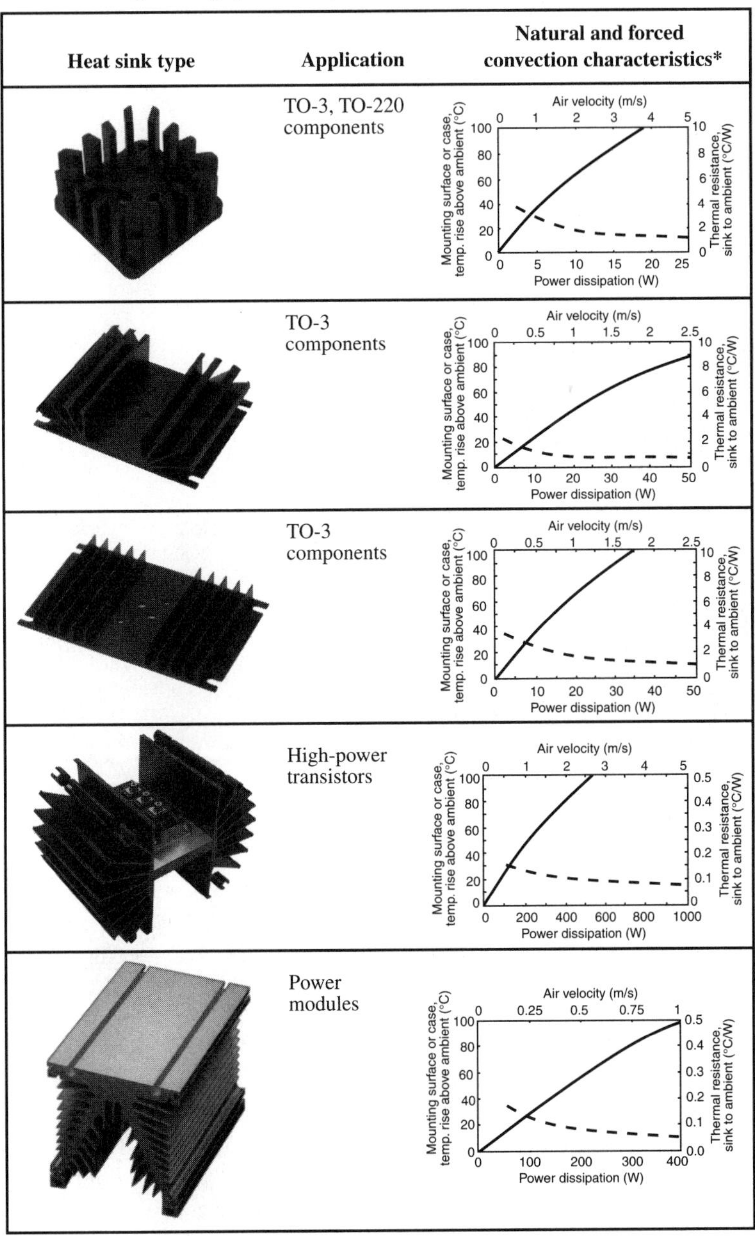

Heat sink type	Application	Natural and forced convection characteristics*
	TO-3, TO-220 components	Air velocity (m/s): 0 1 2 3 4 5 Thermal resistance, sink to ambient (°C/W): 0 2 4 6 8 10 Mounting surface or case, temp. rise above ambient (°C): 0 20 40 60 80 100 Power dissipation (W): 0 5 10 15 20 25
	TO-3 components	Air velocity (m/s): 0 0.5 1 1.5 2 2.5 Thermal resistance, sink to ambient (°C/W): 0 2 4 6 8 10 Mounting surface or case, temp. rise above ambient (°C): 0 20 40 60 80 100 Power dissipation (W): 0 10 20 30 40 50
	TO-3 components	Air velocity (m/s): 0 0.5 1 1.5 2 2.5 Thermal resistance, sink to ambient (°C/W): 0 2 4 6 8 10 Mounting surface or case, temp. rise above ambient (°C): 0 20 40 60 80 100 Power dissipation (W): 0 10 20 30 40 50
	High-power transistors	Air velocity (m/s): 0 1 2 3 4 5 Thermal resistance, sink to ambient (°C/W): 0 0.1 0.2 0.3 0.4 0.5 Mounting surface or case, temp. rise above ambient (°C): 0 20 40 60 80 100 Power dissipation (W): 0 200 400 600 800 1000
	Power modules	Air velocity (m/s): 0 0.25 0.5 0.75 1 Thermal resistance, sink to ambient (°C/W): 0.0 0.1 0.2 0.3 0.4 0.5 Mounting surface or case, temp. rise above ambient (°C): 0 20 40 60 80 100 Power dissipation (W): 0 100 200 300 400

*Solid lines indicate natural convection (use upper and right-hand axes).
Dashed lines indicate forced convection (use lower and left-hand axes).

path also reduces heat transfer (Scott, 1974). Fig. 10.11 illustrates how the relationship and orientation of the heat surfaces in a heat sink affect heat transfer. Fins placed too close together will cause mixing in the convected flows and will shade the radiation transfer; both processes impede heat transfer.

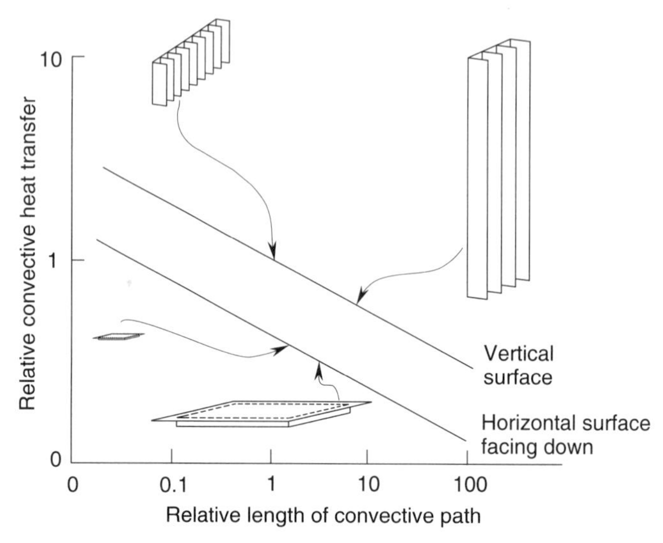

Fig. 10.11 Relationship between heat transfer and heat sink configuration. A vertical orientation for a heat sink is much more effective for transferring heat than a horizontal orientation facing down; a long path for convection flow reduces heat transfer.

10.8 HEAT PIPES AND THERMAL PILLOWS

Heat pipes and thermal pillows can add another wrinkle to heat sink design. These passive devices provide effective conduits for heat if some distance separates the hot components from the cooler environment. Neither device requires moving mechanical parts or electrical power input.

A heat pipe uses the liquid-to-gas phase change to absorb heat (Fig. 10.12). The gas diffuses down the pipe to the cooler end and condenses, releasing the stored thermal energy. The liquid then flows back to the hot end, completing the cycle of heat transfer. The wick is critical to operation because it must overcome the viscous force of the liquid, the vapor pressure of the gas, and gravity (Scott, 1974). The capacity of heat pipes for heat transfer approaches that of evaporative cooling. The disadvantages are the weight of the heat pipe and the need for cooling at the cool end (Swager, 1990).

A thermal pillow transfers heat through natural liquid convection (Fig. 10.13). A plastic bag filled with an inert liquid conducts heat when it is sandwiched between

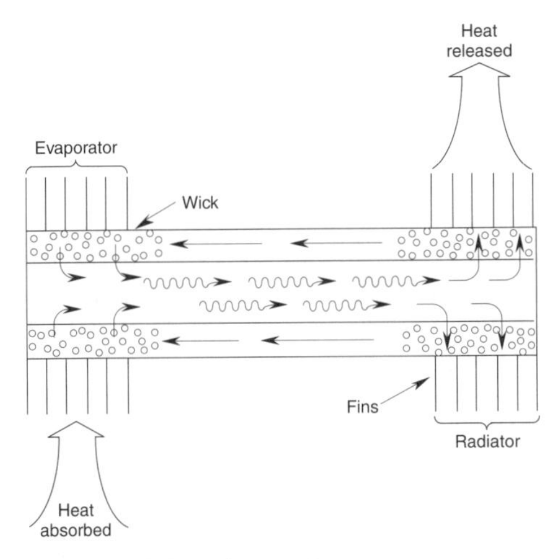

Fig. 10.12 General operation of a heat pipe.

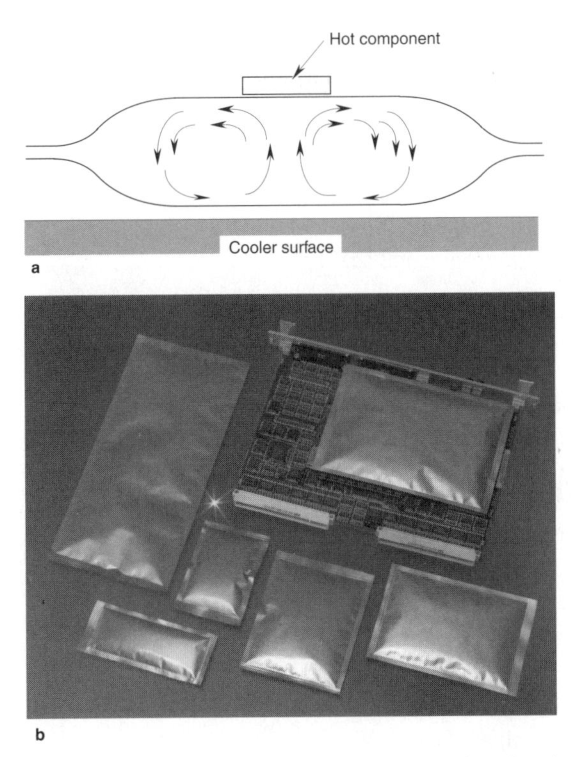

Fig. 10.13 Thermal pillows. Natural convection in the liquid conducts heat in a thermal pillow. (a) Schematic of convective flow in a thermal pillow. (b) Samples of thermal pillows (photograph courtesy of 3M, Inc.).

hot components and a cooler surface. The warmed fluid near the components diffuses away, drawing in cooler fluid to absorb more heat.

10.9 FANS AND FORCED AIR COOLING

Forced air cooling can increase heat transfer by 10 times over natural convection and radiation. Greater system functionality and high processing density result, in turn, from lower operating temperatures. Active heat transfer frequently makes specific operations possible. But forced air cooling has drawbacks:

- Increased system complexity and the attendant reduction in reliability
- Increased size and weight
- Lower system efficiency
- Acoustical noise

Obviously, you will have to decide whether the increased functionality allowed by forced air cooling overcomes the limitations.

10.9.1 *Calculations*

Designing a forced air cooling system involves several considerations:

1. Aerodynamic flow path
2. Choice of fan
3. Heat sink configurations
4. Layout of circuits and modules
5. Filtering and environmental sealing

These considerations are based on the complex, nonlinear relationships of fluid flows. Exact calculations of fluid flows and heat transfer are extremely complex, if not impossible; however, you can make some intuitive and reasonable estimates for choosing and sizing the cooling system.

The increase in air temperature as it flows over warm components and modules is inversely proportional to the volume flow (in cubic meters per hour or cubic feet per minute). Another way of saying this is that the thermal resistance of airflow drops with higher velocity:

$$\theta_{air} = \frac{\Delta T_{air}}{Q} = \frac{k_1}{v},$$

where θ_{air} = thermal resistance (°C/W)
ΔT = increase in air temperature (°C)
Q = dissipated heat (W)
k_1 = constant = 2.94 (°C·m^3)/(W·hr) = 1.73 (°C·ft^3)/(W·min)
v = volume airflow (m^3/hr or ft^3/min)

The thermal resistance of heat transfer between heat sink and air is inversely related to the length of the flow path and fractional powers of the volume airflow and fin geometry and directly proportional to the spacing between fins (Scott, 1974):

$$\theta_{\text{heat sink-to-air}} = \frac{\Delta T_{\text{heat sink-to-air}}}{Q} = \frac{k_2 s}{n^{0.2} h^{0.2} v^{0.8} L},$$

where $\theta_{\text{heat sink-to-air}}$ = thermal resistance (°C/W)
$\Delta T_{\text{heat sink-to-air}}$ = increase in temperature from the air to the heat sink (°C)
Q = dissipated heat (W)
k_2 = constant = 49.2 (°C·m$^{2.8}$)/(W·hr$^{0.8}$)
= 140 (°C·in.$^{0.4}$·ft$^{2.4}$)/(W·min$^{0.8}$)
s = spacing between fins (m or in.)
n = number of channels between fins
h = fin height (m or in.)
v = volume airflow (m^3/hr or ft^3/min)
L = length of channels (m or in.)

From these relationships you can approximate the temperature rise of the heat sink above the ambient for forced air cooling:

$$T_{\text{heat sink}} = (\theta_{\text{ambient}} + \theta_{\text{heat sink-to-air}})Q + T_{\text{ambient}}.$$

An important caveat in these calculations is that airflow is critically dependent on pressure drop and aerodynamic resistance along the flow path. Increased aerodynamic resistance or lowered pressure input diminishes the airflow. Thermal resistance of the heat transfer depends on airflow. Once again, the exact relationships are exceedingly difficult in most cases, but several rules of thumb clarify the necessary issues:

1. Pressure differential through the flow path is proportional to the square of the volumc airflow:

$$\Delta p \propto v^2.$$

2. Pressure differential is inversely proportional to the square of the cross-sectional area:

$$\Delta p \propto \frac{1}{A^2}.$$

3. Pressure differential is proportional to the length of the flow path:

$$\Delta p \propto L.$$

These rules hold for straight, constant-cross-section channels. Bends, junctions, and variation in cross-sectional area of the air-handling ducts increase aerodynamic resistance to flow and, consequently, the pressure drop. Entry and exit ports, dust filters, internal ducting, and cooling fins all contribute to aerodynamic resistance and pressure drop within a forced air cooling system. Fig. 10.14 illustrates some contributions to pressure drop. Fig. 10.15 illustrates pressure drop as a function of airflow.

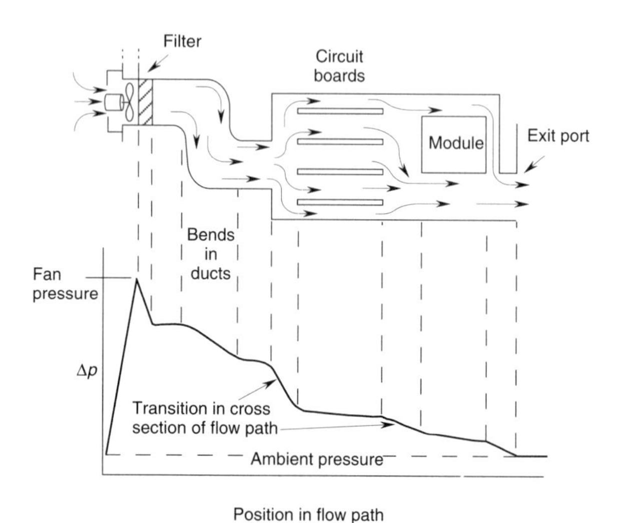

FIG. 10.14 Contributions to pressure drop. Many factors, including entry and exit ports, filters, duct bends, duct size variations, cooling fins, and module placement, contribute.

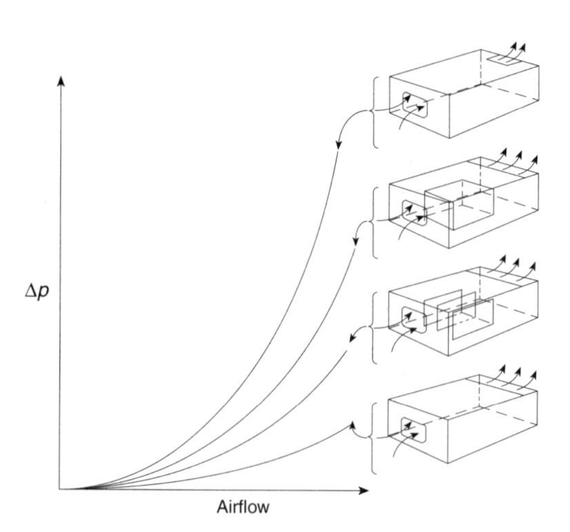

FIG. 10.15 Examples of pressure drop as a function of airflow area.

Further considerations may be important for particular applications. Air density, altitude, and inlet temperatures affect pressure drop, airflow, and heat transfer. Avionics specifically require attention to these factors.

10.9.2 *Fans*

Choice of a fan or blower is the next critical step in specifying a forced air cooling system. Fans and blowers differ according to the following parameters:

1. Pressure as a function of volume airflow
2. Power and efficiency
3. Size and weight
4. Acoustical noise signature

Four common configurations exist for fans: propeller, centrifugal blower, tubeaxial, and vaneaxial. Figs. 10.16 and 10.17 and Table 10.6 illustrate and compare representative examples of fans.

Fig. 10.16 Four common configurations for fans and blowers (photographs courtesy of Rotron, Inc.). (a) Propeller fan; (b) centrifugal blower; (c) tubeaxial fan; (d) vaneaxial fan.

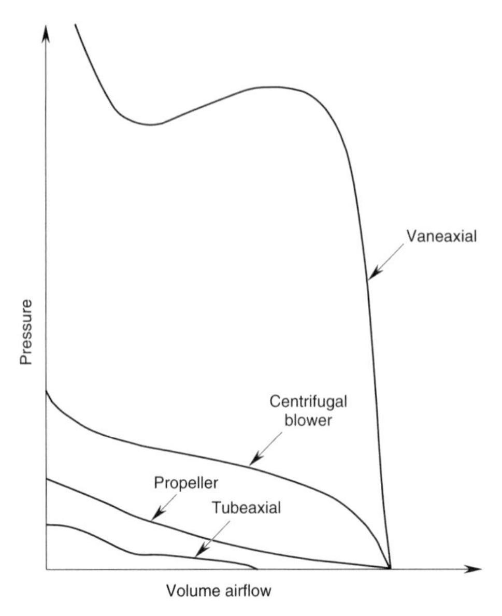

Fig. 10.17 Comparison of pressure and volume airflow characteristics for propeller fan, centrifugal blower, tubeaxial fan, and vaneaxial fan.

Table 10.6 General comparison of fan and blower characteristics

Fan/blower	**Pressure**	**Power required**	**Size**	**Weight**	**Acoustic noise**	**Reliability**	**Cost**
Propeller	Low	Low	Large	Medium	Quiet	High	Lowest
Centrifugal blower	Medium	High	Large	Very heavy	Noisy	Medium	Medium
Tubeaxial	Lowest	Lowest	Small	Light	Quietest	Very high	High
Vaneaxial	Very high	Highest	Smallest	Light	Very noisy	Low	High

Each fan type has characteristics that fit it well to specific applications. Desktop computers often employ tubeaxial fans because they are comparatively quiet. Aircraft avionics use vaneaxial fans because of their small size and high pressure output, which copes well with altitude changes. Racks of electronics often use centrifugal blowers (sometimes called squirrel cage blowers) to maintain sufficient pressure and airflow in a variety of configurations. Power tools and hand appliances have propeller fans attached to one end of the motor shaft because they are the cheapest to manufacture.

10.9.3 *Considerations*

Many details go into the design of a forced air cooling system. Here are a few to consider.

Dust settling on components acts as an insulator and reduces heat transfer, causing temperatures to rise. You may need to filter the inlet air to remove this contamination; however, a filter presents a significant resistance to airflow. Or you may have to seal the circuitry from the environment entirely by using heat exchangers. In that case, hot components are mounted on a heat sink whose base forms the environmental barrier and whose external fins transfer heat to the external, ambient air. Fig. 10.18 is a drawing of a sealed heat exchanger.

For safety's sake a protective grill may be installed over the fan inlet or outlet exhaust. The grill also prevents damage to the fan and the subsequent overheating of the circuitry. But remember that even a grill has aerodynamic resistance that you must factor into your airflow calculations.

You should place temperature-sensitive components near the inlet to reduce temperature variations. As load and operation change, the temperature of the air can vary according to the dissipated heat from different circuits and modules. Alternatively, a temperature sensor and speed control on the fan can vary the airflow to follow the power dissipation and maintain constant temperature.

Designing forced air cooling is tricky business. Table 10.7 summarizes some pitfalls and solutions in system design (Matisoff, 1990).

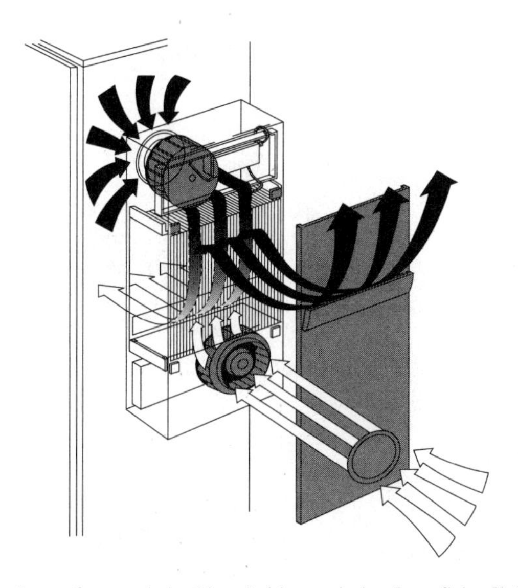

Fig. 10.18 Heat exchange for a sealed cabinet (with permission from Schroff, Inc.).

Table 10.7 Common pitfalls and solutions for forced air cooling systems

Area of concern	Comments	Examples	
		Bad	Better
Inlet and outlet sizing	Restricted openings reduce airflow and require a more powerful fan	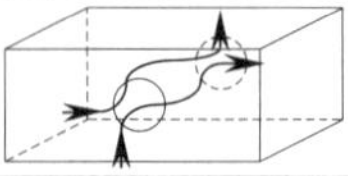	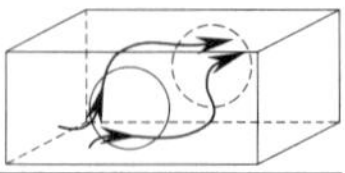
Changes in cross-sectional area	Abrupt changes in duct sizing cause large pressure drops; smooth transitions reduce turbulence and the attendant pressure drop.	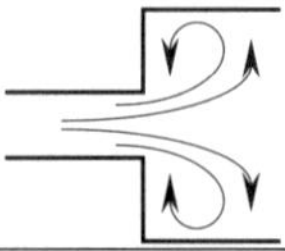	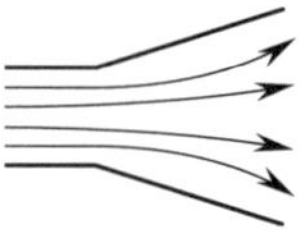
Bends in flow	Reduce the number of bends in the flow path to prevent unnecessary pressure drop.	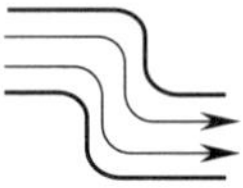	
Shock loss	Avoid directing air into a barrier, causing energy loss.	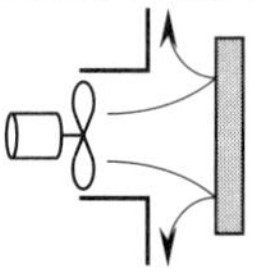	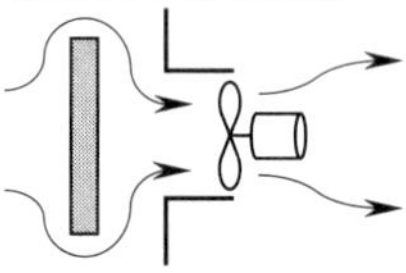
Impeller shroud	Carefully position a fan impeller inside the orifice of a shroud. Avoid turbulent loss around the tips from an orifice that is too large, separated from the impeller, or improperly shaped.	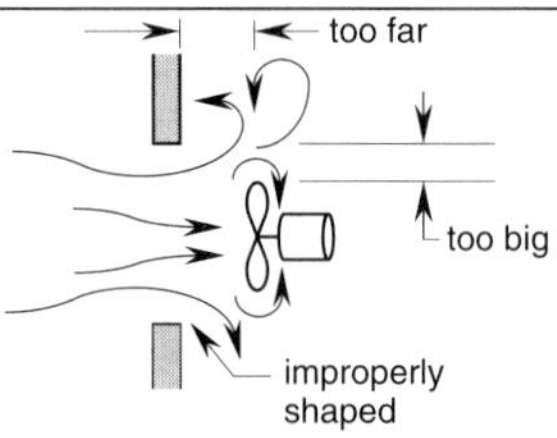 too far; too big; improperly shaped	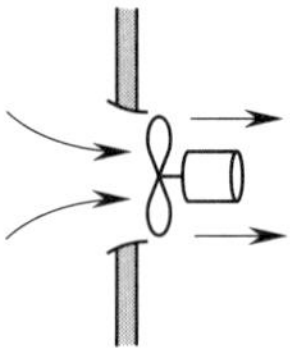
Blower size	Select blowers for maximum efficiency and minimum noise. An undersized blower requires a more powerful motor and is much noisier.	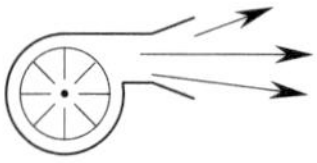	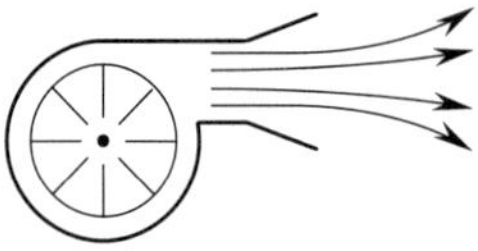
Exposed surface	Maximize the airflow exposure of a surface to be cooled. Minimize turbulent loss.	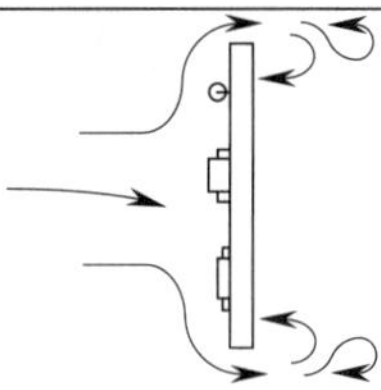	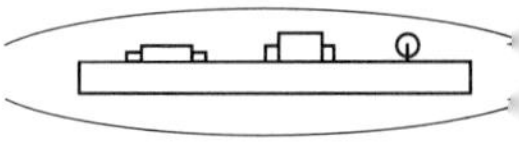
Circuit placement	Temperature-sensitive components should be upstream in the airflow to reduce fluctuations from local heating.	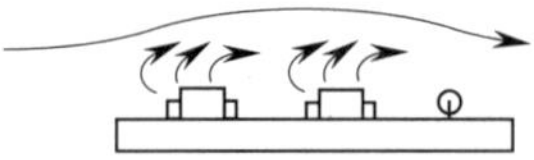	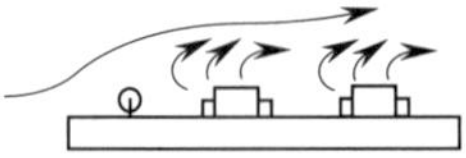

Table 10.7 Common pitfalls and solutions for forced air cooling systems (continued)

Area of concern	Comments	Examples: Bad	Examples: Better
Module placement	Place module blocks to allow smooth transition and minimum pressure drops. Use a finned heat sink to transfer heat and channel airflow.		
Dust	Reduce dust accumulation with filters and positive pressure.		
Safety	Use grills to prevent injury caused by curiosity, accident, or stupidity. Also protects fan from damage.		

10.10 LIQUID COOLING

Liquid cooling can increase heat transfer by 10 times over forced air cooling (but it is expensive). Even as forced air cooling improves system functionality and density over natural convection, so liquid cooling improves the density of operation even further. Fig. 10.19 outlines the components in a liquid cooling system.

The design of a liquid cooling system is almost always a custom effort. If you think that the nonlinear calculations for forced air were complex, wait until you see the extent of the calculations for liquid cooling (Scott, 1974). Liquid cooling requires accounting for laminar, turbulent, and transitional flows; geometry of the pipes; and fluid properties such as viscosity, boiling point, and decomposition.

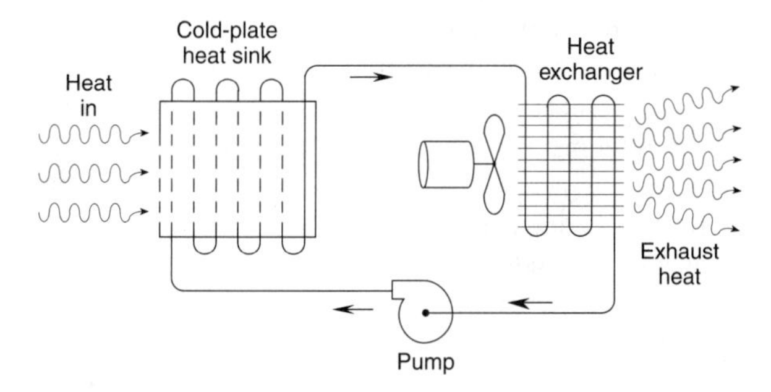

Fig. 10.19 Basic components of a liquid cooling system.

Again, the pressure drop is a nonlinear function of flow rate. You will have to consider the following properties when selecting a coolant liquid:

- Heat capacity
- Temperature limits
- Hydraulic properties such as viscosity
- Corrosive effects on tubing and seals
- Flammability
- Toxicity
- Environmental impact of disposal

Other than the challenge of design, liquid cooling has significant drawbacks such as cost, reliability, and failure modes. The complexity of liquid cooling increases the cost and reduces overall system reliability. Failures range from simple leaks to coolant spills and destruction of the circuitry from overheating.

Clearly, you should consult and involve a cooling specialist or mechanical engineer if you plan to incorporate liquid cooling into your equipment.

10.11 EVAPORATION AND REFRIGERATION

Boiling a liquid and absorbing thermal energy in the phase change to a gas is the most effective way to transfer large amounts of heat. Liquid evaporation can transfer up to 1.24 MW/m^2 (800 $W/in.^2$) of heat.

Where would anyone want that kind of capacity? Three general types of applications exist:

1. Extremely low and stable temperatures
2. Compact, mission-serviceable equipment
3. Surroundings that are too warm for circuit operation

Infrared sensors in missiles and satellites require low, stable temperatures to function. Evaporating liquid nitrogen across the back of the sensor achieves the desired heat transfer and operating temperature. Since a liquid always boils at a fixed temperature, the thermal regulation is inherent to the process.

Some avionics pods, slung beneath the wing or fuselage of an aircraft, may cool by evaporation also. The electronic modules are immersed in a volatile liquid; as the modules dissipate heat, the coolant boils and the gas generated is vented. Again, temperature remains constant at the boiling point of the liquid. This is a very simple and effective mechanism for cooling equipment over a limited duration. Fig. 10.20 illustrates one configuration.

Refrigeration and air conditioning transfer heat through the compression and expansion cycle of a working fluid or gas. Fig. 10.21 outlines the basic components of a refrigeration unit. Some commercial supercomputers use refrigeration to achieve compact cooling of the circuitry. Equipment racks can incorporate air conditioners to cool the enclosed equipment (Fig. 10.22).

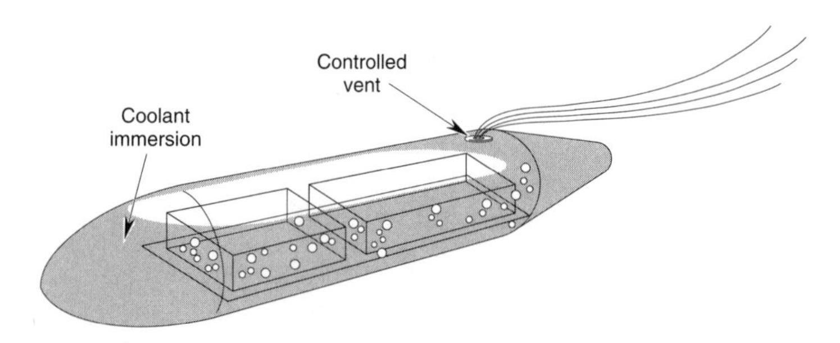

Fig. 10.20 A simple mechanism for dissipating large amounts of heat for a limited-duration mission: liquid immersion, evaporation, and controlled venting.

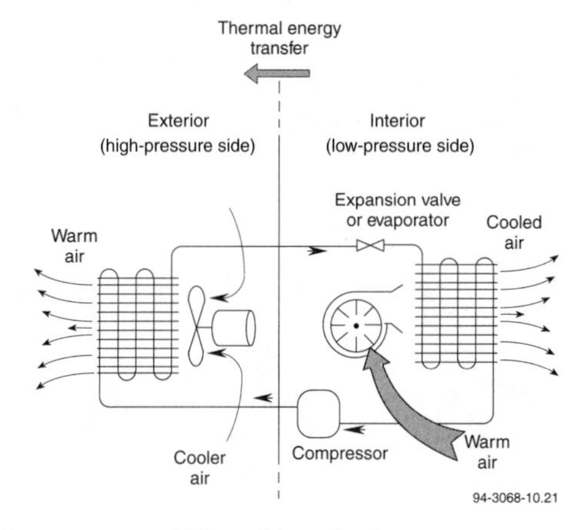

Fig. 10.21 Basic components within a refrigeration loop.

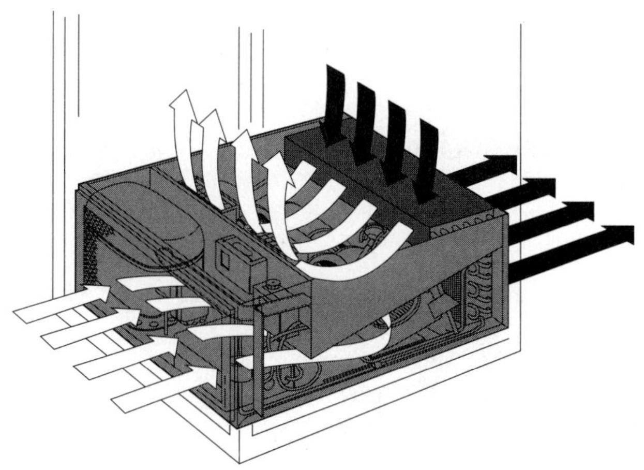

Fig. 10.22 Air conditioner for cooling rack-mounted equipment in a sealed cabinet (with permission from Schroff, Inc.).

Refrigeration and air conditioning have drawbacks:

1. Complexity
2. Lower system reliability
3. Large amounts of power consumed
4. Installation and maintenance costs
5. Environmental impact from leaking or disposal of the working fluid

Thermoelectric devices, however, are solid-state "refrigerators" and avoid some of these shortcomings. Thermoelectric coolers can also be scaled to size for very localized cooling, such as individual chips (Fig. 10.23).

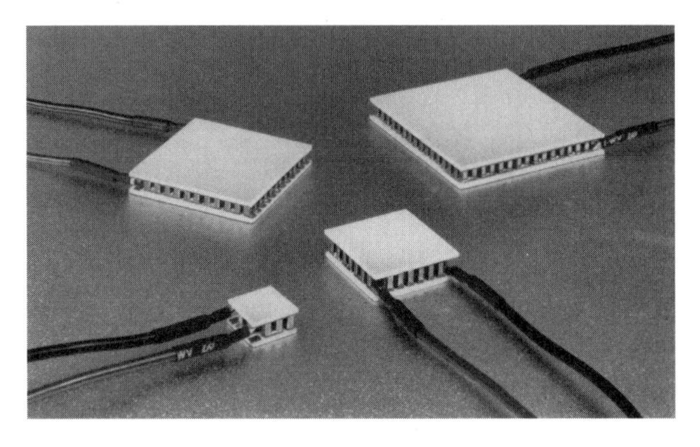

Fig. 10.23 Thermoelectric coolers for integrated circuits (photograph courtesy of Melcor Thermoelectrics).

10.12 TRADE-OFFS IN DESIGN

The variety of cooling configurations available leave you with a lot of design choices. This section is an outline of the steps that you must take to arrive at a solution.

1. Consider the thermal environment in and around your device or equipment:
   a. Calculate total heat transfer (1 W = 3.415 Btu/hr).
   b. Determine the operating temperature range.
   c. List the heat transfer mechanisms: conduction, convection, radiation.
   d. Map out the available thermal paths.

2. Determine what factors may constrain the design of the cooling system:
   a. Availability of an ambient, low-temperature reservoir
   b. Dusty, humid, or corrosive environment that must be sealed out
   c. Acoustical noise requirements
   d. Power required to run the cooling system
   e. Environmental concerns
   f. Safety concerns
3. Calculate the life cycle costs:
   a. Initial cost: design, test, or purchase
   b. Installation
   c. Maintenance
   d. Disposal of coolant
4. Compare the advantages and disadvantages of the various cooling configurations:
   a. Relative capacity (watts per unit area)
   b. Cost
   c. Size and weight
   d. Reliability

Fig. 10.24 shows some of the decision processes.

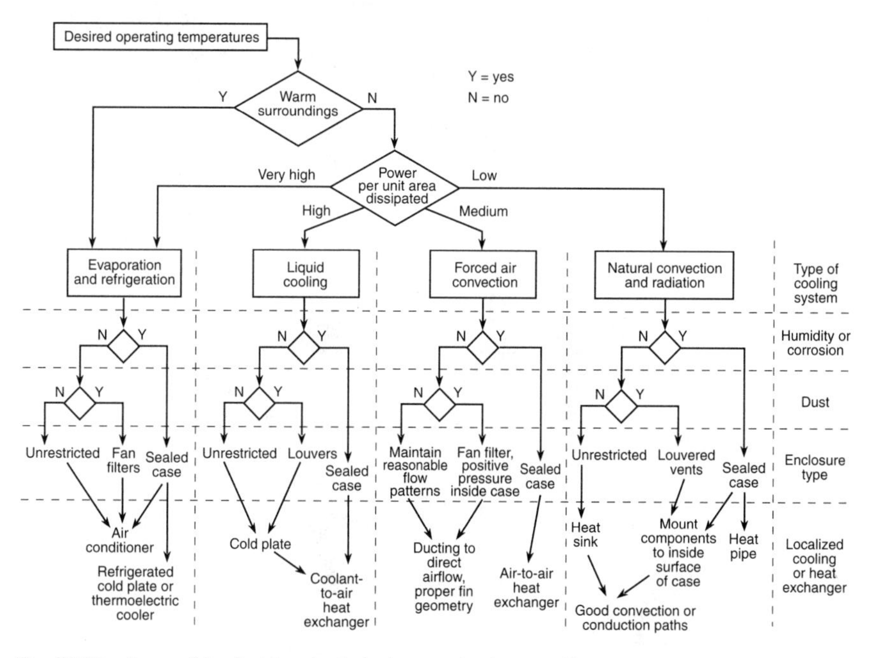

Fig. 10.24 Some of the decisions in designing or selecting a cooling system.

10.13 ANALYSIS AND EXPERIMENTATION

The approximate and nonlinear nature of calculation in heat transfer lends itself to, and perhaps demands, experimentation for each cooling system design. This is a part of the iterative nature of systems engineering in developing electronic products.

The purpose of analysis and experimentation is to converge on a satisfactory cooling design. First establish broad boundaries for the heat transfer in your system with analytical models. Then simulate the transient and steady-state thermal responses with CAD tools or to go directly to experimentation. The results allow you to refine the analytical models and then to proceed with another iteration of analysis, simulation, and experimentation to reduce the margins of error in your understanding of the cooling operation. Fig. 10.25 illustrates the general flow of effort.

Analytical models include calculation of the steady-state response with thermal resistances and calculation of the transient response with differential equations.

Simulation with CAD tools can perform finite element analysis to confirm the two-dimensional and three-dimensional heat flows.

Experimentation uses expected and estimated heat flows within the physical and mechanical configuration of the equipment to determine heat transfer and operating temperatures. You can simulate the equivalent power dissipation from the circuitry with load resistors or heaters. If the environment is warm, you can place the equipment case inside an oven and simulate the hot external sources. Monitor and record temperatures at various levels, from component to external case. From these data, you can plot the transient and steady-state responses.

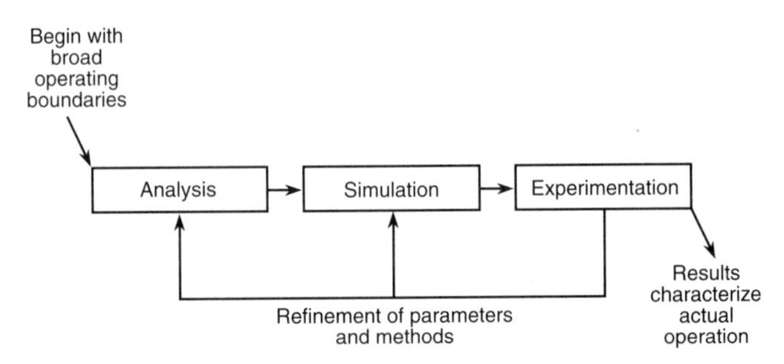

Fig. 10.25 The iterative nature of refining cooling system design.

10.14 CASE STUDIES

Four case studies illustrate several approaches to cooling electronic equipment. From these examples, you may see that natural air convection and forced air convection are the most common mechanisms for cooling.

10.14.1 *Magnetic Tape Unit*

During the 1980s the U.S. Air Force needed some way to quickly change software parameters within the EF-111A aircraft. I designed a custom tape unit to download data from a desktop computer onto a magnetic tape. (See the case study in Chapter 2.)

The custom tape unit, the key to speeding up the process, had to be fairly rugged even though it would usually reside in an office environment. It comprised an electronics module for controlling the tape cassette, a circuit board for communicating with the host computer, and two power supplies (see Figs. 2.16 and 2.17).

The electronics module and power supplies had a maximum operating temperature of 70°C (158°F). However, I chose 60°C (140°F) as the maximum operating temperature to allow for some margin of error. The maximum ambient temperature for the environment surrounding the unit was 40°C (104°F). Radiation and natural convection cooled the external surfaces of the enclosure, while conduction and natural convection transferred heat internally. The enclosure had to seal out dust and sand, so fans were ruled out. In a cramped office or field deployment, papers and other equipment were expected to be placed on top of the unit; therefore, the top and bottom surfaces of the tape unit could not be counted on to transfer heat. For the analysis and experimentation, I initially assumed a worst-case scenario of continuous service. The power dissipated totaled 24 W, broken down as follows:

- Electronics module: 12 W
- +28-V power supply: 10 W (28 V at 1.8 A and 80% efficiency)
- +5-V power supply: 0.9 W (5 V at 300 mA and 40% efficiency)
- Circuit board: 1.5 W (5 V at 300 mA)

Thus far, I have specified the thermal environment of the system:

1. Operating temperature range: 0°C to 40°C (32°F to 104°F)
2. Mechanisms: conduction, natural convection, and radiation
3. Thermal paths: sides of the enclosure
4. Total heat transfer: 24 W, worst case

Next, I used an analytical model to predict the maximum temperatures for the base plates of the electronics module and the +28-V power supply for an ambient temperature of 40°C (104°F):

$$Q = \alpha A_c \Delta T_s + \epsilon \sigma A_e (T_s^4 - T_\infty^4)$$
$$Q = 24\ \text{W}$$
$$\alpha = 10\ \text{W/(m}^2 \cdot {}^\circ\text{C)}$$
$$\epsilon = 0.7$$
$$A_e = 78\ \text{in.}^2 = 0.05\ \text{m}^2$$
$$A_c = 0.05, 0.1, \text{ or } 0.15\ \text{m}^2$$
$$T_\infty = 40^\circ\text{C}$$

I used three different areas, A_c, to calculate cooling from natural convection. The first area, 0.05 m^2, represented a flat plate; the other two areas, 0.1 and 0.15 m^2, represented fins. The radiative area, A_e, remained the same at 0.05 m^2 regardless of the convection area. The model gave the results found in Table 10.8.

Table 10.8 Calculated temperatures for different types of heat sinks ($T_n = 40^\circ$C)

		Temperature (°C)	
Radiator surface	**Convection area (m²), A_c**	**Surface, T_s**	**Differential $\Delta T_s = T_s - T_n$**
Flat panel	0.05	71	31
Fins with double the surface area	0.1	59	19
Fins with triple the surface area	0.15	54	14

To confirm the theoretical analysis, we conducted two experiments to study thermal dissipation from the enclosure. Resistors mounted on flat, polished aluminum plates simulated the electronics module and the +28-V power supply. Styrofoam sheet 2.5 cm (1 in.) thick insulated the top of the enclosure to simulate conditions that would prevent thermal transfer. *These are deceptively simple experiments but excellent for simulating real thermal dissipation.*

The first experiment had a flat back panel. An attached resistor simulated the electronics module by dissipating 14 W (26.5 V across 50 Ω), and another attached resistor simulated the power supply by dissipating 10 W (31.6 V across 100 Ω). The second experiment had cooling fins attached to the back panel. The cooling fins increased the cooling area by a factor of 2.2, to 0.111 m^2 (172 in.^{2}). The two experiments gave the results found in Table 10.9.

Table 10.9 Experimental results for temperatures of different types of heat sinks

Radiator surface	Surface, T_s	Temperature (°C) Ambient, T_n	Differential $\Delta T_s = T_s - T_n$
Flat panel	51	25	26
Cooling fins	46	26	20

Obviously, the experimental results gave a lower final temperature and a smaller differential than the analytical results. Succeeding analysis found that the electronics module and +28-V power supply would need to be operational for only 5 s at a time, corresponding to a duty cycle of less than 1%. Therefore, no cooling fins were needed for the magnetic tape unit, as the average power dissipated was around 2 W; the corresponding temperature rise was about 3°C (6°F).

Scope: process or program
Lessons learned:
1. Experimentation is necessary to confirm analysis.
2. Understanding the actual operation and environment led to a simplified design.

10.14.2 *Commercial Tape Backup*

Consider another example of design philosophy for cooling a tape unit. Colorado Memory Systems (CMS) built a tape drive for backing up hard disks (Leibson, 1989). CMS incorporated a cheap motor to drive a fan that cools the tape cartridges. The motor and fan run only during a backup operation, when cooling air is really needed. By doing this, CMS designed an inexpensive cooling system, extended its reliability by running only on demand, and reduced the average acoustical noise from the unit. In fact, the cooling fan is an audio cue that backup is in progress—a feature! (This is an example of feedback explained in Chapter 4.)

Scope: system or enterprise
Lessons learned: Understanding the actual operation and operating on demand led to an inexpensive design :
1. Reliability was extended.
2. The cost of the cooling fan was reduced.
3. The average acoustical noise was reduced.
4. Fan operation became an audio cue for the tape backup operation.

10.14.3 *Rugged Computer*

Returning to the military arena, C3, Inc., used several heat transfer techniques to cool commercial off-the-shelf technology in a sealed enclosure. A thermal pillow sandwiched between the motherboard and the aluminum case acts as a liquid heat sink and transfers heat through natural convection within the perfluorinated liquid (Hinke, 1992).

Two fans circulate air within the enclosure. One fan creates airflow across the circuit board with the CPU. The other fan draws air from the disk drives through the channels of the surrounding fins that serve as an internal heat sink. The circulating air transfers heat to the internal surfaces of the enclosure; the heat is conducted through the case and dissipates into the ambient environment through radiation and natural convection. Fig. 10.26 shows the internal heat sink of C3's AN/UYK-85A 486 computer.

Fig. 10.26 Internal heat sink for C3's AN/UYK-85A 486 computer. Various forms of cooling can be used within a rugged computer. Note that this is a completely sealed enclosure; there are no fan exhaust ports. All heat is conducted to the outside case through internal fans, thermal pillows, and fins (photograph courtesy of 3M, Inc.).

Scope: system or enterprise
Lessons learned: Heat can be removed from sealed enclosures through a variety of means:
1. Heat sinks
2. Thermal pillows
3. Internal circulation

10.14.4 *Cooling for a Workstation (Boggs, 1994)*

NCR Corporation produced a workstation that uses two CPUs, two system buses, and a high-performance graphics bus. It can accommodate five hard disk drives and peripherals like CD-ROMs and tape drives. Furthermore, it has five expansion slots for add-in boards and 512 Mbytes of dynamic RAM. Fig. 10.27 is a photograph of the workstation.

All these circuits and electronic modules dissipate a lot of heat. The CPUs operate at 55°C (131°F) at an ambient temperature of 40°C (104°F), the maximum specified by NCR. The design team considered heat sinks, heat sinks with built-in fans, and heat pipes for cooling the CPUs. A heat sink with a built-in fan was rated at only 25°C (77°F), below the range needed by NCR. It would also have been noisy. Heat pipes were too expensive. Beyond these concerns, the other circuits needed cooling too.

Fig. 10.27 Workstation that required some careful analysis to cool all circuits. Peripheral cards and hard drives all contribute to the heat load (photograph courtesy of NCR Corporation).

The NCR design team settled on an exhaust fan that draws air through louvers and a baffle to cool the workstation. The fan is single-speed with a capacity of 143 m^3/hr (84 ft^3/min). The position of the fan and the baffle balances airflow so that both CPUs operate at nearly the same temperature. A separate fan, with a capacity of 59 m^3/hr (35 ft^3/min), cools the power supply. Fig. 10.28 illustrates the airflow through the workstation.

But cooling was not the only concern; NCR wanted a quiet workstation as well. This was accomplished by installing a perforated dome over the exhaust and a rubber gasket (Fig. 10.29). The domed exhaust cuts down noise in the air stream, and the gasket prevents transmission of the vibration from the fan to the enclosure cabinet.

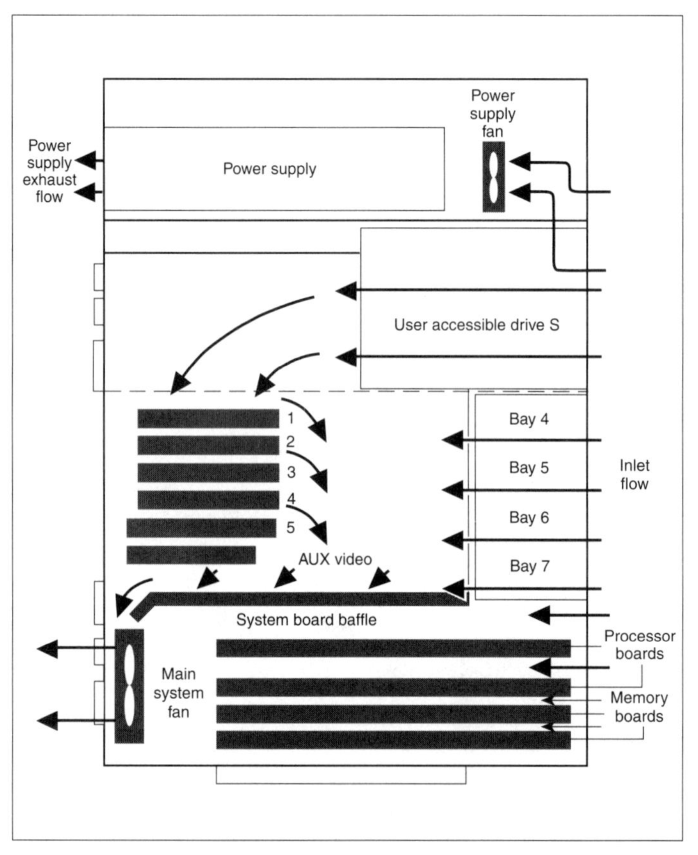

Fig. 10.28 The pattern of airflow through the workstation. Air enters through the front and the main system fan exhausts it. A separate fan cools the power supply (copied with permission from *Design News* magazine, ©copyright 1994).

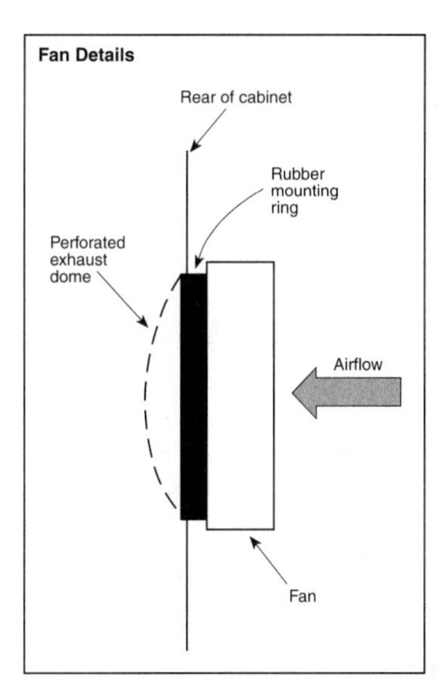

Fig. 10.29 Perforated dome and rubber gasket installed to reduce acoustical noise (copied with permission from *Design News* magazine, © copyright 1994).

Scope: system or enterprise
Lessons learned: Cooling of commercial equipment raises several concerns:

1. The microprocessors generate a lot of heat. An integrated fan and heat sink did not have enough capacity, and the fan was too noisy.
2. Heat pipes were too expensive.
3. An internal baffle balances the airflow over the microprocessors to provide sufficient cooling.
4. Acoustical noise was reduced by using a perforated domed exhaust and a rubber gasket to isolate the vibrations from the fan.

10.15 SUMMARY

Cooling affects reliability and mechanical design as well as the electrical design of a device. Not enough cooling, and the lifetime of the product is shortened. Too much capacity, and you have added extra expense, weight, and bulk to build your instrument.

The cooling system is the noisiest part of operation in most equipment. Quiet is better. You will often have to lower the acoustical signature of your design before it is acceptable to the customer.

You must account for cooling in every project from the beginning. Determine the heat sources and thermal paths. Understand the environmental conditions and operating temperatures. And provide for adequate cooling capacity if expansion in the electronic circuits is necessary.

10.16 RECOMMENDED READING

Incropera, F. P., and D. P. DeWitt. 1990. *Fundamentals of heat and mass transfer*. 3rd ed. New York: Wiley.
A thorough and thoughtfully prepared textbook that provides an in-depth basis for this chapter.

Scott, A. W. 1974. *Cooling of electronic equipment.* New York: Wiley.
An excellently written textbook that is still applicable today for most electronic systems.

10.17 REFERENCES

Boggs, R. N. 1994. Dual processors speed workstation. *Design News*, May 5, pp. 95–96.

Hinke, F. W. 1992. How C3 bags the heat in its military PCs. *Military & Aerospace Electronics,* July, pp. 32–33.

Incropera, F. P., and D. P. DeWitt. 1990. *Fundamentals of heat and mass transfer.* 3rd ed. New York: Wiley.

Leibson, S. H. 1989. The KISS factor. *EDN*, June 15, p. 18.

Matisoff, B. S. 1990. *Handbook of electronics packaging design and engineering.* 2nd ed. New York: Van Nostrand Reinhold.

Scott, A. W. 1974. *Cooling of electronic equipment.* New York: Wiley.

Simpson, C. 1991. The fundamentals of thermal design. *Electronic Design*, September 26, pp. 95–100.

Swager, A. W. 1990. Methods converge to cool fast and dense circuits. *EDN*, December 6, pp. 162–68.

11

SOFTWARE

If architects built buildings the way programmers build programs, then the first woodpecker to appear would destroy civilization.
—Anonymous

11.1 WHY SOFTWARE IN A BOOK ABOUT ELECTRONICS?

Software is pervasive in electronic products. Appliances such as televisions, videocassette recorders, remote controls, microwave ovens, sewing machines, and clothes washers all have embedded microcontrollers, and "software accounts for 50–75 percent of the cost of a microcontroller project" (Gilmour, 1991, p. 37). Automobiles have microprocessors and software built into engine controls and braking systems; even traffic signals are complex systems of distributed processors and software. Other familiar systems that contain embedded software include cellular telephones, the telephone system, automatic cash terminals, cash registers, and environmental controls.

Software is the major cost in many projects. IBM Federal Systems conducted a cost study on a U.S. Army program and found that software cost $8 billion, or half of the total life-cycle cost of $16 billion (Plansky, 1993). Elsewhere the "[software] maintenance phase . . . generally accounts for 60–80 percent of the time and money spent on a program, a far larger share than for hardware" (Murphy, 1990, p. 44).

Unfortunately, problems with software are rampant. Boehm (1981) has studied software productivity and divided it into 14 categories. The most problematic are personnel capability (insufficient capability can extend development by as much as 400%) and product complexity (a complex project can extend development by more than 200%). Generally, poor management of the development effort for software results in escalating field repair and maintenance costs as well as damaged reputation or even bankruptcy.

Leveson and Turner (1993, p. 38) write, "A common mistake in engineering . . . is to put too much confidence in software." Maybe the relative ease of producing and changing code fosters this kind of confidence. In this chapter, I will cover general methods to improve software in the following areas:

- Code generation
- Reliability
- Maintainability
- Correctness

But these techniques are only building blocks in the process to produce software. Reliable and correct software is not inherent in methods; it's the *process* that combines these methods that ensures quality. Your process must plan, measure, and document the development of software.

Within the process, the primary issue is communication between development team members, for engineers performing future modifications, and for satisfying the demands of liability. This chapter will concentrate on good communications and emphasize that it's not *what* you do but *how* you do it.

11.2 TYPES OF SOFTWARE

Software is found in many different types of systems. They include real-time control, data-processing systems such as payroll, and graphical systems such as games and CAD. In this chapter, I focus only on embedded, real-time software and the processes to develop it.

11.2.1 *Algorithms*

Software implements the intent of the designer through lists of instructions, or recipes for action. These lists and recipes are made up of algorithms, which may be very complex or as simple as an equation or a continuous loop of instructions. Algorithms describe the general actions to be taken and consequently are independent of the specific programming language. They are the first thing specified in the design of a program and have the greatest effect on the utility, success, and failure of the software. (Data structures provide another way of designing the processing architecture. They are more applicable to handling huge amounts of data. Most embedded control is better suited to preparing algorithms and letting the data structures sort themselves out.)

You should make a habit of collecting algorithms for your future programming efforts. Be sure to understand each algorithm, its limitations, and its boundary conditions. Jack Ganssle (1992, p. 4) writes, "It's ludicrous that we software people reinvent the wheel with every project. . . . Wise programmers make an ongoing effort to build an arsenal of tools for current and future projects. . . . Make an investment in collecting algorithms for future use. When a crisis hits there is no time to begin research."

11.2.2 *Languages*

Software has many applications within embedded systems:

- Firmware
- Peripheral interface and drivers
- Operating system
- User interface
- Application programs

These applications may use a variety of languages, such as assembly language, Basic, C, C++, and Ada.

Assembly language is peculiar to the processor architecture and closest to the hardware in the processor; it can control individual bits and set registers directly. At the same time, it is the most tedious to program because it requires steadfast attention to exacting detail. Assembly language is best suited for small, simple projects that need minimum memory, the highest execution speed, and precise control of peripheral devices.

High-order languages such as Basic and C make programming easier than assembly language by handling the nitty-gritty details and providing some measure of structure and readability. They are better for larger, more complex projects because they relieve the chore of handling low-level details and make programming faster and easier, but they require more memory and execute the code more slowly. Ada is an example of a high-order language being used in complex projects such as avionics and even medical devices.

Beyond high-order languages, object-oriented programming (OOP) is easing complexity in some tasks. Also, computer-aided software engineering (CASE) tools are beginning to structure and maintain specific tasks to create more reliable software.

11.2.3 *Methods*

Whatever language you choose, your objective will be to reduce complexity and improve understanding of the software. You can use one of several different levels of design architecture: structured, objected-oriented, or CASE. Within each of these disciplines are a number of programming tools.

Webb (1992, p. 37) writes, "Structured design is having a strategy before you start to code. Figure out what you want your program to do, identify its logical parts, and define how those parts interact before you write your first line." Plan and structure the software so that it can be debugged and tested easily. Small modules with clear operational flow will help. Beyond these simple rules, OOP can help by incorporating data abstraction, information hiding, and modularity to aid structured design. Finally, CASE tools provide a blend of environment, tools, and language to further structure and manage the design of the software system.

Tools available for the programming task include compilers, disassemblers, debuggers, emulators, monitors, and logic analyzers. Operating systems and software libraries also ease the task of programming.

11.2.4 *Selection*

The selection of a particular language depends on management directives, the knowledge and expertise of the software team, the hardware, and available tools. Team members' familiarity with the selected language is an important factor because it directly affects productivity.

The speed and data path width (8, 16, 32, or 64 bits) of the processor interact with the software to constrain both function and performance. The amount of memory (RAM and ROM) limits the size of programs. Architectural features such as coprocessors, peripherals (analog-to-digital conversion, timers, pulse-width modulators, interrupt handlers), I/O communications, power-down modes, and the level of integration all affect the choice of language.

Surprisingly, the choice of the language may also depend on the manufacturer. Here are some questions you should ask: Does the vendor provide reasonable documentation and support? Does it provide toll-free telephone support and knowledgeable application engineers? Does it have liability support for life-critical systems?

11.2.5 *Purchase*

Commercial software has utilities and libraries that usually save you from reinventing the wheel. Purchase the software after you have defined your software requirements and surveyed vendors for availability, reputation, and experience. Here are some of the qualifications you may require from a vendor:

- Acceptance testing
- Review of vendor's quality assurance
- Verification testing
- Qualification report

Furthermore, you may want to require documentation from the vendor that shows good process, such as the following:

- Requirements specification
- Interface specification
- Test plans, procedures, results
- Configuration management plan
- Hazard analysis

Don't buy cheap software tools just to save money! You will lose much more money in the long run from wasted time forced by delays and inadequacies of the cheap tools.

11.3 TRADITIONAL SOFTWARE LIFE CYCLE

Software development was once viewed as sequential. The results of one activity flowed into the next. Fig. 11.1 illustrates this view of software development; it's called the *waterfall model.*

While software rarely, if ever, develops according to the waterfall model, the model will acquaint you with the necessary components of the development cycle. Overseeing each step of the model is quality assurance to help you develop reliable and useful software. First, you bound the software problem and project during the concept and analysis stages. The requirements tell *what* the software does, while the design tells *how* the software does it. Programming is the coding stage, in which you actually implement the software program. Testing and verification then ensure that the program performs according to the requirements. Once you have written and released the software, it enters the maintenance phase, in which you update, correct, and modify it for future applications.

For more in-depth study, the recommended text by Pressman (1992) covers these methods in much more detail than I could even hope to introduce in this section.

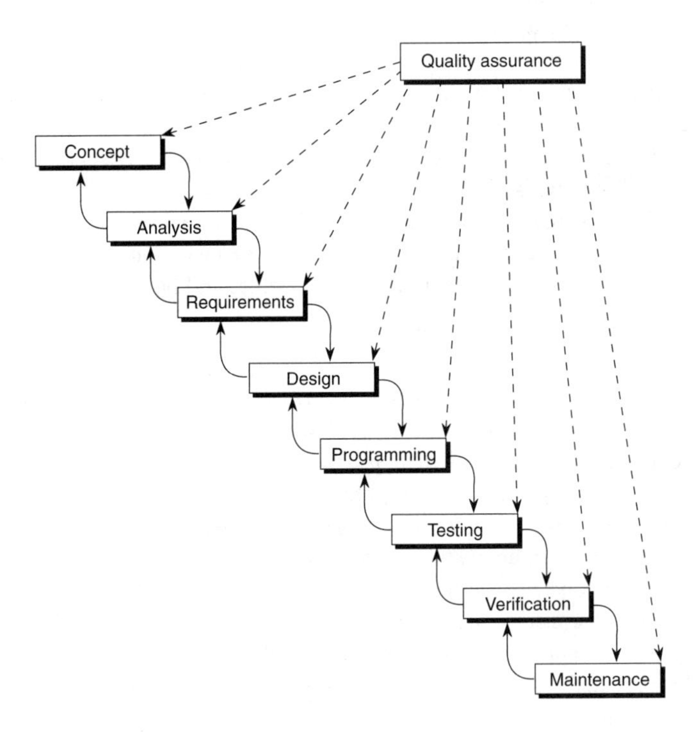

FIG. 11.1 The waterfall model of software development.

11.3.1 *Quality Assurance*

Methods do not guarantee reliable software. They are necessary but not sufficient. Quality assurance is really an attitude toward producing useful, reliable software rather than a collection of methods. With that in mind, you will find that what follows are some examples of methods that may be appropriate for implementing a good process (or practices and procedures).

First, you need to classify your system and its software according to any relevant standards. Second, you need a software development plan. (Writing software without a development plan is like building a house without blueprints. You can construct a shack or shed without plans, but not a large building.) That plan should include hazard and fault-tree analyses if the software implements life- or mission-critical functions, as well as configuration management, documentation, and traceability.

Requirements change and systems evolve, but a software development plan can and should accommodate change. Failure to plan is a plan for failure. Should your product ever encounter a review by a government agency, such as the U.S. Food and Drug Administration reviewing a medical instrument, the reviewers will examine the design and validation documentation for orderly process. A good plan will prevent outright rejection.

The development plan must specify configuration management for standard data formats, any libraries of subroutines, and any code produced. It catalogs developments and changes in the concept and design during the implementation and testing of the software programs. Configuration management ensures that the current or correct version is released.

Built into the plan must be accountability or traceability from the concept through requirements and design to test and maintenance. You must log and document the requirements, any hazards, design, testing, and bug reports.

11.3.2 *Concept and Analysis*

Generally you start out with a problem described in human language. Sometimes you will get a precise mathematical specification that makes translation into code straightforward, but don't count on it. Certainly you will have to model the concept with mathematical formulas and algorithms. Each problem is unique, and I cannot give you any recipes for converting concepts to formats that may be analyzed, translated into requirements, and written into code.

You must analyze the software within the framework of the system description and concept; the software can't be divorced from the system or its hardware. Concentrate your analysis on the interactions and interfaces between the software, the hardware, and the data inputs—this is where most problems occur. Your analysis should include the following:

- Technical trade-offs
- Performance and timing
- Human factors
- Hazard or risk analysis
- Fault-tree analysis

Hazard or risk analysis identifies problems that can result from faulty or incorrect software. What you are trying to do is find areas of operation that must have one of several modes, in descending priority: fail-safe, decreased capacity, identified and labeled, and inconsequential.

Fault-tree analysis is a graphical model of sequential and concurrent events that lead to failures or problems. You use it by simulating failure and then working down the tree for operational impact.

11.3.3 *Requirements*

The requirements follow both concept and analysis and reduce the abstract intentions of the customer into realizable constraints. The requirements tell *what* the software does. They include the standards that the software must adhere to, the development process, the constraints, and reliability or fault tolerance.

Standards can drive the amount of effort expended in developing. If you decide to follow standards that are inappropriate or tighter than necessary, you will waste time and effort. Pick carefully those that apply and understand all that they demand of you. Some standards, like DoD-STD-2167A in the United States can help you by providing a framework for development. Table 11.1 lists some example standards that may fit your software development.

Requirements for the development process may call for several, successive specifications, such as the following:

- General specification
- Functional performance specification
- Requirements specification
- Design specification

Table 11.1 Some standards that apply to software development

Standard	Description
ISO 9000	International standards for quality
DoD-STD-2167A	Software engineering
DoD-STD-2168	Software quality evaluation
ANSI/IEEE STD	
731	IEEE Standard for Software Quality Assurance Plans
828	IEEE Standard for Software Configuration Management Plans
829	IEEE Standard for Software Test Documentation
830	IEEE Standard for Software Requirements Specifications

Within these requirements you need to consider the system constraints such as memory and timing margin, hardware, communications and I/O, and execution speed. Will the selected processor and associated hardware support the requirements? Can the software use these resources and satisfy the requirements?

The requirements may also specify desired reliability, fault tolerance, and failure modes. The hazard and fault-tree analyses can help set the requirements. Reliability may be specified in the frequency and type of errors (bugs) found during testing or maintenance. Fault tolerance specifies how the system responds to failures.

11.3.4 *Design*

The design outlines the path through the system constraints laid down by the requirements. The design tells *how* the software does it. The design defines the preliminary version of the software operation and its procedures, the communications, the development tools, how the software will optimize performance, and how it will manage the memory.

The design specifies how the software will fulfill the requirements. You need to consider these software elements in preparing the design:

1. System preparation and setup
2. Operating system and procedures
3. Communications and I/O
4. Monitoring procedures
5. Fault recovery and special procedures
6. Diagnostic features

As always, the interfaces demand the most attention. Communications have several different formats and concerns:

- Polled I/O
- Interrupt I/O
- Synchronization between tasks
- Intertask signaling
- Combinations of polling and queuing to avoid overrunning events

The design can specify algorithms and techniques that optimize performance. The design can also specify the management of memory within an electronic instrument. The design may reuse modules and libraries in an effort to improve productivity and reliability, but even this course of action requires careful management of the modules.

11.3.5 *Programming*

After all the analyses, requirement specifications, and design plans, you can get down to actually writing code. Two methods of programming—one in assembly language, the other in high-level languages—are currently available to you. CASE tools are on the horizon; they do not yet have general application; however, I would encourage you to research them for each application, as they will become more useful and sophisticated.

An *assembler* takes assembly language mnemonics prepared in a text file and generates an object file, an intermediate step in preparing the final code. A *linker* takes object files, links them to other object files, and produces the binary machine code in a file that maps directly into memory. You may then download the hex file directly into RAM or *burn* the binary machine code from the hex file into a PROM, UVPROM, or EEPROM. Fig. 11.2 illustrates the steps in assembly or compilation of programs.

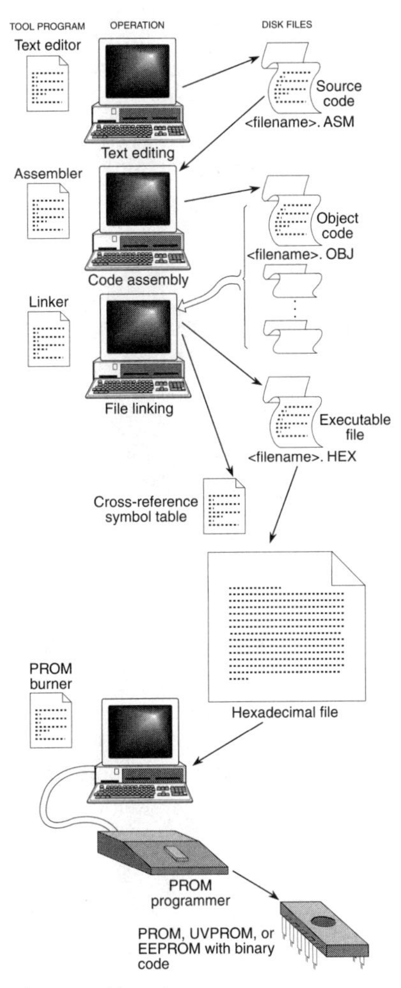

Fig. 11.2 Steps in programming assembly code.

An assembler can produce a cross-referenced symbol table, can include files, and can provide conditional assembly, macros, and structured assembly statements (Gilmour, 1991). A cross-referenced symbol table relates each symbol to the line number of its definition and to each line number where it is used. File inclusion takes the contents of another file and inserts them in specified locations; this avoids rewriting variables and code between files. Conditional assembly assembles code according to an operand expression. Macros are code sequences that are used as in-line subroutines. Structured assembly statements are preprogrammed conditional statements like IF-THEN-ELSE and FOR, DO, and WHILE loops.

Why should I describe assemblers and assembly language? I spent the time because a surprising number of applications need subroutines written in assembly language to achieve performance. High-level languages have compilers that perform all these housekeeping chores and free the programmer from concern for the huge number of details; however, you may find that you must link an assembly language routine to a high-level program to achieve the necessary timing or performance. This is more often true for specialized chips like digital signal processors or microcontrollers than for larger, general purpose CPUs, for which optimizing compilers can produce very compact and efficient code.

Whatever language you choose, you still need to write your programs in a clear and readable format. The source code still is your primary source of documentation. If you use a clear and complete format, all your programs can be more understandable. A good design format uses a consistent header and appropriate comments to describe the function of the code. A header should include the author's name, the date of code creation, revision dates, the document version number, a brief description of the module's function, a brief description of the module's input and output, exceptions, and definitions of the variables used (Fig. 11.3).

Remember the engineering notebook described in Chapter 3? You should use it to log and document all programming and testing. Accurate records will aid later efforts to debug and maintain software, as well as decrease your liability in questions of negligence.

The host system and software integration tools will constrain your effort in programming. You can begin coding on the host system by simulating the program. While slow—because the host has to read, interpret, and imitate each line of code—simulation can allow some concurrent experimentation while hardware is being prepared. A prototype of the final product will further the development effort by providing a much more realistic environment for the software. The final step in development is an engineering or production model that allows testing and verification of the software.

```
;*********************************************************************
;   This is MEDTRON.ASM.                          Author's name: ________

;   CREATED on 7/3/90 from DEVELOP.ASM.
;   REVISED on 7/17/90; massive changes and elimination of code.
;   REVISED on 8/8/90 to add XPROG, CNTPRG, and STARTSE4.
;   REVISED on 8/9/90 to add overflow flag, OVF, rearrange the
;                     position of XPROG, and added MODULATE.
;   REVISED on 8/23/90 to add algorithms for timing corrects: CORRECT1,
;                     CORRECT4, CRTSE1, and CRTSE4.
;   REVISED on 8/24/90 to correct timing values and added TMP flag.
;   REVISED on 8/27/90 to correct timing values.
;   REVISED on 10/10/90 to disable interrupt during programming.

;   FUNCTION:
;         These subroutines are the code for the neurostimulator
;   interface for the Medtronic transmitters.  They program the
;   electrode combinations and generate the desired stimuli -
;   pulses and sequences.  They set frequency, pulse width, pulse
;   amplitude, and electrode polarity.

;   INPUT:
;             Data from variables in the MAIN program.
;   OUTPUT:
;             Stimulus control to selected neurostimulator transmitter.

;   ROUTINES CALLED:
;   SE1:    drives the single channel transmitter.
;   SE4:    drives the four channel transmitter.
;   XTREL: drives the four channel transmitter.
;   ITREL: drives the implanted, powered receiver.
;
;  *******************************************************************
```

FIG. 11.3 An example of a header for a listing of a software module.

11.3.6 *Testing and Verification*

Testing can and should take various forms to increase the chance of catching bugs and verifying correctness. Testing includes internal reviews, black box testing, white box testing, and alpha and beta testing. You really should test and verify software in the context of the entire system because the interaction of hardware, environmental inputs, and software produces complex responses.

Internal reviews by colleagues examine the correctness of your software and can ferret out mistakes and errors in logic. "More than 50 percent of the discrepancies are found by code inspection or code audit" (Haugh, 1991, p. 223).

Black box testing uses typical scenarios to ensure that input and output interfaces are functioning correctly. It ensures the integrity of the information flow into and out of a module without concern for what happens within the software. Most testing is black box testing.

White box testing exercises all logical decisions and functional paths within a software module. It requires an intimate knowledge of the module's software so that you can design the tests to reach all operations and functions. White box testing is far more likely to uncover bugs in special-case processes and seldom-used logical paths. Unfortunately, exhaustive white box testing is impossible for most modules. Some simple programs would have to run for thousands of years to test every possible combination (Pressman, 1992).

Alpha and beta testing are forms of black box testing that usually exercise the complete (or nearly complete) software in actual operating environments. Both naive and expert users can generate inputs and results that a designer and programmer would not anticipate. An alpha test allows the programmer to collaborate with and observe people using the software. A beta test, in contrast, is performed by users isolated from the programmer. A necessary component of beta testing is recording and reporting the results consistently.

All testing must measure and record progress, productivity, complexity, and correctness. The very simplest metric is program size, measured in lines of code. Other measures count and classify the number of logical decisions and define complexity. Analysis tools such as debuggers, logic analyzers, and in-circuit emulators can record when and how often specified operations occur.

11.3.7 *Maintenance*

Once you have tested, verified, and released software, you have to maintain it. As with testing, you cannot separate software maintenance from system concerns.

Proper maintenance requires you to control the software configuration. This includes reports, measurements, personnel, costs, and documentation. You should also have a plan for releasing software upgrades. The aim is to achieve consistency and continuity in your product.

You will need a standard form for bug reports and a plan for corrective action. Each bug report should include the following (Dearden, 1992):

1. Bug number
2. Bug description
3. Severity
4. Date
5. Target device configuration
6. Person responsible for fixing the bug
7. Current status

Each report should become an entry in a database. Beyond the initial reports, a test lab simulator or prototype of your product will help replicate failures, isolate problems, and correct errors more quickly.

11.4 MODELS, METRICS, AND SOFTWARE LIMITATIONS

The essential components of software development interact in different ways in different process models. The process models include waterfall, prototyping, and spiral (Pressman, 1992). These models help plan the development of a project and estimate the effort for it.

11.4.1 *Models*

The waterfall model (Fig. 11.1) is the traditional view of software development. It develops each component sequentially and usually does not iterate through more than two components at a time. In reality, however, software seldom develops according to this model. Part of the problem is that requirements evolve because either the customer doesn't know all of them or the application changes or expands. The other problem is that the software is available only late in the development schedule.

Prototyping the software accommodates the problem of changing requirements and makes a subset of the software available early. The essence of prototyping is a quickly designed model that can undergo immediate evaluation, as shown in Fig. 11.4. A prototype may take one of three forms: (1) a paper model or computer-based simulation, (2) a program with a subset of functions, or (3) an existing program with other features that will be modified for your new product (Pressman, 1992). Its biggest weaknesses are that thorough testing and documentation are easily forgotten. The designer tends to rush the product to market without considering the long-term reliability, maintenance, or configuration control. Another problem is *creeping featurism*—the customer voices new desires after each evaluation, and the project effort balloons.

The spiral model (Fig. 11.5) presents an incremental approach to development that provides a combination of the classical waterfall model and prototyping. Each cycle around the development spiral provides a successively more complete version of the software. The model allows flexibility to manage requirements and control changes.

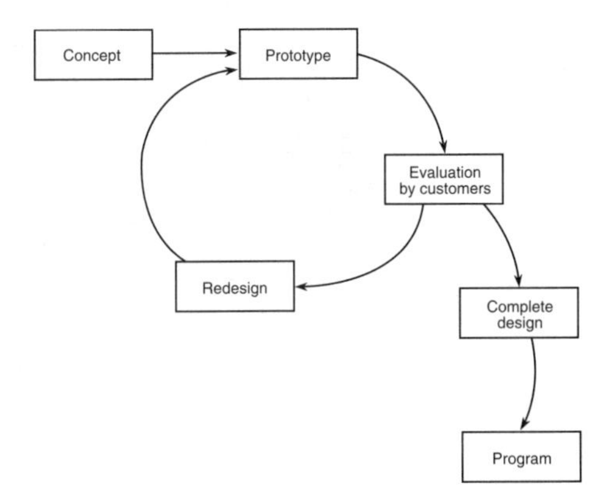

FIG. 11.4 Prototyping model for software development.

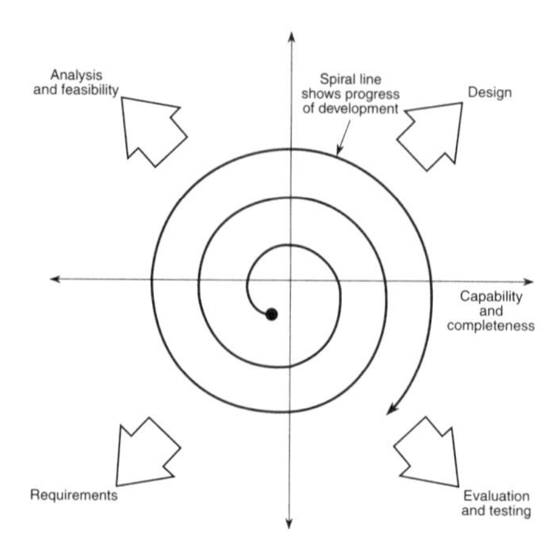

Fig. 11.5 Spiral model of software development.

11.4.2 *Metrics*

Each process model needs metrics to assess performance and progress of the software. Without them, the development team has no objective understanding of the completion of the software at each stage or of its usefulness. Furthermore, software size, development time, personnel requirements, productivity, and the number of defects all interrelate, and metrics can help define those relationships (Putnam and Myers, 1992).

Most software projects are driven by cost and schedule; when the development meets or exceeds a cost milestone or calendar deadline, the software is declared finished. Unfortunately, these are very poor metrics for producing quality software. Table 11.2 lists some metrics that are more suitable for monitoring software development.

Table 11.2 Some metrics that quantify software development

	Type and Orientation of Metric			
Metric	**Technical**	**Quality**	**Productivity**	**Human interface**
Correctness	√	√	—	—
Reliability	√	√	—	—
Efficiency	√	—	—	—
Maintainability	√	—	√	—
Flexibility	√	—	√	√
Testability	√	√	√	—
Portability	√	—	√	—
Reuse	√	√	√	—
Utility	—	√	—	√
Size	—	—	√	—

Note: √ = applicable; — = not applicable.

Measured results provide records of actual performance that can be used to estimate the development time and effort for new projects. First, establish the scope of the project. Second, use past measurements and results to estimate the individual pieces and modules. Remember that personnel capabilities; hardware and software resources; and tools for planning, simulation, and debugging all affect your estimate. To rely on your estimates, you need to track the metrics honestly and record them carefully and consistently. Finally, commitment and involvement by senior management is necessary to ensure good estimates (Putnam and Myers, 1992).

11.4.3 *Process*

Ganssle (1993, p. 85) writes, "Code that cannot be maintained is worse than no code." Process incorporates models of software development to generate useful, reliable, and maintainable software. A good process keeps statistics for feedback and improvement of the software development. In addition, software tools are integral to the software process; don't change them in midstream.

The Software Engineering Institute has defined five levels of process maturity (Card, 1991):

1. *Initial:* chaos or ad hoc process
2. *Repeatable:* design and management defined
3. *Defined:* fully defined and enforced technical practices
4. *Managed process:* feedback that detects and prevents problems, a control process
5. *Optimizing process:* automating, monitoring, and introducing new technologies

Most of us are somewhere in one of the first two levels.

11.4.4 *Software Limitations (Walker, 1994)*

Software has some identifiable limitations:

1. Not all problems can be solved.
2. Specifications cannot anticipate all possible uses and problems.
3. Errors creep into development in a number of ways.
4. Software simulation can predict only known outcomes.
5. Human error can occur in operating the software.

Walker (1994, p. 16) writes, "When problems begin as poorly defined ideas, then much of the work of writing specifications is to clarify carefully what is meant." Specifications fall prey to ambiguous language, assumptions based on experience, inconsistencies, and omissions. Careful review can find and eliminate some problems with assumptions and omissions. CASE tools help with consistency.

Errors in designing software arise from omission, inconsistency, numerical round-off and approximation, typographical errors, and distributed interactions between computing elements. Careful testing, review, verification, and maintenance can reduce these.

Relying on simulation to verify a software design is a trap. Simulation is based on current (and always incomplete) knowledge and can only roughly imitate reality. Walker (1994, p. 128) writes that simulation, (1) "can help in identifying conclusions when models are based upon well established assumptions" and (2) "can provide insights about assumptions when conclusions can be compared with experimental data."

Human error plays a big role in software failure. An operator can pass wrong data or use the software incorrectly. Furthermore, human error has a tendency to compound into a series of consecutive errors that are very difficult to foresee.

All these possibilities illustrate the limitations of software. Beware!

11.5 RISK ABATEMENT AND FAILURE PREVENTION

One of the goals of measurement is to improve the software quality. Once you have established metrics for performance and quality, you can plan for robust software that is fault tolerant. I will focus on safety and reliability, since these usually are real concerns for embedded software systems and since good references already exist for improving productivity (Putnam and Myers, 1992). In particular, I will discuss plans for fault avoidance and reducing the problems resulting from errors and bugs in the software. But once again the issues are bigger than the software alone. "Safety is a quality of the system in which the software is used; it is not a quality of the software itself" (Leveson and Turner, 1993, p. 39).

11.5.1 *Issues*

Before you develop a safe and reliable system, you need to understand the types of faults and failures and how they occur. First, consider the nature of software itself. Software doesn't have inertia or friction; it is nonlinear, discrete, and chaotic. Software doesn't conform to conventional or natural boundaries of stability; it can jump from zero to infinity in a single step. Furthermore, the huge number of possible logic paths within the software leads to confounding complexity.

According to Littlewood and Strigini (1992, p. 62), "software faults are generally more insidious and much more difficult to handle than are physical defects. The problems essentially arise from complexity, which increases the possibility that design faults will persist and emerge in the final product." Moreover, failures do not always have simple, single-point sources. "Accidents are seldom simple—they usually involve a complex web of interacting events with multiple contributing technical, human, and organizational factors" (Leveson and Turner, 1993, p. 38). Safety and reliability are system concerns, not concerns of software alone.

11.5.2 *Development Plan*

Software faults affect safety, fault tolerance, and reliability of the system and product. A careful plan reduces the risk from software failure; for many products it could reduce insurance rates and help settle lawsuits (Sroka and Rusting, 1992). A plan for software should specify how to handle requirements, design and development, reviews, tests, and documentation.

Certain inputs shape the requirements; these include product goals, user approaches, market constraints, prototype results, and hazard or risk analysis. The difficult part is translating the input into specifications for safety, reliability, performance, timing, and the user interface. Care with the risk analysis will ease the translation (Boehm, 1991):

1. Identify the risk.
2. Assess the capability of your personnel.
3. Set priorities for criticality.
4. Prepare management and resolution strategies.
5. Monitor the risk during development.

Reviews of the requirements and code inspections of each module can help abate risk. You also need to review and test module interactions on the system level; this is most difficult and requires a great deal of thought, modeling, and testing. The case study of the Therac-25 at the end of this chapter, for example, shows just how difficult and critical these interactions can be.

Documentation reduces risk by tracking the changes to the requirements, eliminating informal changes that are forgotten, and improving communication. It improves the understanding of the software.

If any of these ideas are to work, upper management must show its commitment to the process by providing tool support, realistic schedules, and adequate finances. Here is a checklist of important steps that a manager may take (Leffingwell and Norman, 1993):

1. Develop a formal requirements document.
2. Perform a hazard analysis.
3. Review quality metrics.
4. Include the quality assurance team in the process.
5. Establish design reviews and a software monitoring plan.
6. Plan for validation and verification testing.

Furthermore, designers and programmers must be committed to establishing effective communications and working knowledge of the process. Within this framework of risk analysis, reviews, tests, and documentation, certain techniques will reduce risk. Follow these guidelines to reduce your risk (Putnam and Myers, 1992):

- Keep the size of the program as small as possible.
- Keep the design of the program as simple as possible.
- Lengthen the development time beyond the minimum.
- Keep the size of the team as small as possible.
- Invest in software tools, methods, and practices that improve the process.
- Add personnel only when new people can be used effectively.

11.5.3 *Safety and Reliability*

Certain techniques and methods can help you write safe and reliable software. First, develop a complete plan, as discussed above. Second, choose safety and reliability requirements that reflect the nature of the application. Next, understand the nature of bugs and errors, both their introduction and their removal (often removing a bug introduces an entirely different fault) (Musa, 1989). Finally, you have three avenues to improve reliability: mathematical formalism, tight management, and limiting the role of software.

Mathematical formalism can prove the reliability and safety of algorithms, but it is tractable only for very small algorithms. Tight management, thorough reviews and tests, and analysis of previous errors check for logical inconsistencies, ambiguities, and incompleteness. Gowen and Yap (1993) have shown that hazard analysis, fault-tree analysis, and event-tree analysis—in addition to inspections, boundary analysis, and testing—help in finding latent faults. Such static methods of verification, however, can't deal with errors in user intent (Hamilton, 1986). The third avenue is to limit the role of software within a system and to use hardware interlocks to increase safety.

Once you have chosen a method of establishing the safety and reliability of the software, you can use the following guidelines to write reliable software (Banks, 1990):

- Make each module independent.
- Reduce the complexity of each task.
- Isolate tasks from external influences, both hardware and timing.
- Communicate through a single, well-defined interface between tasks.

11.5.4 *Fault Tolerance*

Fault tolerance concerns safety and operational uptime, not reliability. It defines how a system prevents or responds to bugs, errors, faults, or failures. One issue of design philosophy you will have to settle is whether to prevent faults or to detect and remedy faults. Above I presented some methods to prevent faults; here I will discuss some ideas for detecting and responding to faults.

Digital systems and software are discrete, nonlinear systems that are not "well behaved." A single, simple operation or decision can drastically alter output. "A single incorrect character in the specification of a control program for an Atlas rocket, carrying the first U.S. interplanetary spacecraft, *Mariner 1*, ultimately caused the vehicle to veer off course. Both rocket and spacecraft had to be destroyed shortly after launch" (Littlewood and Strigini, 1992, p. 63). Obviously, some systems and products need fault tolerance.

Safety-critical systems can have catastrophic consequences during accidents that lead to equipment damage, environmental harm, injury, and even death. Consequently, such devices require fail-safe operation such as detection and correction of faults, fallback modes that limit equipment function, and proper notification of faults. Some examples include cyclic redundancy checks on data communications, check sums on blocks of memory to detect bit flips, and watchdog timers that reset the operation of the device.

Software may detect a fault by measuring a value that is out of bounds or inconsistent. A sensor may be producing a signal that is too large, for instance, or an operator may key in an alphabetic character instead of a number. Careful design of the software can help uncover some faults; furthermore, hardware, such as a watchdog timer, can detect a transient fault and help the software ride it out.

A *watchdog timer* is hardware that monitors a system characteristic to check the control flow and signals the processor with a logic pulse when it detects a fault. The timer expects a signal from the software to reset its count during a preset interval of time; if the software does not reset it, then it times out and generates the logic pulse that indicates a fault. A watchdog timer indicates that the software has hung up in one particular operation—such as an infinite loop. Be careful to select an address and software operation that ensures reset of the watchdog timer within the specified time interval when the software is working correctly (Nygren, 1993). On the other hand, resetting a microcontroller whenever the watchdog timer detects a fault may not be the best choice, because some peripherals may not clear properly; an interrupt service routine that is carefully designed will often suffice.

Once the system detects a fault, it must respond appropriately. Real-time systems can use roll-back recovery (reconstructing previous states) or roll-forward recovery (sort of like filtering out anomalies) that tolerates occasional bad outputs and detects the catastrophic ones (Nygren, 1993). Roll-forward recovery places bounds on responses. A system using a watchdog timer may use roll-back or roll-forward recovery or both to respond to a detected fault.

A systematic approach to fault tolerance incorporates both hardware interlocks and redundancy. An interlock limits the range of outputs that the software generates. In some systems, limiting the role of software is the only way to guarantee an appropriate level of safety. Redundant systems with different software survive faults by reducing the probability that an error will crop up simultaneously in both software programs. Walker (1994, p. 84) writes, "Redundancy can help increase the likelihood of detecting some types of errors and in guarding against hardware and software failures. . . . On the other hand, redundancy increases the cost and the complexity of systems, and redundant systems may contain more logical errors than simpler ones."

11.6 SOFTWARE BUGS AND TESTING

While the last section outlined plans to reduce the problems from errors and bugs, this section concentrates on preventing and eliminating bugs. You need to recognize how bugs get introduced, how to find them, and how to eliminate them. The most serious usually occur in the conceptual design, originate from incorrect assumptions, and are the most difficult to uncover.

11.6.1 *Phases of Bug Introduction*

Bugs can enter software in any of four stages (Table 11.3):

1. Intent
2. Translation
3. Execution
4. Operation

The severity of the problems that bugs present is usually related to how early the bugs are introduced. Errors in developing the requirements are almost always more costly and difficult to diagnose than those introduced later in the process. Hamilton (1986, p. 53) says, "The real problem with large systems . . . is that the user needs to understand the specific application problem before it reaches the state of a 'software problem.'" Leveson and Turner (1993, p. 39) write, "Most computer-related accidents have not involved coding errors but rather errors in the software requirements such as omissions and mishandled environmental conditions and system states." According to Sroka and Rusting (1992, p. 158), "Most of the weaknesses in software-dependent systems can be traced to ill-defined or incomplete system specifications and requirements."

Errors or bugs in translation tend to be less severe but can still produce some baffling problems. Actual operation can introduce errors also. The dynamics of interacting or multitasking modules can cause serious problems; the case study of the Therac-25 in Section 11.10.2 illustrates this. "Even the addition of novel features to a program may produce unexpected changes in existing features" (Littlewood and Strigini, 1992, p. 64).

Example 11.6.1.1 A colleague tells about an error in translation while developing a simulation for tracking radar targets. The simulation modeled radar performance and included sea clutter, a low-level and low-angle effect on radar operation. The requirements called for "sea clutter to cut off at 3° elevation." When the simulation did not work properly, he found that the programmer had integrated the sea clutter effects *above* 3°, not *below* 3°—the opposite of what he wanted.

Table 11.3 Common bugs in software (Van Tassel, 1978)

Category	Types	Comments
Intent	1. Wrong assumptions or misunderstanding	Insufficient grasp of the problem
	2. Problem definition	Correctly solving the wrong problem
	3. Sizing	Limits on operation too broadly or too narrowly defined
	4. Malice	Viruses
Translation	1. Incorrect algorithm	
	2. Incorrect analysis	
	3. Misinterpretation	
Execution	1. Semantic error	Failure to understand how a command works
	2. Syntax error	Failure to follow rules of language, failure to terminate statement or procedure
	3. Logic error	Wrong decision, branch to wrong label, omission of possible conditions, improper initialization or termination of loop
	4. Range error	Failure to predict the possible calculation, incorrect approximation
	5. Truncation error	Incorrect rounding of calculation
	6. Data error	Failure to predict the possible ranges of data or data types, not initializing variable, not initializing or setting size of array, wrong size of character string
	7. Misuse of language	Coding operation inefficiently or incorrectly
	8. Documentation	Wrong or mismatching comment
Operation	1. Changing paradigm	System operating in an unanticipated environment
	2. Interface error	Incorrect type or range of input or output
	3. Performance	Inadequate hardware resources or software timing
	4. Hardware failure	Failure or incorrect functioning of subsystem
	5. Human error	Possible compounding into multiple errors

11.6.2 *The Nature of Bugs*

Bugs or defects are integral to and created by the process of software development. They usually start early in design, and the number of occurrences is inversely proportional to the experience of the software developer. Bugs and complexity go hand in hand; real-time applications, multiple communications, new tools and processes, and the involvement of many disciplines in product development all help spawn bugs in software. While testing is necessary, a single method of testing alone is not good enough; the cheapest and easiest errors are found first. Finally, if you find more bugs than usual, then more bugs usually remain in the software (Leffingwell and Norman, 1993).

11.6.3 *Debugging*

Debugging is a process of building good mental models of the system function; it requires a systems approach (Castaldo, 1990; Lantz, 1990; Mittag, 1993). Often problems arise in the interaction between hardware and software, so you need to foster good communication between team members. Good documentation is just as necessary; some of the best documentation consists of comments in the source code that give the intent of the code. Finally, and most important, good design and programming will facilitate debugging (see Section 11.7).

Debugging is an art, and experience is often the best resource. Even if you are just starting out, however, these few steps will help you debug effectively and efficiently:

1. Collect your wits. Don't get stuck in a mental infinite loop.
2. Work on one problem at a time.
3. Examine the intent of the software module.
4. Outline the events that led to the fault.
5. Isolate the problem.
6. Develop a set of reasonable hypotheses.
7. Run tests likely to disprove each hypothesis to single out the real cause.
8. Record the results and your thoughts. You'll need them for documentation and to sharpen your debugging skills.

When the going gets tough try the following (Ziegler et al., 1992):

9. Take a break to interrupt the frustration of the mental infinite loop.
10. Have someone else look over the problem (but don't get too much help—a crowd only leads to confusion).

Several techniques are the stock-in-trade of software debugging; they are print statements, breakpoints, and conditional compilation. Each allows you to view intermediate values computed internally to a software routine. Print statements are the simplest, but they can clutter a program quickly and must be removed after you finish with them. A breakpoint stops program execution so that you can view what has happened in the software up to that point. Breakpoints can be complex or conditional upon logic decisions made by the software.

Software debuggers and simulators allow you to exercise and debug your program on the target hardware. Hardware tools also provide invaluable assistance for debugging software; they include in-circuit emulators (ICEs), logic analyzers, and oscilloscopes (Fig. 11.6) (Mittag, 1993). An ICE replaces the processor with a pod that simulates it, remains transparent to the ancillary circuitry, operates at the same speed, and gives access to the internal registers. An ICE is most useful for solving logic problems. A logic analyzer records and displays the digital values and their time of occurrence. It provides accurate timing information. An oscillo-

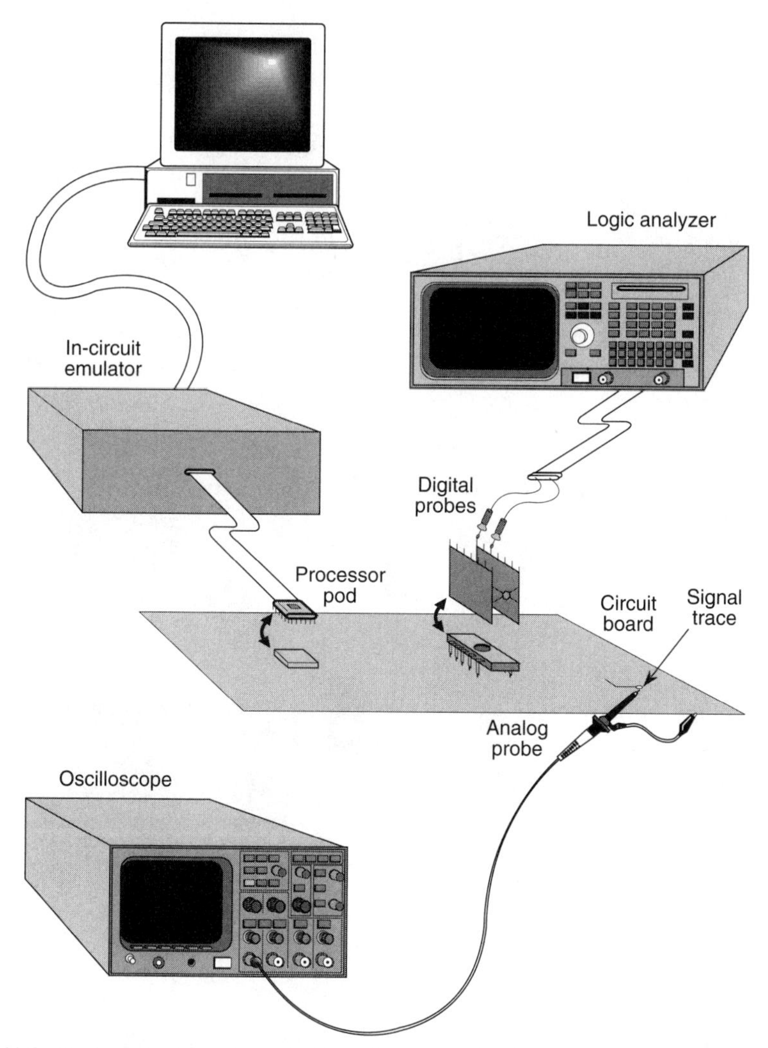

FIG. 11.6 Hardware for debugging embedded software: an ICE, a logic analyzer, and an oscilloscope.

scope measures and displays the real-time analog values of signals and is useful for debugging glitches, ringing, and metastability within circuitry.

11.6.4 *Inspections and Reviews*

Inspections and reviews increase the quality and reliability of your software because multiple pairs of eyes are more likely to catch errors. Scrutiny by peers also motivates a programmer to remove errors (Olivier, 1992). Informal "beer checks" and formal reviews complement each other to improve the development process for software, and they are much cheaper than fixing defects after customers find them.

Beer checks are an informal challenge to your peers to peruse your design and code and try to find defects. Make your design and code available along with a list of requirements. You owe a beer to anyone who finds an error. The relaxed atmosphere doesn't mean that the checking isn't valid.

Otherwise, formal code inspections will foster quality and productivity. The process for a formal code review includes the author of the software, a moderator, a recorder, and several reviewers. The moderator runs the review, the recorder documents all comments and recommendations, and the author and reviewers discuss the design and code. A formal review includes the following steps (Blakely and Boles, 1991):

1. Choose complex software modules. They have a greater likelihood of errors.
2. Prepare packets for the review that include these items:
   - A cover sheet
   - Listings
   - Module descriptions
   - Pseudo code
   - Design overview documents
3. Give the packets out 1 week before the meeting.
4. Limit the meeting to 1 hour—any longer and effectiveness plummets.
5. Use these rules to guide conduct and avoid personality clashes:
   - Do not critique coding style.
   - Identify problems; do not offer solutions.
   - Keep all comments civil, and focus on the code.
6. Have the recorder keep an inspection log, and have the reviewers follow a checklist like that in Fig. 11.7.
7. Follow up the review meeting within 2 weeks or 1 month, and discuss the changes implemented to accommodate the critiques.

11.6.5 *Testability*

Testability predicts whether existing faults will be revealed during testing and depends on reachability (the faults are executed), necessity (the data become infected), and sufficiency (the infected data propagate to output) (Voas et al., 1993). Testability helps verify the reliability of a system, increases confidence that your product or system will operate correctly, and reduces your liability.

Design	Yes	No	Comments
❑ Are assumptions correct?	❑	❑	
❑ Is the intent or purpose clear?	❑	❑	
❑ Does the code embody the intent?	❑	❑	
❑ Is any code unused?	❑	❑	
❑ Are appropriate standards followed?	❑	❑	
Style			
❑ Do comments follow a standard format?	❑	❑	
❑ Does the prologue clearly state purpose?	❑	❑	
❑ Do comments avoid paraphrasing code?	❑	❑	
Control flow			
❑ Are the interfaces clear?	❑	❑	
❑ Are loops properly terminated?	❑	❑	
❑ Do all modules have a single entry and return?	❑	❑	
❑ Are all cases considered at each branch?	❑	❑	
Interfaces			
❑ Are the interfaces initialized?	❑	❑	
❑ Are parameters consistent?	❑	❑	
❑ Are values bounded?	❑	❑	
❑ Are all possible values considered?	❑	❑	
❑ Are registers and stacks restored upon exit?	❑	❑	
Logic			
❑ Is the precision of calculation sufficient?	❑	❑	
❑ Are performance requirements addressed?	❑	❑	
❑ Are all error conditions explicitly addressed?	❑	❑	
❑ Are all counters, arrays, and defaults initialized?	❑	❑	
❑ Are all complex expressions necessary?	❑	❑	
Data			
❑ Are all variables initialized?	❑	❑	
❑ Are input ranges checked?	❑	❑	
❑ Are types declared?	❑	❑	
❑ Are variable names meaningful?	❑	❑	
❑ Are the approximations useful for the desired range?	❑	❑	
❑ Is the precision appropriate (neither too high nor too low)?	❑	❑	

Fig. 11.7 A sample checklist for code inspection.

11.7 GOOD PROGRAMMING PRACTICE

According to Van Tassel (1978, p. 33), "Programs are to be read by humans." *For programs to be useful, reliable, and maintainable, you must make them readable and understandable.* Good design and programming practices can make programs more readable.

11.7.1 *Style and Format*

You can vastly improve the utility of your software by merely commenting it appropriately. A brief header before each routine should describe its objective and summarize its operation (Fig. 11.3). Comments should be clear, concise, and correct; a wrong comment is worse than none at all. An appropriate and consistent naming convention for variables will clarify the software function. Finally, keep routines and modules short to reduce complexity. Fig. 11.8 summarizes some of these techniques.

A program has two reasons for existence: to *do* something and to *communicate* the designer's intent to others. The structure of the program and the comments

Design
- ❑ Start documentation at the beginning of design.
- ❑ Write each module out in words before you begin coding (pseudo code).
- ❑ Keep routines short. (Keep source code to one page for the sake of comprehension; comments and headers may add extra pages.)
- ❑ Write clearly. Don't sacrifice clarity for "efficiency."
- ❑ Make the routine right before you make it faster.
- ❑ Make the routine clear before you make it faster.
- ❑ Make the routine simple to make it faster.
- ❑ Keep the routine right when you make it faster.

Comments
- ❑ Comments should explain and clarify the intent, not paraphrase the.code.
- ❑ Incorrect comments are worse than no comments.
- ❑ Write a prologue for each routine.
- ❑ Describe what a function does and its parameter(s).
- ❑ Improve readability with spaces and indention.
- ❑ Provide more comments than you think you need.

Variables
- ❑ Name variables to improve code clarity.
- ❑ Don't use global variables, because they can be altered anywhere.
- ❑ Don't pass pointers, either.
- ❑ Pass only intact values.

Fig. 11.8 A checklist of style guidelines for better programs.

within it are the primary means of communicating the designer's intent. Good style and format will help that communication.

11.7.2 *Structured Programming*

Good style dovetails into a discipline called *structured programming,* which establishes a framework for generating code that is more readable and consequently more useful, reliable, and maintainable. That framework is based on clearly defined modules or procedures, each doing one task well in a variety of situations. Modules can isolate device-dependent code for simplicity and reuse. Modules also can be divided up among members of the programming team for a more productive parallel effort. Furthermore, libraries of modules and procedures load faster and resist inadvertent changes (Webb, 1992). Finally, structured programming encourages the installation and testing of one module at a time to simplify the verification of the software (it's better to vary only one parameter at a time).

Ganssle (1992) expands on these points with the following advice:

1. "Most studies find that 90% of a processor's time is spent executing 10% of the code. Identify this 10% in the design (*before* writing code) and focus your energies on this section" (p. 15).
2. Listen to the customer when you develop the specifications.
3. Prototype complex tasks on the host computer and investigate their behaviors.
4. Design the architecture for debugging and testing.
5. Code small modules that you can "test and forget."
6. Code a single entry and exit in routines.
7. Document and comment carefully—in fact, do it first.

11.7.3 *Coupling and Cohesion*

Coupling and cohesion can help you define your tasks and design your modules. Modules should have a minimum of communication or coupling. If two tasks or processes communicate heavily, they should reside in the same module. Cohesion means that everything within a module should be closely related—that is, they should stick together. One guideline is that a module should correspond to a physical object—perhaps a device or process that it controls.

11.7.4 *Documentation and Source Control*

While documenting individual modules is straightforward, a document that describes the overall system function and how modules interrelate is even more important, as well as more difficult to prepare (Moon, 1993). This should be one of the first documents that you begin and one of the last that you finish, to ensure completeness and veracity.

Finally, you should back up all source files to manage configuration properly. If you work in a distributed development environment, copy the server files to tape regularly. You should lock up multiple copies of the source code, both disk and paper printout, in separate locations so that a fire or magnetic erasure in one location won't cause an irreparable loss. Keep copies of the source code for a long time. You won't know how long ago a file was corrupted, and you will need to trace the event and restore the file. Remember, file storage is cheap, but reconstructing lost data is expensive or impossible.

11.7.5 *Scheduling*

One of the biggest problems anyone faces with software is estimating development time. You should keep a record of all effort expended in current jobs to help you estimate future jobs. This is another good reason to keep metrics of progress. Besides accounting for the known activities, such as meetings, planning, designing, testing, and debugging (and no one ever schedules enough time for debugging), make allowances for the unexpected. Ganssle (1992, p. 248) writes, "'The unexpected' is frequently cited for delays. Tools fail, prototype hardware is flaky, and unknown coding issues arise. . . . But why do we continue to schedule projects with no room or plan for dealing with these problems?"

11.8 USER INTERFACE

The user interface is a major concern from a systems viewpoint, and software plays a principal role in the interface. "Good user interfaces require extraordinary attention to detail" (Constantine, 1993, p. 46). Unfortunately, they are often left as an afterthought for the programmer to handle. The user interface can make or break an instrument; for instance, *half of all hospital accidents are due to improper use of correctly operating equipment* (Kriewall, 1993).

"Strangely, the higher the cost of the equipment, the lower the quality of the human-interface design" (Constantine, 1993, p. 44). This situation may be cultivated by a small cadre of *wizards* (skilled, knowledgeable users), or it may be a trade-off between development time for the interface and training time for the users and operators. Neither explanation is a good excuse for a conscientious designer.

Common design issues for a user interface include response time, error handling, and help facilities. The response time should have a reasonable interval and consistent variation applicable to the task. Error handling should be clear and give remedial action (see the case study on the Therac-25 for how *not* to handle errors). Help facilities should be on line and context sensitive. Command sequences should be useful and consistent.

11.8.1 *Guidelines*

While user interfaces are extremely varied, some common guidelines outlined in Table 11.4 will help in developing effective instruments. These are a distillation of principles from Chapter 4.

In essence, the interface should be correct and understandable. Make the software of the interface correct; don't wait to fix it in the documentation or in the next release, because you can't undo the damage from users' poor perceptions. Design for the users rather than the technology. Just because it's easy to code doesn't mean you should throw in a lot of extra features; the resulting confusion from the complexity will increase the frequency of errors both by the software developers and by the users. Don't spice up the interface with unneeded graphics. "Norm Cox, former designer of the Xerox Star, . . . thought sexy graphics are like 'lipstick on a bulldog'" (Trower, 1993, p. 16).

Table 11.4 Some guidelines for user interfaces (Constantine, 1993; Pressman, 1992)

Action or concern	Comments
Tune dialogue to user.	Make it smooth and consistent. Use logical rather than visual thinking.
Make error messages meaningful.	Let the user know what is going on.
Provide help facilities.	Let the user know what is going on.
Verify critical actions.	Help user understand consequences.
Permit reversal of actions.	Forgive mistakes. Don't label input as right or wrong.
Reduce memory load.	Don't compromise simple operations by extending them for infrequent ones.
Display only relevant information.	Remove static or redundant information. Show the dynamic data.
Deactivate commands.	"Fade out" or clear from screen unused or inappropriate selections.
Use good layout techniques.	Use a modular format with indention, columns, and mixed-case lettering.

11.8.2 *Development*

Storyboarding and rapid prototyping are particularly suited for developing user interfaces because they are informal and fast. (You could try to develop the user interface with a top-down structured approach, but it relies on both paper analysis and the designer's complete understanding of user requirements—not usually possible.) A prototype can take any of three forms:

1. A paper prototype depicting user interaction
2. A working prototype with a subset of functions
3. An existing program that has all the features but needs modification

While storyboarding and rapid prototyping are flexible tools, they are difficult to document. Moreover, creeping featurism can sidetrack development because prototyping concentrates on short-term results and can miss long-term concerns that may require substantial reworking in the future. Consequently, prototyping needs close supervision to constrain growth and avoid creeping featurism. Coupling the prototype studies with systematic analysis and documentation is most effective for achieving a complete development and satisfying the requirements of software quality assurance (Mallory, 1992).

Finally, you need cooperation from both the customer and management. Senior management must support the goals of prototyping and understand its impact on development schedules. The customer must be committed to both evaluation and refinement of the prototype and make timely decisions for you (Mallory, 1992).

11.9 EMBEDDED, REAL-TIME SOFTWARE

Most of the software that runs or controls instruments and equipment is embedded, real-time software. Real-time software is code that responds to current events in a timely manner (Laplante, 1993). Embedded software is hardware specific; often a user interacts with a portion of the software system but does not have complete control over the source code. Generally it is inside an instrument and not obvious. Two concepts appear in every project and require attention from the software design: occurrence and loading.

Occurrence is the timing of events in real-time software. Events occur either synchronously or asynchronously. *Synchronous* events occur at predictable times in the control flow of the software. *Asynchronous* events are controlled by external sources and cannot be predicted in the control flow.

Loading is a measure of processor capacity. Two metrics for loading are utilization and throughput. *Utilization* is the amount of useful processing measured against the processor's capability. *Throughput* is a unit measure of the number of instructions or I/O operations that a processor can perform.

11.9.1 *Real-Time Software*

Real-time software has two components—the operating system and the device drivers—that affect every project. They are intertwined with system issues and queuing; network models will help you address these issues in your systems analysis (Laplante, 1993; Pressman, 1992).

The operating system controls the flow of events in the software and usually incorporates a priority-scheduling mechanism. Often called a real-time operating system (RTOS), the system may be either a dedicated operating system or a general-purpose one enhanced for real time. Most general-purpose operating systems are too large to fill the role of an RTOS. Generally, you should purchase a commercial RTOS rather than develop your own.

The kernel is the smallest portion of the operating system that schedules, dispatches, and communicates with tasks. Scheduling within an RTOS may be clock driven, event driven, or a combination of the two.

A clock-driven scheduler uses a polled loop. It is simple to write, debug, and analyze and good for high-speed data channels. A clock-driven scheduler, however, fails to handle bursts of events and is wasteful of CPU cycles (Laplante, 1993).

An event-driven scheduler uses event interrupts to control program flow. It services interrupts by context switching or saving CPU register states. The context switch is a major contributor to response times, so you should save only the minimum necessary information. An event-driven scheduler is easy to write, but it is vulnerable to timing variations, race conditions, and hardware failures, and it has difficulty in providing advanced services (Laplante, 1993).

The kernel calls software modules, or device drivers, to operate peripheral devices. These device drivers make the attached equipment work and insulate the rest of the operating system from the complexities of the immediate task (Tuggle, 1993). Sometimes you may have to mix assembly language routines with high-level language programs to achieve the necessary timing in a device driver (Brown, 1990). When a device driver is part of an interrupt service routine, you should follow the guidelines set forth in Fig. 11.9.

- ❑ Assign an ISR to each device.
- ❑ Assign a unique ISR to each periodic activity.
- ❑ Assign a unique ISR to each distinct activity.
- ❑ Group closely related functions into one ISR.
- ❑ Separate functions with different schedules into unique ISRs.

General format

- ❑ Save the CPU registers that will be used in the ISR (push them onto the stack).
- ❑ Service the device.
- ❑ Reenable the interrupt controller if one is used.
- ❑ Restore the CPU registers (pop them from the stack).
- ❑ Enable interrupts.
- ❑ Return.

Fig. 11.9 A checklist for interrupt service routines (ISRs) (Ganssle, 1994; Moon, 1993).

11.9.2 *Concerns*

Performance, fault tolerance, and reliability are major concerns for embedded, real-time software. You can measure and define these attributes in several ways. You can define performance for embedded software with the following metrics:

- Execution speed of the processor
- Response time of the system
- Data transfer rate
- Interrupt handling: context switching and interrupt latency
- Memory size

Synchronization of tasks will often reduce idle time of the processor or interrupt conflicts. Improving intertask communications with queues, mailboxes, and semaphores will also improve performance.

You measure performance with various methods. At the most basic level you can count instruction cycles of the source code, but this is not always practical or possible. A good step up from counting cycles is to use software simulators that provide indications of performance. But to measure the actual performance of the system, you must time I/O and bus signals with an oscilloscope, logic analyzer, or performance analyzer.

Fault tolerance defines how the software deals with misused resources and outright errors. Fault tolerance has various degrees, listed below in an increasing order of capability (Laplante, 1993):

1. Limit on downtime of the system
2. Absence of catastrophic errors
3. Predictableness
4. Robustness (ability to recover gracefully from errors)

Greater fault tolerance in the system requires more system design at the beginning of the project and more system testing and verification.

Some types of failure that affect reliability, beyond the bugs and catastrophic errors already mentioned, include missed or incomplete tasks, deadlock, spurious interrupts, and stack overflow. A lack of computing resources will cause the software to miss tasks or not complete them. Deadlock occurs when two tasks compete for the same resources and each has one resource already in its possession that it will not relinquish to the other task. Sensors can overload a processor with spurious interrupts and stack overflows. You need to address all of these concerns when you consider fault tolerance.

11.10 CASE STUDIES AND DESIGN EXAMPLES

This section presents several ways that software may be used in an embedded system. The two case studies discuss how software can be improved in function and quality. The design example shows some applications of Ada, a high-level language.

11.10.1 *Sprinkler System (Ducas, 1992)*

A sprinkler system for golf courses is a good example of an embedded, real-time system.

The sprinkler system has a Macintosh central computer that communicates with distributed controllers using the Motorola 6803. Each controller handles 36 sprinklers; somewhere between 10 and 60 controllers manage an entire golf course. Each controller has an 80-key membrane keypad, an LCD display, and control for 36 triac solenoid drivers. The system can display messages in several foreign languages. Fig. 11.10 outlines the hardware of the sprinkler system.

During development, the original programmer began the software in assembly language. A replacement engineer converted it to C, a high-level language. The original configuration had several major problems that caused the sprinklers to start watering at the wrong times and occasionally shower some surprised golfers. The problems stemmed from unreliable communications between the central computer and the controllers and from power dropouts and noise spikes. In addition, the assembly code in the controllers was not complete, difficult to maintain, and still buggy.

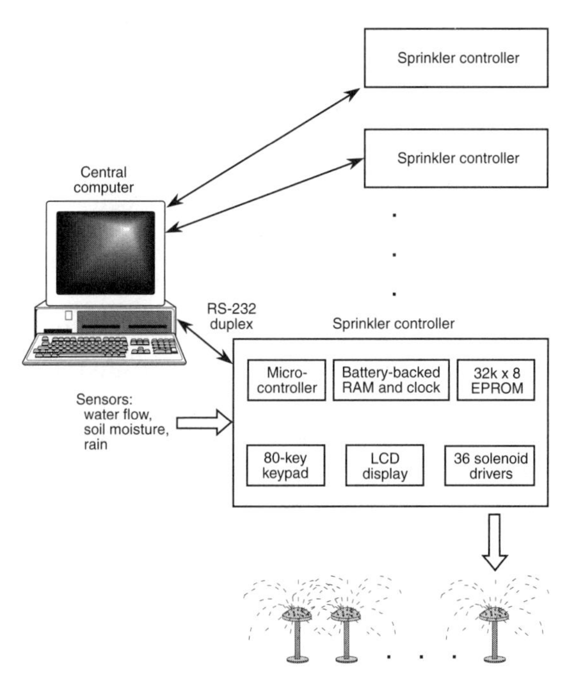

Fig. 11.10 Outline of the golf course sprinkler system.

The revised configuration avoided major hardware changes. The C code supposedly provided faster development and better documentation, but I suspect that the replacement engineer was more conscientious and better equipped to do a thorough system design than the original developer. He improved the communications by changing the protocol and adding error checking to remove induced errors, and he added noise suppression filtering to the controller hardware. (This is an important point where a thorough system design solves problems with both hardware and software.) After 12 weeks of intense effort, the revised system functioned reliably with a robust measure of fault tolerance.

Scope: system or enterprise
Lessons learned:
1. System design and documentation build a better product.
2. Complementary components (noise filter hardware, software protocol, and error checking to improve communications) make a design more robust.

11.10.2 *Therac-25 (Leveson and Turner, 1993)*

The Therac-25 was a medical linear accelerator developed during the 1970s and early 1980s to treat tumors with high-energy beams. Between 1985 and 1987, Therac-25 accelerators had several accidents that seriously and fatally overdosed patients. This case study is just a brief overview of the complex nature of failure in this embedded system.

Six patients accidentally received massive overdoses of radiation from the Therac-25 that resulted in serious injuries and death. In each of these cases the Therac-25 delivered up to 250 Gy of radiation (5 Gy of whole-body radiation can kill) yet displayed "no dose given." How could this happen? Leveson and Turner (1993) provide a revealing, in-depth study of the possibilities; I'll just summarize some thoughts here.

The problems arose from a custom operating system that had a real-time executive using a preemptive scheduler. Systems analysis showed that dynamic issues in the multitasking environment led up to the catastrophic overdoses. *While each task ran fine and without failure in isolation, asynchronous operation and communicating in the multitasking environment set up conditions for failure of the system.*

First, a race condition caused by an operator's quick editing at the keyboard passed wrong data to various tasks. Then a cryptic error message appeared with no follow-up explanation in the user manual, so operators did not know the remedial action. The Therac-25 depended on software safety mechanisms; it had no adequate mechanical interlocks. For instance, it had a 1-byte status register that rolled over every 256 passes; a zero indicated that it was okay to proceed when that was not necessarily the case.

What allowed these circumstances? Unfortunately, little information exists on the software development, management, or quality control; apparently very little of the software was ever documented. The picture that emerges is that a single programmer developed the code without having audits of the source code or testing of the design protocol.

Lessons learned from the Therac-25 accidents are many. Any safety-critical system should have a general plan for system development that includes analyses and inspections of the software, system tests, and hardware safeguards. The analyses include failure mode, effect, and criticality analysis (FMECA) and fault-tree analysis (FTA). A failure mode and effect analysis describes the system response to all failure modes of the individual components. A fault-tree analysis shows the failure paths and the relative importance of different failure mechanisms. Hardware safeguards can limit the catastrophic nature of software faults by constraining outputs.

A safe system comes from a good design—not from testing for faults. Leveson and Turner (1993, p. 29) emphasize that "focusing on particular software bugs is not the way to make a safe system. Virtually all complex software can be made to behave in an unexpected fashion under certain conditions. . . . The particular coding error is not as important as the general unsafe design of the software overall."

Scope: system or enterprise
Lessons learned:
1. Safety-critical systems need a general plan for system development.
2. Hardware safeguards can limit software faults.
3. A good design, not extensive testing, makes a safe system.

11.10.3 *Some Applications for Ada and Some Arguments Against Using It*

Ada was developed to ease the burden of designing and programming large military systems. It is a high-level language that is currently enjoying some use beyond military applications.

Ada is being used in the avionics of the F-22 advanced tactical fighter and the RAH-66 Comanche helicopter for the U.S. military, in commercial aircraft such as the Boeing 747 and the 777, and in numerous other embedded systems for companies such as Shell Oil, General Electric, and Alcoa. Surprisingly, Nippon Telephone and Telegraph in Japan is the world's largest user of Ada (Mayer, 1992a). Even some medical companies are incorporating Ada into their medical instruments because the Ada compilers are certified and that can reduce liability concerns (Roundtable discussion, 1993).

Proponents argue several cogent reasons for using Ada. It provides packages, strong typing, structural constructs, and data abstraction. It strives for maintainability, reusability, and readability. Both legal and financial incentives emphasize using certified compilers and reuse of software, purposes for which Ada was developed (Mayer, 1992a).

Others argue that C++ is a much better choice. They claim ease of use, efficiency, better run-time and compile performance, and support for object-oriented programming. Furthermore, C++ can integrate with commercially available software fairly easily (Mayer, 1992a). On the other hand, Mayer (1992b, p. 20) writes, "C gives programmers tremendous opportunities to make mistakes." Your experience and preferences should guide your choice between Ada and C++.

Development of the avionics software for the Boeing 777 illustrates some good system and software practices. Boeing ironed out problems by setting up meetings between the vendors of Ada compilers and the avionics suppliers during 1986 to 1989. "Getting those guys to head in one direction was a challenge and, frankly, the only way we could do it was by getting the compiler vendors with the suppliers and forcing them to discuss the problems" (Mayer, 1992b, p. 19). Even small programs such as the cabin management system for the Boeing 777, which has 70,000 lines of code, could be reasonably developed in Ada because the tools were in place, the code was compatible with other systems, and the workforce had experience with Ada.

11.11 SUMMARY

I have presented some methods to help you design and write good, reliable software. But methods and tools will not ensure correct software. Development of quality software requires good communication; documentation; a process plan; and a commitment to inspect, test, and verify the operation of software.

Correct and reliable software requires a good process. A good process can implement one or a combination of several models of development: waterfall, prototyping, and spiral. Each has its particular strengths and weaknesses. A good process defines metrics and measures progress, reliability, and correctness. A good process also manages risk; identifies the nature and introduction of bugs; and incorporates good style, testing, and inspections.

Beyond process, good system design implements and documents maintainable, reliable, and fault-tolerant software. The system design considers the user interface, performance issues, and safeguards and relates these concerns to the development of both the software and the hardware.

11.12 RECOMMENDED READING

Ganssle, J. G. 1992. *The art of programming embedded systems.* San Diego: Academic Press.
An outstanding and practical textbook that is must reading for anyone doing electronics development.

Laplante, P. A. 1993. *Real-time systems design and analysis: An engineer's handbook.* New York: IEEE Press.
A practical textbook for developing real-time systems.

Pressman, R. S. 1992. *Software engineering: A practitioner's approach.* 3rd ed. New York: McGraw-Hill.
A thorough textbook for software development.

Walker, H. M. 1994. *The limits of computing.* Boston: Jones and Bartlett.
A very readable and interesting book for software development.

The journal *Embedded Systems Programming*, published by Miller Freeman Inc., is an excellent source of current issues and practical solutions to embedded, real-time systems.

11.13 REFERENCES

Banks, W. 1990. Designing with tolerance. *Embedded Systems Programming,* June, pp. 60–68.

Blakely, F. W., and M. E. Boles. 1991. A case study of code inspections. *Hewlett-Packard Journal,* October, pp. 58–63.

Boehm, B. W. 1981. *Software engineering economics.* Englewood Cliffs, N.J.: Prentice-Hall.

Boehm, B. W. 1991. Software risk management: Principles and practices. *IEEE Software,* January, pp. 32–41.

Brown, R. 1990. Mix C and assembly language for fast real-time control. *EDN*, September 17 (special issue), pp. 41–50.

Card, D. 1991. Understanding process improvement. *IEEE Software,* July, pp. 102–103.

Castaldo, A. M. 1990. Lessons in debugging. *Embedded Systems Programming,* June, pp. 28–34.

Constantine, L. L. 1993. User interface design for embedded systems. *Embedded Systems Programming,* August, pp. 44–58.

Dearden, S. 1992. Develop large-scale embedded designs. *Electronic Design*, June 11, pp. 96–104.

Ducas, L. 1992. Archimedes C computerizes an irrigation system. *Archimedes Time-Saver* (a newsletter published by Archimedes Software), Winter, pp. 2–5.

Ganssle, J. G. 1992. *The art of programming embedded systems.* San Diego: Academic Press.

Ganssle, J. G. 1993. Defensive programming. *Embedded Systems Programming*, January, pp. 85–87.

Ganssle, J. G. 1994. Coding ISRs. *Embedded Systems Programming,* July, p. 82.

Gilmour, P. S. 1991. How to select tools for microcontroller software. *IEEE Spectrum,* February, pp. 37–39.

Gowen, L. D., and M. Y. Yap. 1993. Traditional software development's effects on safety. *Proceedings: Sixth Annual IEEE Symposium on Computer-Based Medical Systems, June 13–16, 1993, Ann Arbor, Michigan,* pp. 58–63. Los Alamitos, Calif.: IEEE Computer Society Press.

Hamilton, M. H. 1986. Zero-defect software: The elusive goal. *IEEE Spectrum*, March, pp. 48–53.

Haugh, J. M. 1991. Never make the same mistake twice: Using configuration control and error analysis to improve software quality. In *Proceedings: IEEE/AIAA 10th Digital Avionics Systems Conference, 14–17 October 1991, Los Angeles, California,* p. 220. New York: Institute of Electrical and Electronics Engineers.

Kriewall, T. J. 1993. Opening remarks at the Sixth Annual IEEE Symposium on Computer-Based Medical Systems, June 14, Ann Arbor, Mich.

Lantz, A. 1990. Debugging guidelines facilitate software development. *EDN,* September 17 (special issue), pp. 21–27.

Laplante, P. A. 1993. *Real-time systems design and analysis: An engineer's handbook.* New York: IEEE Press.

Leffingwell, D. A., and B. Norman. 1993. Software quality in medical devices: A top-down approach. In *Proceedings: Sixth Annual IEEE Symposium on Computer-Based Medical Systems, June 13–16, 1993, Ann Arbor, Michigan,* pp. 307–311. Los Alamitos, Calif.: IEEE Computer Society Press.

Leveson, N. G., and C. S. Turner. 1993. An investigation of the Therac-25 accidents. *IEEE Computer*, July, pp. 18–41.

Littlewood, B., and L. Strigini. 1992. The risks of software. *Scientific American,* November, pp. 62–75.

Mallory, S. R. 1992. Using prototyping to develop high-quality medical device software. In *Designer's handbook: Medical electronics.* Santa Monica, Calif.: Canon Communications.

Mayer, J. H. 1992a. Is Ada ready for the 90s? *Military & Aerospace Electronics,* August, pp. 21–24.

Mayer, J. H. 1992b. Ada takes off in non-DoD applications. *Military & Aerospace Electronics,* October 15, pp. 19–22.

Mittag, L. 1993. Debugging with hardware. *Embedded Systems Programming,* October, pp. 42–47.

Moon, M. F. 1993. Guidelines for tasking design. *Embedded Systems Programming,* October, pp. 28–34.

Murphy, E. E. 1990. Software R&D: From an art to a science. *IEEE Spectrum,* October, pp. 44–46.

Musa, J. D. 1989. Tools for measuring software reliability. *IEEE Spectrum,* February, pp. 39–42.

Nygren, D. C. 1993. Error detection in real-time. *Embedded Systems Programming,* October, pp. 36–54.

Olivier, D. P. 1992. Inspections: A successful approach to achieving high-quality software. In *Designer's handbook: Medical electronics*, pp. 163–165. Santa Monica, Calif.: Canon Communications.

Plansky, P. 1993. Software drives military microprocessor designs. *Military & Aerospace Electronics,* February 15, p. 17.

Pressman, R. S. 1992. *Software engineering: A practitioner's approach.* 3rd ed. New York: McGraw-Hill.

Putnam, L. H., and W. Myers. 1992. *Measures for excellence: Reliable software on time, within budget.* Englewood Cliffs, N.J.: Yourdon.

Roundtable discussion. 1993. Sixth Annual IEEE Symposium on Computer-Based Medical Systems, June 15, Ann Arbor, Mich.

Sroka, J. V., and R. M. Rusting. 1992. Medical device computer software: Challenges and safeguards. In *Designer's handbook: Medical electronics,* pp. 158–162. Santa Monica, Calif.: Canon Communications.

Trower, T. 1993. The seven deadly sins of interface design. *Microsoft Developer Network News,* July, p. 16.

Tuggle, E. 1993. Writing device drivers. *Embedded Systems Programming,* August, pp. 42–67.

Van Tassel, D. 1978. *Program style, design, efficiency, debugging, and testing.* 2nd ed. Englewood Cliffs, N.J.: Prentice-Hall.

Voas, J., K. Miller, and J. Payne. 1993. A software analysis technique for quantifying reliability in high-risk medical devices. In *Proceedings: Sixth Annual IEEE Symposium on Computer-Based Medical Systems, June 13–16, 1993, Ann Arbor, Michigan,* pp. 64–69. Los Alamitos, Calif.: IEEE Computer Society Press.

Walker, H. M. 1994. *The limits of computing.* Boston: Jones and Bartlett.

Webb, J. 1992. Make structured programming work for you. *BasicPro*, January–February, pp. 37–42.

Ziegler, J., T. Hornback, and A. Jordan. 1992. The ten commandments of debugging. *Electronic Design,* September 3, pp. 61–68.

12

DEBUGGING AND TESTING

. . . real life is twice as bad as worst case.
—Mike Fults, *Hughes Missile Systems Newsletter*,
December 1993

12.1 UNDERSTAND YOUR COMPONENTS AND CIRCUITS

This chapter introduces a voluminous amount of material. I had a difficult time writing it and can only hope to encourage you to investigate each component and its characteristics before using it. The chapter provides a foundation of proven principles used in debugging and testing. It also introduces you to some components, their tests, and their circuits. You need to understand how components function and fail to develop effective and reliable products.

Debugging is the art of tracking down and correcting undesirable system behavior. Testing characterizes components and verifies that they meet requirements. At times the two may become indistinguishable as you struggle to solve problems. Debugging comprises a set of guidelines to focus the search for a solution. Testing involves a variety of methods that verify operation within a specified environment. The range of methods encompasses inspections, destructive and nondestructive parts analyses, and static and dynamic evaluation.

As with everything else, you must plan for testing early in development and identify the critical components and subsystems that need specific tests to certify reliable operation. This chapter will detail these concerns and illustrate some examples of tests that you may use.

12.2 STEPS TO DEBUGGING

Think first! Unfortunately, many people take a scatterbrained approach to debugging and waste far too much time solving a problem.

Use strong inference in debugging; that is, hypothesize, isolate, and test. Five steps of the scientific method will help you solve debugging problems efficiently:

1. Observe and characterize.
2. Brainstorm.
3. Hypothesize.
4. Test.
5. Repeat steps 1 through 4 until the problem is solved.

Start by collecting observations and characterizing the problem. Brainstorm all possible causes for the problems and list them. Build a hypothesis that explains the observations and eliminates all causes but one. Design experiments that test the hypothesis and try to *disprove* it—you want to develop the strongest possible case for the hypothesis by using the most stringent experiments that might possibly eliminate it. Repeat the four steps until a workable hypothesis is found and the problem is solved.

Some simple observations and tests will go a long way to debugging most problems. Technicians have told me that 60% of all problems can be found with a visual inspection—problems such as missing components, broken wires, and backward installation. Check continuity to confirm connections. Understand your test equipment. Verify your design, both the schematic and the software. Sometimes only a break from debugging or another perspective helps to solve the problem. You can be blinded by your own familiarity with the situation and become mired in a narrow and misaligned focus. In particular, don't just swap parts to fix a problem; you may only substitute another component that is on the other side of a marginal threshold and waiting for a change in conditions to fail. Fig. 12.1 gives a checklist of simple tests that will organize and speed debugging (Jackson, 1991; Ziegler et al., 1992).

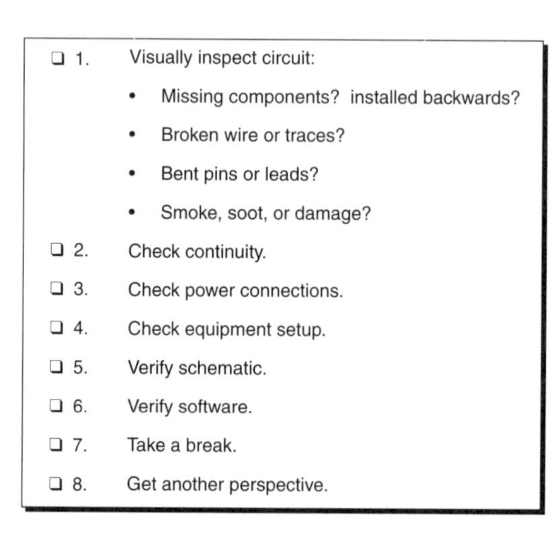

FIG. 12.1 Checklist for debugging.

12.3 TECHNIQUES FOR TROUBLESHOOTING

While the previous section set out guidelines for debugging, this section gives some specifics for troubleshooting and debugging. You need the proper diagnostic equipment to collect the necessary observations. This section provides some tips for using equipment to probe circuits.

12.3.1 *Equipment*

Generally you will need a dual-trace oscilloscope (analog or digital), a digital multimeter (DMM), a function generator, a power supply, a logic analyzer or an in-circuit-emulator (ICE), a soldering iron, and a supply of components (Fig. 12.2).

A good dual-trace oscilloscope should have a sensitivity of 1 or 2 mV/cm and a 100-MHz bandwidth. (Higher-frequency circuits demand wider bandwidths; use

a

b

Fig. 12.2 Equipment for troubleshooting electronics. (a) A minimal setup with oscilloscope, DMM, function generator, power supply, and soldering iron. (b) A more sophisticated setup that also includes a logic analyzer, multiple power supplies, multiple function generators, and highly accurate DMM.

an oscilloscope with a bandwidth that is 5 to 10 times the highest fundamental frequency in your circuit.) High bandwidth is needed for race conditions, oscillations, and pulse edges. You also need good probes—more on them later. For digital circuits, you will find either a logic analyzer or an ICE indispensable, but an oscilloscope is absolutely necessary, regardless.

You should have a high-resolution DMM with at least 5 digits of accuracy. Sometimes you may need to place a 100-kΩ resistor in series with the probe to prevent its input capacitance from loading the circuit under test. Don't disparage analog meters; they are good for showing trends, especially in power supplies (Pease, 1989a).

A good function generator will generate the necessary stimulus for testing your circuits. You should get one that provides at least sine, square, and triangular waves. Read the manuals and understand the limitations of your test equipment. Otherwise, you may attribute its problems to your circuit.

Rounding out a minimum inventory of necessary equipment is a stable power supply with coarse and fine adjustment knobs. It should be well filtered and capable of withstanding short circuits and step load changes. Sometimes you may need to use batteries for supplying power in low-noise applications.

A soldering iron, rosin-core solder, and a solder sucker facilitate component removal and replacement. *A good temperature-controlled soldering iron is well worth its cost.* It provides enough heat to melt solder quickly and reduce the spread of heat into the component and board. A cheap, low-wattage iron will take too long to melt the solder and heat too large an area around the joint; consequently, it will lift the pads of PCB traces and delaminate the PCB. *Never use acid-core solder or flux (usually found in plumbing supplies), because the acid will rapidly corrode leads, traces, and components.* Finally, practice your soldering technique for good joints; they should be smooth, shiny, and properly wetted. Avoid the dull, grainy globs of cold solder junctions.

A can of freeze mist and a hot air gun (or hair dryer) will occasionally help the debugging process. Directing a cold or hot stream of air at a component will sometimes force marginal operation to cross into an unacceptable region of performance.

You should keep a supply of components handy to substitute into circuits to vary operation and test the margins and limits of performance. A good stock of precision resistors, from 0.1 Ω to 10 MΩ, stable capacitors, from 10 pF to 1 μF, and working ICs will usually suffice. In addition, a working circuit is always useful for comparison.

12.3.2 *Probing*

You may sometimes find it difficult to probe without disturbing the circuit operation; it's a kind of Heisenberg uncertainty principle for troubleshooting electronic circuits. You can't avoid it, but you can reduce its effect. Consequently, you need to know what causes disturbances and use the appropriate probes and equipment to minimize the effects. In particular, probes can introduce disturbances through load capacitances, ground lead problems, and cables.

The oscilloscope probe is an "integral part of the circuit under test" (Williams, 1991, p. 167). Its input resistance and capacitance load a circuit and can affect or change its operation. A 10-pF capacitance from a probe at a summing junction of an op-amp can add significant lag in the feedback and lead to ringing or outright oscillation. A probe with an attenuation network often needs compensation to achieve a flat response to pulses without long tails or overshoots on the edges of the pulse. Fig. 12.3 illustrates how turning the screw can adjust a 10× probe's input capacitance.

For circuits with high impedance, a field-effect transistor (FET) input can reduce the capacitance to 1 or 2 pF and maintain a high input resistance. If you build your own, be sure to keep the signal paths short, use a low-dielectric material to support the circuit (like glass epoxy circuit board with the copper peeled off), and match impedance to the cable (Pease, 1989a; Williams, 1991). Fig. 12.4 is a block diagram of an FET input probe.

The ground lead on a probe can cause big problems; at high frequencies it becomes inductive and causes ringing in the signal. People often make the ground

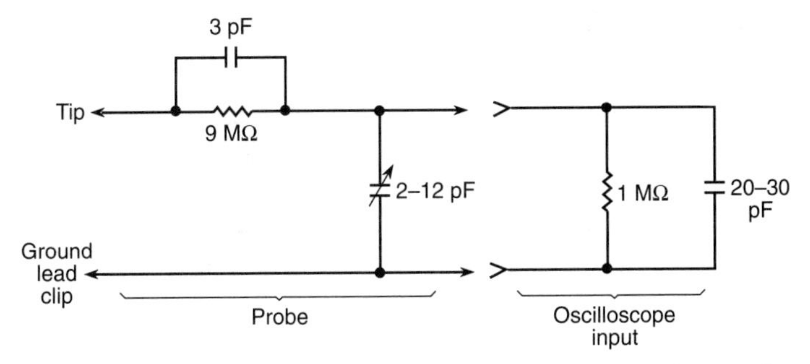

Fig. 12.3 Adjusting the compensation on a 10× oscilloscope probe by turning the screw that controls the variable capacitor.

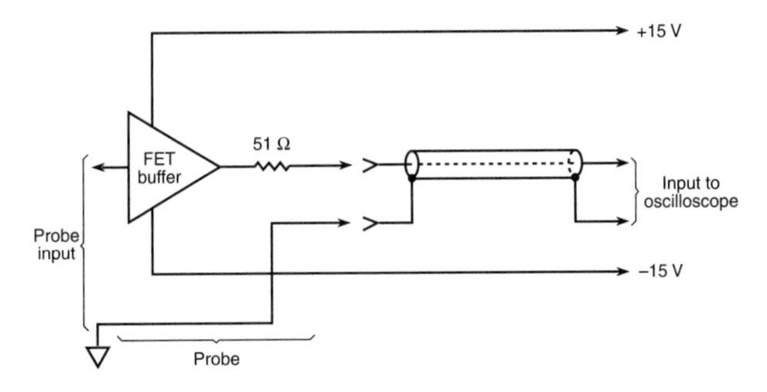

Fig. 12.4 An FET buffer built into the probe to provide very high input impedance.

leads on probes too long; always try to keep them as short as possible. If touching or wiggling the ground lead changes the waveform, then the probe grounding is not acceptable and you need to change it.

Cabling can significantly alter signals in probed circuits. Capacitance loading, inductive loops, and improper termination will degrade the rise time of signal pulses. You should match impedance and use carbon or metal-film resistors with short leads for termination.

12.3.3 *General Tips on Power*

Often schematics or diagrams do not show power supplies, because all circuits need power and the designer wanted to avoid unnecessary clutter in the drawing. You do need them, though, so make sure to connect power supplies correctly to your circuits.

Never use the absolute maximum operating parameters for power on your components. Any variation in the margin of the components will lead to their destruction. Furthermore, avoid damage to ICs by properly sequencing power. Replace chips only when the power supplies and signal generators are off. Clamp diodes on the power supply lines will prevent accidental polarity reversal (Fig. 12.5). Fig. 12.6 summarizes tips for powering circuits.

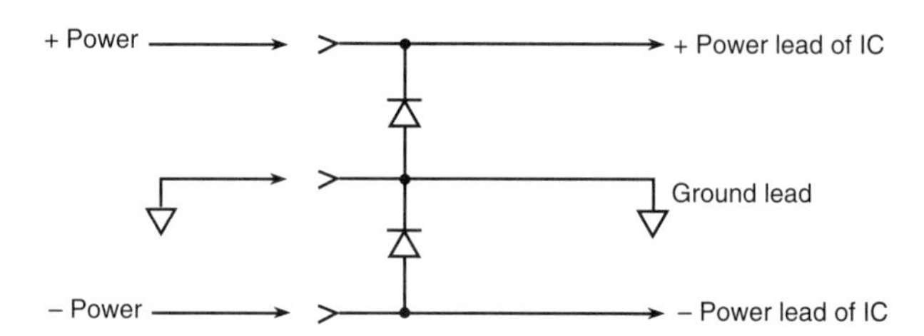

Fig. 12.5 Clamp diodes to prevent damage to components from accidental polarity reversal on DC power inputs.

- ❑ Provide power to all components and circuits (spec sheets don't always show them).
- ❑ *Never* use the absolute maximum operating parameters for integrated circuits.
- ❑ Put clamp diodes on power supply lines to prevent polarity reversals.
- ❑ Replace chips in circuits and test sockets with power and signal generators *off*.
- ❑ Sequence power properly.

  When powering up:

  1. Turn on power first.
  2. Turn on signal generators last.

  When powering down:

  3. Turn off signal generators first.
  4. Turn off power to components last.

Fig. 12.6 Checklist for powering circuits during debugging and troubleshooting (Frederiksen, 1988).

12.3.4 *Tips for Analog Circuits*

Basically, keep leads short, use the right oscilloscope probe, and lay out the circuit and grounds (or returns) carefully. Resistors should be located right next to the op-amp to reduce noise from stray capacitance. Compensate the oscilloscope probe; use an FET input probe for high-impedance circuits. Don't overdrive the scope; long settling times will distort the picture of the real signal.

12.3.5 *Tips for Digital Circuits*

The tools of the trade are the oscilloscope (to catch glitches and runt pulses) and the logic analyzer (for timing and disassembly). One of your challenges in tracking down problems will be triggering to catch glitches.

Software in the target system can aid debugging of the hardware. A built-in test uses a minimum of hardware and yet assures you that some portion of the circuitry is working. ROM diagnostics, such as the beep in the Apple Macintosh, can isolate problems by indicating what is operating. A simple ROM monitor allows you to change or examine memory and to drive I/O. You can even write simple looping routines to toggle an I/O port to see whether it is functioning (Ganssle, 1990a, 1990b)

Ultimately, you should design the hardware and software in a modular fashion that makes problems easy to isolate during debugging and testing. Partition the circuits according to function during design. This helps isolate signals for debugging.

12.3.6 *Tips for Software*

Review Section 11.6 on software bugs and testing to guide your debugging efforts. Debugging software probably requires more paper-and-pencil work than hardware does, but some hardware tools and techniques will help. The tools include LEDs on power and I/O lines, an ICE for logic problems, and even an oscilloscope for capturing pulses. A simple LED often can tell you whether the program is toggling a signal or not (Mittag, 1993).

Debugging software efficiently really requires a modular design (so plan for this from the beginning). You should work on one problem at a time and isolate the problem by reducing the system to the minimum configuration that still evidences the bug. As always, good documentation is an invaluable aid to isolating the problem because it explains the purpose and operation of each module. Apply the five steps of debugging in Section 12.2 to explain the cause of the bug. Castaldo (1990, p. 32) says, "Debugging is a process of building a good mental model of a system's particular aspect."

12.4 CHARACTERIZATION

The first step in debugging is to characterize the component or problem. Moreover, you can test components, circuits, and systems in a variety of environments to evaluate their performance and margins of performance. The results of these tests are necessary to establish the utility and reliability of your product; they also provide a basis for the models used in simulations.

Simulation can help you understand some component characteristics, but don't always believe its results. Simulations are based on simplifying assumptions that at some point must break down. In addition, it is difficult if not impossible to guarantee that the software in a simulation is error free. Consequently, final verification requires testing of the actual equipment, not merely a simulation of its response.

Simulations of components aid the development of concepts and circuit designs by narrowing the areas of concern. But simulations will never be able to fully describe the leading edge of technology; they are based on models developed from previous experience and tests. Testing, therefore, will never go out of style, and you will always need to do it.

Tests of individual components characterize parameter tolerances and variation and provide for calibration of their function. The test environments include electrical stimulation and variations in temperature and vibration.

You can use hot and cold thermal environments, for instance, to challenge and characterize components. Power consumption of most ICs and the conversion accuracy of ADCs vary with temperature. (Environmental chambers that provide the high and low temperatures for testing circuits require some special care; leads can contribute stray capacitance or thermocouple-induced voltages in connectors. Chapter 13 further describes the problems associated with chambers for testing components and circuits [Frederiksen, 1988].)

Example 12.4.1 I once designed a transistor circuit for driving LEDs. I tested a prototype breadboard in an environmental chamber, varied the temperature, and found that, unfortunately, its gains decreased at low temperature and no longer turned on the LEDs. I changed the resistor bias to provide gain margin over temperature; then the LEDs lit at all temperatures.

Chapter 13 covers in more detail some of the temperature and vibration tests of systems that you can use to stress both subsystem components and systems. The remainder of this chapter outlines some electrical tests you can conduct to characterize the components in your designs.

12.5 ELECTROMECHANICAL COMPONENTS

Every electronic instrument needs some electromechanical components to provide the electrical interface. Each component has its characteristic properties and failure modes. I have selected a small sample of mundane mechanical parts to illustrate certain properties that you will encounter and have to consider during design and test. You should experiment with characterizing components before you include them in a circuit; you may be surprised at how often your experimentation reveals anomalous or undocumented behavior.

12.5.1 *Connectors*

First among electromechanical components are connectors. They are ubiquitous in electronic products and have a wide variety of configurations (see Fig. 5.4). Their purpose is to provide the mechanical link that completes the electrical connection.

The main concerns about connectors focus on the metal contacts: How well do they maintain their electrical, mechanical, and chemical properties? The resistance of the contacts and their current capacity change with wear, dust, and force. At high frequencies, proper mating becomes important; you want to avoid insertion losses in poorly mated connectors. Insertion force and the number of insertion cycles affect the system's reliability. If the connector is removed and mated frequently, you will need a connector with low insertion force to handle many insertion cycles. The range of possible environments will determine the rate of corrosion in the contacts. Obviously you will have to specify these requirements up front in the design and possibly perform environmental testing to verify their reliability.

12.5.2 *Fuses*

A fuse prevents damage to a circuit from a fault that draws excessive current. A fuse has a metallic conductor with a specified cross section and resistance that will melt when the current density becomes too high. (Be aware that fuse resistance becomes significant at lower values of rated current; as cross section decreases, the resistance increases.) Fig. 12.7 illustrates some forms for fuses.

You will find several types of fuses that can fit a variety of situations. Normal-blow (U.S.) or class F (European) fuses open within 10 ms for currents 10 times the rated capacity or within 100 ms for currents 2 times the rated capacity. Slow-blow (U.S.) or type T (European) fuses target systems that normally sustain large, momentary surge currents; they open within 100 ms for currents 10 times the rated capacity or within 20 s for currents 2 times the rated capacity.

If you design products for international markets, you should understand the differences in fuses, protection philosophies, current capacity, and size. The European design philosophy is to protect against large overloads caused by short circuits, whereas the U.S. design philosophy is to protect against overload current

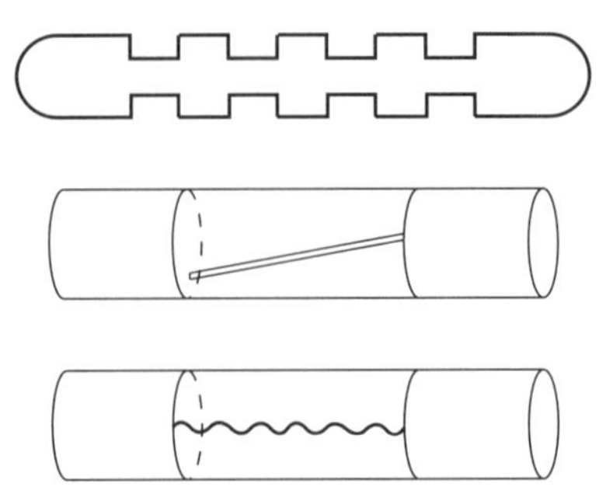

FIG. 12.7 Several different cross sections for fuses.

that slightly exceeds the fuse rating. Consequently, an 8-A normal-blow fuse in the United States is equivalent to a 6-A Eurofuse. Furthermore, a 10-A fast-blow fuse in the United States is equivalent to a 9-A Eurofuse. Fuse size varies across international boundaries as well. In Europe, the standard IEC 127 specifies that fuses should be 20 mm (0.79 in.) long by 5 mm (0.20 in.) in diameter; the standard UL198G in the United States and its equivalent standard CSA22-2 no. 59 in Canada specify that fuses should be 31.8 mm (1.25 in.) long by 6.4 mm (0.25 in.) in diameter (Sinclair, 1991).

Approval testing for fuses comprises three tests. One measures the statistical variation in time to open the circuit for high currents of 10 times the rating. A second test measures the statistical variation in time to open the circuit for medium currents of 2 times the rating. The final test determines whether the terminals of a blown fuse will sustain a high voltage (1.5 kV) without arcing.

12.5.3 *Switches*

Switches provide the mechanical means to make and break electrical connections. Of immediate concern to a user is the ergonomics in the handles and pads of switches. Are they obvious in their function? Do they operate easily with low probability of accidental activation? Besides the ergonomics, you need to understand the basic operation and parameters of each switch.

The ideal properties for a switch are low contact resistance, low force to maintain contact, no corrosion, durability, and low cost. Usually the biggest concerns are the configuration, electrical resistance, capacity, and durability of the switch contacts. Contacts can be normally open (NO) or normally closed (NC), as shown in Fig. 12.8. The resistance between contacts is inversely proportional to the area of the contact; furthermore, wiping action between contacts gives lower resistance than press action.

Current-capacity ratings depend on the application. A DC rating will typically be much lower than an AC rating; for example, a switch might have a 24-VDC rating but a 240-VAC rating (assuming a resistive load—derate for reactance).

Fig. 12.8 One possible configuration for switch contacts: NO = normally open, NC = normally closed.

Inductance within a DC circuit can generate huge potentials across opening switch contacts and cause devastating arcs that quickly erode the contacts. AC current tends to extinguish arcs across contacts because both the voltage and the current pass through 0 once every 8.3 ms (at 60 Hz) or 10 ms (at 50 Hz).

The interaction between the force required to move a switch armature and its inertial mass results in bounce when the contact is made or broken. Contact bounce generates oscillations in the voltage output from the switch and can trigger logic multiple times if you have not accounted for it. Contact bounce typically lasts between 5 and 20 ms (Fig. 12.9). See Section 7.10 and Fig. 7.23 for recommended circuits to debounce a switch.

The durability, current capacity, corrosion resistance, and cost of a switch are primarily functions of the metal used in the contacts. Precious metals such as gold and platinum give the best performance overall but are terribly costly; therefore, manufacturers usually electroplate a precious metal onto the contact surface to get the desired performance at far less expense. Table 12.1 lists some metals used in electroplating contacts.

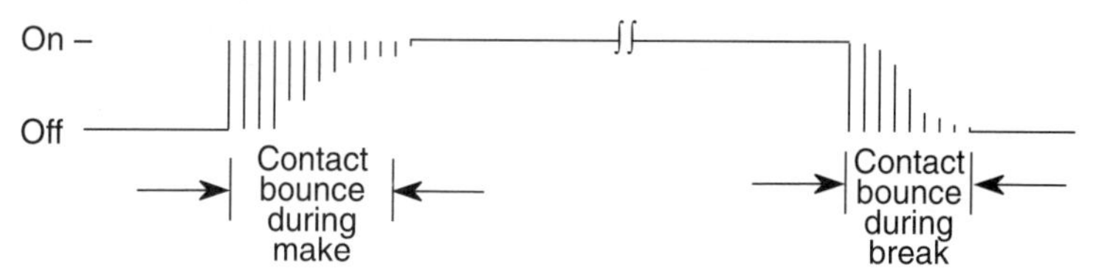

Fig. 12.9 Timing of contact bounce when a switch makes or breaks contact.

Table 12.1 Comparison of some materials that may be electroplated onto switch contacts

Plating material	Applications	Comments
Gold	Electronics	Resists corrosion, low currents only
Silver	High current	Low resistance, corrodes easily
Platinum	Electronics	Resists corrosion, low currents only
Silver-nickel	General	Resists burning, high resistance
Palladium-silver	General	Resists corrosion, high resistance
Tungsten	Power switching	High melting point, oxidizes easily

The fabrication of switches requires many considerations. The electrical connections to the outside world within a switch may be soldered, crimped, friction fit, or even welded for ease of manipulation by robots. The switch enclosure may have to be sealed to reduce atmospheric corrosion. "Explosion-proof" switches are sealed to prevent arcs generated during switching from igniting volatile gases.

12.5.4 *Keyboards*

Keyboards are a special application for switches. All the concerns voiced above for switches also apply to keyboards. In addition you have to understand the frequency and pattern of use. Durability is a major concern; all keyboards are subject to degradation by dust and spills. Keyboards in industrial environments, moreover, may have to withstand a corrosive atmosphere as well as pokes from screwdrivers.

12.6 PASSIVE COMPONENTS

Simple as they may seem, resistors, capacitors, and inductors need attention during product development. They form the majority of passive components used in circuit design. Each has its characteristic properties, mainly arising from the manufacturing processes and materials used for each type of component.

12.6.1 *Resistors*

Resistors have so many uses: pull-up, pull-down, impedance matching, voltage division, bias set, gain control, and even the purposeful generation of heat, to name just a few applications. Their primary function is to resist or restrict current. Most resistors have their resistance encoded in three or four colored bands around the body of the component (Table 12.2).

Resistance varies, however, with manufacturing tolerances, temperature, humidity, and voltage stresses. Manufacturing tolerances arise from the accuracy of trimming (removal of resistive material to tailor for the desired value), package materials used, and soldering of leads to the resistive conductor. Manufacturers specify resistors by the percentage variation in resistance; the ranges used are 20%, 10%, 5%, and 1%. Precision resistors have tolerances of 1% or less. Temperature variations attributed to soldering, self-heating, and gradients in the environment affect resistors. Humidity can change resistance of some resistors; stability is measured in parts per million per year. Did you know that fingerprints on the package of high-value resistors ($> 10\ M\Omega$) can reduce their resistance?

Resistors also change impedance with frequency. The lead inductance and internal capacitance become significant at high frequencies. Fig. 12.10 shows a circuit model of a resistor that illustrates its frequency dependence.

Table 12.2 Resistor color code

Color	First band (tens digit)	Second band (ones digit)	Third band (power of 10 exponent)	Fourth band (% tolerance)
Black	0	0	0 = 1	
Brown	10	1	1 = 10	
Red	20	2	2 = 100	
Orange	30	3	3 = 1,000	
Yellow	40	4	4 = 10,000	
Green	50	5	5 = 100,000	
Blue	60	6	6 = 1,000,000	
Violet	70	7	7 = 10,000,000	
Gray	80	8	8 = 100,000,000	
White	90	9	9 = 1,000,000,000	
Gold			−1 = 0.1	5
Silver			−2 = 0.01	10

Note: Tolerances of 1% and better are printed on the resistor, not indicated by color.

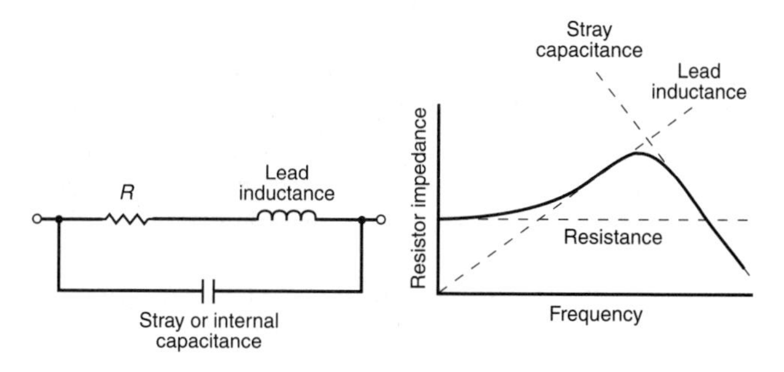

Fig. 12.10 Model of a resistor used in high-frequency applications.

Resistors can contribute a significant amount of thermal noise to the operation of some circuits. The noise is proportional to resistance, temperature, and bandwidth (see the equation for Johnson or thermal noise in Section 7.7).

If resistors are used at elevated temperatures, you will have to derate the amount of power that they dissipate. You can use a graph similar to that in Fig. 12.11 to derate a resistor's power dissipation. If cooling is used, the resistor's dissipation rating may be increased (Dorf, 1993).

Variable resistors, or potentiometers, were once common in circuit design, but they have fallen into disfavor because of their unreliability and bulk. Try to eliminate all trimming potentiometers, if possible; they are very unreliable and sensitive to moisture, dirt, and vibration (Upham, 1985). Instead, use solid-state devices, such as EEPROMs, to store calibration values. These can drive analog switches or DACs to generate analog values.

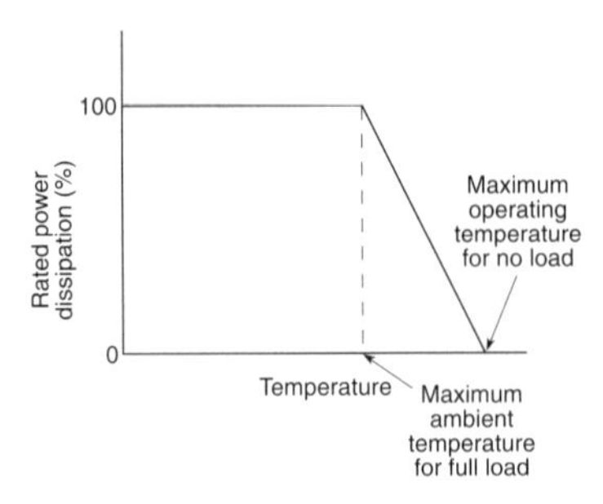

Fig. 12.11 Power derating for resistors as a function of temperature.

12.6.2 *Types of Resistors*

Table 12.3 lists the seven major types of resistors. Each type has advantages and disadvantages. Some trade cost for lower accuracy; others trade precision and low temperature coefficient of resistance (TCR) for lower maximum resistance.

A carbon-composition resistor contains graphite, ceramic, and resin compressed under high temperature into a rod that forms its resistive element. Metal end caps on the rod attach the leads, and then plastic or ceramic insulation coats the rod and end caps. High-resistance components have long resistive elements to reduce leakage, while high-dissipation components are thick for greater thermal dissipation. Carbon-composition resistors (or "carbon comps") have poor TCRs (about 1200 ppm/°C) and poor stability over both temperature and shelf life. Cycling between heating and cooling quickly reduces the reliability of carbon-composition resistors. In their favor, they are very cheap.

Carbon-film resistors use carbon vacuum-deposited onto a ceramic core as the resistive element. They are more accurate than carbon-composition resistors and have lower variation over temperature, humidity, and shelf life. They tend to

Table 12.3 Comparison of types of resistors (Ginsberg, 1981; Kerridge, 1992; Sinclair, 1991)

Type	Maximum resistance (MΩ)	Maximum voltage (V)	Temperature coefficient of resistance (±ppm/°C)	Accuracy of resistance (±ppm)	Shelf life (ppm/yr)	Stability	
						Soldering (%)	Humidity (%)
Carbon composition	22	500	1,200	10,000	5,000	2	15
Carbon film	50	500	250–1,000	5,000	2,000	0.5	3
Metal film	10	350	15–100	100	100	0.2	0.5
Thick film	2,000	250	25–300	2,000	100	0.15	1
Thin film	1	200	10–100	50	50	0.02	0.5
Wire wound	1	1,000	1–10	20	20	0.1	0.5
Foil	0.25	500	0.5–2	5	5	0.002	0.02

be less expensive than other types of resistors. You can use them in general low-power circuits.

Metal-film resistors have the resistive element formed by sputtering metal onto ceramic. They are now replacing carbon-composition resistors in many applications because of their good accuracy and low variability. Thick-film resistors are very similar in that they have sputtered metal oxides on ceramic. Both types are useful in general low-level electronic applications.

Thin-film resistors are tantalum nitride on a silicon substrate. Processing sputters the tantalum nitride onto silicon and then photoetches the conducting path into the tantalum nitride. Thin-film resistors have a low TCR, as low as 10 ppm/°C, and are good for resistance networks and surface-mount technology.

Wire-wound resistors are the original high-precision or high-power-dissipation components. Usual processing winds nickel-chromium wire on a beryllium oxide core. They have excellent TCRs that can be as low as 1 ppm/°C and can have high power ratings (to 2 kW). The wound core is inductive, however, and limits operation to less than 50 kHz.

Foil resistors have a thin foil of nickel-chromium that is bonded to a ceramic substrate and then its pattern is laser trimmed on the substrate. They have the best TCRs (< 1 ppm/°C) because the differential expansion between the foil and ceramic substrate balances the rise in resistance with temperature by compressing the foil and decreasing the resistance. Foil resistors have the highest precision of any resistor with both low inductance (< 100 nH) and low capacitance (< 1 pF) and allow bandwidths up to 500 MHz. They are good for RF circuits.

12.6.3 *Capacitors*

The primary function of capacitors is to store charge. They have a huge variety of applications; they may filter, couple, tune, block DC, pass AC, decouple (or bypass), compensate, isolate, store energy, or suppress noise.

A capacitor's impedance varies with frequency. The lead inductance, called the equivalent series inductance (ESL), becomes significant at high frequencies, while the internal leakage and resistance, called the equivalent series resistance (ESR), spoil the I-V quadrature phase relationship, or Q. Use high-quality capacitors with low ESL and ESR for the best performance. Fig. 12.12 gives a circuit model of a capacitor that illustrates how its impedance depends on frequency.

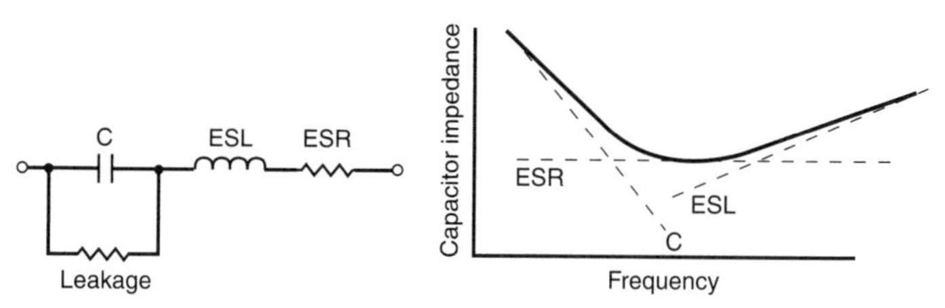

Fig. 12.12 Model of a capacitor and its impedance characteristics.

Capacitors have three parameters that may affect their capacitance (Fig. 12.13). Processing during manufacture may alter any one or all three to tailor the capacitance value:

$$C = \epsilon_0 \, k A / d ,$$

where ϵ_0 = permittivity of free space (8.85×10^{-12} F/m)
d = thickness of dielectric
A = area of plates
k = dielectric constant of material between plates

Breakdown voltage in a capacitor is directly proportional to *d,* the thickness of the dielectric. Thinner dielectric requires lower operating voltages. Etching the plate foils increases the area of the plates, but the capacitor then requires more dielectric and generally has a higher ESR. Heavily etched foils are fragile and difficult to handle. Different materials can have dielectric constants between 3 and 18,000, but increased permittivity is usually more sensitive to temperature and voltage fluctuations (Rappaport, 1982).

Capacitors suffer from a memory effect, called dielectric absorption, in which the internal electric dipoles align to raise capacitance and reduce discharge capability. This means that a residual voltage, sometimes larger than you would expect, may remain on a capacitor after discharge. You can measure the coefficient of dielectric absorption, or DA, for a capacitor in the following way (Frederiksen, 1988):

1. Charge the capacitor for 5 min.
2. Discharge the capacitor through a 5-Ω resistor for 10 s.
3. The DA is the ratio of the remaining voltage to the charging voltage.

Teflon and polypropylene capacitors have the lowest DA, followed by ceramic COG types. Mica and Mylar have the worst memory effect or largest DA.

Capacitors can fail in several ways. The most common failure mode is open circuit, but short circuit and leakage failures do occur. Arcing between the plates can cause all three; an arc can evaporate metal contacts, opening the circuit, or it can penetrate the dielectric, shorting the circuit.

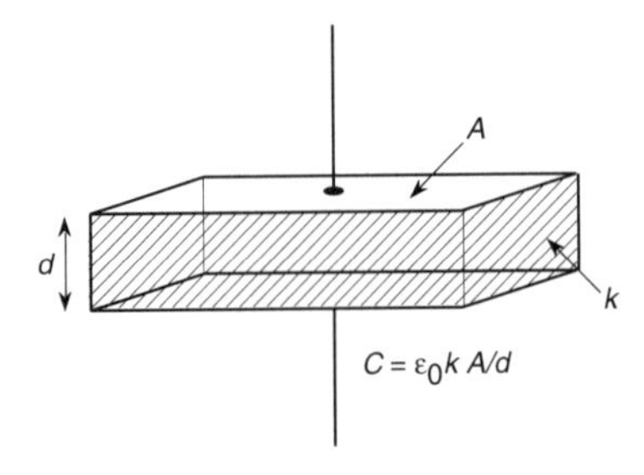

FIG. 12.13 Parameters that affect capacitance.

The main causes of failure are excessive voltage, current, or temperature and sudden charging or discharging. Mica and ceramic capacitors are least affected by abuse, plastic dielectrics are more sensitive, and electrolytics are the most fragile and should always be derated conservatively (Sinclair, 1991).

Certain capacitors require your close attention. Filter capacitors in switching power supplies, for instance, can suffer self-heating from large ripple currents if the ESR is not very low. The ESR dissipates heat in the form of I^2R power within the capacitor. Ceramic and electrolytic capacitors are not very stable over temperature, either.

12.6.4 *General Characteristics of Capacitors*

Capacitors, like resistors, have several different manufacturing processes that set their properties. Sinclair (1991) and Ginsberg (1981) detail these properties. Table 12.4 compares some characteristics of different capacitors (Dorf, 1993; Ginsberg, 1981; Rappaport, 1982).

The most common type of capacitor that you will encounter is the multilayer ceramic. It is frequently used in bypass or coupling circuits. A multilayer ceramic capacitor has stacks of metallized ceramic slabs sintered into a monolithic block, electrodes and leads added to the edges of the block, and a conformal coating. They have the highest dielectric coefficients ($k > 3000$), so they can be much smaller than other types of capacitors. Special ceramic capacitors have good temperature stability; once called negative-positive-zero (NP0) because they do not change with temperature, they are now called C0G type. Multilayer ceramic capacitors also suit automatic pick-and-place machinery. Furthermore, DIP capacitors mount like ICs and do not restrict cooling airflow.

Metallized-film (Mylar, polypropylene, polycarbonate, Teflon) capacitors have alternating layers of metal foil and flexible plastic dielectric. They have excellent thermal stability and can be self-healing because shorts often have high resistance with localized heating that clears the short circuit. While they are handicapped by larger sizes, they are suitable for high-voltage applications.

Mica capacitors have alternating layers of metal foil and mica insulation that have fairly low capacitance. They are useful in high-frequency circuits.

Electrolytic capacitors are found in the filters of power supplies because of their large capacitance and reasonable cost. They have an oxide film electrochemically formed on a metal surface; the metal is the anode (positive terminal), the oxide film is the dielectric, and the conducting liquid or gel is the cathode (negative terminal). Electrolytic capacitors have some significant limitations, though; the leakage current is proportional to temperature, and the ESR can cause internal heating. You should study the specification for ripple current in filter applications before picking a capacitor. The performance is limited by the internal temperature and resistance.

Table 12.4 Comparison between types of capacitors

Type	Dielectric constant	Capacitance range	Maximum frequency (Hz)	Maximum voltage (V)	Temperature coefficient (± ppm/°C)	Tolerance (%)	Leakage	Applications
Ceramic (C0G)	35	10–220 pF	10^9	35–100	±30	±1	Medium	Small, inexpensive
Ceramic	35–6,000	100pF–100nF	10^9	35–20,000	+500, −10,000	−35, +80	Medium	Decoupling
Mylar	3.0	1nF–50μF	10^9	50–750	+200, +800	±5	Low	Industrial, low-voltage telecommunications
Polypropylene	3.3	100pF–50μF	10^9	50–1,500	−200	−2	Very low	High-voltage applications
Polycarbonate	2.7	100pF–30μF	10^9	50–600	±50	±1	Low	Military, telecommunications
Mica	6.8	2.2pF–100nF	10^9	100–600	±20	±1	Low	RF applications
Teflon	2.0	1nF–2μF	10^7	50–200	−100	±0.25	Lowest	High reliability and stability
Tantalum	11	100nF–500μF	10^3–10^7	6–100	+1,000	±5	Medium	Power filtering, PCB power entry filtering
Aluminum electrolytic	7.0	100nF–1.6F	10^3	3–600	+10,000	−10, +50	Highest	Power supply filtering

Aluminum electrolytics comprise aluminum foil, aluminum oxide, and an acid solution of ammonium perborate. The foil is the positive terminal, while the ammonium perborate forms the negative terminal. Unfortunately, the solution outgasses at high temperatures (such as during reflow soldering) and must be sealed within the capacitor enclosure. The life of the capacitor depends on the rate of escape of the volatile fluids. Electrolytic capacitors are prone to explosion under reverse or overvoltage conditions because the resistance drops and the capacitor quickly overheats.

Tantalum electrolytic capacitors have tantalum pentoxide for their dielectric, which has a higher dielectric constant than aluminum. Tantalum electrolytics also have better volumetric efficiency, better low-temperature operation, longer life, lower leakage current, and lower ESR. If cost of parts is not the only priority in manufacturing an instrument, there is no reason not to use tantalum electrolytics in place of aluminum electrolytics.

12.6.5 *Decoupling Capacitors*

Decoupling capacitors are frequently used in circuit design to reduce the impedance of the power distribution system. They are most effective near or in the IC package. You can calculate the capacitance needed to decouple a circuit by specifying the average current pulse, the duration of the pulse, and the maximum voltage drop:

$$C = i\,\mathrm{d}t/\mathrm{d}v = (I_{\mathrm{transient}})\,(t_{\mathrm{transient}})/(V_{\mathrm{drop}}).$$

Example 12.6.5.1 An IC generates a current pulse that averages 100 mA for 10 ns. If you want to maintain a voltage drop of 0.25 V or less, choose a decoupling capacitor with greater than 4 nF of capacitance.

If $I_{\mathrm{transient}} = 100$ mA, $t_{\mathrm{transient}} = 10$ ns, and $V_{\mathrm{drop}} < 0.25$ V,
then $C > 4$ nF. Therefore choose $C = 10$ nF.

Don't choose a capacitor that is too large, or the ESL will lower the resonance frequency and cause ringing in the voltage of the power supply.

You can find a variety of types of decoupling capacitors: ceramic, leadless chips to reduce inductance, and capacitors that fit under IC DIP or PGA packages or are built into the IC package. All of these strive to reduce the inductive loop by locating the capacitor as close to the IC as possible (see Chapter 6). Good-quality capacitors have low ESL and cause the least voltage drop; the best have ESL < 20 nH and ESR < 0.5 Ω (Keenan, 1987).

Example 12.6.5.2 You can calculate the voltage drop caused by a current pulse through the lead and circuit inductance of the decoupling capacitor. If you have a total inductance of 20 nH and a current pulse of 100 mA with a duration of 10 ns, then the drop will be about 0.2 V. Use the following equation to calculate voltage drop:

$$V_{\text{drop}} = L\,\mathrm{d}i/\mathrm{d}t\,.$$

For L = ESL + trace inductance = 20 nH, di = 100 mA, and dt = 10 ns, V_{drop} = 0.2 V.

12.6.6 *Inductors*

The role of inductors in circuit design is significant but much smaller than that of resistors and capacitors. Inductors are usually found in power supplies and high-voltage generators. An inductor generates a changing magnetic field that opposes any change in current.

The inductance depends on the number of turns, the current, and the core material that can raise the relative permeability. They are nonlinear devices (unless they have an air core or equivalent) with hysteresis (Fig. 12.14), saturation, and eddy current losses. Laminated or ferrite cores reduce the loss from eddy currents. Most inductors are custom produced for the application; you will need to spend time with the manufacturer's application engineer to get the right inductor for your design.

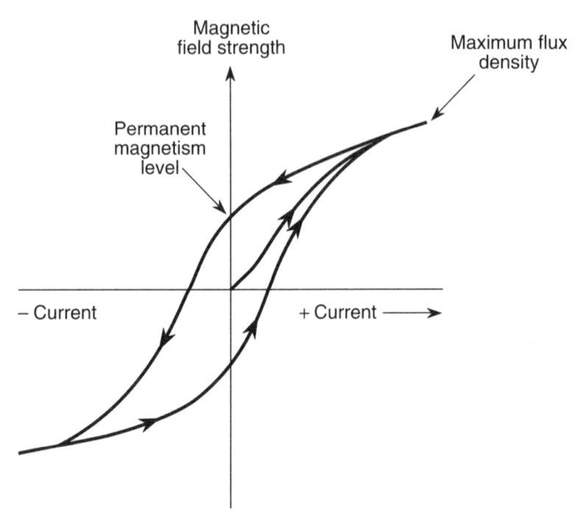

Fig. 12.14 The hysteresis curve for an inductor. The loss from energy dissipation is proportional to the enclosed area. A narrow figure is low loss; a wide figure is large loss.

12.7 ACTIVE DEVICES

A great many components are classified as active devices—those that amplify or control greater power with lesser power. I introduce three types of these devices: the mechanical relay, the solid-state relay, and semiconductor transistors (with an emphasis on MOSFETs). You can extend these general principles of specification and testing to other components that you will use.

12.7.1 *Mechanical Relays*

Mechanical relays use a magnetic coil to actuate a mechanical armature and electrical contacts to switch electrical connections. The low contact resistance of a relay approximates a perfect switch fairly simply and cheaply. Relays provide isolation between primary and secondary circuits and a fail-safe mode in which the secondary circuit remains open when a fault occurs.

Fig. 12.15 illustrates the parts of a relay. Current through the coil establishes the magnetic field necessary for moving the armature and contacts. The magnetic core provides a high-permeability path for the magnetic field.

A relay is a nonlinear, switched device. The power required to pull in the armature is greater than the power required to hold the armature closed; the ratio of pull-in power to holding power can be as much as 10:1. By exploiting this fact, switching it with a larger current and holding it with a lower current, you can reduce the power dissipation in a relay. Moreover, the maximum power dissipation is limited by overheating within the coil. High temperature forces derating of the power dissipated. (Latching relays, of course, require no holding power; they have a permanent magnet core to attract and hold the armature.)

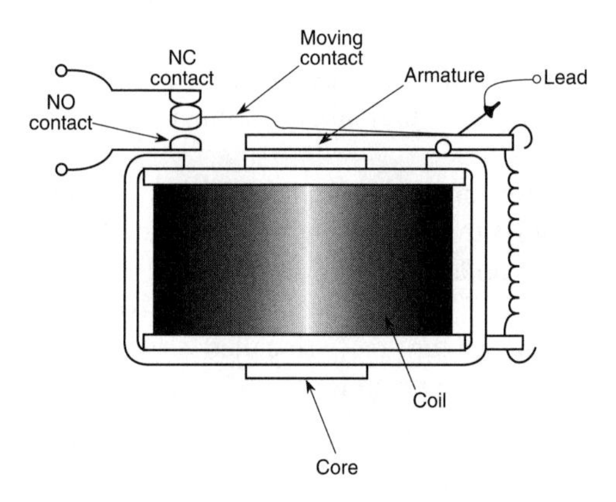

FIG. 12.15 The anatomy of a relay.

You can characterize relay coils by their resistance and inductance. The resistance sets the I^2R dissipation and how much voltage you must supply to attain the necessary current to pull in the armature. The inductance determines how much spike suppression you will need (see Section 7.11).

Some specifications and tests that you may require to certify a particular relay include a high-voltage test for the contacts, assurance of insulation resistance, and a test of life-cycle reliability.

In a typical high-voltage test, 2 kV is applied for 1 min across the contacts to see whether they arc over. To test the insulation resistance, use a test voltage of 500 V between one terminal of the coil and a probe to the insulated coating of the coil wire; the resistance should be greater than 100 MΩ (Sinclair, 1991).

The operational life of a relay is specified by the number of open-closed cycles the armature and contacts can complete without failure. Beware of manufacturers specifying reliability as 10^7 cycles for the mechanical life when a relay may function correctly for only 10^5 to 10^6 cycles while current flows in the contacts and armature connections.

If you have an environment with shock or vibration, you will have to specify and test for the maximum shock that a relay may sustain and still operate correctly. The shock is measured in each axis as acceleration while the relay is operating.

The relay contacts have the same wear and arc considerations that switches have. The duty cycle will also affect how long a relay lasts. Usually contacts are either closed nearly all the time or open nearly all the time.

12.7.2 *Solid-State Relays*

Solid-state relays use magnetic or optical isolation between the primary and secondary circuits. They are inherently faster (by a factor of 10 or more) and more resistant to shock and vibration than their mechanical cousins, and they have no contact bounce. They are also more reliable (by a factor of 10 or more), generate far less EMI because there is no contact arcing or bounce, and generally have lower voltage and lower power requirements than mechanical relays. Since solid-state relays do not have contacts, they also are immune to contamination and oxide formation.

Solid-state relays use either optocoupling or transformer coupling to achieve isolation. Fig. 12.16 shows two block diagrams of solid-state relays.

Solid-state relays excel at switching currents quickly and cleanly. They emit low-EMI radiation from AC circuits because of zero-voltage turn-on and zero-current turn-off. Zero-voltage turn-on (also called zero crossover or synchronous switching) forms the conducting path for the AC load current whenever it passes through zero voltage; this action causes only a small step change in power to the load and reduces the inrush current to lamps and extends lamp life. Zero-current crossing at turn-off breaks the electrical connection to the load whenever the current

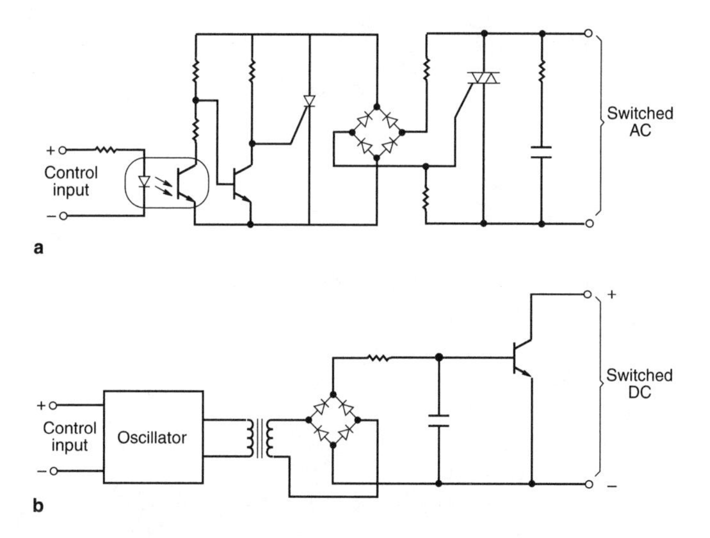

Fig. 12.16 Two examples of solid-state relays. (a) An AC relay that uses an optocoupler for electrical isolation. (b) A DC relay that uses transformer coupling for electrical isolation.

passes through zero; this action reduces the inductive kick or flyback in reactive loads (Ormond, 1989).

Manufacturers specify the current rating of solid-state relays for resistive loads, for reactive loads, and for temperature effects. Usually the AC rating can sustain a current surge for one cycle that is 10 times the rms current rating.

You must deal with three types of specifications when choosing a solid-state relay: input, output, and isolation. Input specifications include the following:

- Current sinking at a specified voltage level (such as +5 VDC)
- Voltage range at worst-case V_{cc} (power supply voltage) over temperature

Output specifications include the following:

- Steady-state current and thermal derating (see Fig. 12.17)
- Surge current as a function of time (see Fig. 12.18)
- Leakage current
- Output voltage, both rms and peak

The current ratings for solid-state relays hold only for devices mounted on heat sinks; you must derate the load ratings for free-air operation. Semiconductor outputs, drive circuitry, and snubber networks within the solid-state relay all contribute to the leakage current; high-power devices can have tens of milliamperes of leakage. Leakage current is specified at the maximum value for a particular load. The output voltage specifies both a maximum rms voltage and a peak blocking or breakdown voltage. Isolation specifications include the following:

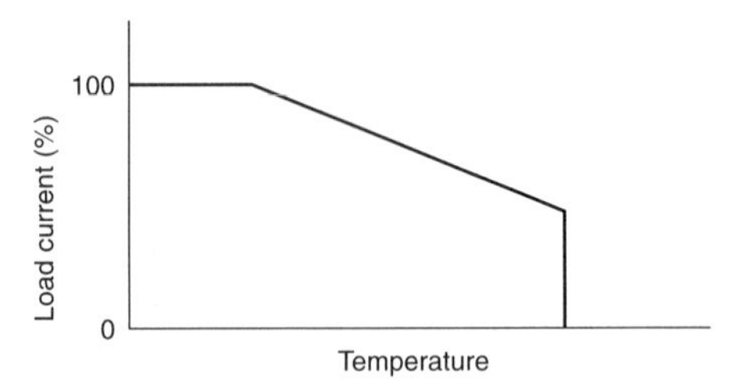

Fig. 12.17 An example of a thermal derating curve for a solid-state relay.

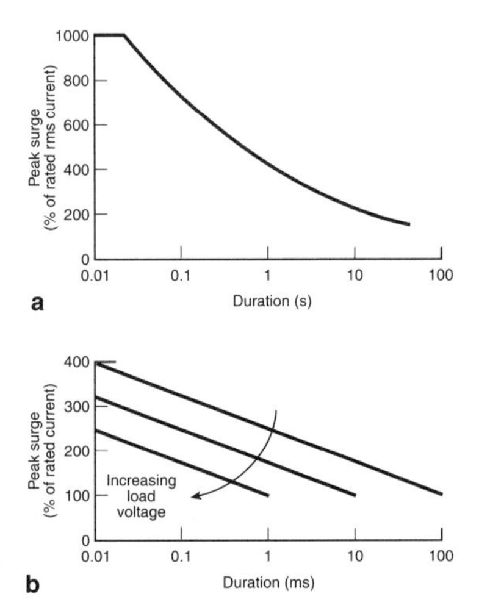

Fig. 12.18 Examples of peak current surges for solid-state relays. (a) Peak surge rating for an AC relay. (b) Peak surge for a DC relay.

- Leakage resistance at 500 VDC between input and output terminals
- Isolation voltage
- Capacitance between input and output

The leakage resistance typically is greater than 1 GΩ. Isolation voltage is measured between the input terminal and the case, as well as between the output and the case. A typical rating for isolation voltage is 3750 V rms. A low capacitance between input and output reduces feed-through; it typically is less than 20 pF. You may need to test your solid-state relays to certify that they meet these specs.

Solid-state relays have some significant disadvantages as well. Their contact configuration is usually limited to single pole, single throw, and normally open, and their contact on-resistance is higher than that of electromechanical relays. Finally, solid-state relays tend to be twice as expensive as comparable electromechanical relays.

12.7.3 *Semiconductors*

Semiconductor components—such as diodes, transistors, and optoisolators—are nonlinear devices that amplify or control electrical power. When incorporating semiconductors into your products, it is especially important to do the following (Pease, 1989b):

- Test and qualify each component.
- Develop a good relationship with each manufacturer.
- Maintain an alternative source for each component.
- Have manufacturers notify you before they change their process.

Some of the common problems and concerns with semiconductor devices are leakage current, reverse current, safe operating areas, thermal effects, ESD, and parameter variations. *One of the simplest and most common problems is incorrect orientation—putting diodes in backward, reversing leads on transistors, or rotating the packages.* You can reduce these problems by marking the orientation on both the components and the circuit board.

Always design your circuits to keep components within their safe operating ranges. Don't exceed voltage, current, or temperature ratings. Some of these ratings are as follows:

- Voltage versus pulse width
- Surge current
- Derating of current versus time

Temperature is always a concern in the operation of semiconductor devices. Temperature affects the transconductance of bipolar transistors that have negative temperature coefficients. If you place several transistors in a parallel arrangement, one transistor can sustain more current and dissipate more heat than the others and slip into positive thermal feedback. Eventually, it hogs current, goes into thermal runaway, and self-destructs. Always understand the operational limitations of your devices and use the appropriate heat sinks.

A signature analyzer or curve tracer can help you identify good and bad components. It applies an AC signal to the component under test and graphs the voltage (on the x-axis) versus the current (on the y-axis). A short circuit causes a vertical line, while an open circuit causes a horizontal line on the display. Normally operating devices have unique curves (Fig. 12.19). A signature analyzer can evaluate devices either alone or in circuit. If the device is in circuit, you must have a known good board on hand for comparison of signatures.

Finally, beware of trusting a prototype for testing a device and characterizing its circuit. A prototype may work fine because it has typical values for its device parameters, but it may not test the full range of parameter values. Some parameters can vary significantly; for instance, Pease (1989c) provides the following cautions:

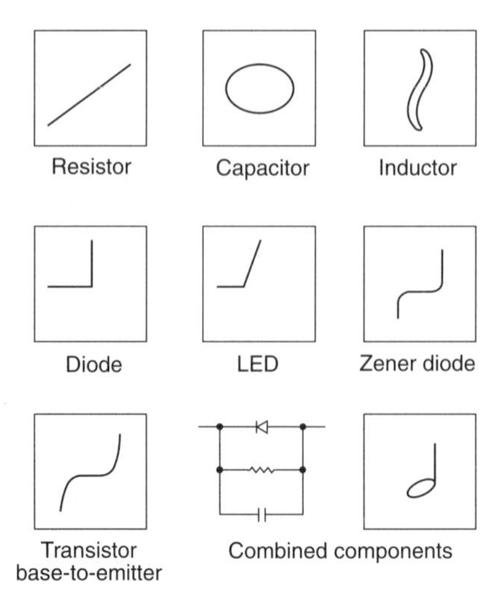

FIG. 12.19 Some examples of component signatures on a signature analyzer. You must adjust the frequency and voltage for the best display (Kral, 1993).

1. The gain of transistors can vary widely from transistor to transistor.
2. Fabrication affects switching speed and ruggedness.
3. Current gain and frequency response of optocouplers can vary widely.

An epitaxial base architecture leads to faster transistors, while single-diffused transistors are more rugged. They may have the same part number but different fabrication processes, so their actual parameters may vary significantly.

A resistor connected to the base of the output transistor in an optocoupler will drastically increase the frequency response. This is an example of how circuit design can affect the performance of your components.

12.7.4 *MOSFETs*

MOSFETs have a number of advantages over bipolar transistors. They are low-noise devices that have high input impedance at low frequencies and positive temperature coefficients. Consequently, they don't experience thermal runaway (except at high voltage and low current). Here are some of the parameters that you will have to consider:

- Drain current versus drain-to-source voltage
- Noise
- Gate capacitance

Allow margins of tolerance on the parameters of MOSFETs when you design them into a circuit. As an example, you may need to insert some resistance or a ferrite bead on the gate lead to prevent high-frequency oscillations (Pease, 1989c).

MOSFETs can fail in several ways. Defects in the silicon die, such as gate oxide defects, can cause gate-to-source short circuits, or distortions in the field can increase leakage and cause thermal runaway. Packaging faults can also destroy MOSFETs. Thermal and current stress between the wire bonds and die cause cracking, separation, and ultimately an open circuit. Water vapor that invades the enclosure can corrode the aluminum metallization and open contacts on the silicon die (Clemente et al., 1985).

Electrical and temperature testing can reveal defects in the silicon die and stress the wire bonds in the package (Fig. 12.20). High temperature and humidity will accelerate corrosion from leaky enclosures and force these failures.

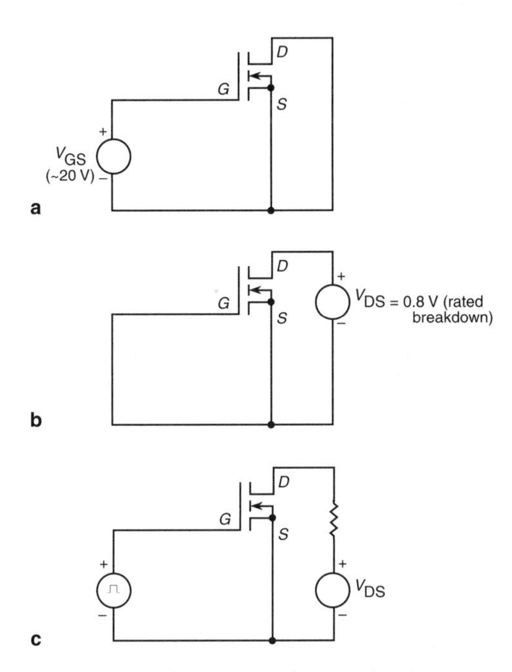

FIG. 12.20 Some test circuits for MOSFETs. (a) A forward-biased gate stress will reveal gate oxide defects that cause a gate-to-source short. (b) A drain-to-source voltage during high temperature will detect increased leakage from distortions in the field. (c) Temperature cycling or pulsed gate voltage with drain current will stress wire bonds and force failure of weak connections.

12.8 OPERATIONAL AMPLIFIERS

Operational amplifiers (op-amps) provide a nearly ideal gain component for use in amplifiers, filters, oscillators, and control systems. But nuances in their functions can cause much trouble if you aren't aware of or don't test for them. You will find that more complex integrated components have more complex parameters of operation and specifications.

"'Oscillators won't. Amplifiers will.'—oscillate, that is" (Pease, 1989e, p. 151). You can prevent a number of potential problems when using op-amps; these include oscillation, 50-Hz or 60-Hz noise, and saturation in the output. Besides these concerns, beware of the idiosyncrasies of op-amps and their circuits (Pease, 1989d):

1. Passive components around the op-amp cause most problems.
2. Don't rely on characteristics that aren't guaranteed by the manufacturer.
3. Virtual ground at the input is only a first-order model that applies to DC and low frequencies; at high frequencies, inputs have a significant error voltage to drive the output.

With these cautions in mind, you will be ready to do some real troubleshooting. First, break problems into regimes of operation. Is the problem an oscillation? Then consider the AC operation. Fig. 12.21 lists some of the diagnostics and solutions for treating oscillations. Is the output stuck at one voltage rail? Then consider the DC operation of the op-amp. Other areas to consider are bad offset voltage, long overload or short-circuit recoveries, long settling times, and a significant response to a changing thermal environment.

If the fundamental component of oscillation is *not* near the unity-gain frequency of the open loop gain, then the problem is *not* feedback loop stability. If the feedback loop is the culprit, then larger gain in the closed loop will stabilize the oscillation. Otherwise, a small or stray capacitance at the inverting input of an op-amp will introduce a feedback pole and lead to instability. There are several preventive measures you can take to yield proper operation in op-amp circuits (Frederiksen, 1988):

DIAGNOSTIC: What is the frequency of oscillation?

Frequency range	**What to check for**
200–1000 MHz	Circuit layout, parasitics
20–100 MHz	Stray feedback
1–40 MHz	Improper power supply bypassing
50–1000 kHz	Improperly damped op-amp loop, linear regulator oscillating
<1 kHz	Electrochemical delays
<10 Hz	Thermal delays

SOLUTIONS: Reduce oscillations and ringing.
Place power-supply bypass capacitors close to the op-amp.
Reduce capacitive loads; they add phase shift.
Use a small feedback capacitor to prevent oscillations.

Fig. 12.21 Some troubleshooting tips for finding and preventing oscillations in op-amp circuits (Pease, 1989d, 1989e).

1. Ensure that ground and return planes are correct and that the layout is good.
2. Use capacitors to bypass all power supply lines next to each op-amp.
3. Make sure the problem is not related to the power line (noise is 50 or 60 Hz) or EMI.

Once you have solved a problem, be sure to test the circuit carefully with a variety of loads, large-amplitude input signals, and frequencies.

Finally, if earlier versions of an op-amp circuit worked but newer ones don't, ask what changed. Were suppliers or components switched? Similar components from different suppliers can differ, and a small change in the fabrication by one manufacturer will change the characteristics of the component.

12.8.1 *Characteristics*

The remainder of this section introduces test circuits for some of the important parameters of an op-amp. These parameters can characterize an op-amp and help you reduce uncertainties in operation. Your particular application will probably not require that the op-amp be characterized fully but will focus on one or two critical parameters.

12.8.2 *Offset Voltage*

The offset voltage is a small input voltage that makes the output go to zero for DC applications. Offset voltage, a key specification for most op-amps, results from internal circuit imbalances and is a function of the IC fabrication process (Frederiksen, 1988). Fig. 12.22 shows a test circuit for measuring offset voltage, which is the output divided by 1000:

$$V_{\text{offset}} = V_{\text{out}}/1000.$$

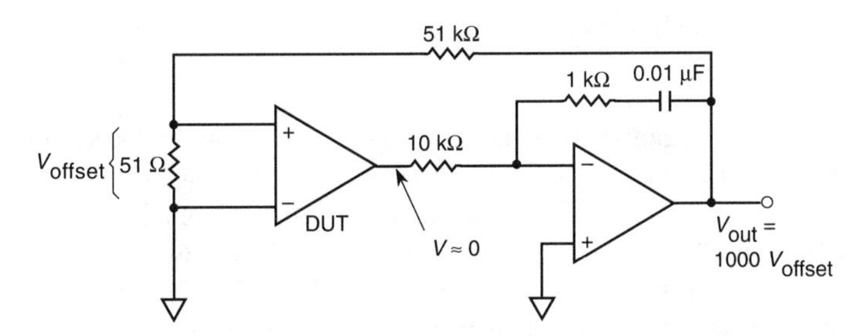

Fig. 12.22 A test circuit for measuring the offset voltage of an op-amp as suggested by Frederiksen (1988). The additional op-amp ensures that the output of the device under test (DUT) is very near 0 V for measuring the offset voltage.

Keep the input resistor small so that the input current will not cause significant measurement error.

12.8.3 *Offset Current*

The offset current is the sum of the inverting and noninverting input currents when the output of the op-amp is zero for DC applications. These currents are small, so you can't discount the offset voltage in the measurement. For op-amps with FET inputs, you will have to integrate the offset current for a time and then divide the resulting output by the integration time. Frederiksen (1988) has developed two separate circuits to measure the inverting and the noninverting input currents.

12.8.4 *Dynamic Loop Gain*

You can analyze the stability of feedback networks by plotting loop gain as a function of frequency. The phase at the unity-gain frequency indicates stability. A stable feedback loop has a positive difference between the phase (at unity gain) and −180°. Fig. 12.23 shows an example of phase margin in a stable circuit.

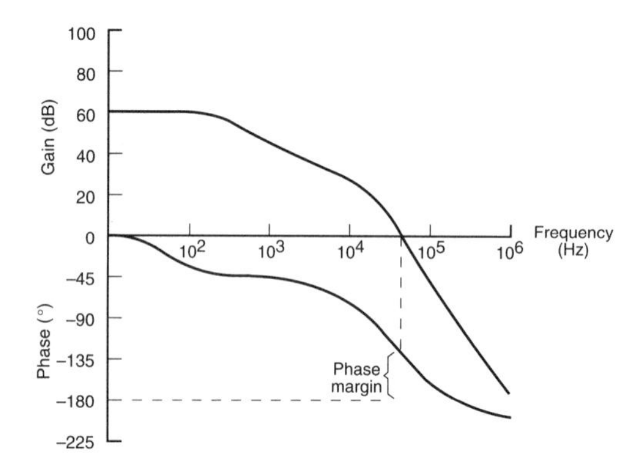

FIG. 12.23 An example of a stable phase margin in a Bode plot.

Measuring loop gain dynamically can be tricky. Circuits tend to saturate when the feedback is disconnected, and the circuit impedances can swamp the circuit's loop gain at high frequencies. Consequently, you can use both a buffer and some simulated load impedances in the test circuit to break the current and voltage loops and eliminate the errors from impedance loading. Fig. 12.24 diagrams the test circuit. Once you have the test circuit in place, sweep the frequency of the input stimulus and plot the resulting gain and phase (Caldwell, 1990).

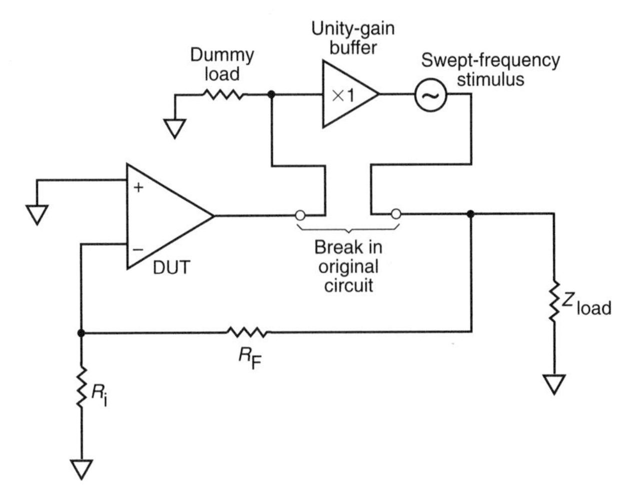

FIG. 12.24 Circuit for dynamically measuring loop gain. The unity-gain buffer and swept-frequency stimulus reduces loading errors. The buffer breaks the current loop, and the dummy load emulates the actual load.

12.8.5 *CMRR*

The common-mode rejection ratio (CMRR) defines how well an op-amp rejects common-mode voltages (that is, identical voltages presented simultaneously to the inverting and noninverting inputs). CMRR is not the same as the gain-bandwidth product, as some think. It is a nonlinear curve of the ratio between the change in offset voltage and common-mode voltage (ΔV_{offset} versus V_{cm}). You should, however, design common-mode noise out of a circuit rather than rely on the op-amp to eliminate it. Pease (1990) presents a circuit that measures CMRR at frequencies of interest rather than at DC.

12.8.6 *Power Supply Rejection Ratio*

The power supply rejection ratio defines how well the op-amp rejects the variations in voltage. If you use the test circuit for offset voltage in Fig. 12.22 and vary the supply voltages, you can determine the power supply rejection ratio.

For the op-amps that use a +15-V and –15-VDC supply, follow this procedure. Set the negative supply to –15 VDC and toggle the positive supply between +10 VDC and +20 VDC. The output should vary less than 1.0 V, which corresponds to 80 dB of rejection. Likewise, set the positive supply to +15 VDC and toggle the negative supply between –10 VDC and –20 VDC. Again, the output should vary less than 1.0 V.

12.8.7 *Slew Rate*

The slew rate of an op-amp is the limiting rate of change in output voltage for an input pulse that is faster than the op-amp. The slew rate is the smaller of the positive and negative edges. Fig. 12.25 illustrates the test circuit for measuring the slew rate.

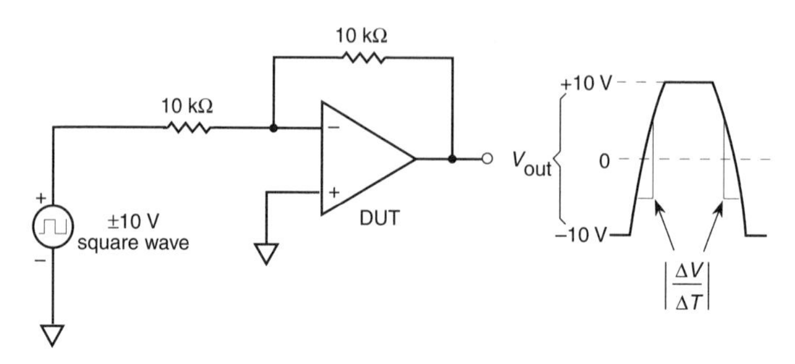

Fig. 12.25 Test circuit for measuring slew rate (Frederiksen, 1988).

12.8.8 *Output Noise*

You can measure the thermal noise of an op-amp circuit by integrating the rms noise voltage over a specified bandwidth. Fig. 12.26 illustrates the test circuit for measuring output noise.

Another type of noise in op-amps is *popcorn noise*. It is characterized by abrupt offset step voltages (which sound like popcorn popping in audio circuits) lasting milliseconds and is associated with surface contamination on the IC die

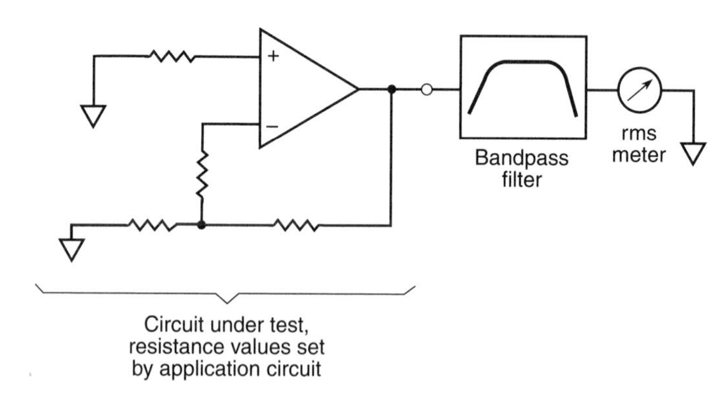

Fig. 12.26 A test circuit for measuring the noise voltage of an op-amp circuit (Whitehead, 1989).

during fabrication. You can screen op-amps for popcorn noise by measuring the spot noise between 10 and 100 Hz and discarding op-amps with pops that have higher amplitude than typical (Whitehead, 1989). But screening parts greatly increases cost; you should use a better part anyway.

12.9 ANALOG-DIGITAL CONVERSION

Components that transform signals between the analog and digital domains are necessary to computer control (Fig. 12.27). Certifying the fidelity of conversion is critical to achieving the specified performance. Unfortunately, no single test can completely characterize an ADC or a DAC. Tests are peculiar to the application; for instance, radar and video require good transient responses, digital audio requires linearity, and data acquisition requires high resolution. A good, complete reference for standard definitions and testing is *IEEE Standard 1057: Standard for Digitizing Waveform Recorders*.

You will have to verify the performance and accuracy of a converter at the desired conversion speed and not rely on the spec sheets. All converters are subject to errors from many sources, including architecture, quantization, fabrication, and circuit layout. Fig. 12.28 outlines the equipment you will need to evaluate converters. It should have proper amplifier bandwidth and settling time, filtering to remove harmonics and broadband noise, a low-jitter clock, a clean power supply with decoupling, good layout, and controlled temperature (Giacomini, 1992a). Many manufacturers of converters supply PCB-based evaluation boards to reduce concerns about layout.

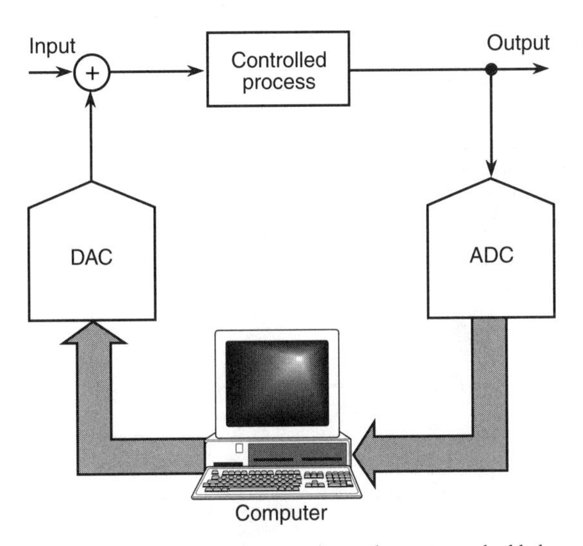

Fig. 12.27 ADCs and DACs. They are components integral to many embedded computer systems.

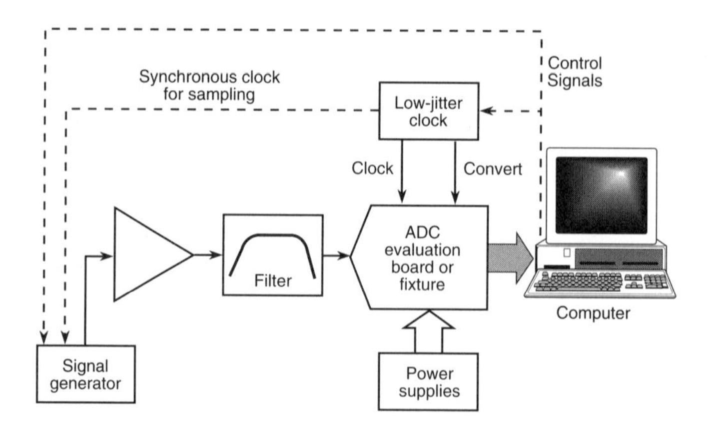

Fig. 12.28 General test configuration for evaluating ADCs.

12.9.1 *Basic Definitions*

I will begin with some basic definitions for converters so that you can better understand the rationale for testing them. The analog domain is continuous in time and signal magnitude, while the digital domain is discrete in time and magnitude. Fig. 12.29 illustrates how an ADC converts analog signals to discrete values represented by binary code. A single binary value represents a range of analog values in the quantization band around its code center point. Analog values not exactly at the code center point have an associated amount of quantization error. The step to the next binary code occurs at the code transition (Zuch, 1987a).

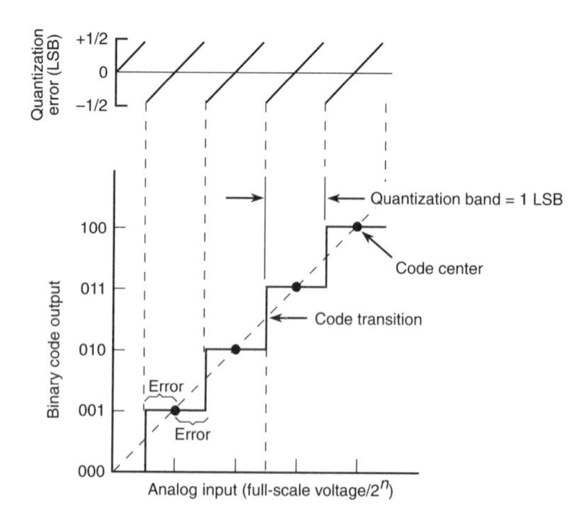

Fig. 12.29 An ideal transfer function for an ADC.

The resolution of an ADC defines its capability to represent analog values; in particular, the number of bits in the binary output sets the resolution. Tables 12.5 and 12.6 give a range of resolutions for ADCs as well as the associated quantization band, full-scale analog output, and theoretical SNR.

ADCs may represent analog values in either a single voltage range such as 0 to +5 V (unipolar converters) or in a dual voltage range such as –10 to +10 V (bipolar converters).

Fig. 12.30 illustrates how a DAC converts binary code representing discrete values into analog signals. Each binary code maps to a unique analog output, but it represents a range of analog values in the quantization band around its code center point.

Table 12.5 Resolution, binary weights, and analog scale values for unipolar converters

			Quantization band		Analog full-scale output		
Resolution (n = bits)	**Number of states (2^n)**	**Binary weight (2^{-n})**	**(+10 V range)**	**(+5 V range)**	**(+10 V range)**	**(+5 V range)**	**Theoretical SNR (dB)**
4	16	0.0625	0.625 V	0.3125 V	9.3750	4.6875	25.8
6	64	0.01563	0.1563 V	78.13 mV	9.8438	4.9219	37.9
8	256	3.906×10^{-3}	39.06 mV	19.53 mV	9.9609	4.9805	49.9
10	1,024	9.765×10^{-4}	9.765 mV	4.883 mV	9.9902	4.9951	62.0
12	4,096	2.441×10^{-4}	2.441 mV	1.221 mV	9.9976	4.9988	74.0
14	16,384	6.104×10^{-5}	610.4 μV	305.2 μV	9.9994	4.9997	86.0
16	65,536	1.526×10^{-5}	152.6 μV	76.29 μV	9.9998	4.9999	98.1

Table 12.6 Resolution, binary weights, and analog scale values for bipolar converters

			Quantization band		Analog full-scale output		
Resolution (n = bits)	**Number of states (2^n)**	**Binary weight (2^{-n})**	**(+10 V range)**	**(+5 V range)**	**(+10 V range)**	**(+5 V range)**	**Theoretical SNR (dB)**
8	256	3.906×10^{-3}	78.13 mV	39.06 mV	9.9219	4.9609	49.9
10	1,024	9.765×10^{-4}	19.53 mV	9.765 mV	9.9805	4.9902	62.0
12	4,096	2.441×10^{-4}	4.883 mV	2.441 mV	9.9951	4.9976	74.0
14	16,384	6.104×10^{-5}	1.221 mV	610.4 μV	9.9988	4.9994	86.0
16	65,536	1.526×10^{-5}	305.2 μV	152.6 μV	9.9997	4.9998	98.1
18	262,144	3.815×10^{-6}	76.29 μV	—	9.999924	—	110.1
20	1,048,576	9.536×10^{-7}	19.07 μV	—	9.999981	—	122.2
22	4,194,304	2.384×10^{-7}	4.768 μV	—	9.999996	—	134.2
24	16,777,216	5.960×10^{-8}	1.192 μV	—	9.999999	—	146.2

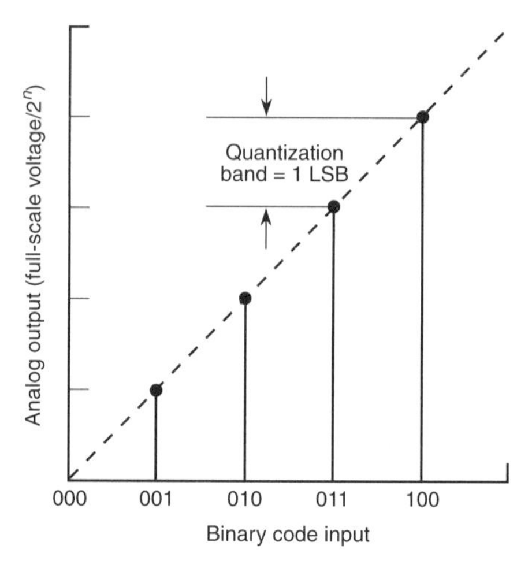

FIG. 12.30 An ideal transfer function for a DAC.

While most of us are familiar with the straight binary code, converters can use a variety of binary codes. For unipolar converters, straight binary code is the most popular; complementary binary code, however, is convenient for many DAC architectures and binary-coded decimal is specific for decimal digit displays (otherwise it is inefficient). Table 12.7 illustrates these codes for a 12-bit unipolar converter (Zuch, 1987b).

Bipolar converters may use offset binary code, two's complement, or sign-magnitude binary. Computation (digital arithmetic) can use two's complement, but it is not natural for DAC architectures. Table 12.8 illustrates these codes for a 12-bit bipolar converter (Zuch, 1987b).

Table 12.7 Examples of unipolar codes for a 12-bit converter

Fraction of	**Analog voltage range**		**Straight**	**Complementary**	**Binary-coded**
full scale (FS)	**+10 V**	**+ 5 V**	**binary**	**binary**	**decimal**
+FS–1 LSB	9.9976	4.9988	1111 1111 1111	0000 0000 0000	1001 1001 1001
+3/4	7.5000	3.7500	1100 0000 0000	0011 1111 1111	0111 0101 0000
+1/2	5.0000	2.5000	1000 0000 0000	0111 1111 1111	0101 0000 0000
+1/4	2.5000	1.2500	0100 0000 0000	1011 1111 1111	0010 0101 0000
1 LSB	0.0024	0.0012	0000 0000 0001	1111 1111 1110	0000 0000 0001
0	0	0	0000 0000 0000	1111 1111 1111	0000 0000 0000

Table 12.8 Examples of bipolar codes for a 12-bit converter

Fraction of full scale (FS)	Analog Voltage Range ±10 V	±5 V	Offset binary	Sign-magnitude binary	Two's complement
+FS−1 LSB	+9.9952	+4.9976	1111 1111 1111	1111 1111 1111	0111 1111 1111
+3/4	+7.5000	+3.7500	1110 0000 0000	1110 0000 0000	0110 0000 0000
+1/2	+5.0000	+2.5000	1100 0000 0000	1100 0000 0000	0100 0000 0000
+1/4	+2.5000	+1.2500	1010 0000 0000	1010 0000 0000	0010 0000 0000
+0	0	0	1000 0000 0000	1000 0000 0000	0000 0000 0000
−0	0	0	1000 0000 0000	0000 0000 0000	0000 0000 0000
−1/4	−2.5000	−1.2500	0110 0000 0000	0010 0000 0000	1110 0000 0000
−1/2	−5.0000	−2.5000	0100 0000 0000	0100 0000 0000	1100 0000 0000
−3/4	−7.5000	−3.7500	0010 0000 0000	0110 0000 0000	1010 0000 0000
−FS + 1 LSB	−9.9952	−4.9976	0000 0000 0001	0111 1111 1111	1000 0000 0001
−FS	−10.0000	−5.0000	0000 0000 0000	—	1000 0000 0000

12.9.2 *Nonideal Behavior and Errors*

Seven types of error are prominent in converter designs: quantization error, differential nonlinearity, integral nonlinearity, missing codes, aperture jitter, input noise, and settling error. Some of these occur in both ADCs and DACs:

1. *Quantization error:* range of analog values around the code center point.
2. *Differential nonlinearity:* differences in the quantization band between each digital code; it can vary between −1 LSB and greater than +1 LSB (Fig. 12.31). You should strive for differential nonlinearity of less than ±0.5 LSB in your designs.

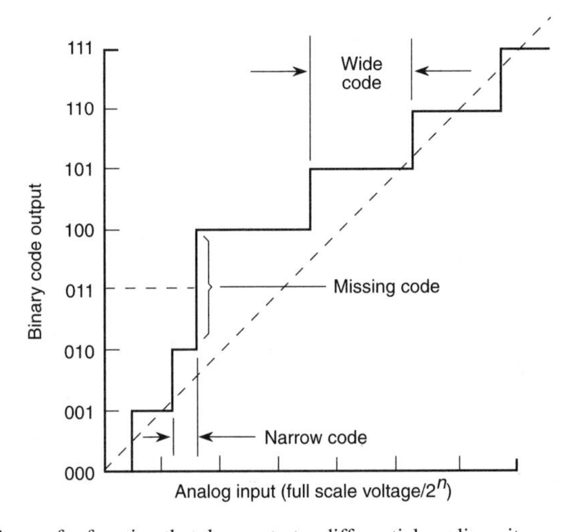

Fig. 12.31 A transfer function that demonstrates differential nonlinearity.

3. *Integral nonlinearity:* the deviation from a linear transfer function. For AC signals, it causes harmonic distortion (Fig. 12.32).

Other errors are specific to ADCs:

4. *Missing codes:* a special case of differential nonlinearity in which specific binary codes aren't produced at the output regardless of the value of the analog input (see Fig. 12.31). A missing code is a differential nonlinearity of −1 LSB. An ADC should have no missing codes at the desired resolution; you might use an ADC with more advertised resolution than you need to eliminate missing codes.
5. *Aperture jitter:* the uncertainty in the times at which samples are taken. This is a particular concern for high-speed ADCs.
6. *Input noise:* the equivalent noise contributed by the internal circuitry.

One type of error is specific to DACs:

7. *Settling error:* time for the output to reach the final analog output value.

Quantization error sets the noise floor for a converter. It limits the theoretical SNR. Assuming a sine wave input, it can be calculated as follows:
Taking the rms signal = signal amplitude/ $\sqrt{2}$ = (full scale/2) / $\sqrt{2} = 2^{n-1}\, q/\sqrt{2}$, where q is the value of the LSB, and the rms noise $= q/\sqrt{12}$ (it is the rms value of the quantization error in Fig. 12.29), then SNR = (rms signal)/(rms noise) $=(2^{n-1} q/\sqrt{2})/(q/\sqrt{12}) = 2^{n-1}\sqrt{6}$.

Therefore, the SNR for an ideal ADC, with n bits of resolution, is

$$\mathrm{SNR(dB)} = 20\log_{10}(2^{n-1}\sqrt{6}) = 6.02n + 1.76.$$

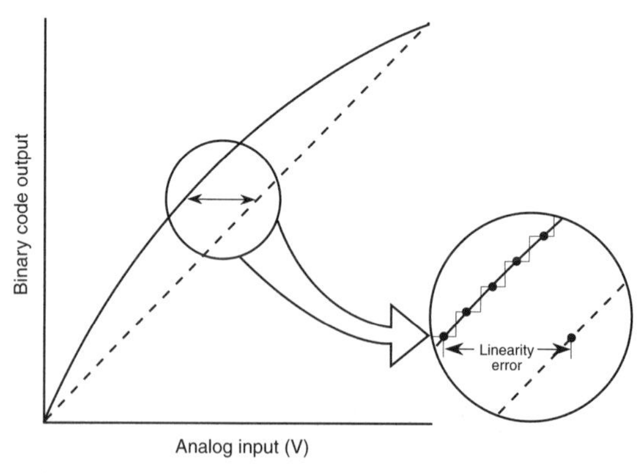

Fig. 12.32 A transfer function that demonstrates integral nonlinearity.

This formula calculated the SNRs found in Tables 12.5 and 12.6. Consequently, the dynamic range of a converter is $20 \log_{10}(2) = 6.02$ dB/bit of resolution.

Manipulating the SNR gives the effective number of bits, which is a measure of the signal power versus noise power. Any noise or distortion in the ADC will reduce the number of effective bits (Grove, 1992):

$$\text{Effective bits} = n = [\text{SNR (dB)} - 1.76]/6.02.$$

If you can measure the SNR, you can calculate the effective bits and compare that value to the specified resolution of an ADC to define its performance. Fig. 12.33 gives an example of a spectrum from an ADC with a signal fundamental, distortion harmonics, and noise. The actual SNR includes distortion products and noise and is less than the theoretical SNR (which is due to quantization error alone). The effective bits, therefore, will be less than the specified resolution of the converter.

The limitations of device physics and converter architectures show up in slew rate, settling time, recovery time from large voltage inputs, and comparator hysteresis. Variations in fabrication lead to differential and integral nonlinearities, gain variation, offset, aperture jitter, lack of amplitude flatness, and nonlinear phase. Circuit design can contribute spurious noise, clock and conversion jitter, and power supply noise and variation.

Understanding that errors are inevitable, you should prepare an error budget that will estimate the degradation in resolution of the converter. You take the root sum square of the errors to get a statistical error. You can use the following list to prepare an error budget (Zuch, 1987a):

1. Quantization (bits of resolution)
2. Linearity error (% of full scale)
3. Drift over temperature (in fraction of LSB)

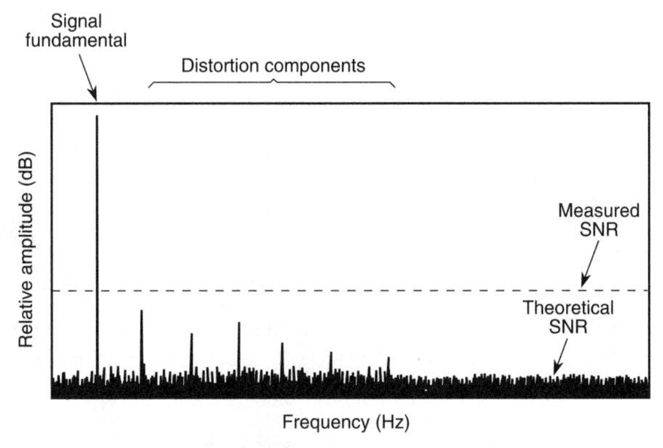

Fig. 12.33 An example of a spectrum produced by an actual converter.

4. Gain or full-scale error (% of full scale)
5. Differential nonlinearity (in fraction of LSB)
6. Offset error (in fraction of LSB)
7. SNR (dB)
8. Power supply sensitivity (dB)
9. Other errors (dependent on your application):
   - Aperture jitter
   - Amplitude flatness
   - Nonlinear phase
   - Random noise (thermal, 1/*f* or flicker, shot)
   - Spurious (unwanted pickup)

Example 12.9.2.1 A 16-bit ADC has the following specs:

Error	**Units**	**% error**
Quantization	16 bits	0.00076
Linearity error	0.001% of full scale	0.001
Drift over temperature	±0.25 LSB	0.00038
Gain or full-scale error	0.003% of full scale	0.003
Differential nonlinearity	±0.5 LSB	0.00076
Offset error	±1 LSB	0.00152
SNR	90 dB	0.003
Power supply sensitivity	84 dB	0.006

The total statistical error = 0.0077%, which corresponds to 82 dB, or somewhere between 13 effective bits (80 dB) and 14 effective bits (86 dB) of resolution. The worst-case error is a linear sum = 0.016%, which corresponds to 76 dB, or 12 effective bits of resolution—a highly unlikely case.

12.9.3 *ADC*

Ideally, an ADC assigns a specific digital code for each increment of analog input voltage. Physical limitations, manufacturing variation, component tolerance, and system noise conspire, as mentioned above, to prevent perfectly sized increments and assignments of digital codes. The transfer function of an ADC may change significantly under dynamic operating conditions. Consequently, you must test the ADC under conditions similar to the intended operating environment to characterize its transfer function.

12.9.4 *High-Accuracy ADCs*

High-resolution ADCs require accuracy and precision, which means good integral and differential linearity. These properties are generally not achieved at high conversion speed. Two tests form a basic paradigm that can characterize high-resolution ADCs. The first test is a histogram test for differential nonlinearity (DNL). The second test is a frequency response test for integral nonlinearity (INL).

12.9.4.1 The Histogram Test for DNL

The width of the quantization band can deviate greatly from the ideal and create nonideal bands called *differential nonlinearity errors*. You quantify these errors, expressed in LSBs, by measuring the actual width of the quantization band and subtracting the ideal width. Missing codes are a special case of DNL. The quantization band shrinks to 0, and the DNL becomes −1 LSB. A few localized DNLs and missing codes may be inconsequential when you digitize full-scale signals, but when the signal becomes a fraction of full scale, these errors can greatly distort a measurement (Neil and Muto, 1982).

A simple test that generates a histogram will reveal DNL and missing codes. You can use the test configuration shown in Fig. 12.28 to test for DNL in an ADC. Input a clean, accurate signal that is noncoherent with the sampling rate of the ADC. (*Noncoherent* means not synchronous with any harmonic of the sampling frequency.) Have the ADC digitize the signal, and count the number of occurrences for each output code. Deviation from the average number of occurrences indicates a DNL error. Codes with a larger-than-average number of counts must have quantization bands wider than ideal. Codes with less than the average must be narrower than ideal. Codes that have zero counts are missing codes (Harris, 1987).

You find the DNL at any code (or point) by dividing the counts (actual number of code occurrences) by the ideal value for that code and then subtracting 1. This calculation scales and offsets the data so that a perfect code (count equals ideal value) gives a value of 0 LSB for the DNL error. A code that is 0.5 LSB too wide gives a +0.5 LSB DNL error. Conversely, a code that is 0.5 LSB too narrow gives a −0.5 LSB DNL error. A missing code gives a −1 LSB DNL error. Fig. 12.34 illustrates how the numbers from the range of possible codes may be plotted as DNL versus code.

The selection of the type of input signal and its generator will affect your test for DNL. A triangular wave is simple and useful for ADCs with 14 bits of resolution or fewer in most situations, but the fidelity of the waveform generator is critical for producing a good, undistorted signal. A triangular wave produces a flat probability density function for the number of occurrences per code bin in the histogram. Therefore, you can calculate the DNL for a triangular wave input as follows:

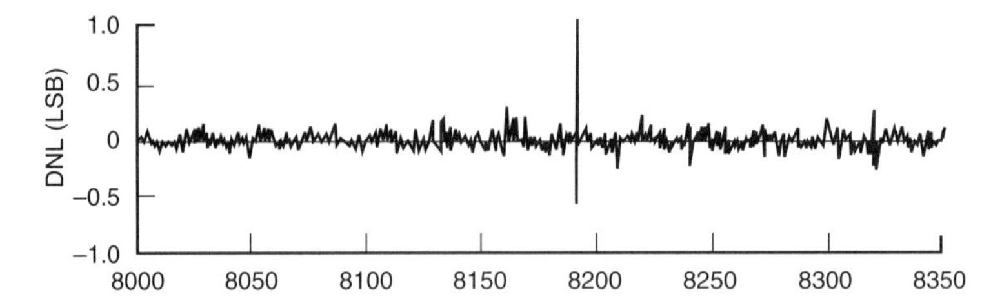

Fig. 12.34 An example plot of DNL error from a histogram test of a 14-bit ADC. Note that only 350 codes out of a total of 16,384 are shown.

The ideal count per code bin is

$$C_{\text{avg}} = N/(2^n),$$

and the error for the *i*th bin is

$$\text{DNL} = (C_i / C_{\text{avg}}) - 1,$$

where N = total number of samples taken
n = bits of resolution in the ADC
C_{avg} = the average (or ideal) count per bin
C_i = the count in the *i*th bin

How many samples (N) should you take? That depends on the variation of the data. If it is large, you must take a large number of samples. Generally, somewhere between 50 and 200 samples per code bin will suffice for an accurate calculation of DNL.

On the other hand, only a pure sine wave is practical or even possible for ADCs with 16 bits of resolution or more. A sine wave requires more intense computation because it has a saddle-shaped probability density function (Fig. 12.35). This test is relatively insensitive to even-order distortion. You can calculate the DNL for a sine wave input as follows (Neil and Muto, 1982):

The ideal count for the *i*th code bin is

$$P_i = \left\{\sin^{-1}[V(I_i - 2^{n-1})/A2^n] - \sin^{-1}[V(I_i - 1 - 2^{n-1})/A2^n]\right\}/\pi,$$

and the error for th *i* th bin is

$$\text{DNL} = (C_i/NP_i) - 1,$$

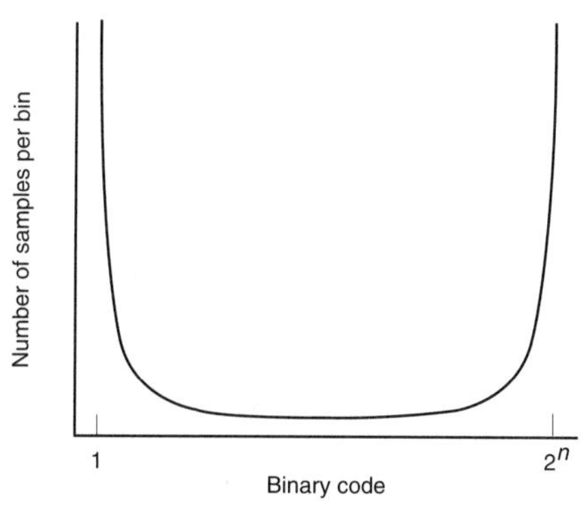

FIG. 12.35 The probability density function for a sine wave used in a DNL histogram test.

where
V = full-scale analog range of the ADC
A = positive peak analog amplitude
n = bits of resolution in the ADC
N = total number of samples taken
P_i = the ideal count in the ith bin
C_i = the count in the ith bin
I_i = ith code value

Generally, somewhere between 400 and 800 samples per code bin will suffice for an accurate calculation of DNL. The codes in the middle of the range have many fewer counts than those at the ends of the range (remember the saddle-shaped probability density function) and require many more total samples to be taken.

12.9.4.2 Testing INL versus Frequency

Another way of characterizing dynamic errors is INL. While DNL measures how far a converter's quantization bands are from the ideal, INL measures the curvature in the converter's overall transfer function. Generally, INL becomes problematic when digitizing a full-scale input, since a lower-amplitude signal could fall into a part of the transfer function that is relatively linear (Neil and Muto, 1982).

You can use the test configuration shown in Fig. 12.28 to test for INL in an ADC. You apply a very pure sine wave at nearly full-scale amplitude to the ADC and perform a fast Fourier transform (FFT) analysis on its output. The resulting spectrum should ideally have just one component at the frequency of the input sine wave, and the rest of the spectrum should be noise. Any other components represent nonlinearities in the transfer function of the ADC and reduce the SNR of the converter.

Unfortunately, data samples with finite length cause problems for the FFT analysis because the partial cycles of the sine wave signal at the beginning and end of the sample interval cause discontinuities in the data record (Fig. 12.36b). (The

FFT assumes that the data are repetitious and continuous for all time; any recording discontinuities appear to be part of the real signal.) The errors that result from the discontinuities in the data record are called *side lobes* or *leakage* and cause the spectral line from the input sine wave to spread into a series of spectral lines. You can use a window function on the incoming data samples to reduce the leakage significantly. A window function weights the data in the middle of the record more heavily and smoothly suppresses the points toward both ends of the sample interval (Fig. 12.36c).

You will find many window functions available for FFT processing; the main trade-off between them is frequency resolution versus distortion effects introduced by the window. A window function whose distortion side lobes are low will also spread out the frequency components. I prefer the Blackman-Harris window (Harris, 1987; Pinkowitz, 1986):

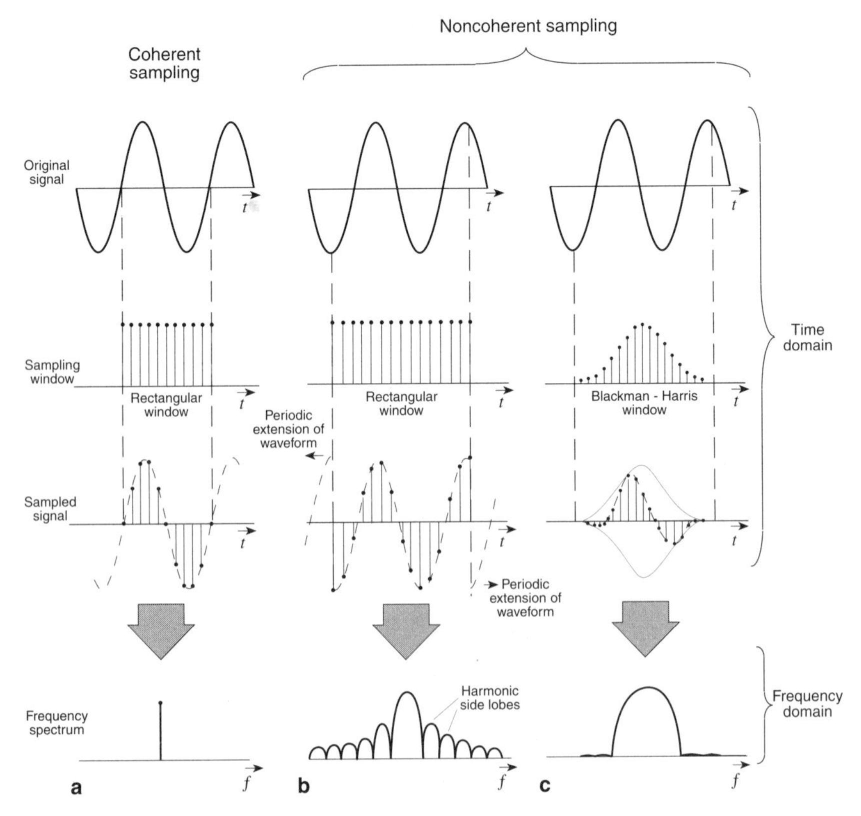

Fig. 12.36 Problems with sampling. Coherent sampling can produce an artificially clean spectrum with a single frequency as shown in (a). The sampling window function drastically affects the computed spectrum for noncoherent sampling as shown in (b) and (c). The trade-off in window function is between larger harmonic side lobes and a wider peak at the fundamental frequency.

$$W_i = 0.35875 - 0.48829 \cos\frac{2\pi(i-1)}{N} + 0.14128 \cos\frac{4\pi(i-1)}{N} - 0.01168 \cos\frac{6\pi(i-1)}{N},$$

where W_i = weighting of the ith data sample
N = number of samples in the data record

The value of the ith weighted data sample then becomes W_iD_i where D_i is the value of the original data sample (Fig. 12.36c).

Once the FFT has produced the spectrum, you can calculate the actual SNR to characterize the INL of the ADC. First, calculate the magnitude of the input sine wave (the signal) by root sum square (RSS) of the energy contained in the bin with the largest spectral line and the three bins on either side (a total of seven bins contain the energy of the windowed signal fundamental). Next, calculate the magnitude of the noise floor by RSS of the remaining bins. Finally, take the ratio of the signal magnitude to the RSS value of the noise floor. Any difference between the theoretical SNR and the measured SNR represents INL. (Strictly speaking, this test is not an INL test because it also includes noise, jitter, and spurious effects and it does not give INL versus code.)

$$S = \sqrt{\sum_{i=n}^{n+7} M_i^2}$$

$$\mathrm{RSS}_{\mathrm{noise}} = \sqrt{\sum_{i=1}^{n-1} M_i^2 + \sum_{i=n+8}^{N} M_i^2}$$

$$\mathrm{SNR} = S/\mathrm{RSS}_{\mathrm{noise}}$$

or

$$\mathrm{SNR\ (dB)} = 20 \log_{10}(S/\mathrm{RSS}_{\mathrm{noise}}),$$

where S = input fundamental signal
$\mathrm{RSS}_{\mathrm{noise}}$ = noise plus harmonics
M_i = magnitude of ith spectral bin
n = first spectral bin of the fundamental
N = number of samples in the data record

Fig. 12.37 illustrates some results from testing a 16-bit ADC.

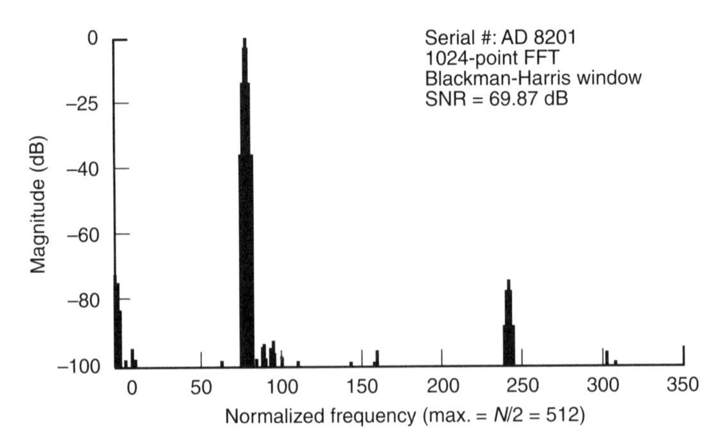

Fig. 12.37 A spectral plot for calculating INL for a 16-bit ADC.

12.9.4.3 Input Noise

Input noise is the equivalent noise contributed by the internal circuitry (it is over and above the noise contributed by the quantization error). Often comparators are the source of input noise. If the noise exceeds 1 LSB, then testing can be as simple as shorting the inputs and measuring the distribution of output codes. If noise is less than 1 LSB, then a more sophisticated test may be needed. You can sweep the input signal through the desired point with a ramp generator and record the output codes. Then calculate the probability density functions of the codes around the desired point. Less noise will narrow the distribution. More noise will widen the distribution.

12.9.5 *High-Speed ADCs*

High-speed ADCs, primarily flash converters, have a special set of concerns aside from measuring DNL and INL. Aperture uncertainty (or jitter) in the S/H block becomes a significant source of error. Applications such as radar and video require good transient responses, differential gain, and differential phase. The following tests can measure and quantify these parameters.

12.9.5.1 Aperture Uncertainty

Aperture uncertainty or jitter affects the performance of ADCs at high frequencies. Fig. 12.38 quantifies the effect for different values of jitter (Kester, 1990).

Ushani (1991b, p. 158) writes, "Getting an exact measurement of aperture uncertainty is impossible—parameters such as test-setup timing jitter, noise on the input signal, and the A/D converter's noise are not distinguishable from aperture

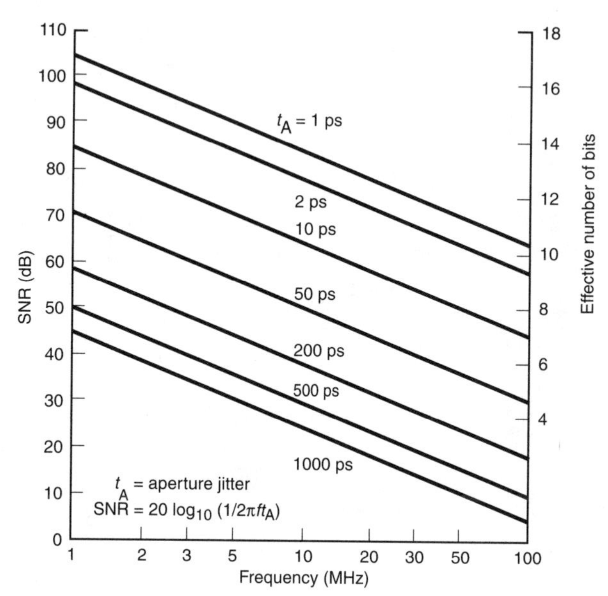

Fig. 12.38 Aperture uncertainty or jitter increases the noise over frequency in an ADC.

uncertainty." But you can get an estimate of the uncertainty by assuming that the noise is constant over frequency; then noise resulting from the aperture uncertainty is proportional to the slew rate. An indirect method for measuring noise due to aperture uncertainty drives the output from the S/H block into a spectrum analyzer to eventually derive the noise voltage. Then aperture uncertainty is the ratio of the noise voltage to the slew rate. Giacomini (1992a) details the equipment and procedure to measure the aperture uncertainty for ADCs that have an output from the S/H block.

12.9.5.2 Intermodulation Products

A nonlinear device such as an ADC produces intermodulation products when it processes multiple tones or frequencies; the energy within the intermodulation products sets the system's dynamic range. Testing for intermodulation products is very similar to testing for INL. The only difference is that you generate and combine pure, dual-frequency sine waves as inputs and then calculate the dynamic range. Fig. 12.39 shows how to modify the general test configuration in Fig. 12.28 to measure intermodulation products. (Beware. Summing two signals without distortion can be difficult [Giacomini, 1992b].)

The clock jitter must be low; otherwise it will look like aperture jitter in the S/H block and adds intermodulation products. Sampling ADCs are more susceptible to random noise in the clock than delta-sigma converters that use oversampling,

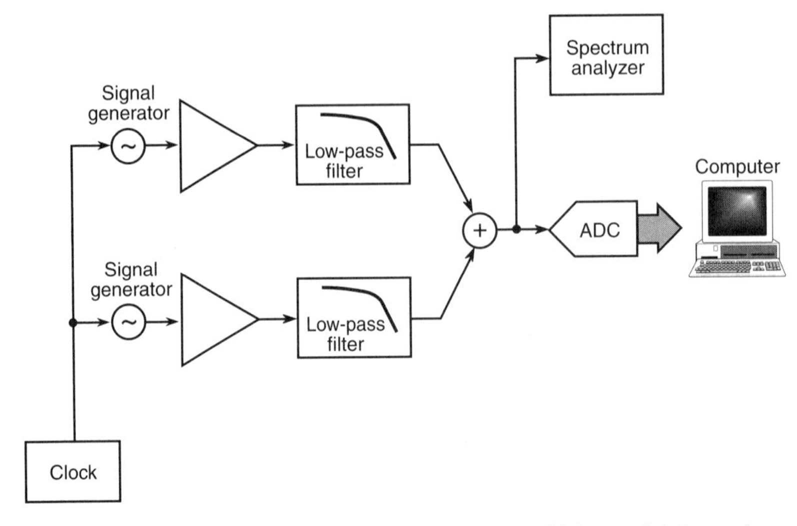

FIG. 12.39 Test configuration for measuring the dynamic range with intermodulation products.

noise shaping, and digital filtering to generate their digital codes. The filtering reduces the clock jitter (Harris, 1990).

12.9.5.3 Transient Response

The transient response of an ADC is particularly important for high-speed applications. According to Kester (1990, p. 107), "The response of a flash converter to a transient input such as a square wave is often critical in radar applications." The input must be a good, clean, fast pulse to test the transient response of an ADC. Fig. 12.40 illustrates a test configuration that tests the delay for a 0–100% rise time. It produces a pulse by using an analog multiplexer to switch between two voltage levels and then measures the transient response (Ushani, 1991b).

You begin by dialing a programmable delay into the clock circuit to start a conversion. If the delay is less than the transient response time of the ADC, it will generate intermediate values between the reference values. You increase the delay until the converter's output produces only the reference values, then you calculate the transient response time by summing the delay time and the ADC's conversion time.

12.9.5.4 Overvoltage Recovery

You can use the setup for transient response (Fig. 12.40) to measure the ADC's recovery time to overvoltages. Adjust one of the reference voltages to its maximum overrange value, and set the other reference voltage to a value near full scale in the

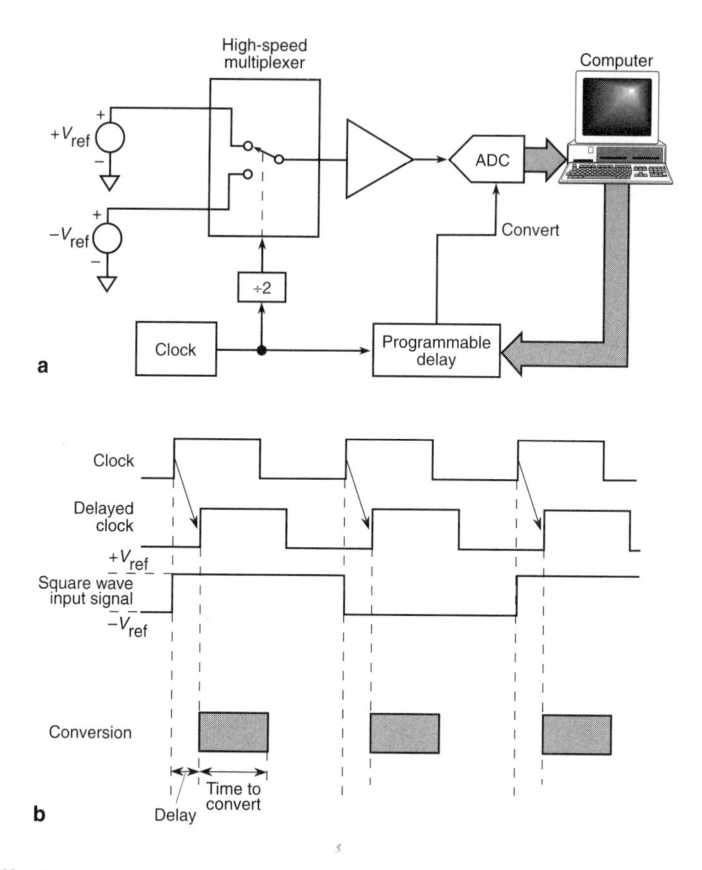

Fig. 12.40 Test configuration for measuring the transient response time of an ADC. (a) Equipment block diagram. (b) Timing diagram for calculating the response time.

opposite polarity or direction. Adjust the conversion delay until the output gives the valid value. The delay equals the overvoltage recovery time (Ushani, 1991b).

12.9.5.5 Differential Gain and Differential Phase

If you deal with composite video signals, you should measure the differential gain and differential phase. Differential gain is the percentage difference between the digitized amplitude of two signals. Differential phase is the phase difference between the digitized values of the same two signals. The input signal is a low-level, high-frequency sine wave (to represent a color subcarrier frequency) superimposed on a low-frequency sine wave. An ADC should process the input without distortion so that neither the amplitude nor the phase of the chrominance signal is altered as a function of the luminance signal (Kester, 1990; Ushani, 1991b).

12.9.6 *Manufacturing Calibration*

Incorporating ADCs into mass-produced instruments requires a balance between calibration accuracy and the cost of testing. The tests for DNL and INL, described above, take too long and cost too much for large quantities. If you need an alternative, you should investigate a number of possible tests for manufacturing calibration, including beat frequency, servo loop, crossplot, sine wave curve fit, frequency response, and basis set factorization.

12.9.6.1 Beat-Frequency Test

The beat-frequency test gives a quick, qualitative look at DNL and INL. The test uses a full-scale sine wave whose input frequency is slightly greater than the sampling frequency of the ADC:

$$\Delta f = f_s / 2^N \pi ,$$

where f_s = sampling frequency
N = resolution of ADC in bits

The output code should change by only 1 LSB; jumps in the output sine wave indicate missing codes (called *sparkle codes* in the TV industry), while flat spots and variations from the sine wave indicate INL. Fig. 12.41 illustrates an example of these nonlinearities.

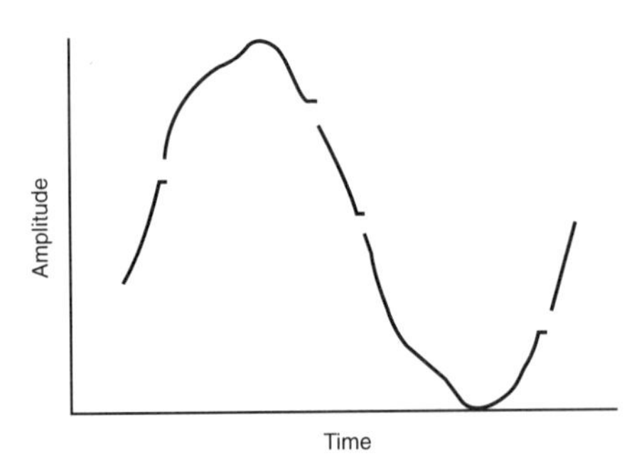

Fig. 12.41 A beat-frequency test can give a quick indication of nolinearities. Gaps represent missing codes; flat spots and dips indicate INL (Neil and Muto, 1982).

12.9.6.2 Servo Loop

A servo loop locates code transitions and gives the average value at a point. It is a good test for ADCs used in low-speed applications that require high precision and resolution. Fig. 12.42 illustrates the test setup for a servo loop.

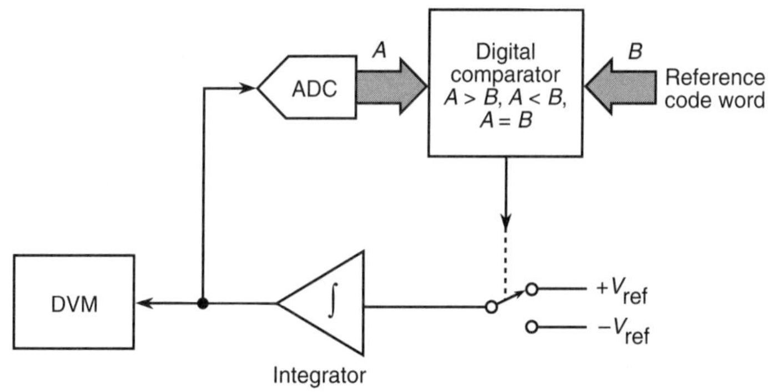

FIG. 12.42 The servo loop tests for code transitions around the point specified by the reference code word.

12.9.6.3 Crossplot

Crossplot also gives a qualitative measure of errors including DNL, INL, hysteresis, and comparator oscillation. Crossplot determines the actual transfer function of the ADC at a specific point by setting a DC voltage input, superimposing a dithered analog signal across a range of several LSBs, and displaying the ADC codes versus the dithered analog input. Calibrate the oscilloscope display to 1 division per LSB for the display that gives a stair-step output (Knapp, 1991).

Fig. 12.43 illustrates some errors found on the crossplot display. A shift in the display represents INL. A variation in step width represents DNL. Overlap indicates hysteresis; that is, the transitions are a function of the direction of slew. Finally, transition pulses indicate comparator oscillation (Knapp, 1991).

12.9.6.4 Sine Wave Curve Fit

Sine wave curve fit is a nonlinear least-squares fit of a sine wave to the data. You use the residue to give total error and effective bits. Unlike an FFT test for INL, the test needs neither an integral number of samples nor windowing. It is good for parameter estimates, but using too few periods (< 5) gives too good a fit and too many effective bits.

12.9.6.5 Frequency Response

The frequency-response test provides a measure of amplitude and phase as a function of frequency. You could use a swept sine wave, but you would need a very good, distortionless generator, and it would be difficult to get phase information. Otherwise, you could use the step response to estimate the frequency response by sweeping the sampling through a step input. Then take the discrete derivative (first-forward difference) to get impulse response and perform an FFT to get the spectrum (Souders and Flach, 1987).

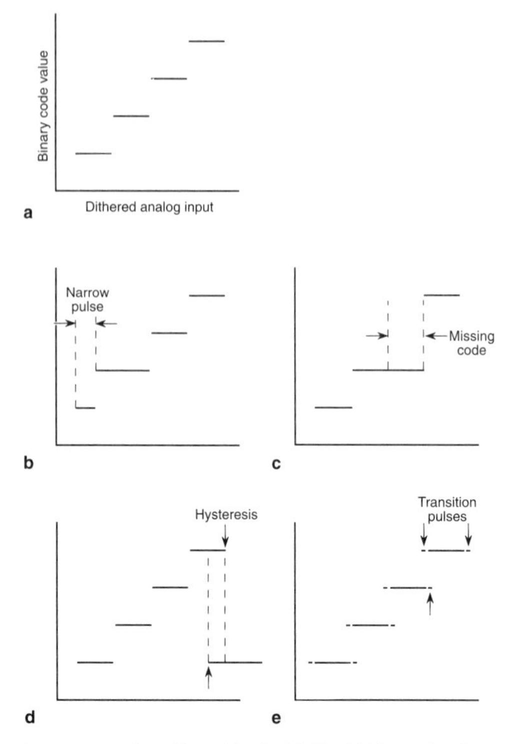

Fig. 12.43 Crossplots can reveal nonlinearities in ADCs. (a) Example of a normal crossplot. (b) and (c) Narrow or missing pulses indicate DNL. (d) Overlap indicates hysteresis in the slew. (e) Transition pulses indicate comparator oscillation.

The advantage of this method is speed of measurement, ease of getting the phase, and adaptability to automatic test equipment. You will need a precision step generator and a high enough sampling rate to avoid alias error.

12.9.6.6 Basis Set Factorization

Finally, you could use a sort of basis set factorization to test quantities of ADCs. First, you test a small sample of ADCs to develop a basis vector that characterizes the ADCs. Afterward, you confirm the ADC model with sample testing instead of exhaustive tests of each unit. Souders and Stenbakken (1991) detail the test.

12.9.7 *DACs*

DACs have four primary limitations in generating analog signals: resolution, linearity, settling time, and glitches. Resolution depends on the number of binary bits converted: The more bits there are, the greater the resolution. You may select a current-output DAC for your design because it settles faster than a comparable voltage-output DAC. For optimum settling, the DAC design can switch output to ground with a MOSFET, thereby minimizing the load impedance (Ushani, 1991a).

One popular way to test for linearity is to compare outputs from the production DACs with the output of a precise converter as a reference (Fig. 12.44). The reference converter should be more accurate and have at least 2 or 4 bits higher resolution than the production DACs. The comparison may be performed by a differencing circuit such as an op-amp or an instrumentation amplifier (Sheingold, 1986).

The settling response is the time the output of the DAC takes to approach a final value within the limits of a defined error tolerance. It becomes important for high-speed applications. One concern in testing for settling response is to avoid overdriving the differencing circuit (such as an oscilloscope preamplifier). You can design an input stage with voltage clamping for the differencing circuit. Or you can use a tracking loop scheme similar to the servo loop for testing ADCs. In the tracking loop, a programmable pulse generator drives the DAC under test. The DAC drives a fast comparator that toggles an integrator. The output of the integrator is a slowly changing voltage that represents the settling time. Fig. 12.45 illustrates the test setup for a tracking loop to test DACs (Sheingold, 1986).

Glitches occur during transitions of the digital input signals; they are caused by three mechanisms (Michaels, 1981):

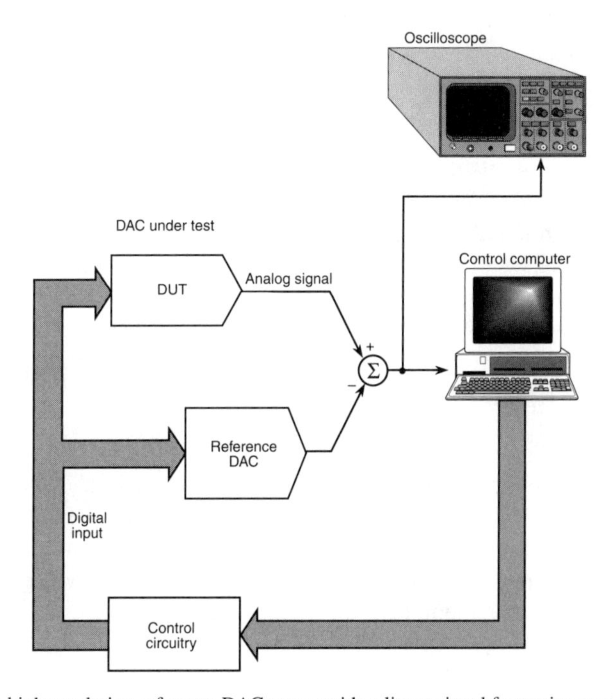

Fig. 12.44 A high-resolution reference DAC can provide a linear signal for testing production units.

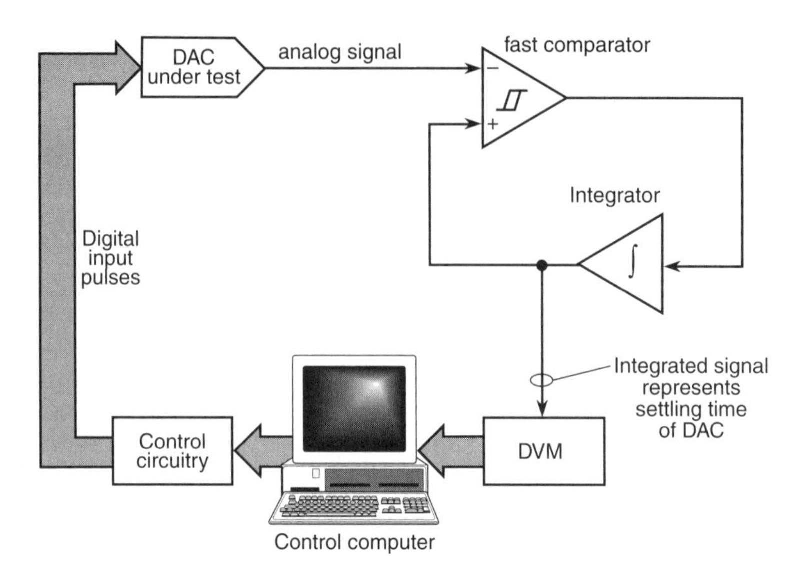

FIG. 12.45 A tracking loop used to measure the settling response of a DAC.

1. Unequal delays in logic driving converter (skew in data)
2. Unequal switching times of current drivers in converter
3. Feed-through or external coupling of logic signals into the analog output

The largest glitches occur at the midrange transition between 1000 and 0111. They disrupt low signal levels the most. The way around glitches is to choose either a low-glitch DAC or a deglitched DAC. Low-glitch DACs remove internal problems by careful switch design, but the glitch amplitude still varies with input code. Deglitched DACs follow the converter with a track-and-hold amplifier; consequently, they generate very small glitches at all codes with equal amplitude (Moscovici, 1983; Pinkowitz, 1985). You can test for glitches by probing the analog signal when the binary code changes at major and semimajor transition points.

12.10 DIGITAL COMPONENTS

Digital components deal with more than high and low logic levels. They are plagued with problems similar to those of analog components, particularly at high frequencies. I will introduce four problems—ground bounce, phase jitter, metastability, and transmission line concerns—and explain how you might go about diagnosing and solving them. I will also introduce design for test.

12.10.1 *Ground Bounce*

When the output changes state, it momentarily draws a surge of current through the pin and bond wire inductances between the chip and board and causes the chip's ground to bounce. (Ground bounce is introduced in Section 7.5 and Fig. 7.9.) Ground bounce, also called simultaneous switching noise, can cause unswitched outputs to exceed the threshold of the driven gate, thereby forcing a logic transition when none really occurred. Otherwise, ground bounce shifts the input level with a negative ground excursion that adds to low-level input voltage, which then exceeds the input threshold. Factors that contribute to ground bounce are signal rise time (faster rise times accentuate the problem), load capacitance (larger is worse), and the number of simultaneously switched outputs (more switching draws more current) (Shear, 1989). Fig. 12.46 illustrates how logic gates can generate ground bounce.

You can avoid ground bounce by using good layout that reduces the inductive loop, by using a center-pin package (about 4 nH inductance) or even MCM packaging instead of the traditional end-pin package (about 15 nH inductance), or by testing chips from different manufacturers. You will need a good test fixture, as shown in Fig. 12.47.

One caution: All logic families have thresholds that are a function of pulse width; as the width narrows, the pulse has to have a higher amplitude to trigger the logic input (Shear, 1989).

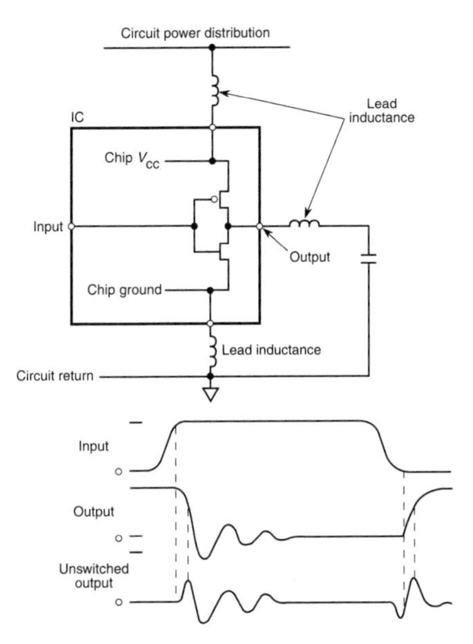

Fig. 12.46 How logic gates can generate ground bounce. Lead inductance, load capacitance, and simultaneous switching all contribute to voltage transients on unswitched outputs.

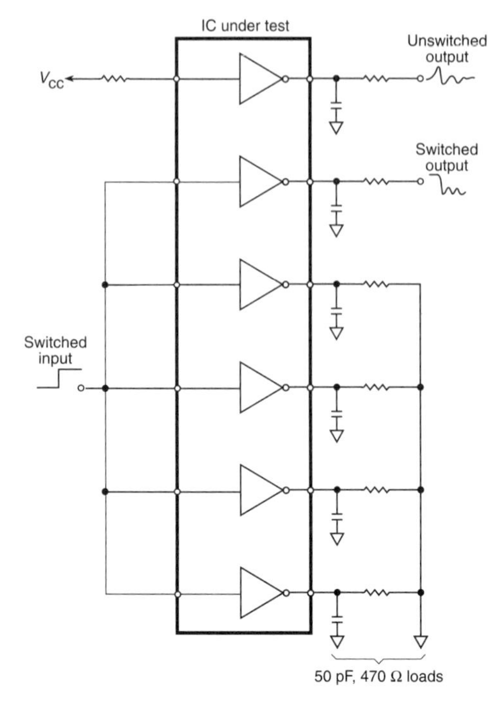

Fig. 12.47 Test fixture for measuring ground bounce in digital components.

12.10.2 *Phase Jitter*

Phase jitter is the variation in arrival times of later signal edges from a reference edge. (Phase jitter is introduced in Section 7.5 and Fig. 7.13.) You can measure the jitter by collecting a number of observations of signal edges with the test setup shown in Fig. 12.48. The sampled edge is a long time interval from the trigger. Sampling over long intervals requires a digital storage oscilloscope that can delay trigger and display multiple edges without adding appreciable jitter of its own. (Long delays generally increase jitter.)

Jitter filters high-frequency content from the signal edge of the mean waveform. If you can plot a histogram of the arrival times, you can define the jitter and its "smear width."

12.10.3 *Metastability*

Metastability occurs when the output from a logic component assumes an illegal level (between logic high and logic low) for a duration. Flip-flops are particularly susceptible. They are found in discrete packages, PLDs, PALs, and other programmable components.

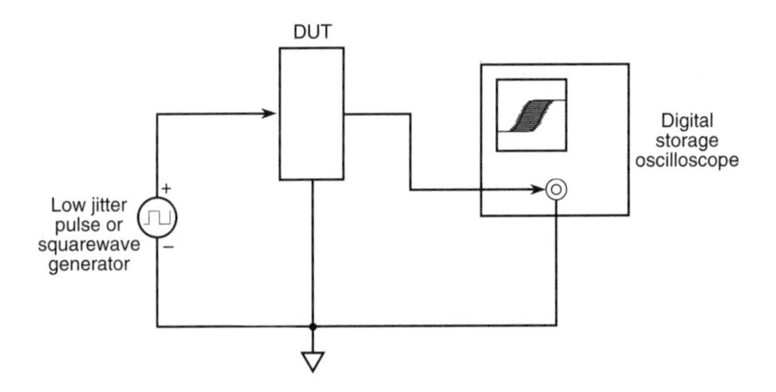

Fig. 12.48 Test configuration to measure phase jitter.

Metastability causes problems such as long propagation delay, runt pulses, oscillation, and intermediate voltages between logic levels (Shear, 1992). Fig. 12.49 points out some of these problems.

Illegal input levels cause metastability, as do uncorrelated clock and data signals that violate setup and hold time criteria at asynchronous interfaces. Fig. 12.50 shows a timing diagram to illustrate the necessary criteria for setup and hold time. Setup and hold time criteria, and therefore metastability, depend on processing during fabrication and the architecture of the device. Components from the same manufacturer can vary by a factor of 2, and components from different manufacturers, by a factor of 5 (Shear, 1992).

Metastability is a statistical problem that results in propagation delay within a flip-flop. Fig. 12.51 graphs the propagation delay versus the setup time for a component entering metastability. You can quantify metastability with a test configuration similar to that shown in Fig. 12.52. It will help you estimate the MTBF of a component as described by Shear (1992).

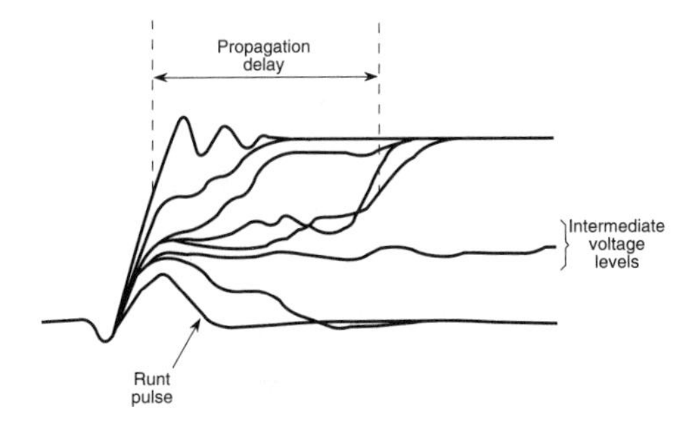

Fig. 12.49 Example of outputs caused by metastability.

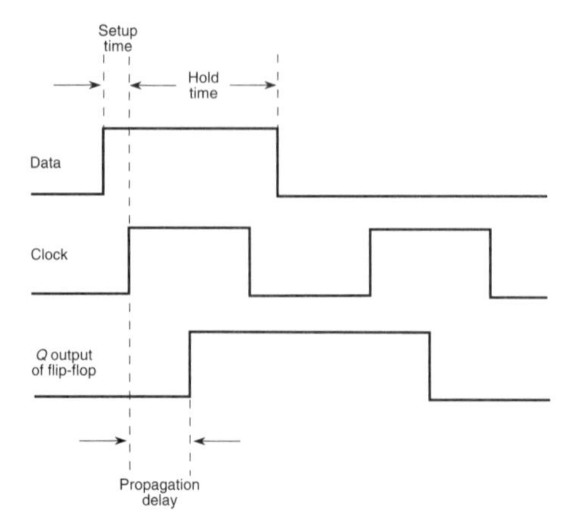

FIG. 12.50 Timing diagram that defines the location of setup and hold times.

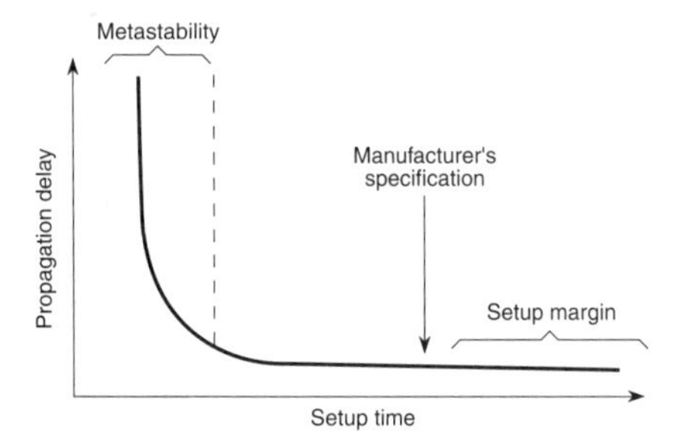

FIG. 12.51 Relationship between propagation delay and setup time. Decreased setup time increases the probability of metastability and propagation delay.

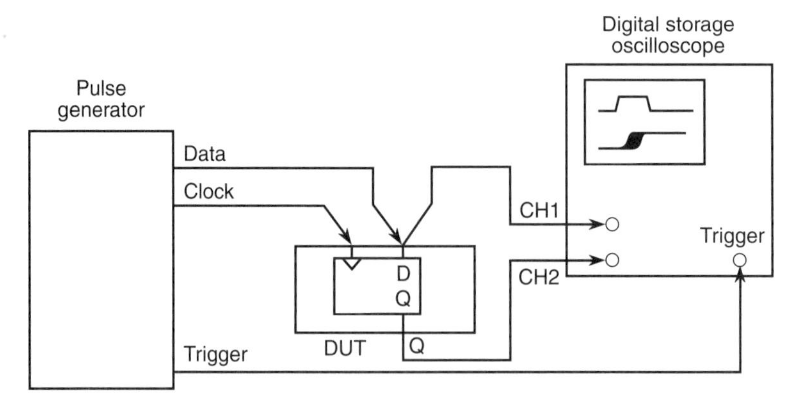

FIG. 12.52 Configuration for testing metastability. The proper equipment and test fixture will yield matched delays and accurate edge measurements.

You can reduce the probability of metastability in one of three ways: Add delay to ensure that the output resolves to a legal state, use faster parts, or use a multistage synchronizer. A multistage synchronizer interfaces asynchronous inputs to synchronous circuitry (Fig. 12.53). It greatly increases MTBF, but it does delay data.

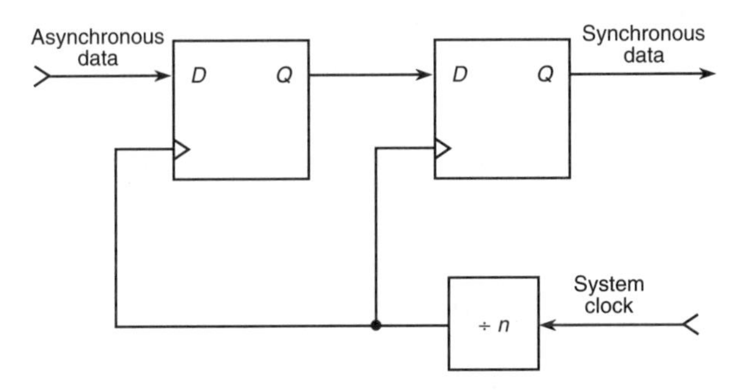

Fig. 12.53 A multistage synchronizer. It will greatly reduce the probability of metastability, but it also delays the data.

12.10.4 *Transmission Line Diagnosis*

One way you can analyze transmission line limitations in circuitry is with time-domain reflectometry (TDR). TDR elucidates impedance matching, packaging interconnections, crosstalk, and differential concepts.

Lueker (1989, p. 171) writes, "Time-domain reflectometry is analogous to radar in that a signal is sent into a component and its reflection is detected; the characteristics of the reflection reveal the characteristics of the component. The signal sent into the component is usually a voltage step, and the shorter the step's rise time, the better. An oscilloscope displays the step's reflection." Fig. 12.54 diagrams input, transmitted, and reflected energy. Fig. 12.55 shows how the reflection coefficient in a transmission line affects whether the energy is reflected or transmitted.

A typical TDR system (Fig. 12.56) measures the transmitted and reflected energy to determine the changes in impedance. It can also determine the length of the transmission from the propagation delay. Multiple discontinuities in impedance cause multiple reflections (Fig. 12.57). A TDR system needs special software to calculate impedances from the reflected energy because the TDR trace becomes difficult to interpret with multiple reflections. Moreover, the finite rise time of a signal pulse degrades the resolution of the TDR.

Fig. 12.58 shows how even distributed and lumped impedances, such as lumped capacitance or inductance, can be measured and modeled by TDR. Using these

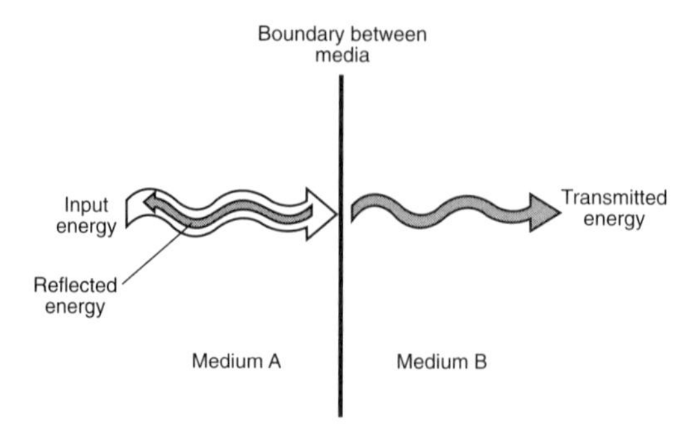

Fig. 12.54 Effect of a boundary between media. The change of material (or impedance) will transmit and reflect fractions of the original input energy.

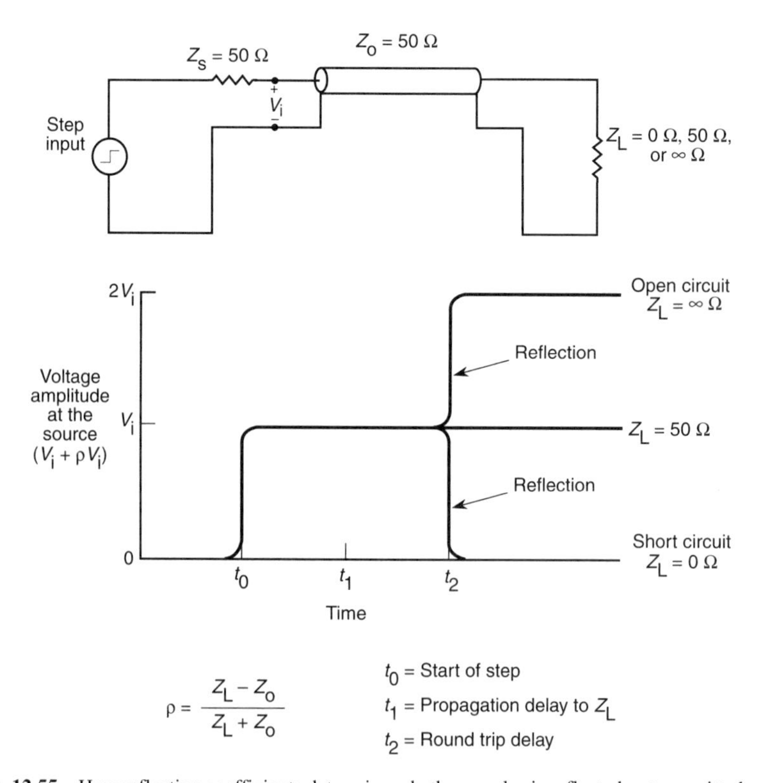

Fig. 12.55 How reflection coefficients determine whether a pulse is reflected or transmitted.

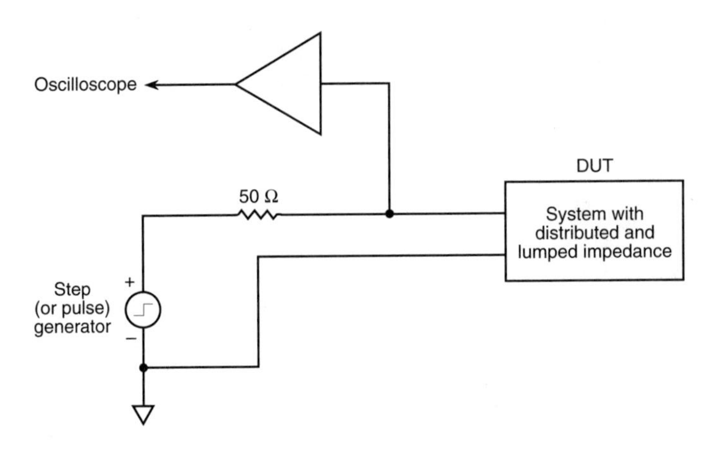

Fig. 12.56 Configuration for a typical TDR.

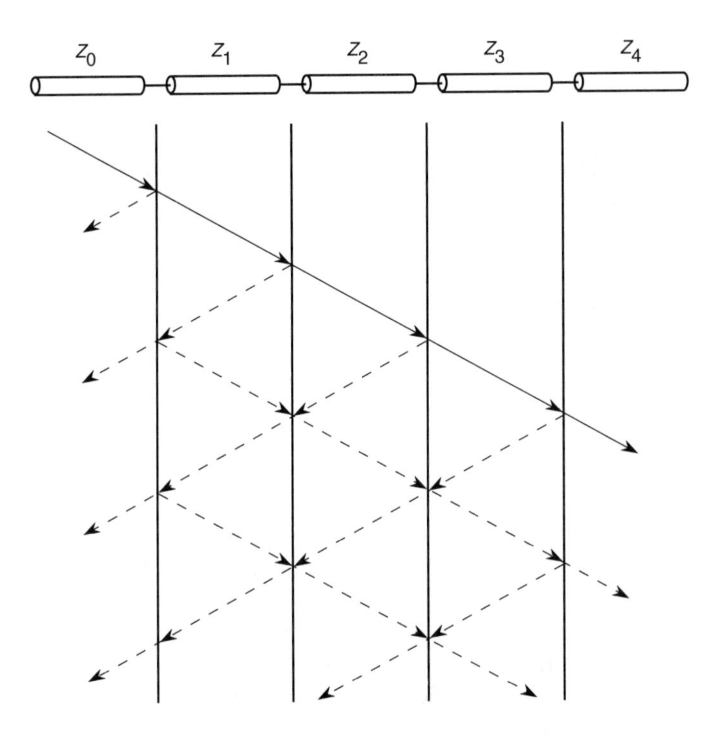

Fig. 12.57 Effect of multiple discontinuities in impedance. Multiple reflections and transmissions make the calculation of impedance very difficult.

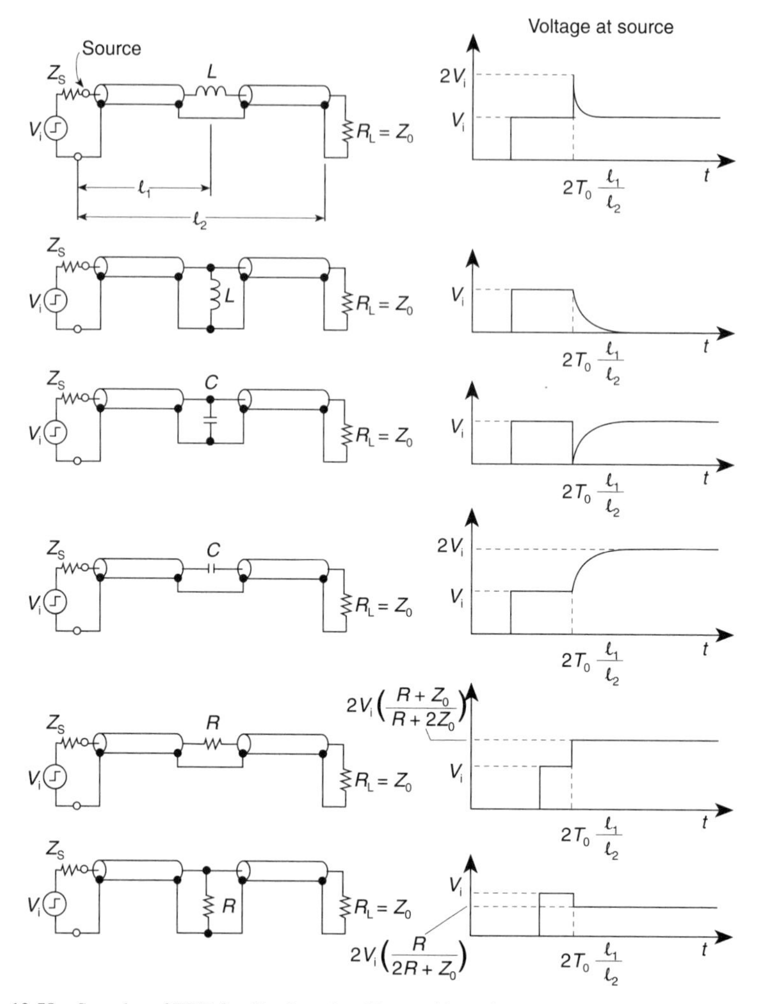

Fig. 12.58 Samples of TDR for distributed and lumped impedances.

concepts, TDR can help model the packaging interconnections within ICs. Fig. 12.59 shows an example of modeling and iteration from Tektronix advertising literature.

TDRs can have multichannel and differential applications in automatic test equipment that measures crosstalk and common-mode rejection in differential video amplifiers (Lueker, 1989). A large number of lines can allow you to detect pulse skew and crosstalk between signal lines. With differential TDR, you can measure the differential input capacitance of a differential video amplifier, its output response, and its common-mode rejection (Lueker, 1989).

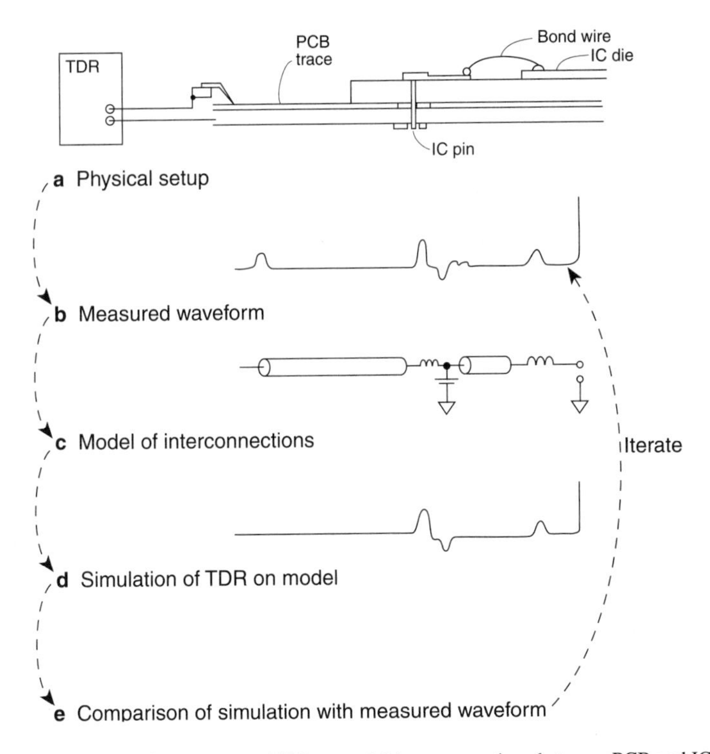

FIG. 12.59 Sequence of events to use TDR to model interconnections between PCB and IC chip die (Tektronix, 1993).

12.10.5 *Tests for Complex Logic*

Complex logic circuits and ICs need a test configuration that stimulates the circuitry and records the results. Fig. 12.60 illustrates a general scheme for testing complex logic.

Miczo (1992, p. 25) says, "Most solutions simulate electrical failures at each node, and then search forward through the logic for an unblocked path to the output (observability) and backward for a path from the input (controllability). The real difficulty comes in trying to navigate the heavily interdependent, clock-sensitive

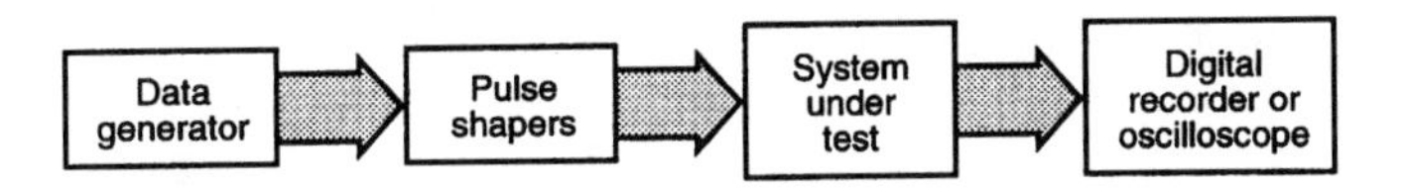

FIG. 12.60 A general configuration for stimulating complex logic and recording the results. It can provide both pattern and time stimuli.

logic as the software looks for these pathways. For most designs, the search space approaches infinity, making unguided probes impossible. For example, an IC with 400 flip-flops and 100 I/Os has 10^{150} possible logic paths, which is orders of magnitude greater than the number of particles in the universe. . . . As a result, users . . . end up adding nearly total design-for-test structures to the sequential logic in their designs."

Design for test usually warrants the extra cost and effort up front because of these advantages:

- Eliminating rework
- Reducing bad inventory
- Reducing service calls
- Improving the image of quality in the marketplace

IEEE 1149.1 defines testing of ICs by boundary scan; generally, boundary scan uses three or more I/O pins to daisy-chain all the storage elements in circuit for testing. It requires 10% to 30% more silicon, which increases die costs and power consumption and decreases speed. Static tests, however, can miss manufacturing flaws that show up only in dynamic testing, such as boundary scan at operating speeds (Miczo, 1992).

12.11 CASE STUDY: SENSOR CALIBRATION

Recall the data acquisition system described in the case study in Chapter 9 and Fig. 9.31, which involved sensors distributed around the hull of a submarine. Each sensor converts an analog signal to digital format with a 16-bit ADC. The ADC needed calibration to certify the fidelity of conversion.

We used a histogram test for DNL to certify that each sensor had a resolution of 14 bits with no missing codes. We used a pure sine wave, FFT test for INL to determine the available dynamic range of each sensor. We measured the temperature sensitivity of the ADC and its circuit board by testing for DNL in an environmental chamber at high and low temperature extremes.

We simulated the operating environment by performing these tests on each ADC circuit board and its DC-DC power converter as a unit, powered by the actual DC power supply used on the submarine. This approach allowed us a realistic measure of system noise and confidence in verifying the final operation.

Lessons learned: Carefully selected tests that simulate the actual environment can characterize components and subsystems well. Here we combined the ADC with its power supply and tested in an environmental chamber at various temperatures.

1. Testing for INL provided a calibration for dynamic range.
2. Testing for DNL ensured the resolution of the system.

12.12 SUMMARY

Testing and debugging electronic products requires a thoughtful approach to diagnose problems and correct them. Work on one problem at a time. Check the obvious first. You need to understand the nuances of test equipment and probing circuits. Furthermore, you need a basic understanding of the physical limitations of electronic devices and of the tests that can characterize them.

This chapter just begins to introduce some devices and methods of testing. You may extrapolate and extend these tests to fit your applications. Ultimately, your own experience will be the best guide.

12.13 RECOMMENDED READING

IEEE standard 1057: Standard for digitizing waveform recorders. New York: Institute of Electrical and Electronics Engineers.
Details the terminology and testing methods for ADCs.

Pease, R. A. 1991. *Troubleshooting analog circuits.* Boston: Butterworth-Heinemann.
Don't let the title fool you digital types. It's a well-written, very useful textbook and reference for everyone.

12.14 REFERENCES

Caldwell, D. J. 1990. Minimize loading errors in loop-gain measurements. *EDN*, May 24, pp. 165–170.

Castaldo, A. M. 1990. Lessons in debugging. *Embedded Systems Programming*, June, pp. 28–34.

Clemente, S., D. Lidow, and A. Lidow. 1985. Armed with a battery of data, designers can reinforce MOSFET circuit reliability. *EDN*, June 6, pp. 117–124.

Dorf, R. C., ed. 1993. *The electrical engineering handbook.* Boca Raton, Fla.: CRC Press.

Frederiksen, T. M. 1988. *Intuitive operational amplifiers: From basics to useful applications.* New York: McGraw-Hill.

Ganssle, J. G. 1990a. The Zen of diagnostics. *Embedded Systems Programming*, June, pp. 81–84.

Ganssle, J. G. 1990b. The Tao of diagnostics. *Embedded Systems Programming*, July, pp. 83–86.

Giacomini, J. D. 1992a. High performance ADCs require dynamic testing. *Electronic Design*, August 6, pp. 55–65.

Giacomini, J. D. 1992b. Most ADC systems require intermodulation testing. *Electronic Design*, August 20, pp. 57–65.

Ginsberg, G. L. 1981. *A user's guide to selecting electronic components.* New York: Wiley.

Grove, M. B. 1992. Measuring frequency response and effective bits using digital signal processing techniques. *Hewlett-Packard Journal*, February, pp. 29–35.

Harris, S. 1987. Dynamic techniques test high-resolution ADCs on PCs. *Electronic Design*, September 3, pp. 109–112.

Harris, S. 1990. The effects of clock jitter on Nyquist sampling analog-to-digital converters, and on oversampling delta sigma ADC's. *Journal of the Audio Engineering Society*, July, pp. 12-101–12-112.

Jackson, R. M. 1991. Patience and reason solve digital-system debugging problems. *EDN*, May 9, pp. 135–140.

Keenan, R. K. 1987. *Decoupling and layout of digital printed circuits*. Pinellas Park, Fla.: TKC.

Kerridge, B. 1992. Elegant architectures yield precision resistors. *EDN*, July 20, pp. 86–92.

Kester, W. 1990. Measure flash-ADC performance for trouble-free operation. *EDN*, February 1, pp. 103–114.

Knapp, R. 1991. Evaluate your ADC by using the crossplot technique. *EDN*, April 11, pp. 251–262.

Kral, R. 1993. Using the signature analyzer for component test. *Evaluation Engineering*, May, pp. 62–67.

Lueker, J. 1989. Differential techniques move TDR into the mainstream. *EDN,* August 17, pp. 171–178.

Michaels, S. R. 1981. D-A glitch: causes abound but solutions are rare. *Electronic Design*, April 6, pp. 183–187.

Miczo, A. 1992. Design for test. *Computer Design*, July, pp. 25–26.

Mittag, L. 1993. Debugging with hardware. *Embedded Systems Programming*, October, pp. 42–47.

Moscovici, A. 1983. Meet the deglitcher. *Electronic Products,* March 4, pp. 71–74.

Neil, M., and A. Muto. 1982. Tests unearth A-D converter's real-world performance. *Electronics,* February 24, pp. 127–132.

Ormond, T. 1989. Solid-state relays satisfy a wide range of switching needs. *EDN*, July 20, pp. 190–196.

Pease, R. A. 1989a. The right equipment is essential for effective troubleshooting. *EDN*, January 19, pp. 157–166.

Pease, R. A. 1989b. Active-component problems yield to painstaking probing. *EDN,* August 3, pp. 127–134.

Pease, R. A. 1989c. Rely on semiconductor basics to identify transistor problems. *EDN,* August 17, pp. 129–138.

Pease, R. A. 1989d. Keep a broad outlook when troubleshooting op-amp circuits. *EDN,* September 1, pp. 131–142.

Pease, R. A. 1989e. Troubleshooting techniques quash spurious oscillations. *EDN,* September 14, pp. 151–160.

Pease, R. A. 1990. What's all this CMRR stuff, anyhow? *Electronic Design*, November 8, pp. 125–130.

Pinkowitz, D. C. 1985. Deglitched DACs improve vector-stroke displays. *EDN,* April 18, pp. 239–242.

Pinkowitz, D. C. 1986. Fast Fourier transform speeds signal-to-noise analysis for A/D converters. *Digital Design* 16(6):64–66.

Rappaport, A. 1982. Capacitors. *EDN*, October 13, pp. 105–118.

Shear, D. 1989. EDN's advanced CMOS logic ground-bounce tests. *EDN,* March 2, pp. 88–114.

Shear, D. 1992. Exorcise metastability from your design. *EDN*, December 10, pp. 58–64.

Sheingold, D. H., ed. 1986. *Analog-digital conversion handbook*. 3rd ed. Englewood Cliffs, N.J.: Prentice-Hall.

Sinclair, I. R. 1991. *Passive components: A user's guide*. 2nd ed. Boston: Newnes.

Souders, T. M., and D. R. Flach. 1987. Accurate frequency response determinations from discrete step response data. *IEEE Transactions on Instrumentation and Measurement* IM-36(2):433–439.

Souders, T. M., and G. N. Stenbakken. 1991. Cutting the high cost of testing. *IEEE Spectrum,* March, pp. 48–51.

Tektronix, Inc. 1993. Bridging the gap between measurement and simulation: The IPA 310 interconnect parameter analyzer. Advertising literature, May.

Upham, A. F. 1985. Failure analyses and testing yield reliable products. *EDN,* August 8, pp. 165–173.
Ushani, R. K. 1991a. Subranging ADCs operate at high speed with high resolution. *EDN*, April 11, pp. 139–152.
Ushani, R. K. 1991b. Classical tests are inadequate for modern high-speed converters. *EDN*, May 9, pp. 155–166.
Whitehead, R. 1989. Operational amplifier noise prediction. *Harris Analog App Note No. 519,* pp. 10-24–10-27.
Williams, J. 1991. The mysteries of probing. *EDN*, October 10, pp. 165–176.
Ziegler, J., T. Hornback, and A. Jordan. 1992. The ten commandments of debugging. *Electronic Design*, September 3, pp. 61–68.
Zuch, E. L. 1987a. Interpretation of data converter accuracy specifications. In *Data acquisition and conversion handbook,* edited by E. L. Zuch, pp. 40–48. Mansfield, Mass.: Datel.
Zuch, E. L. 1987b. Know your converter codes. In *Data acquisition and conversion handbook*, edited by E. L. Zuch, pp. 34–39. Mansfield, Mass.: Datel.

13

INTEGRATION, PRODUCTION, AND LOGISTICS

The end of a matter is better than its beginning,
and patience is better than pride.
—Ecclesiastes 7:8

13.1 PUTTING IT ALL TOGETHER

After all the analysis, concepts, development plans, designs, and characterization tests introduced in the previous 12 chapters, you still have to build the product, maintain it, and finally dispose of it. Building, moreover, involves procurement of parts, manufacturing and assembly, various levels and types of tests, and integration. Fig. 13.1 shows how these activities interact.

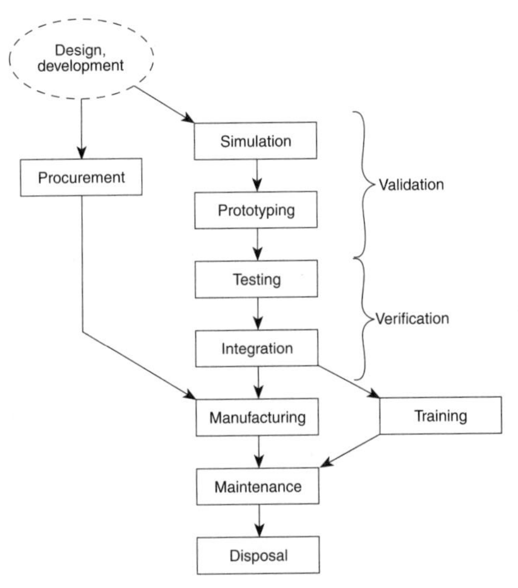

Fig. 13.1 The general flow of the main activities that complete the manufacture of a product. (Iterations in design and development have been left out for the sake of clarity.)

13.2 INSPECTION AND TESTS OF COMPONENTS

A reliable product requires reliable components and subsystems. You will need to perform a wide variety of tests throughout manufacture to ensure reliability. These tests include inspections, static tests such as destructive parts analysis for IC packages, stress tests such as burn-in to weed out weak components, and dynamic tests such as electronic stimulus-response tests for circuit boards. You apply these tests at various stages of assembly: component, board, module, and subsystem.

13.2.1 *Philosophy of Testing*

Testing will detect failures in production and reduce costs while improving your company's reputation. If you miss failures at any stage, the costs for repair can go up as much as 10-fold for each level of integration: component, board, system, and fielded system. Regaining your reputation is even more costly.

A good test strategy can increase your control over manufacturing and ultimately reduce costs by providing the necessary information to alter the manufacturing process, device specifications, and product design. Designing your product for testing will decrease production costs as well. For instance, you need to consider the mechanical constraints of test fixtures when designing your circuit boards and components. Other helpful hints for design for test include the following (Heide, 1986):

1. Disable the clocks so that the board tester can drive the signals.
2. Make the part number, revision level, and check sum available in ROM.
3. Drive bidirectional buses with tristate buffers.
4. Tie control pins to ground or power through resistors so that the tester can take control.
5. The order for BIT should be ROM, RAM, I/O, timers, and displays, with an indication after each to confirm its operation.

You should test each stage of assembly to cover the range of faults and improve quality. The stages of testing include inspection, burn-in (or environmental stress screen), component, board, and assembly.

13.2.2 *Incoming Inspection*

Do you always trust the vendor? By keeping accurate records of failures in your products, you can identify and avoid vendors that distribute unreliable components. You can either sample incoming parts or do 100% testing. Statistical sampling to accept or reject lots usually is more economical for production quantities, but 100% testing may be satisfactory for small production quantities or for highly reliable custom systems. Regardless of choice, inspection is just the first step in testing.

13.2.3 *Burn-In*

Burn-in stresses components to remove premature failures in components and subsystems. It is a combination of high temperature and electrical stimulation that induces weak or faulty devices to fail sooner. Survivors tend to exhibit very low failure rates for years after (Mayerfeld, 1987).

Traditionally, burn-in uses the simple and somewhat arbitrary criterion of soaking components at 125°C (257°F) for 168 hr to simulate 1 year of normal operation (Kuo and Kuo, 1983). At the same time, the test gear generates complex patterns of stimulation to reach all circuits, monitors the outputs, and continuously computes the failure rate. Burn-in can be applied to components, circuit boards, or an entire system. (An appropriately designed environmental stress screen, however, may be more suitable for subsystems or systems; see below.)

Burn-in attempts to weed out the weak components early in production. Components that fail during burn-in are called *infant mortalities*. The graph in Fig. 13.2 is called the *bathtub curve* and shows how burn-in can speed the early failures caused by infant mortality.

Infant mortality is caused by problems with the fabrication of the components such as anomalies, workmanship, and ESD. Surface anomalies on an IC die due to corrosion, contamination, electromigration, and inversions can cause failure. Workmanship and process faults are quality defects that can fail; excessive handling can cause ESD and damage ICs (Kuo and Kuo, 1983).

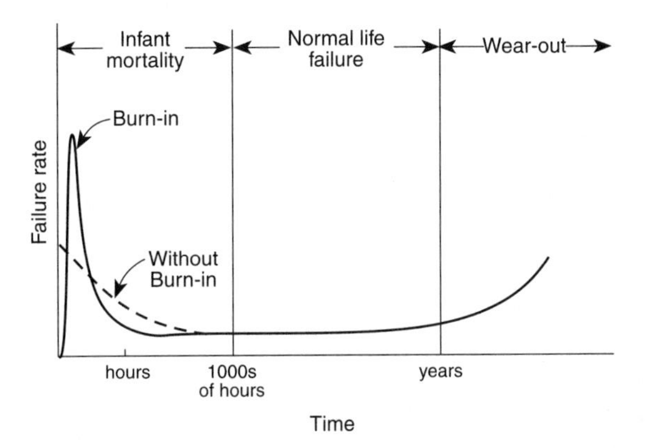

Fig. 13.2 The *bathtub curve* that predicts failure rates. Burn-in accelerates early failures or infant mortality. (Suggested by Jensen and Petersen [Kuo and Kuo, 1983].)

13.2.4 *Environmental Stress Screen*

Environmental stress screen (ESS) stresses systems to accelerate premature failures and remove them before delivery. ESS cycles temperature and vibration to simulate thermal and vibration transients that stress components, solder, and bond junctions. About 80% of failures can be attributed to temperature cycling, while vibration shakes out 20% of the failures (Klass, 1984). ESS is different from burn-in because vibration and temperature cycling can find some defects that cannot be found by high temperature alone.

Typically ESS varies temperature between −50°C (−58°F) and 75°C (167°F) and uses random vibration. You must determine the sensitivities of your product first; these include temperature range, ramp time between extremes, soak time at the extremes, vibration bandwidth and amplitude, and magnitude of shock (O'Shea, 1993).

Here are some general guidelines that you may follow to prepare for ESS (O'Shea, 1993):

1. Reduce the forcing functions of the stress environments to the lowest common denominator to maintain cost.
2. Use concurrent testing to find the maximum number of faults in a single screening.
3. Find a balance between too much stress and not enough so that the system is stressed without permanent damage.

ESS can have problems. The oven that bakes and freezes the components can condense moisture on the tested systems. During tests, you may have to turn off the oven temporarily to avoid disturbing measurements. The leads need careful routing to isolate stray capacitance, and the connectors should be placed in the oven to avoid thermocouple-induced voltages in the connectors. Finally, too much handling can cause damage from ESD (Frederiksen, 1988; Kuo and Kuo, 1983).

13.2.5 *Component Level Tests*

Aside from burn-in and ESS described above, you can screen components with electrical tests. The electrical tests stimulate components to their design parameters. You, the vendor, or an independent testing lab can conduct these tests. You may choose the vendor to test the components for reduced lead times, handling, and possibly costs (they may not have tooling costs). An independent testing lab provides support for a broad range of components, customization of tests, and independent verification; the down side is that an independent testing lab can have significant tooling costs. Ultimately, you will have to balance unit cost, the minimum purchase quantity, and tooling costs (Heide, 1986).

If your product uses custom ICs, you will have additional tests to perform. A destructive parts analysis examines the IC die and its package for fabrication quality.

In addition to static tests, contact and electronic beam probers provide dynamic certification for custom ICs. Even though they are expensive, they may provide the only direct verification of the IC (Morris and Sekulic, 1993).

If vendor screening is unavailable or the tooling charges are very high, you might skip screening in favor of later board or assembly tests. Be sure that no significant degradation can occur in the final product because you skipped screening. (Review Chapter 12 for further tests of components.)

Blank circuit boards need testing too. First, a visual inspection will find missing pads or traces. The test coupon allows you to test for solderability, trace foil adhesion, and plating in the vias with a microscopic examination. You or the vendor should test electrical continuity of every board because rework is very expensive (Heide, 1986).

13.2.6 *Circuit Board Level Tests*

Once the electronic circuit board has been assembled and soldered, you can enter the next level of testing.

First, a visual inspection will usually find 70% of the defects. These include solder quality and splashes; short circuits; open circuits; and missing, wrong, or backward-inserted components.

Next, special testers and fixtures can speed testing (Fig. 13.3). A probe tester checks for continuity and short circuits in a PCB. A very large-scale integration (VLSI) tester verifies chips. These are good for small production runs.

For larger production runs, a bed-of-nails tester using a vacuum test fixture can isolate and test each electrical node. It allows analog measurements of the signals at the electrical node. A bed-of-nails tester is easy to program, versatile, and fairly comprehensive in its coverage. On the other hand, it is expensive, doesn't detect speed- or drive-related problems, and wastes time on good boards without defects (Heide, 1986).

A functional board tester uses the edge connector to exercise the circuit board. It simulates real use by testing at speed and has comprehensive coverage of the circuit operation. While a functional board tester is less expensive, it requires increased programming time, has poorer fault resolution, and is less flexible because it is customized to the application (Heide, 1986).

13.2.7 *Assembly Tests*

Once you have a completed module or subsystem, you should fully test it to the guaranteed specifications. Assembly tests are specific to the module or product, but they can include ESS and a test bench simulator (see the case study). BIT has a low tooling cost and may serve as one of the assembly tests. Assembly testing should precede field and integration tests.

a

b

Fig. 13.3 Two examples of testers and their fixtures. (a) A VLSI tester. (b) A tester for circuit boards (photographs courtesy of Hewlett-Packard).

13.3 SIMULATION, PROTOTYPING, AND TESTING

Engineering development, systems integration, and training require simulation, prototyping, and parametric tests. Simulation supports engineering development and training, while prototypes help find unforeseen problems and validate the purpose of the system. Finally, parametric tests verify that the system meets the requirements.

13.3.1 *Simulation*

Simulation emulates the product's function in software to illustrate the bounds of operation. It usually is much faster and cheaper than building the actual hardware, but it is limited to known properties of the materials and configuration. Consequently, it cannot provide a conclusive verification of system function.

You can use simulation in engineering development, training, and later test support. The larger the system, the more important simulation becomes. A flight simulator aided the development of the B-2 bomber. "More than 6,000 hr. of manin-the-loop simulation were completed before the first B-2 ever flew. The FMS (flight mission simulator) aided by detailed data from the B-2 wind tunnel program, was a significant engineering tool for developing flight controls" (B-2 flight mission simulator, 1992, p. 44).

Simulation cannot provide thorough testing on parameter variations, nor can it provide integration testing, because it cannot describe the entire system.

13.3.2 *Prototyping*

Prototyping validates the purpose of the system and the intent of both users and customers, as well as finding unforeseen problems (see Chapter 4). Prototypes can vary from mock-ups to scaled-down functional models of the final system. Clear examples of why prototype testing is so necessary come from aircraft testing, as shown in the next two examples.

Example 13.3.2.1 A crash of the YF-22A advanced tactical fighter in April 1992 illustrates how a prototype can find problems. "The airplane was performing a planned go-around using afterburner with thrust vectoring activated when it entered a pitch oscillation. . . . After a severe series of pitch oscillations, the YF-22 crashed onto the runway and slid to a stop in flames" (Report pinpoints factors, 1992, p. 53). The pilot had induced oscillations in flight by overreacting to aircraft responses during a time when some software in the avionics [called the flight control laws] changed operational paradigm. The oscillations occurred when the landing gear retracted, thereby commanding a large and instantaneous change in the flight control laws that interacted with logic for the control surfaces on the wing and stabilizers. The designers had not realized before the prototype demonstrated it that this particular problem could occur.

Example 13.3.2.2 The program for the F-117 stealth fighter built two subscale prototypes in the late 1970s to test the concepts of absorbing and reflecting radar waves. "The designers had to grapple with 1,001 details that could have derailed the program. 'It was very good that we did Have Blue prototypes [F-117 stealth fighter demonstrators] first,' said Mr. Brown [Alan Brown, the first F-117 program manager]. 'That gave us a real head start'" (Lynch, 1992, p. 25).

But prototypes usually are not fully representational of the final product. If software is the primary component in the prototype, the trap is to make just a few more tweaks and deliver without full parametric testing. Finally, human nature can run afoul of prototypes and alter results in mysterious ways. Be prepared for occasional strange behavior when field testing a prototype, as seen in the next example.

Example 13.3.2.3 I once designed some electronics that monitored switch gear onboard a submarine (described in the arc fault detection case study in Chapter 2). During the technical evaluation, we installed instrumentation on top of a tall cabinet to record the system's operation within the submarine. One of the measured parameters was cabin pressure, which at random intervals dropped to one-half atmosphere for several seconds at a time (life cannot be sustained by pressure this low for extended periods). After inquiry and some head scratching on our part, a sailor pulled a colleague aside and showed him how one of the crew would sometimes pull himself up on the cabinet, put his mouth over the vent for the pressure sensor, and suck—just to confound the results from the data collection!

13.3.3 *Parametric Testing*

Testing the variations on parameters will verify that your system meets the requirements. You need to collect and store all test results so that you can retrieve and examine them at any time. If you design medical devices, a carefully conceived database is necessary because, as Göring (1991, p. 49) writes, "regulatory agencies may request [these] data, even years after release."

When you test boards and modules, you can streamline the process by identifying all inputs and outputs, combining those inputs that produce the same output, and designing test cases to produce unique outputs. You may have access to automatic test equipment (ATE) that stimulates the circuits, stores the responses, and then compares the outputs to the stored values. ATE generally needs a known good board to store the correct values for later comparisons. Fig. 13.4 is a block diagram of one example of such ATE.

13.3.4 *Calibration*

Eventually all measurements have to trace back to a physical standard—volt, joule, second, or kilogram. If your finest measurements are large compared with the theoretical limit, your measuring instrument may be sufficiently accurate and precise. Measuring logic levels with an oscilloscope to confirm the presence of a signal, for instance, may be sufficient without having a paper trail of traceability back to a calibration standard. A very sensitive electrometer, on the other hand, may need periodic calibration. Some measurements, like integral nonlinearity for

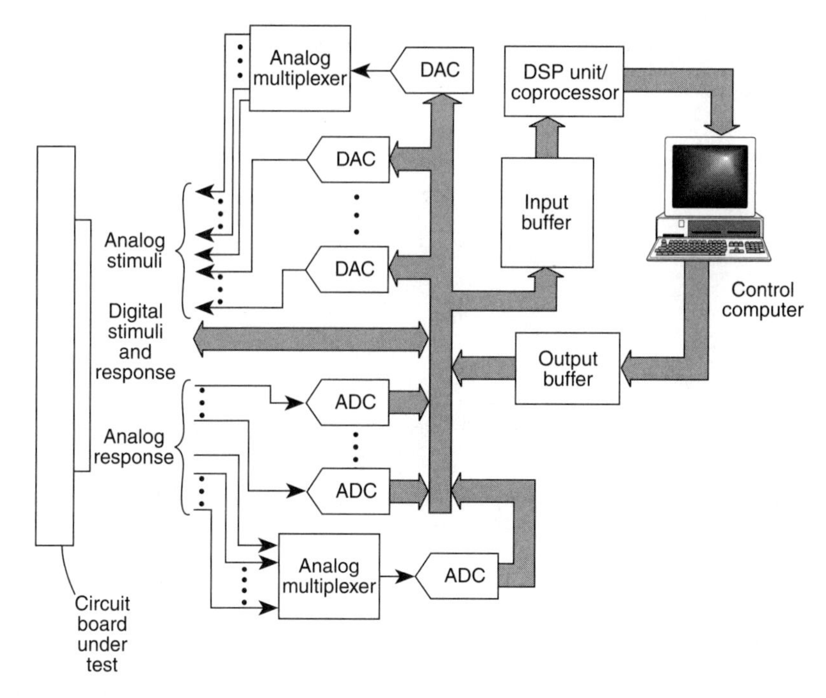

FIG. 13.4 General components and modules within ATE.

an ADC, are more art than science; calibration is less the issue than how you set up and conducted the tests. Most important, you should document what is calibrated to reduce any liability in developing and testing a product.

13.4 INTEGRATION

Integration confirms whether the components, modules, and subsystems work together in the desired way. It is the final stage of tests and assembly to confirm that the system design and implementation meet the requirements (verification). Fig. 13.5 illustrates the general sort of test structure and flow that you might expect throughout a project.

Thus far I have described tests on components and modules; these are the first steps in integration. Next, you need to check the fit of the components and test the system end to end. You may have to calibrate the system as well. If your product must conform to specific standards such as thermal environment, shock and vibration, and EMI/RFI, you will have to test accordingly. These are the more common types of tests; your application may also require tests for liquid spills, acoustical noise, humidity, or salt spray.

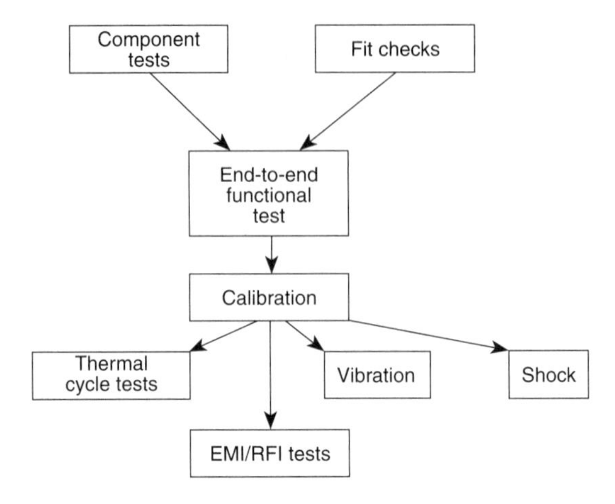

Fig. 13.5 Some tests that constitute integration. The tests may be tailored to specific applications and products.

13.4.1 *Fit Checks*

A fit check simply proves mechanical clearances between components and mechanisms. Cable connections and harnesses are a special concern for many instruments and yet are often overlooked. Currently, software that performs three-dimensional modeling can do most of the checking, particularly for a complex project with many components. Cardboard and wooden mock-ups, however, still can be a quick and efficient method to check fit.

13.4.2 *End-to-End Tests*

End-to-end testing is absolutely critical to verifying the operation of any system. The more complex the system is, the more extensive and necessary the testing. End-to-end testing is a carefully designed blend of tests that look like prototyping and parametric testing. You should examine and document every region of operation, if possible, to verify that the system meets the requirements.

End-to-end testing is specific to the application and product. You must prepare thorough and detailed procedures before beginning integration to avoid mistakes and pitfalls. You will find enough unforeseen problems without creating some because of lack of preparation. Remember, prevent stupidity and control ignorance. Mistakes will occur in even the most carefully planned procedures—which shows just how important end-to-end testing can be. The next two examples point out the importance of end-to-end testing and its hazards.

Example 13.4.2.1 In the previous example, a colleague installed the electronic equipment that monitors a submarine's power switch gear. He tested its ability to detect arcing faults in the switch gear by simulating an arc and detecting the closure of relay contacts. (The relay trips circuit breakers to extinguish an arc and protect the switch gear from the searing heat of the arc.) At one point in the end-to-end test, he overlooked a step in the procedure and forgot to disconnect the relay contacts from the trip circuit of a circuit breaker. When he simulated the next arc, the system operated correctly, tripped the circuit breaker, scrammed the nuclear reactor (dropped the control rods into the core), and shut down power to the submarine—an effective test and demonstration, but not planned!

Example 13.4.2.2 The Hubble Space Telescope is one of the most expensive scientific instruments ever built. It orbits the earth above atmospheric disturbances to record celestial phenomena from the farthest reaches of space. Scientists discovered that it could not focus properly once it was in space because its main mirror had been ground wrongly. The laser measurements of the curvature of the mirror reflected off the wrong portion of a calibrating rod and gave an erroneous curvature. No end-to-end test on the mirror optics was performed; such a test would have caught the improper curvature in the mirror.

13.4.3 *Thermal Tests*

Sometimes an electronic instrument has to operate under extremes of temperature. Thermal tests during integration confirm that your product will meet the environmental requirements—but note that this is not burn-in or ESS. Thermal tests are verification of operation, while burn-in and ESS are stress screens to reduce infant mortality.

Thermal tests cycle the ambient temperature between extremes around the system under test. Some applications, such as satellite instruments, may require concurrent vibration or vacuum.

13.4.4 *Vibration and Shock Tests*

If your product will operate in an environment that is vibrating or mechanically shocked, such as in an industrial plant or on a moving vehicle, then you will have to test the completed system for survival through a series of shock and vibration tests. The tests should be designed according to the environment.

Each test may deliver a mechanical impulse or vibration in one of the three axes of orientation. Vibration may be either random to present a wide spectrum of frequencies for a duration or swept frequency to identify the resonance peaks and determine whether they are below the maximum allowed.

13.4.5 *EMI/RFI Tests*

If you are concerned about electrical noise radiation and interference, then you will have to identify and test to the applicable standards (see Chapter 6). These tests usually are valid only for the complete system, since subsystems may interact in complicated ways that increase susceptibility to noise or interference. Often an independent testing lab can do the EMI/RFI testing for you because they have the specialized support equipment and expertise. They can work with you to customize the tests for your product and certify it to governmental or military requirements.

13.4.6 *Scheduling*

Integration can take either of two paths: an incremental approach or a complete assembly test. The incremental approach connects subsystems one at a time and tests them thoroughly—much like the spiral model of software development. Some systems do not lend themselves to an incremental approach and must be tested in nearly final form. You will probably find that a combination of these two approaches best fits your application.

Just as integration requires careful procedures, so you should schedule it very carefully. Avoid late changes to your project, because they are difficult to test comprehensively.

> **Example 13.4.6.1** A recent satellite did not deploy correctly because a mechanical change made late in the program was not thoroughly tested. The Italian-U.S. satellite was a tethered experiment from the U.S. space shuttle and it couldn't deploy fully because the tether reel hit a protruding bolt. The bolt was a change to the mechanism and was added late in the program. An inquiry after the failure showed that the program should have had an end-to-end test and more careful discipline in the engineering development (Inadequate tests, 1992).

13.5 VALIDATION AND VERIFICATION

Kaufman (1989, p. 26) writes, "No amount of design verification, as it is done today, can accomplish more than provide assurance that we have perfectly implemented a possibly imperfect specification. . . . Design verification focuses narrowly on assuring that a design meets its specification. . . . Design validation, on the other hand, focuses on ensuring that the system actually performs the way the designer and the users want it to. . . . Achieving early design validation . . . can at the very least have a high payback in reduced development costs."

The testing described in the last two sections concerns either validation or verification of your product. *Validation* certifies that the product meets the customer's intent or desires. *Verification* confirms that the product meets the requirements. Unfortunately, intent is not always properly translated into the requirements; consequently, validation is a necessary step and should precede verification.

You should carefully record the results from validation and verification. These documents will become an integral part of your product.

13.6 PROCUREMENT

Procurement is one of those areas that is consistently overlooked and underestimated. You will appreciate the process more once you include the purchasing agents, manufacturers' representatives, vendors, and suppliers as part of your design team.

Early in the development of an instrument, plan to work with your purchasing department: the buyers, expediters, and shipping and receiving. Early discussions with them may reveal some surprising complications in procuring components, parts, and services. Some new components, for instance, are allocated among very large customers and distributors, making them very hard to find and buy. Purchasing can help you understand the process and schedule the procurement more effectively. You can extend the team to include vendors, distributors, and subcontractors for similar reasons.

You need to allow sufficient time in the calendar for purchasing and delivering components and subsystems. Seemingly little things can turn out to be big obstacles. Paperwork and signatures all take time. Production delays in procurement can delay your project as well. Even telephone calls to distributors take time. I have found that I can complete only about seven telephone calls an hour to get price and delivery information from distributors for common components.

You may need to qualify and certify vendors as one step in maintaining quality in components. Build a history of dealings with vendors, distributors, and subcontractors to determine who will give you consistently good service and quality components. Manufacturers can help you if they certify their process to the international standards of ISO 9000. Otherwise, you may have to institute a program to certify each vendor with on-site audits and qualification testing.

13.7 MANUFACTURING

If your product is mass produced, you must enter the world of manufacturing after proving the design in integration. Of course you must account for manufacturing concerns from the beginning and design for manufacturing. According to Boothroyd and Knight (1993, p. 53), "80 percent of all product costs are fixed at the design stage." Design for manufacturing includes design for assembly, design for disassembly, design for environment, and design for test.

I have introduced some of these concerns in previous chapters. Circuits should be testable to ensure quality and remove failures. Design for disassembly eases maintenance cost and effort (see Section 13.8). Design for environment accounts for recycling and materials that are more friendly to the environment (see Section 13.10).

13.7.1 *Personnel*

Beyond your own experience and knowledge, you will find that employees and suppliers can provide invaluable insight into manufacturing concerns and problems. Work with them to improve the process of fabricating and assembling a product.

Some forward-looking companies give considerable responsibility to the individual worker, encourage initiative, and help workers recover and learn from mistakes. One such company claims that "no one has lost his or her job for making a mistake. After all, people who do not make mistakes will not be seeking new ways to succeed" (Luciano, 1993, p. 37).

Providing vision for each product will help individuals manage their own responsibilities and better understand how they may contribute. All this slashes development time and leads to a higher-quality product.

13.7.2 *Six-Sigma Design*

One way to achieve high quality is to strive for it. *Six-sigma design* sets a goal for production of less than 3.4 failures in 1 million units. Everyone in the design, development, manufacturing, and support process must be involved to do this (Smith, 1993).

Smith (1993, p. 43) writes, "First, total customer satisfaction means more than satisfaction with the product; it is meeting or exceeding every requirement of every customer. . . . A failure to satisfy a customer, who is everyone from the next person in the process to the end-user, is a defect. . . . Reducing total defects per unit will not only reduce cycle time per unit, but will also result in fewer defects in the delivered products and fewer early-life failures. The net result is a better-satisfied customer, plus lower warranty and manufacturing costs per unit."

What you are trying to do is meet the customer's expectations. Motorola has found that those expectations are as follows (Smith, 1993):

1. Delivery of the product when the customer wants it
2. Installation that is "plug and play" or has clear and simple instructions
3. No infant mortality
4. Reliability in use
5. Swift repair

13.7.3 *Robust Design*

A robust design has margins in all aspects of operation that will result in a final product that is easier to produce and has lower defects and better customer satisfaction. You can prepare a more robust design by the following means (Smith, 1993):

- Reducing the number of parts
- Simplifying the manufacturing process with fewer steps
- Reducing the number of key characteristics that require testing

Accurate and continuous communication between all parties involved contributes to a good design. Networking the CAD and computer-aided manufacture (CAM) systems is just one way to improve communication and access to information.

13.8 MAINTENANCE AND REPAIR

Once an instrument has been delivered to a customer, its support has just begun. If it is reliable, it should need minimum maintenance. Unexpected use or abuse, however, can break even the most reliable equipment. Common problems and failures (many mechanical) include the following:

- Cable flexure stress
- Connectors and contacts
- Potentiometers
- Switches
- Physical abuse
- Reversed polarity
- Overvoltage

Maintenance is either planned or unscheduled. Planned maintenance includes software modifications and upgrades and hardware changes. Unscheduled maintenance includes fixes to remove bugs in software and correct damage to hardware.

The prevalence of software often requires a corrective action plan for when problems arise. First, it needs a method for reporting the problem, then an analysis to diagnose the problem, and finally, a procedure to fix it.

Good manuals, as covered in Chapter 3, are an important part of any product. They must be clear, concise, and thorough.

You will need to decide the level of repair and maintenance that suits your product. This starts with the maintenance philosophy for the product. Is it to be repaired or replaced? What level of replacement—component, circuit board, module, or the entire instrument? The level of replacement will determine what spare parts and components inventory you will have to carry. Obviously the level of expertise needed for repair will affect both the cost of repair and the training of personnel.

Commonality of design will reduce the maintenance inventory and the training to maintain your product. Moreover, if you involve maintenance personnel early in development you will design an easier-to-maintain system. "Northrop managers attributed the B-2's current and projected reliability and maintainability design (R&M) figures to having logistics and maintenance personnel involved early in the design process" (B-2 reliability focus, 1992, p. 52). Even if you're building a one-of-a-kind prototype, ask a technician for advice. You will design and build a better product.

13.9 TRAINING

One aspect of product support is the training necessary to install, use, and repair an instrument. First, you must define who will need the training and understand their capabilities. Is it the buyer or owner? Will your product need skilled, knowledgeable operators?

Next, you need to determine the level of training that you will provide. Maybe a manual is sufficient. If the instrument is complicated, you will certainly need good manuals and possibly need to supplement them with training classes. Is support staff available to do the training? For complex equipment, demonstrator models are good instructional aids.

Provide a fundamental understanding of your product. Thorough training as well as seeking advice from the users to improve your product will ultimately help them own the solution and commit to your product (see Example 4.9.2).

13.10 DISPOSAL AND ENVIRONMENTAL CONCERNS

Bendz (1993, p. 64) writes, "Design-for-environment (DFE) programs call for the careful inclusion of environmentally safe attributes in the early design stages of new products, as opposed to re-engineering them late in the product cycle. These attributes are being defined every day by global legislators, by technology advances, and by customer perception. So DFE includes monitoring of worldwide events and providing appropriate engineering responses."

More and more products are being designed with consideration for the entire life cycle. Not only is cost of initial production a concern for manufacturers, but disposal costs are becoming more important. Environmental consciousness is a factor in what is used and what is discarded. Consider batteries for handheld devices. Rechargeable ones ultimately use fewer natural materials and produce less waste for disposal. Furthermore, batteries that avoid heavy metals are much preferred for disposable applications.

Some companies are pursuing a zero-waste goal. This kind of policy strives for the following (Bendz, 1993):

- Product simplification (compatible with reduced assembly costs)
- Design for disassembly (compatible with maintenance concerns)
- Reduced material diversity
- Use of recyclable materials
- A consistent recycling strategy, including inventory management and testing of used components.

Recycling reduces both the volume of raw materials and the cost of disposal. Moreover, many countries are enacting "take-back" laws that require manufacturers to reclaim or recycle their products (Bendz, 1993).

Example 13.10.1 Digital Equipment Corporation is recycling the plastic from monitors and computer enclosures by retrieving the old plastic cases and grinding them into chips that are used to mold new components (Billings, 1993).

Other environmental concerns include eliminating hazardous chemicals and materials from processing, fabrication, and operation. Manufacturers of cooling and air-conditioning units, for instance, are replacing chlorofluorocarbon (CFC) working fluids with more environmentally inert chemicals.

Example 13.10.2 After soldering, the solder flux must be cleaned from circuit boards. Once, the only cleaning agents used were CFC solvents. New water-based systems not only eliminate the CFC solvents but are reducing the wastewater in the process (Bendz, 1993).

Cost, as always, plays a major part in design and manufacturing decisions, but DFE can actually work in your favor to contain costs. Plastic enclosures that follow design-for-disassembly principles, such as molded-in snaps, can lower process costs and facilitate recycling because of the all-plastic construction and the ease of disassembly. "Xerox shows that many design-for-assembly principles are congruent with those that underlie disassembly and asset recovery management" (Bendz, 1993, p. 66).

13.11 CASE STUDY: TEST SYSTEM FOR AVIONICS

Developing an aircraft is a monumental undertaking. The avionics alone are an enormously complex system that is difficult to test and verify. Boeing developed a flexible test system to verify each line-replaceable unit (LRU) in the avionics of the Boeing 777 airliner. The system provides all of the electrical signals normally seen by an LRU when it is installed in the aircraft and records its response to them (Bryan et al., 1992).

The test system has four sections: a host computer, a hardware simulator, an avionics test bench, and a flight-deck mock-up (Fig. 13.6). The host computer simulates the operation and flight characteristics of the aircraft with mathematical models of flight characteristics of the 777. The hardware simulator interfaces the LRU under test to the host computer and acts like the aircraft hardware. "[The] airborne hardware simulator can simulate almost all of the approximately 350 data-producing nodes on the Boeing 777 airplane" (Bryan et al., 1992, p. 16). The avionics test bench holds the LRU and provides the necessary support and test equipment. Finally, the flight-deck mock-up allows a test engineer to interact with the LRU as a pilot would.

The test system has a VMEbus architecture rather than a proprietary one. It is flexible and extensible and allows a wide selection of board vendors. Since it uses standard modular units, both hardware and software, it can be dismantled and reused on future projects. It uses a modified real-time operating system to control the software execution. (Bryan et al., 1992).

The test system collects, stores, and analyzes all digital data and analog signals. Since it captures all the data, it can verify previous results by running the same test again.

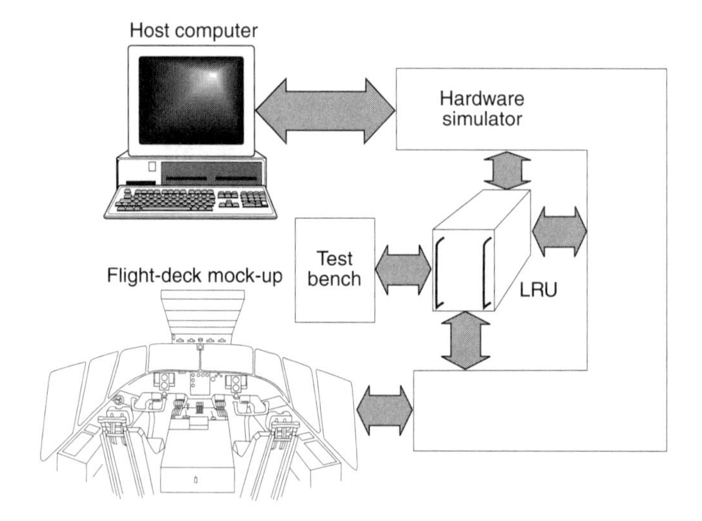

FIG. 13.6 Test system for testing LRUs in a simulated aircraft environment.

Obviously, this is a complex and capable test system for verifying the electrical characteristics of LRUs. It does have limitations: It doesn't simulate the environment of a flying aircraft, including temperature, vibration, and humidity. LRU response to such conditions must be handled by separate tests and equipment setups.

Scope: process or program
Lessons learned: A properly planned test system speeds the development of complex systems. The test system for new avionics on the Boeing 777 capitalizes on a modular and open architecture. The system has some advantages:

1. It is flexible and extensible for numerous functions and future projects.
2. It allows a wide selection of circuit boards.

It does have limitations: It cannot simulate the entire environment of an aircraft. So it does not satisfy all requirements of testing and integration.

13.12 SUMMARY

This chapter rounds out the life cycle of an electronic instrument. It includes validation, verification, procurement, production, training, maintenance, and disposal. Each of these contributes to a cost-effective and high-quality instrument. While procurement, production, and training all have directly assignable costs, the others affect that intangible and subjective asset—your company's reputation.

Today's market is demanding reliable products with good performance that have minimal impact on the environment. A systems perspective that accounts for the entire life cycle of an electronic product is one step toward satisfying that demand.

13.13 REFERENCES

B-2 flight mission simulator serves as key resource for flight control, avionics, and systems development. 1992. *Aviation Week & Space Technology,* May 18, p. 44.

B-2 reliability focus pays early dividends. 1992. *Aviation Week & Space Technology,* October 12, pp. 52–53.

Bendz, D. J. 1993. Green products for green profits. *IEEE Spectrum,* September, pp. 63–66.

Billings, S. L. 1993. Enclosures that ease service burdens. *Design News,* April 5, pp. 76–78.

Boothroyd, G., and W. Knight. 1993. Design for assembly. *IEEE Spectrum,* September, p. 53.

Bryan, J., B. Cornelius, P. McCusker, and D. Wilson. 1992. Building a test system for the Boeing 777. *VMEbus Systems* 9(5):11–19.

Frederiksen, T. M. 1988. *Intuitive operational amplifiers: From basics to useful applications.* New York: McGraw-Hill.

Göring, D. 1991. An automated test environment for a medical patient monitoring system. *Hewlett-Packard Journal,* October, p. 49.
Heide, P. W. 1986. Manufacturing test strategies for microcomputer based medical instrumentation. Association for the Advancement of Medical Instrumentation annual meeting, April 12–16, Chicago.
Inadequate tests cause tether woes. 1992. *Aviation Week & Space Technology,* November 16, p. 65.
Kaufman, P. 1989. Validation must replace verification. *High Performance Systems,* September, p. 26.
Klass, P. J. 1984. Guides issued for avionic stress screening. *Aviation Week & Space Technology,* October 15, pp. 99–103.
Kuo, W., and Y. Kuo. 1983. Facing the headaches of early failures: A state-of-the-art review of burn-in decisions. *Proceedings of the IEEE,* November, pp. 1257–1266.
Luciano, V. P. 1993. Display terminals as you like them. *IEEE Spectrum,* September, p. 37.
Lynch, D. J. 1992. How the skunk works fielded stealth. *Air Force Magazine,* November, p. 25.
Mayerfeld, P. 1987. Semiconductor reliability: The burn-in question. *Defense Electronics,* September, pp. 88–92.
Morris, S. R., and Z. Sekulic. 1993. IC fault analysis using contact. *Evaluation Engineering,* May, pp. 46–53.
O'Shea, P. 1993. ESS: Practical advice and stellar results. *Evaluation Engineering,* May, pp. 36–38.
Report pinpoints factors leading to YF-22 crash. 1992. *Aviation Week & Space Technology,* November 9, pp. 53–54.
Smith, B. 1993. Six-sigma design. *IEEE Spectrum,* September, pp. 43–47.

APPENDIX A: OUTLINE OF MILESTONES AND DOCUMENTS FOR DEVELOPING ELECTRONIC INSTRUMENTS

Activity	**Documents**
Concepts	Engineering notes, memos
Conceptual design review (CoDR)	Memo
Development plan	Design plan, testing plan, validation and verification plan, integration plan, production plan
Analysis	Engineering notes, memos
Specifications	Specifications document
Preliminary design review (PDR)	Memo
Design	Drawings, engineering notes, memos
Prototypes	Drawings, engineering notes, memos
Critical design review (CDR)	Memo
Validation and verification (V&V)	Memo or document
Initial production review	Memo
Production	Drawings, engineering notes, memos
Sales	Marketing plan, brochures, advertisements
Training and customer service	Training manuals, training booklets and pamphlets, operations manuals, owner's manual
Maintenance and disposal	Service manuals, repair manuals

APPENDIX B: DESIGN REVIEW CHECKLISTS

Conceptual Design Review (CoDR)

❐ Have all necessary people been involved?
- Management
- Design architect, R&D
- Engineering: electrical, software, mechanical, industrial
- Manufacturing
- Service and maintenance personnel
- Customers
- Purchasing
- Marketing
- Legal
- Financial

❐ Has the operational environment been identified?
- Temperature ranges, cycles, and transients
- Humidity and moisture
- Vibration and shock
- Handling and transport
- Storage
- Special considerations such as abuse, misuse, and radiation

❐ Has the environmental impact study been completed?

❐ Is a development plan being prepared?
- Documentation, specifications, guidelines, heuristics
- Design process
- Testing and debugging
- Integration
- Validation and verification
- Marketing, maintenance, support
- Disposal

❒ Has analysis covered all assumptions?
- Need
- Function: size, weight, packaging, human interface
- Operation: cooling, grounding, shielding, software, circuit design
- Risk, hazard, fault tolerance
- Feasibility
- Resources

❒ Have the trade-offs been assessed and documented?

❒ Is the design testable?

❒ Are the failure modes identified and understood?
- Electrical transients and over ranges
- Mechanical shock and vibration
- Software/hardware interactions
- Abuse and operator error

❒ Have the parameters of operation been identified?
- Observable
- Controllable

❒ What variations in parameters can be tolerated? What are the allowable margins?
- Signal ranges
- Noise and distortion
- Mechanical dimensions: package size, cable lengths
- Feedback stability
- Power distribution, variation, and stability

❒ What heuristics apply?
- Design criteria
- Experience with previous designs

Preliminary Design Review (PDR)

❒ Has a development plan been completed?
- Design plan
- Testing plan
- Validation and verification plan
- Integration plan
- Production plan

❒ Is the development plan being followed? Are the metrics useful for improving your processes?

❒ Does the analysis reveal the intent and desires of the customer and user?

❒ Do the specifications clearly state the intent?
- Unambiguous language
- Clear goals, such as mathematically precise statements

❒ Do the specifications cover the scope of the project?
- Human interface design
- EMI, grounding, and shielding
- Size, weight, and power consumption
- Manufacturing, disassembly, maintenance, and disposal
- Cost and time to market

❒ What is the baseline for assessing progress and instrument function? What are the limitations?
- Models
- Simulations
- Prototypes

❒ Are the packaging requirements clear and achievable?
- Accommodation of power distribution and cooling
- Manufacturing and production plan
- Recycling of materials

❒ Are guidelines for software established?
- Development plan: model, metrics, and process
- Establishing the limits of software function
- Establishing inspections and reviews
- Analyses: hazard, risk, and fault tree
- Specified fault tolerance and error recovery methods
- Defining modules and interfaces
- Documentation and source control
- Structured programming
- Comments

❒ Are guidelines for grounding and shielding established?
- Applicable standards
- Grounding options
- Characterization of the sources, receivers, and transmission: source and load impedances, frequency bandwidth
- Possible mechanisms: conductive, capacitive, inductive, or electromagnetic
- Design procedures: reduce bandwidth, balance currents, route signals for self-shielding, and add shields if necessary

❐ Has an error budget been completed?
- Sensor noise, drift, and linearity
- Interference in transmission lines
- Amplifier linearity, CMRR, SNR
- Conversion errors: quantization, differential linearity, integral nonlinearity, aliasing
- Round-off errors in calculations
- Variations in component values
- Pass-band and stop-band ripple
- Phase linearity

❐ Are guidelines for high-speed circuits needed?
- Control of rise time on the edges of pulses
- Reducing bandwidth
- Decoupling capacitors where needed
- Power and ground (return) planes
- Keeping circuit board traces short and perpendicular on opposing layers
- Terminating transmission lines and keeping stubs short
- Grouping circuits according to power, speed, and function (analog versus digital)

❐ Are guidelines for low-power circuits needed?
- Lower the clock frequency.
- Lower the supply voltage (from +5 VDC to +3.3 VDC or lower if possible).
- Shut down unused circuits.
- Select the slowest low-power logic.

❐ Has the power distribution and consumption been analyzed?
- Define the physical constraints of volume and weight.
- Define the efficiency.
- Define the acceptable noise.
- Establish the reliability.
- Establish the margin for future growth in power consumption.

❐ Have the cooling needs been analyzed?
- Establish the thermal environment.
- Define sources and appropriate ranges of dissipation.
- Define the desired mechanisms: heat sinks, thermal pillows, fans, liquid cooling, and refrigeration.

❐ Have the mechanical concerns been addressed?
- Package will withstand the expected use and abuse.
- Cables are routed for access and ease of installation and repair.
- Cables, wires, and modules are clearly labeled.

- Cables and components are clamped and restrained for shock and vibration.
- Connectors are keyed and clearly labeled.
- Types and numbers of fasteners are kept to a minimum.
- Heavy components are mounted near the edges of circuit boards or chassis supports.
- Cantilever mounting is avoided.

❒ Have the maintenance needs been assessed?
- Responsibilities and skill levels have been identified.
- Level of repair has been determined: throw away, replace modules, replace components.
- Facilities have been identified.
- Necessary components are accessible to repair.
- Adjustments and alignments are kept to a minimum.

❒ Is the design testable?
- Access and probe points
- Ports and connections for test equipment
- Test fixtures and equipment
- BIT

Critical Design Review (CDR)

❒ Has the development plan been followed?
- Design plan
- Testing plan
- Validation and verification plan
- Integration plan
- Production plan

❒ Are all the necessary people still involved?
- Management
- Design architect, R&D
- Engineering: electrical, software, mechanical, industrial
- Manufacturing
- Service and maintenance personnel
- Customers
- Purchasing
- Marketing
- Legal
- Financial

❒ Were metrics kept that will improve your processes?

❒ Is documentation up-to-date and consistent?

❒ Did validation confirm the intent and desires of the customer and user?

❒ Did verification cover all the specifications?
- Human interface
- EMI, grounding, and shielding
- Size, weight, and power consumption
- Power
- Cooling
- Software: functionality, fault tolerance
- Circuitry: high-speed design, low-power design, noise
- Manufacturing, disassembly, maintenance, and disposal
- Cost and time to market

❒ Has the human interface been completely explored, defined, and implemented?
- Ease of operation
- Defined user skills
- Logical and intuitive layout and operation
- Adequate lighting

❒ Have the safety factors been established?
- Plan prepared and followed
- Protection from high voltages
- Interlocks on hazardous operations
- Sharp edges and pointed corners avoided

❒ Does the packaging meet requirements?
- Package image acceptable to customers
- Accommodation of power distribution and cooling
- Manufacturing and production plan
- Recycling of materials

❒ Does the software work?
- Development plan: model, metrics, and process
- How many reviews?
- How many bugs? Were all resolved? How critical were they?
- How widely have the limitations and operational constraints been tested?
- Does the software display specific modes in fault tolerance and error recovery?
- Documentation and source control
- Structured programming
- Comments

❒ Are guidelines for grounding and shielding established?
- Applicable standards
- Testing results

❒ Are errors and noise within budget?
- Sensor noise, drift, and linearity
- Interference in transmission lines
- Amplifier linearity, CMRR, SNR
- Conversion errors: quantization, DNL, INL, aliasing
- Round-off errors in calculations
- Variations in component values
- Pass-band and stop-band ripple
- Phase linearity

❒ Do all circuits meet the specifications for power consumption, timing, and noise?

❒ Were calibration standards met?

❒ Is cooling adequate?
- Heat-dissipating components are not located near heat-sensitive components.
- The temperature contour is mapped for each circuit board and the instrument.
- Airflow is adequate with margin for future growth in heat dissipation.

❒ How testable is the design?
- Access and probe points
- Ports and connections for test equipment
- Test fixtures and equipment
- BIT

Initial Production Review

❒ Can the design be manufactured? Are the desired technologies available?

❒ Is the production plan in place?

❒ Has all testing, validation, and verification been completed? Are there any surprises? If so, are they "show stoppers"?

❒ Does the design still meet the need?
- Prototype testing
- Alpha and beta tests
- Focus groups

❒ Is the design stable?

❐ Is the documentation adequate?
- Drawings, schematics, procedures
- CAD/CAM data files
- Software configuration control of modules

❐ Has analysis been made for reliability?
- Minimum, standard components
- Derating factors
- Stressed in various realistic circumstances

❐ Does the design use available tooling?
- Avoid nonstandard tooling.
- Automate manufacture to reduce expensive labor.
- Allow reasonable tolerances on dimensions: radii, fillets, curves, holes.

❐ Does the design ease the effort of manufacturing?
- Use logical layout and wire routing.
- Use sequential assembly.
- Reduce many or specialized fasteners.
- Consider snap fits.
- Strive for simplicity.
- Use clear markings to guide assembly.

❐ Have suppliers been consulted regularly? Have their requirements been established?

APPENDIX C: SOFTWARE DESIGN TOOLS

Many software packages are available to aid the design and development process. From system design software to schematic capture to word processing, new programs are introduced daily. I cannot even begin to list all the vendors and programs on the market today; many will be obsolete by the time this book is published anyway. What I hope to do is introduce the areas where software can help you design and develop electronic instruments.

The main areas where software programs can aid the design process include the following:

1. Project management
2. System analysis and architecture
3. Circuit design
4. Chip design
5. Circuit analysis and simulation
6. PCB placement and routing
7. Software development
8. Integration
9. Reliability and maintainability

Software tools for managing a project include word processors, desktop publishing software, spreadsheets, and project planners. Project management must have documentation, which obviously requires word processors and desktop publishing. Spreadsheets and project planners help in estimating effort as well as handling accounting and finances.

Software tools for system analysis allow the architect to begin defining the project by simulating performance and dynamics. An architect can play out what-if scenarios to bound the problem. The results of these analyses will help form the requirements for the system.

Tools for circuit design include schematic capture and programmable logic tools. IC design requires special CAD tools for designing custom or semicustom chips.

After you design the circuits, software tools abound for analyzing and simulating them. You may be familiar with analysis programs like Spice. Many more programs provide various degrees of sophistication, size, and speed. They rely on physical models, macromodels, and behavioral models of circuit components. Software also exists to simulate, trace, and analyze possible faults in the circuits. Furthermore, tools can analyze signal quality—such as transmission-line effects, crosstalk, and propagation delay—and verify signal timing.

Software tools are often necessary for circuit board design. Place-and-route programs help the designer to locate components and traces on the PCB. Tools that perform thermal analysis help designers place hot, cool, or sensitive components on the PCB.

Software particularly needs software tools. These include editors, assemblers and compilers, debuggers, configuration management tools, and CASE tools.

Integration and logistics require software tools that keep parts lists, provide configuration control, and predict and follow maintenance. Reliability and maintainability tools include databases of resources, spares and parts lists, predictive programs for reliability and maintainability, and analysis tools for stress, failure modes and effects, and sneak circuits.

INDEX

Numbers in *italics* refer to figures or tables.